Practical Guide to LTE-A, VoLTE and IoT

Practical Guide to LTE-A, VoLTE and IoT

Paving the Way Towards 5G

Ayman Elnashar
Emirates Integrated Telecommunications Company (EITC)
Dubai Media City
Dubai
UAE

Mohamed El-saidny
MediaTek
Dubai Internet City
Dubai
UAE

This edition first published 2018
© 2018 John Wiley & Sons Ltd.

The right of Ayman Elnashar and Mohamed El-saidny to be identified as the authors of this work has been asserted in accordance with law.

Registered Offices
John Wiley & Sons, Inc., 111 River Street, Hoboken, NJ 07030, USA
John Wiley & Sons Ltd, The Atrium, Southern Gate, Chichester, West Sussex, PO19 8SQ, UK

Editorial Office
The Atrium, Southern Gate, Chichester, West Sussex, PO19 8SQ, UK

For details of our global editorial offices, customer services, and more information about Wiley products visit us at www.wiley.com.

Wiley also publishes its books in a variety of electronic formats and by print-on-demand. Some content that appears in standard print versions of this book may not be available in other formats.

Library of Congress Cataloging-in-Publication Data applied for
Hardback ISBN: 9781119063308

Cover design by Wiley
Cover image: © jamesteohart/Shutterstock

Set in 10/12pt WarnockPro by SPi Global, Chennai, India

This book is dedicated to the memory of my parents (God bless their souls). They gave me the strong foundation and unconditional love, which remains the source of motivation and is the guiding light of my life.

To my dearest wife, my pillar of strength, your encouragement and patience has strengthened me always.

To my beloved children Noursin, Amira, Yousef, and Yasmina. You are the inspiration!

I want to offer my sincerest appreciation to the innovation and vision of UAE. It has provided me with a fulfilling career, an unmatched lifestyle and the inspiration to author this book.

– Ayman Elnashar, PhD

I would like to dedicate this book to my amazing family for their continuous support and encouragement. To my beloved wife, you have guided and inspired me throughout the years. To my beautiful daughter, you always surprise me with your motivational spirit and hard work.

"The scientific man does not aim at an immediate result. He does not expect that his advanced ideas will be readily taken up. His work is like that of the planter—for the future. His duty is to lay the foundation for those who are to come, and point the way". – Nikola Tesla

Mohamed El-saidny

Contents

About the Authors

Ayman Elnashar, PhD, has 20+ years of experience in telecoms industry including 2G/3G/LTE/WiFi/IoT/5G. He was part of three major start-up telecom operators in the MENA region (Orange/Egypt, Mobily/KSA, and du/UAE). Currently, he is Head of Infrastructure Planning: ICT and Cloud with the Emirates Integrated Telecommunications Co. "du", UAE. He is the founder of the Terminal Innovation Lab and UAE 5G innovation Gate (U5GIG). Prior to this, he was Sr. Director – Wireless Networks, Terminals and IoT where he managed and directed the evolution, evaluation, and introduction of du wireless networks including LTE/LTE-A, HSPA+, WiFi, NB-IoT and is currently working towards deploying 5G network in UAE.

Prior to this, he was with Mobily, Saudi Arabia, from June 2005 to Jan 2008, as Head of Projects. He played key role in contributing to the success of the mobile broadband network of Mobily/KSA. From March 2000 to June 2005, he was with Orange Egypt.

He published 30+ papers in the wireless communications arena in highly ranked journals and international conferences. He is the author of *Design, Deployment, and Performance of 4G-LTE Networks: A Practical Approach* published by Wiley & Sons, and *Simplified Robust Adaptive Detection and Beamforming for Wireless Communications* to be published in May 2018.

His research interests include practical performance analysis, planning and optimization of wireless networks (3G/4G/WiFi/IoT/5G), digital signal processing for wireless communications, multiuser detection, smart antennas, massive MIMO, and robust adaptive detection and beamforming.

Mohamed El-saidny, M.Sc., is a leading technical expert in wireless communication systems for modem chipsets and network design. He established and managed the Carrier Engineering Services Business Unit at MediaTek, the department responsible for product business development and strategy alignment with network operators and direct customers. He has 15+ years of technical, analytical and business experience, with an international working experience in the United States, Europe, Middle East, Africa, and South-East Asia markets.

Mohamed is the inventor of numerous patents in CDMA and OFDM systems and the co-author of *Design, Deployment and Performance of 4G-LTE Networks: A Practical Approach* book by Wiley & Sons. He published several international research papers in IEEE Communications Magazine, IEEE Vehicular Technology Magazine, other IEEE Transactions, in addition to contributions to 3GPP specifications.

Preface

This book is a practical guide to the design, deployment, and performance of LTE-A, VoLTE/IMS and IoT. A comprehensive practical performance analysis for VoLTE is conducted based on field measurement results from live LTE networks. Also, it provides a comprehensive introduction to IoT, 3GPP NB-IoT and 5G evolutions. Practical aspects and best practice of LTE-A/IMS/VoLTE/IoT, plus LTE-Advanced features such as Carrier Aggregation (CA), are presented. In addition, LTE/LTE-A network capacity dimensioning and analysis are demonstrated based on live LTE/LTE-A networks KPIs. A comprehensive foundation for 5G technologies is provided including massive MIMO, eMBB, URLLC, mMTC, NGCN and network slicing, cloudification, virtualization and SDN.

Chapter 1 provides an overview of LTE/LTE-A networks. This chapter is the foundation for the chapters 2 to 6. In Chapter 2, we will introduce the IP Multimedia Subsystem (IMS), which is the core network of Voice over LTE (VoLTE) and other advanced voice evolutions. The IMS architecture, core network elements, call and signaling flow between different network elements are comprehensively presented. The chapter provides the foundation for VoLTE performance analysis highlighted in chapter 3 and chapter 4. Other IMS services such as Voice over Wi-Fi (VoWiFi), Video over LTE (ViLTE), and Web Real-Time, Communication (WebRTC) are not discussed in detail. In addition, practical deployment scenarios for IMS-network-based on real use cases from a live network deployment are presented.

Chapter 3 presents practical performance analysis including an end-to-end assessment of call setup delay in different radio conditions, main challenges impacting the in-call performance, and performance aspects of Single Radio Voice Call Continuity (SRVCC) and its evolution releases. Therefore, Chapter 3 provides comprehensive analysis for call setup delay including CSFB and VoLTE with different scenarios (stationary and mobility), and handover analysis including SRVCC in terms of data interruption time. Finally, recent topics in handover performance and data interruption reduction during handover are presented. Synchronized handover is introduced as a potential solution to reduce data interruption time during handover.

Chapter 4 presents comprehensive practical analysis of VoLTE performance based on commercially deployed 3GPP Release 10 LTE networks. The analysis in chapter 4 demonstrates VoLTE performance in terms of RTP error rate, RTP jitter and delays, BLER, and VoLTE voice quality in terms of MOS. In Chapter 4, we will also evaluate key VoLTE features such as RoHC, TTI bundling, and SPS.

Chapter 5 analyzes the new features in LTE-A including carrier aggregation (CA), LAA, downlink 256QAM, uplink 64QAM modulation, uplink data compression (UDC), and eVoLTE.

Chapter 6 evaluates LTE counters collected for a commercially deployed 3GPP Release 10 LTE network. The analysis in this chapter includes LTE users (connected and active), LTE scheduling, LTE traffic downlink/uplink, TTI utilizations, physical resource block utilization, modulation and codec scheme, channel quality indicator (CQI), MIMO, and CSFB performance.

Chapter 7 will cover the evolution of the Internet of Things (IoT) from different aspects. The aim is to provide the reader with a holistic overview of IoT evolution and a guide on how to build, design, and customize a successful IoT use case from all perspectives including technical and commercial aspects. We will focus on 3GPP cellular IoT evolution for connectivity, i.e., narrowband IoT (NB-IoT). However, we will initially summarize the IoT evolution from an end-to-end perspective, including the IoT platform, IoT protocols, connectivity, and sensors layer. The IoT evolution is different from the regular mobile evolution; the latter is focusing on connectivity only, while IoT evolution should be addressed from an end-to-end prospective. This is because the IoT connectivity is only 5% to 10% of the IoT value chain and therefore, the service provider should offer an end-to-end use case. We will present in this chapter practical IoT use cases along with dashboards to demonstrate the value of IoT.

Chapter 8 will provide a comprehensive introduction to 5G access and core networks. Advanced 5G technologies such as massive MIMO, 5G flexible frame design, URLLC, mMTC, NGCN and network slicing are also presented. Finally, virtualization and software defined network are summarized along with service provider roadmap for converged native cloud.

Acknowledgments

Many people volunteered their time and talent to make this project a reality, and we thank each and every one of them for their invaluable support. We acknowledge the huge contribution of Mohamed Yehia from du for chapters two and six. We also appreciate the support of Mohanad ElSakka from du for reviewing chapters two and six. We acknowledge the support of our colleagues at du from different sections for providing feedback, practical results and conducting testing scenarios. Without their support, this book could not be produced with such practical scenarios from live network. du is a vibrant and multiple award-winning telecommunications service provider in UAE, serving nine million individual customers with its mobile, fixed line, broadband Internet, and Home services. du also caters to over 100,000 UAE businesses with its vast range of ICT solutions. Finally, we would like to thank the organizations that provided permission for use of their copyrighted material which significantly improved the presentation of this book.

1

LTE and LTE-A Overview

1.1 Introduction

Cellular mobile networks have been evolving for many years. As the smartphone market has expanded significantly in recent years and is expected to grow more in the years to come, network evolution needs to keep up with the pace of users' demands. This chapter provides an overview for network operators and interested others on the evolution of cellular networks, with particular focus on 3GPP for the main technologies of WCDMA/UMTS and LTE. In addition, it highlights the interaction of 3GPP with non-3GPP technology (i.e. Wi-Fi).

The initial networks are referred to collectively as the First Generation (1G) system. The 1G mobile system was designed to utilize analog; it included AMPS (Advanced Mobile Telephone System). The Second Generation (2G) mobile system was developed to utilize digital multiple access technology: TDMA (Time Division Multiple Access) and CDMA (Code Division Multiple Access). The main 2G networks were GSM (Global System for Mobile communications) and CDMA, also known as cdmaOne or IS-95 (Interim Standard 95). The GSM system still has worldwide support and is available for deployment on several frequency bands, such as 900 MHz, 1800 MHz, 850 MHz, and 1900 MHz. CDMA systems in 2G networks use a spread-spectrum technique and utilize a mixture of codes and timing to identify cells and channels. In addition to being digital and improving capacity and security, these digital 2G systems also offer enhanced services such as SMS (Short Message Service) and circuit-switched data. Different variations of the 2G technology have evolved to extend the support of efficient packet data services and to increase the data rates. GPRS (General Packet Radio System) and EDGE (Enhanced Data Rates for Global Evolution) systems have evolved from GSM. The theoretical data rate of 473.6 kbps enables operators to offer multimedia services efficiently. Since it does not comply with all the features of a 3G system, EDGE is usually categorized as 2.75G.

The Third Generation (3G) system is defined by IMT2000 (International Mobile Telecommunications). IMT2000 requires a 3G system to provide higher transmission rates in the range of 2 Mbps for stationary use and 348 kbps under mobile conditions. The main 3G technologies are [1]:

WCDMA (Wideband CDMA): This was developed by the 3GPP (Third Generation Partnership Project). WCDMA is the air interface of the 3G UMTS (Universal Mobile Telecommunications System). The UMTS system has been deployed based on the existing GSM communication core network (CN) but with a new radio access technology in the form of WCDMA. Its radio access is based on FDD (Frequency Division Duplex). Current deployments are mainly in 2.1 GHz bands. Deployments at lower frequencies are also possible, such as UMTS900. UMTS supports voice and multimedia services.

TD-CDMA (Time Division CDMA): This is typically referred to as UMTS TDD (Time Division Duplex) and is part of the UMTS specifications. The system utilizes a combination of CDMA and TDMA to enable efficient allocation of resources.

TD-SCDMA (Time Division Synchronous CDMA): This has links to the UMTS specifications and is often identified as UMTS-TDD Low Chip Rate. Like TD-CDMA, it is also best suited to low-mobility scenarios in micro or pico cells.

CDMA2000 (C2K): This is a multi-carrier technology standard which uses CDMA. It is part of the 3GPP2 standardization body. CDMA2000 is a set of standards including CDMA2000 EV-DO (Evolution-Data Optimized) which has various revisions. It is backward compatible with cdmaOne.

WiMAX (Worldwide Interoperability for Microwave Access): This is another wireless technology which satisfies IMT2000 3G requirements. The air interface is part of the IEEE (Institute of Electrical and Electronics Engineers) 802.16 standard, which originally defined PTP (Point-To-Point) and PTM (Point-To-Multipoint) systems. This was later enhanced to address multiple

Practical Guide to LTE-A, VoLTE and IoT: Paving the Way Towards 5G, First Edition. Ayman Elnashar and Mohamed El-saidny.
© 2018 John Wiley & Sons Ltd. Published 2018 by John Wiley & Sons Ltd.

issues related to a user's mobility. The WiMAX Forum is the organization formed to promote interoperability between vendors.

Fourth Generation (4G) cellular wireless systems have been introduced as the latest version of mobile technologies. 4G technology is defined as meeting the requirements set by the ITU (International Telecommunication Union) as part of IMT Advanced (International Mobile Telecommunications Advanced).

The main drivers for the network architecture evolution in 4G systems are: all-IP based, reduced network cost, reduced data latencies and signaling load, interworking mobility among other access networks in 3GPP and non-3GPP, always-on user experience with flexible Quality of Service (QoS) support, and worldwide roaming capability. 4G systems include different access technologies:

LTE and LTE-Advanced (Long Term Evolution): This is part of 3GPP. LTE, as it stands now, does not meet all IMT Advanced features. However, LTE-Advanced is part of a later 3GPP release and has been designed specifically to meet 4G requirements.

WiMAX 802.16m: The IEEE and the WiMAX Forum have identified 802.16m as the main technology for a 4G WiMAX system.

UMB (Ultra Mobile Broadband): This is identified as EV-DO Rev C. It is part of 3GPP2. Most vendors and network operators have decided to promote LTE instead.

The evolution and roadmap for 3GPP 3G and 4G are illustrated in Figure 1.1.

The standardization in 3GPP Release 8 defines the first specifications of LTE. The Evolved Packet System (EPS) is defined, mandating the key features and components of both the radio access network (E-UTRAN) and the core network (Evolved Packet Core, EPC). Orthogonal Frequency Division Multiplexing (OFDM) is defined as the air interface, with the ability to support multi-layer data streams using Multiple-Input, Multiple-Output (MIMO) antenna systems to increase spectral efficiency. LTE is defined as an all-IP network topology differentiated over the legacy circuit switch (CS) domain. However, Release 8 specification makes use of the CS domain to maintain compatibility with 2G and 3G systems by utilizing the voice calls Circuit-Switch Fallback (CSFB) technique for any of those systems. Other significant aspects defined in this initial 3GPP release are Self-Organizing Networks (SONs) and Home Base Stations (Home eNodeBs), aiming to revolutionize heterogeneous networks. Moreover, Release 8 provides techniques for smartphone battery saving, known as Connected-mode Discontinuous Reception (C-DRX).

LTE Release 9 provides improvements to Release 8 standards, most notably enabling improved network throughput by refining SONs and improving eNodeB (eNB) mobility. Additional MIMO flexibility is introduced with multi-layer beamforming. Furthermore, CSFB improvements have been introduced to reduce voice call-setup time delays.

The International Telecommunication Union (ITU) has created the term IMT-Advanced (International Mobile Telecommunications-Advanced) to identify mobile systems whose capabilities go beyond those of IMT2000. In order to meet this new challenge, 3GPP's partners have agreed to expand specification scope to include the development of systems beyond 3G's capabilities. Some of the key features of IMT-Advanced are: worldwide functionality and roaming, compatibility of services, interworking with other radio access systems, and enhanced peak data rates to support advanced services and applications with a nominal speed of 100 Mbps for high mobility and 1 Gbps for low-mobility users.

Release 10 defines LTE-Advanced (LTE-A) as the first standard release that meets the ITU's requirements for Fourth Generation, 4G. The increased data rates up to 1 Gbps in the downlink and 500 Mbps in the uplink are enabled through the use of scalable and flexible bandwidth allocations up to 100 MHz, known as Carrier Aggregation (CA). Additionally, improved MIMO operations have been introduced to provide higher spectral efficiency. The support for heterogeneous networks and relays added to this 3GPP release also improves capacity and coverage. Lastly, a seamless interoperation of LTE and WLAN networks is defined to support traffic offload concepts.

Release 11 continues the evolution towards the LTE-A requirements. Enhanced interference cancellation and CoMP (Coordinated Multi-Point transmission)

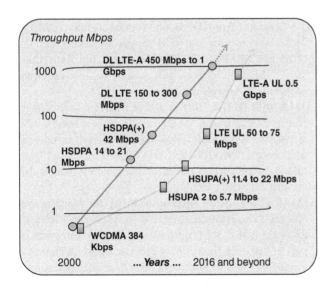

Figure 1.1 3G and 4G roadmap and evolution.

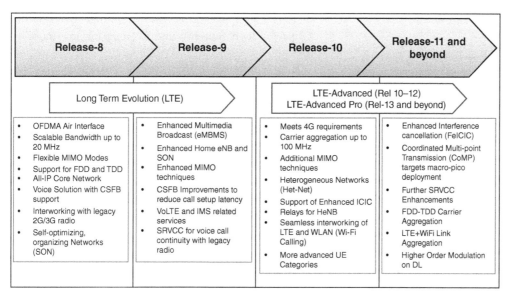

Figure 1.2 3GPP LTE releases.

are means for further improving the capacity in 4G networks.

Key features of 3GPP LTE releases are outlined in Figure 1.2.

1.2 Link Spectrum Efficiency

The Shannon–Hartley theorem states the channel capacity, meaning the theoretical tightest upper bound on the information data rate that can be sent with a given average signal power through a communication channel subject to noise of power:

$$C = B \log_2 \left(1 + \frac{S}{N}\right) \tag{1.1}$$

where

C is the channel capacity in bits per second

B is the bandwidth of the channel in hertz

S is the average received signal power over the bandwidth, measured in watts

N is the average noise or interference power over the bandwidth, measured in watts

S/N is the signal-to-noise ratio (SNR)

Therefore, channel capacity is proportional to the bandwidth of the channel and to the logarithm of the SNR. This means that channel capacity can be increased linearly either by increasing the channel bandwidth given a fixed SNR requirement or, with fixed bandwidth, by using higher-order modulations that need a very high SNR to operate.

Spectral efficiency refers to the information rate that can be transmitted over a given bandwidth in a specific communication system, measured in bits/sec/Hz

As mentioned above, the LTE targets higher capacity by using fixed and high bandwidth in a cell and using higher-order modulations that need high SNR to operate, which is achieved by using different MIMO techniques.

As the modulation rate increases, the spectral efficiency improves, but at the cost of the SNR requirement, which makes LTE scalable at different radio conditions.

Then, the spectral efficiency can be defined as follows:

Efficiency = the number of information bits/the total number of symbols.

The number of information bits + parity bits = total number of bits = total number of symbols * modulation order:

Spectral efficiency

     = Coded bit rate ∗ Modulation order   (1.2)

For example, with LTE, CQI index 1, QPSK, a modulation order of 2, code rate = $78/1024 = 0.0762$, efficiency = $0.1523 = 0.0762 * 2$. More specifically, $78/1024 = 0.076$ is the ratio of the information symbols (78) to the total number of symbols (1024). Then, the efficiency is equal to 0.076 * 2 (QPSK modulation, one symbol occupies two information bits)= 0.152. Modulation order = 2 (QPSK), 4 (16QAM), 6 (64QAM). Table 1.1 summarizes the different values of CQI, code rate, and modulation, and the corresponding efficiencies [1].

Table 1.1 Spectrum efficiencies of the LTE system.

LTE CQI	Code Rate	Modulation	Efficiency
1	0.08	2	0.1523
2	0.12	2	0.2344
3	0.19	2	0.3770
4	0.30	2	0.6016
5	0.44	2	0.8770
6	0.59	2	1.1758
7	0.37	4	1.4766
8	0.48	4	1.9141
9	0.60	4	2.4063
10	0.46	6	2.7305
11	0.55	6	3.3223
12	0.65	6	3.9023
13	0.75	6	4.5234
14	0.85	6	5.1152
15	0.93	6	5.5547

1.3 LTE-Advanced and Beyond

The International Telecommunication Union (ITU) has created the term IMT-Advanced to identify mobile systems whose capabilities go beyond those of IMT2000. Table 1.2 provides the IMT-Advanced requirements.

In order to meet this new challenge, 3GPP's partners have agreed to expand the specification scope to include the development of systems beyond 3G's capabilities. Some of the key features of IMT-Advanced are: worldwide functionality and roaming, compatibility of services, interworking with other radio access systems, and enhanced peak data rates to support advanced

services and applications with a nominal speed of 100 Mbps for high mobility and 1 Gbps for low-mobility users.

The requirements for further advancements for Evolved Universal Terrestrial Radio Access E-UTRA (LTE-Advanced) are defined in TR 36.913. The reports that include the requirements of IMT-Advanced and the basis for evaluation criteria were approved in an ITU-R (International Telecommunication Union) Study Group 5 meeting in November 2008. LTE-Advanced targets are provided in Table 1.3. Therefore, LTE-A naturally meets the ITU requirements for IMT-Advanced.

Average spectrum efficiency is defined as the aggregate throughput of all users normalized by bandwidth and divided by the number of cells. This requirement is essential for operators in terms of capacity and cost per bit. Cell edge user throughput is defined as the 5% point of CDF of average user throughput normalized by bandwidth, assuming ten users in a cell. Average and cell-edge requirements are summarized in Table 1.4. Base coverage urban (shown in bold font in the table) is used as a benchmark, which is a similar model to 3GPP "Case 1" (see ITU-R IMT.EVAL).

In order to support higher peak data rates in the uplink and downlink, LTE-Advanced improved many of the features introduced in Release 8 and Release 9 and introduced new features such as carrier aggregation, enhanced Inter-Cell Interference Coordination (eICIC) for Het-Nets, and relay nodes. Carrier aggregation extends the maximum bandwidth in the downlink/uplink by aggregating two to five carriers. Each carrier to be aggregated is referred to as a Component Carrier (CC). Since Release 8 carriers have a maximum bandwidth of 20 MHz, CA allows for a maximum transmission bandwidth of 100 MHz (5 × 20 MHz). A pre-Release 10 UE can access

Table 1.2 IMT-Advanced requirements.

Requirements	IMT-Advanced
Spectrum bandwidth	40 MHz
Downlink peak spectrum efficiency	15 bps/Hz (4 streams)
Uplink peak spectrum efficiency	6.75 bps/Hz (2 streams)
Control plane latency	< 100 ms
User plane latency	< 10 ms

Where:
Peak spectrum efficiency is defined in 3GPP as the highest theoretical data rate normalized by the spectrum bandwidth.
Control plane latency is defined in 3GPP as the transition time from Idle mode to Connected mode.
User plane latency is defined in 3GPP as the one-way transit time in RAN for a packet being available at the IP layer.

Table 1.3 LTE-Advanced targets versus LTE targets.

Requirements	LTE Targets	LTE-A Targets
Spectrum bandwidth	1.4–20 MHz	Wider than LTE, e.g. 100 MHz
Downlink peak data rate	300 Mbps	1 Gbps
Downlink peak spectrum efficiency	15 bps/Hz (4 streams)	30 bps/Hz (8 streams)
Uplink peak data rate	75 Mbps	500 Mbps
Uplink peak spectrum efficiency	3.75 bps/Hz (1 stream)	15 bps/Hz (4 streams)
Control plane latency	< 100 ms	<50 ms
User plane latency	5 ms	5 ms

Table 1.4 Average and cell-edge requirements.

Scenario	Downlink and Uplink	Antenna Configuration	LTE Targets [a]	LTE-A Targets [a]	IMT-Advanced [b]
Average spectrum efficiency (bit/s/Hz/cell)	DL	2 × 2	1.69	2.4	—
		4 × 2	1.87	2.6	[3, 2.6, **2.2**, 1.1]
		4 × 4	2.67	3.7	—
	UL	1 × 2	0.74	1.2	—
		2 × 4	—	2.0	[2.25, 1.8, **1.4**, 0.7]
Cell-edge user spectrum efficiency (bit/s/Hz)	DL	2 × 2	0.05	0.07	—
		4 × 2	0.06	0.09	[0.1, 0.075, **0.06**, 0.04]
		4 × 4	0.08	0.12	—
	UL	1 × 2	0.024	0.04	—
		2 × 4	—	0.07	[0.07, 0.05, **0.03**, 0.015]

a) Based on radio environment of "Case 1": Inter-cell distance: 500 m, carrier frequency: 2 GHz, bandwidth: 10 MHz, DL Tx power: 46 dBm, penetration loss: 20 dB: mobility speed: 3 km/h (see 3GPP TR 25.814).

b) [Indoor, Microcellular, Base coverage urban, High speed].

one of the component carriers, while CA-capable UEs can operate with multiple component carriers. The CCs can be of the same or different bandwidths, adjacent or non-adjacent CCs in the same or different frequency band, or CCs in different frequency bands. Carrier aggregation can also benefit from both TDD and FDD joint operation. Enhanced MIMO provides higher spatial gains, increased peak data rates, higher spectral efficiency, and increased capacity.

The main goal of heterogeneous networks is to manage traffic between macro and small cell networks. Controlling the interference scenarios within these heterogeneous networks due to different power levels of macro and small cells can be managed with features

such as ICIC/eICIC/feICIC and CoMP (Coordinated Multi-Point transmission), which further improve capacity at cell edges.

The Release 8 specification makes use of the CS domain to maintain compatibility with 2G and 3G systems by utilizing the voice calls circuit-switch fallback (CSFB) technique for any of those systems. As LTE has evolved into an all-IP network and IMS implementation has matured, Voice over IP over LTE has been implemented to carry voice packets natively over the LTE network. At an LTE cell edge, the ability to fall back into a circuit-switched network is possible with features such as Single Radio Voice Continuity (SR-VCC). Figure 1.3 summarizes the key features for LTE and LTE-Advanced [1].

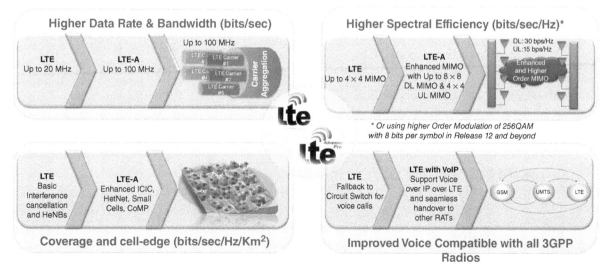

Figure 1.3 Key features for LTE and LTE-Advanced.

1.3.1 LTE and Wi-Fi

There is plenty of spectrum in the 5 GHz band, which is especially suited to small-cell deployments. For the last few years, Wi-Fi has been actively used to offload cellular traffic, and several operators are using it as part of mobile plans. The idea of LTE/Wi-Fi aggregation arose so that smartphones could receive data from a cellular network and a Wi-Fi network at the same time. Therefore, different phases for increasing carrier aggregation in the 5 GHz band have been studied in 3GPP to utilize 5 GHz solely for LTE or jointly with Wi-Fi. The amount of spectrum available in the 5 GHz band per region is summarized in Table 1.5.

The two main options for LTE to use the 5 GHz band are as follows:

LAA (Licensed Assisted Access): A standalone LTE operation in unlicensed spectrum. Several challenges occur with this implementation. In particular, there is a need for eNB and access points (AP) to be collocated, which requires new devices. This would allow operators to benefit from the additional capacity available from the unlicensed spectrum, particularly in hotspots and corporate environments. With LAA, the extra spectrum resource, especially on the 5 GHz frequency band, can complement licensed-band LTE operation to provide additional data plane performance. The use of this technology has prompted

Table 1.5 5 GHz available spectrum per region.

Region	Estimated Bandwidth Available (MHz)
America	580
Europe and Japan	455
China	325

regulators to study the feasibility of this deployment and the impact of LTE on Wi-Fi spectrum. However, it is expected that the feature will gain strong momentum in the upcoming years, especially as it brings a substantial capacity boost from the unlicensed band, if it is proven that the quality of service for the end user is not impacted by the interference situation between the two bands.

LWA (LTE–Wi-Fi Aggregation): Dual connectivity between LTE and Wi-Fi. LWA can be enabled at the radio level and can split the data plane traffic so that some LTE traffic is tunneled over Wi-Fi and the rest runs over LTE. The Wi-Fi data rate can go up to 867 Mbps in 802.11ac and the LTE currently being deployed with 300 to 450 Mbps (two- or three-carrier aggregation), and when both are aggregated, the total aggregated throughput can go beyond 1 Gbps. This is, in some networks, called Giga LTE (G-LTE).

LAA and LWA are summarized in Figure 1.4, and comparisons are provided in Table 1.6.

Other forms of LTE/Wi-Fi aggregation are Multi-Path TCP (MPTCP) and a simple HTTP range retrieval request, but neither is necessarily 3GPP standardized:

Multi-Path TCP (MPTCP): TCP subflows are created for each different network. Flow control is operated for each subflow. The advantages of this method are good aggregation performance and the ease of configuration in the network (it may require a proxy server to apply multiple kinds of services which are non-MPTCP capable). An MPTCP scheduler selects a path for each packet based on the network condition. Each subflow controls each congestion window size and the packet loss on one network does not affect the quality of other networks. The disadvantage of this implementation is that not all kinds of applications can benefit from it, especially UDP-type applications.

Figure 1.4 LTE with LAA and LWA.

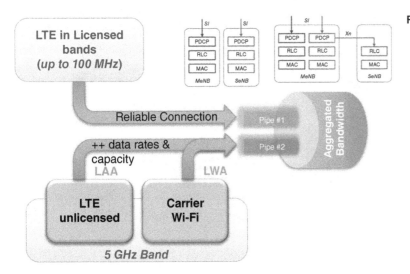

Table 1.6 LAA versus LWA.

Targets	LAA (Licensed Assisted Access)			LWA (LTE–Wi-Fi Aggregation)
3GPP	R10 based	R13	R14	R13
Unlicensed physical layer	Uses LTE on 5 GHz band for higher data rates			Uses Wi-Fi on 5 GHz band for higher data rates
Deployment	Collocated or through backhaul		Collocated or dual-connectivity	Dual-connectivity
Listen before Talk (LBT)	eNodeB Best effort	eNodeB New design	UE listen before talk	802.11
Regulation	LBT not needed in US, Korea, China	LBT needed in Europe and Japan	UE listen before talk fully complied	Wi-Fi widely accepted

HTTP range retrieval request: A client on a device requests half of some data be sent to one network and the rest to another network. This implementation is considered a quick implementation that does not require modification on the device side, but it is only applicable to HTTP traffic.

1.3.2 Wi-Fi Calling

Voice over Wi-Fi is becoming an integral part of the evolution of VoIP in general. One of the advantages set for VoWiFi is its seamless handover with LTE and potentially with CS networks (3G/2G). This is because the 3GPP defines interfaces between the LTE core network (EPS) and the Wi-Fi core network.

The benefits of VoWiFi are as follows:

Wi-Fi is a long-living ecosystem: Smartphone users are still comfortable using Wi-Fi; Wi-Fi is already carrying the bulk of smartphone data consumption and is easy to deploy in Wi-Fi hotspots.

It complements LTE services: Once IMS is being invested in, VoWiFi will be easy to integrate. VoLTE and VoWiFi can work together over EPC. There is added value in cases where VoLTE coverage is not continuous.

There is significant business potential: VoWiFi helps to solve the challenge of indoor access for mobile users when it can extend the coverage over LTE and 2G/3G networks. It also reduces or eliminates roaming costs because calls may be treated as being local.

It provides new service opportunities: Mobile operators need the offering of Wi-Fi services as a differentiator from over-the-top (OTT) clients. VoWiFi is now available to users almost exclusively from over-the-top (OTT) clients for smartphone users. Figure 1.5 illustrates VoWiFi for trusted and untrusted Wi-Fi networks.

We summarize the untrusted and trusted VoWiFi services as follows:

Untrusted access to EPC for VoWiFi services:
The device establishes a dedicated IPSec tunnel to an evolved Packet Data Gateway (ePDG) element located at the edge of the EPC.
The ePDG establishes a PMIPv6 or GTPv2 tunnel (i.e. on the S2b interface) to the P-GW in the EPC.
This IPSec tunnel transfers both signaling and media related to the operator services.

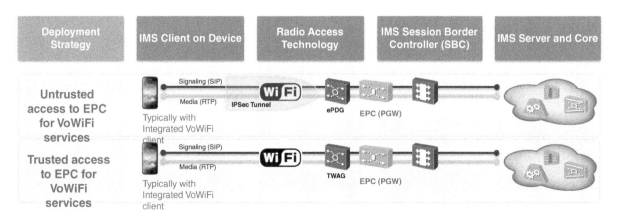

Figure 1.5 VoWiFi for trusted and untrusted 3GPP networks.

Enables mobility between Wi-Fi and LTE access networks whether or not a call is in progress.

Mobility is performed at the EPC level, preserving the IP address allocated for the device.

Trusted access to EPC for VoWiFi services:

Wi-Fi access points are connected to a Trusted WLAN Access Gateway (TWAG) network function that connects to EPC services via a standard interface (S2a interface).

Supports multiple APN/PDN,

Allowing APN-specific traffic to be offloaded directly to the Internet without routing it to the EPC and only IMS APN is sent to the EPC.

Enables mobility between Wi-Fi and LTE access networks.

EPC mobility scenarios could enable subsequent call continuity to CS networks with Single Radio Voice Continuity (eSRVCC), or directly to a CS network through Dual Radio VCC (DR-VCC). The VoWiFi deployment choice should address preferences such as the existing network architecture and interface (including security), mobility requirements, and device implementations.

1.3.2.1 QoS Challenges in VoWiFi

In 3GPP radio access (2G/3G/4G), the voice quality can be guaranteed to some extent, but Wi-Fi access brings new challenges:

Wi-Fi is operating on unlicensed radio bands, which means that resources may not be allocated and higher interference can therefore be experienced.

Multiple users and applications sharing the same Wi-Fi network cause congestion and potentially low-bandwidth broadband connectivity.

Wi-Fi access to the operator voice core may be over ISP networks and therefore not managed by the 3GPP network operator.

Therefore, packet losses, jitter, or latency in the packet delivery can cause degradation of the voice quality. QoS architecture in Wi-Fi access is based on packet prioritization and not on resource reservations as in the case of VoLTE. It requires additional IP DSCP marking in the transport layer, and can be available with trusted/untrusted solutions only.

1.3.3 Internet of Things (IoT)

The transition to 5G has started and will continue until 2020. During the cellular evolution to 5G, aspects of the intermediate defining steps have become important. The Internet of Things (IoT) and the migration between multiple radio access technologies (3GPP and non-3GPP) are defining the path towards 5G.

The Internet of Things defines the method for intelligently connected devices and systems to leverage and exchange data between small devices and sensors in machines and objects. IoT concepts and working models have started to spread rapidly, which is expected to provide a new dimension for services that improve the quality of consumers' lives and the productivity of enterprises. The IoT effort started with the concept of machine-to-machine (M2M) solutions to use wireless networks to connect devices to each other and through the Internet, in order to deliver services that meet the needs of a wide range of industries. The IoT must deal with several challenges involving massive numbers of cheap devices providing low energy consumption and connected in a wider range, referred to as a Low-Power Wide Area (LPWA). Therefore, the IoT is typically classified into:

1) IoT connectivity in unlicensed spectrum.
2) Cellular IoT in licensed spectrum.

Many technologies are emerging to deal with these two categories, including:

Unlicensed networks: For short-range scenarios, we have technologies such as Bluetooth Low Energy, Wi-Fi, IEEE802.11ah, IEEE802.15.4, ZigBee, and Z-wave. For long-range scenarios there are Sigfox, Weightless, OnRamp, LoRa, and ETSI LTN.

Cellular IoT in licensed spectrum: LTE evolution for Machine-Type Communication (MTC), narrowband LTE, and the GSM evolution of IoT referred to as Extended-Coverage GSM (EC-GSM).

The main targets for low-power, wide-area (LPWA) networks are:

Enhanced coverage (path loss ~164 dB).

Very low power consumption (battery life > 10 years).

Low data rate and high capacity core network (signaling and network entity optimization).

Massive numbers of very low-cost devices.

We can summarize the IoT networks as follows:

Wide area

    Unlicensed spectrum LPWA (non-3GPP)

        Sigfox (uplink only)

        Semtech LoRa (uplink, downlink)

        Neul Weightless

    Licensed spectrum (3GPP but not C-IoT)

        GSM (7B GSM connections today)

    Short range in unlicensed spectrum

    Bluetooth Low Energy, Wi-Fi, IEEE802.11ah, ZigBee, Z-Wave.

Some IoT technologies are based on standard protocols supported by industry alliances like the LoRa Alliance and Weightless SIG, some are based on proprietary protocols, and some are standards in progress. The forthcoming IoT networks under 3GPP can be summarized as follows:

Cellular IoT in licensed spectrum
3GPP eRAN (Release 12/13).
LTE evolution for MTC (machine-type communication).
Category 1 but it does not meet the IoT requirement (battery/cost/range).
Release 12 with Category 0.
Release 13 to meet LPWA requirement (Category M).
NB-CIoT and NB-LTE.
Will be evolved into NB-IoT as per latest 3GPP RAN meeting, and is expected to be released with 3GPP Release 13.
3GPP GERAN (Release 13).
GSM evolution: upgrade of GSM by using one carrier for IoT with Extended-Coverage GSM (EC-GSM) is expected with 3GPP Release 13.

For the unlicensed networks, some of the highlighted technologies have already been deployed and meet the four factors for LPWA (long range, very low power, low data rate, and very low cost). . For 3GPP evolution of cellular IoT, the LTE-MTC is defined with the first version released with 3GPP Release 8 based on Category 1 but it does not meet the IoT requirement (battery/cost/range). This idea took an additional turn by providing a new Release 12 Category 0. The ongoing enhanced version (eMTC) is under evaluation in Release 13 to meet the LPWA requirement (i.e. Category M).

On the other hand, narrowband LTE introduces two underlying technologies being discussed in 2015/2016 3GPP Release 13: NB-CIoT and NB-LTE, where the main difference lies in the physical layer. It is expected that the two will merge and provide a final version referred to as NB-IoT. This technology is targeting three different modes such as utilizing the spectrum currently being used by GERAN systems as a replacement for one or more GSM carriers (standalone operation). The second mode utilizes the unused resource blocks within an LTE carrier's guard band (guard-band operation). The final mode utilizes resource blocks within a normal LTE carrier (in-band operation). The NB-IoT should support the following main objectives:

180 kHz UE RF bandwidth for both downlink and uplink.
OFDMA on the downlink with either 15 kHz or 3.75 kHz subcarrier spacing.
For the uplink, two options will be considered: FDMA with GMSK modulation, and SC-FDMA (including single-tone transmission as a special case of SC-FDMA).
A single synchronization signal design for the different modes of operation, including techniques to handle overlap with legacy LTE signals while reducing the power consumption and latencies.

Utilization of the existing LTE procedures and protocols and relevant optimizations to support the selected physical layer and core network interfaces targeting signaling reduction for small data transmissions.

EC-GSM has been introduced as cellular system support for ultra-low-complexity and low-throughput Internet of Things. It targets the following:

Re-using existing designs: Only changing them when necessary to comply with the study item objectives; a reduction in functionality in the GERAN specification to minimize implementation effort and complexity.

Backward compatibility and co-existence with GSM: Multiplexing traffic from legacy GSM devices and CIoT devices on the same physical channels. No impact on the radio units already deployed in the field. Speed the same as supported today (in normal coverage).

Achieving extended coverage: Provide EC by using control channels with blind repetitions and data channels: blind repetitions of MCS-1 (lowest MCS in EGPRS) and HARQ retransmissions. EC has different coverage classes (CCs). The total number of blind transmissions for a given CC can differ between different logical channels.

Figure 1.6 summarizes the LTE UE categories with differing 3GPP evolutions.

With the developments in cellular networks, the doors have been opened wide for the development of 5G architecture and its requirements. The envisioned market space for 5G technology is driven by requirements to enhance the mobile broadband smartphone, with a massive range of machine-type communication and LPWA growth, and the need to provide ultra-reliable, low-latency communications. 3GPP expects the evolution to start from Release 14/15 and is aiming to meet the IMT2020 requirements in Release 16. Figure 1.7 summarizes the main targets for 5G, and the main use cases are summarized in Figure 1.8.

1.4 Evolved Packet System (EPS) Overview

3GPP Release 8 is the starting point when defining the standardization for the Long Term Evolution (LTE) specifications:

The Evolved Packet System (EPS) is defined, mandating the key features and components of both the radio access network (E-UTRAN) and the core network (Evolved Packet Core, EPC).
Orthogonal Frequency Division Multiplexing (OFDM) is defined as the air interface.

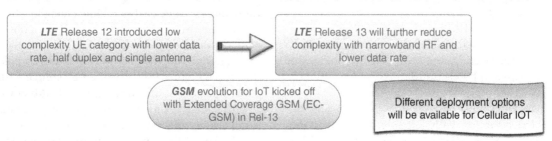

	Release 8	Release 12	Release 13	Release 13
	UE Category 1	UE Category 0	UE Category M	NB-IoT
Downlink peak rate	10 Mbps	1 Mbps	1 Mbps	200 kbps
Uplink peak rate	5 Mbps	1 Mbps	1 Mbps	100 kbps
Number of antennas	2	1	1	1
Duplex mode	Full duplex	Half duplex	Half duplex	Half duplex
UE receive bandwidth	20 MHz	20 MHz	1.4 MHz	0.2 MHz

LTE Release 12 introduced low complexity UE category with lower data rate, half duplex and single antenna → **LTE** Release 13 will further reduce complexity with narrowband RF and lower data rate

GSM evolution for IoT kicked off with Extended Coverage GSM (EC-GSM) in Rel-13

Different deployment options will be available for Cellular IOT

Figure 1.6 Different LTE UE categories for IoT.

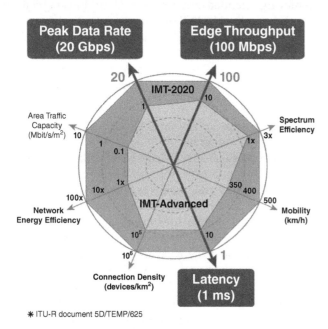

✳ ITU-R document 5D/TEMP/625

Figure 1.7 Main targets for 5G compared to IMT-Advanced.

In OFDM, the carriers are packed much closer together (subcarriers).

This increases spectral efficiency by utilizing a carrier spacing that is the inverse of the symbol or modulation rate.

LTE uses a variable channel bandwidth of 1.4, 3, 5, 10, 15, or 20 MHz.

LTE radio access is designed to operate in two main modes of operation: FDD (Frequency Division Duplex) and TDD (Time Division Duplex).

In *FDD*, separate uplink and downlink channels are utilized, enabling a device to transmit and receive data at the same time.

TDD mode enables full-duplex operation using a single frequency band and time division multiplexing for the uplink and downlink signals.

Figure 1.9 summarizes FDD and TDD operation. The FDD system:

Can operate in full-duplex or half-duplex modes.

Half-duplex FDD is where the mobile can only transmit or receive; i.e.

The user cannot transmit and receive at the same time.

There is reduced mobile complexity since no duplex filter is required.

TDD mode enables full-duplex operation using a single frequency band and time division multiplexing of the uplink and downlink signals. The basic principle of TDD is to use the same frequency band for transmission and reception but to alternate the transmission direction in time. This is a fundamental difference compared to FDD, where different frequencies are used for continuous UE reception and transmission. Like FDD, LTE TDD supports bandwidths from 1.4 MHz up to 20 MHz depending on the frequency band. One advantage of TDD is its ability to provide asymmetrical uplink and downlink allocation. Depending on the system, other advantages include dynamic allocation, increased spectral efficiency, and the improved use of beamforming techniques. This is due to having the same uplink and downlink frequency characteristics.

Since the bandwidth is shared between the uplink and downlink and the maximum bandwidth is specified to be 20 MHz in Release 8, the maximum achievable data rates are lower than in LTE FDD. Therefore, to improve LTE TDD performance, it can be seen that FDD + TDD joint operation (carrier aggregation) will be useful in markets using TDD.

Figure 1.8 Main use cases for 5G.

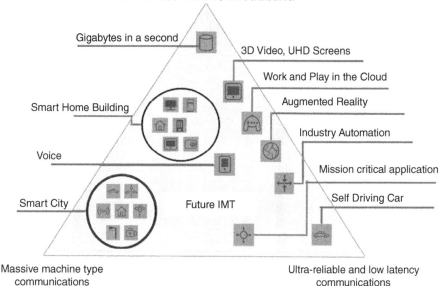

Enhanced Mobile Broadband

Gigabytes in a second

3D Video, UHD Screens

Work and Play in the Cloud

Smart Home Building

Augmented Reality

Industry Automation

Voice

Mission critical application

Smart City

Future IMT

Self Driving Car

Massive machine type communications

Ultra-reliable and low latency communications

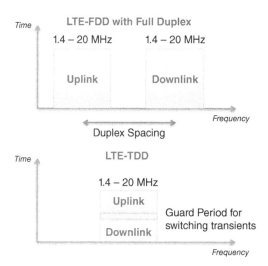

Figure 1.9 FDD and TDD operation.

It is worth noting that the same receiver and transmitter processing capability can be used with both TDD and FDD modes, enabling faster deployment of LTE. However, in an FDD UE implementation, this normally requires a duplex filter when simultaneous transmission and reception is facilitated. In a TDD system, the UE does not need such a duplex filter.

One of the main factors in any cellular system is the deployed frequency spectrum. 2G, 3G, and 4G systems offer multiple band options. The frequency band choice depends on the regulator in each country and the availability of spectrum sharing among network operators in the same country. The device's support for the frequency bands is driven by the hardware capabilities. Therefore, not all bands are supported by a single device. The demands of a multi-mode and multi-band device depend on the market where the device is being sold.

LTE uses a variable channel bandwidth of 1.4, 3, 5, 10, 15, or 20 MHz. Most common worldwide network deployments are in 5 or 10 MHz, given the bandwidth available in the allocated spectrum for the operator. LTE in 20 MHz is being deployed increasingly, especially in bands like 2.6 GHz as well as 1.8 GHz after frequency re-farming.

LTE-FDD requires two center frequencies, one for the downlink and one for the uplink. These carrier frequencies are each given an EARFCN (E-UTRA Absolute Radio Frequency Channel Number). In contrast, LTE-TDD has only one EARFCN. The channel raster for LTE is 100 kHz for all bands. The carrier center frequency must be an integer multiple of 100 kHz.

Tables 1.7 and 1.8 summarize the LTE band allocation for FDD and TDD, respectively.

As was the case with UMTS, LTE supports both FDD and TDD modes. FDD frequency bands are paired, which enables simultaneous transmission on two frequencies: one for the downlink and one for the uplink. The paired bands are also specified with sufficient separations for improved receiver performance. TDD frequency bands are unpaired, as uplink and downlink transmissions share the same channel and carrier frequency. The transmissions in uplink and downlink directions are time-multiplexed. The unpaired bands used in TDD mode start from band 33. Note that band 6 is not applicable to LTE (a UMTS-only band) and bands 15 and 16 are dedicated to ITU Region 1.

1.5 Network Architecture Evolution

Figure 1.10 illustrates the network topologies for 3G/Evolved HSPA/LTE. In a 3G network, prior to the

Table 1.7 LTE-FDD band allocation.

Operating Band	Band Name	Uplink Operating Band (MHz)	Downlink Operating Band (MHz)
1	2100	1920–1980	2110–2170
2	1900	1850–1910	1930–1990
3	1800	1710–1785	1805–1880
4	AWS	1710–1755	2110–2155
5	850	824–849	869–894
7	2600	2500–2570	2620–2690
8	900	880–915	925–960
9	1800	1749.9–1784.9	1844.9–1879.9
10	AWS	1710–1770	2110–2170
11	1500	1427.9–1452.9	1475.9–1500.9
12	700	698–716	728–746
13	700	777–787	746–756
14	700	788–798	758–768
17	700	704–716	734–746
18	800	815–830	860–875
19	800	830–845	857–890
20	800	832–862	791–821
21	1500	1447.9–1462.9	1495.9–1510.9
22	3500	3410–3490	3510–3590
23	2000	2000–2020	2180–2200
24	1600	1626.5–1660.5	1525–1559
25	1900	1850–1915	1930–1995

Table 1.8 LTE-TDD band allocation.

Operating Band	Band Name	Uplink and Downlink Operating Band (MHz)
33	1900	1900–1920
34	2000	2010–2025
35	PCS	1850–1910
36	PCS	1930–1990
37	PCS	1910–1930
38	2600	2570–2620
39	1900	1880–1920
40	2300	2300–2400
41	2500	2496–2690
42	3500	3400–3600
43	3700	3600–3800
44	700	703–803

Location Register (VLR), and the Gateway MSC. The packet-switched elements are the Serving GPRS Support Node (SGSN) and the Gateway GPRS Support Node (GGSN).

Furthermore, the control plane and user plane data are forwarded between the core and access networks. The Radio Access Technology (RAT) in the 3G system uses Wideband Code Division Multiple Access (WCDMA). The access network includes all of the radio equipment necessary for accessing the network, and is referred to as the Universal Terrestrial Radio Access Network (UTRAN).

UTRAN consists of one or more Radio Network Sub-systems (RNSs). Each RNS consists of a Radio Network Controller (RNC) and one or more NodeBs. Each NodeB controls one or more cells and provides the WCMDA radio link to the UE.

After the introduction of HSPA and HSPA+ systems in 3GPP, some optional changes have been added to the core network as well as mandatory changes to the access network. On the core network side, an evolved direct tunneling architecture has been introduced whereby the user data can flow between the GGSN and the RNC or directly to the NodeB. On the access network side, some of the RNC functions, such as the network scheduler, have been moved to the NodeB side for faster radio resource management (RRM) operations.

Figure 1.11 provides the LTE Evolved Packet System (EPS) network topology. In summary, EPS consists of:

The E-UTRAN access network (Evolved UTRAN).
The E-UTRAN consists of one or more eNodeBs (eNBs).
An eNB consists typically of three cells.
The eNBs can, optionally, interconnect to each other via the X2 interface.
The EPC core network (Evolved Packet Core).
The EPC core network consists of the main network entities: MME, S-GW, and P-GW.
The EPS can also interconnect with other radio access networks: 3GPP (GERAN, UTRAN) and non-3GPP (e.g. CDMA, Wi-Fi).

The EPC includes the MME, S-GW, and P-GW entities. They are responsible for different functionalities during a call or registration process. The EPC and the E-UTRAN interconnect with the S1 interface. The S1 interface supports a many-to-many relationship between MMEs, serving gateways, and eNBs.

The MME connects to the E-UTRAN by means of the S1 interface. This interface is referred to as S1-C or S1-MME. When a UE attaches to the LTE network, UE-specific logical S1-MME connections are established. This bearer, known as an EPS bearer, is used to exchange the necessary UE-specific signaling messages between the UE and the EPC.

introduction of the HSPA system, the network architecture is divided into circuit-switched and packet-switched domains. Depending on the service offered to the end user, the domains interact with the corresponding core network entities. The circuit-switched elements are the Mobile services Switching Center (MSC), the Visitor

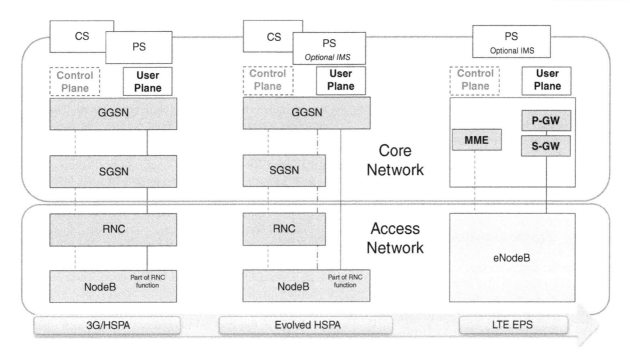

Figure 1.10 3G/Evolved HSPA/ LTE network topology.

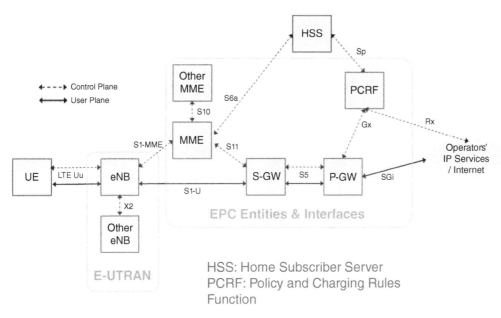

Figure 1.11 LTE/EPS network topology.

Each UE is then assigned a unique pair of eNB and MME identifications during S1-MME control connection. The identifications are used by the MME to send the UE-specific S1 control messages and by the E-UTRAN to send messages to the MME. The identification is released when the UE transitions to Idle state, where the dedicated connection with the EPC is also released. This process may repetitively take place when the UE sets up a signaling connection for any type of LTE call.

The MME and E-UTRAN handle signaling for control plane procedures established for the UE on the S1-MME interface, including:

Initial context setup/UE context release
E-RAB setup/release/modify
Handover preparation/notification
eNB/MME status transfer
Paging UE capability information indication.

The HSS is considered to lie outside the EPC entities and is used to update new EPS subscription data and functions to the EPC. The HSS is located within the HLR of the mobile network and has recently become part of the converged database (CDB) for the entire subscriber services. The PCRF provides QoS policy and charging control (PCC), similar to the 3G PS domain.

The main features of EPS are summarized in Table 1.9.

S1-MME uses S1-AP over SCTP as the transport layer protocol for guaranteed delivery of signaling messages between the MME and the eNodeB. One logical S1-AP connection per UE is established and multiple UEs are supported via a single SCTP association. The following functionalities are conducted in S1-AP:

Setup, modification, and release of E-RABS.
Establishment of an initial S1 UE context.
Paging and S1 management functions.
NAS signaling transport functions between the UE and the MME.
Status transfer functionality.
Trace of active UEs and location reporting.
Mobility functions for UE to enable inter- and intra-RAT HO.

MMEs can also periodically send MME loading information to the E-UTRAN for mobility management procedures. This is not UE-specific information. The S-GW is connected to the E-UTRAN by means of the S1-U interface. After the EPS bearer is established for control plane information, the user data packets start flowing between the EPC and the UE through this interface.

1.6 LTE UE Description

Like that in UMTS, the mobile device in LTE is referred to as the UE (User Equipment) and is comprised of two distinct elements: the USIM (Universal Subscriber Identity Module) and the ME (Mobile Equipment). The ME supports a number of functional entities and protocols including:

RR (Radio Resource): This supports both the control and user planes. It is responsible for all low-level protocols including RRC, PDCP, RLC, MAC, and PHY layers. The layers are similar to those in the eNB protocol layer.

EMM (EPS Mobility Management): A control plane entity which manages the mobility states of the UE; LTE Idle, LTE Active, and LTE Detached. Transactions within these states include procedures such as TAU (Tracking Area Updates) and handovers.

ESM (EPS Session Management): A control plane activity which manages the activation, modification, and deactivation of EPS bearer contexts. These can either be default or dedicated EPS bearer contexts.

The physical layer capabilities of the UE may be defined in terms of the frequency bands and data rates supported. Devices may also be capable of supporting adaptive modulation including QPSK (Quadrature Phase Shift Keying), 16QAM (16 Quadrature Amplitude Modulation), and 64QAM (64 Quadrature Amplitude Modulation). Modulation capabilities are defined separately in 3GPP for the uplink and downlink. The UE is able to support several scalable channels including 1.4 MHz, 3 MHz, 5 MHz, 10 MHz, 15 MHz, and 20 MHz

Table 1.9 Main features of EPS entities.

EPS Element	Element	Basic Functionality
EPC	MME Mobility Management Entity S-GW Serving Gateway P-GW Packet Data Network (PDN) Gateway	• Signaling and security control • Tracking area management • Inter core network signaling for mobility between 3GPP access networks • EPS bearer management • Roaming and authentication • Packet routing and forwarding • Transport level Quality of Service mapping • IP address allocation • Packet filtering and policy enforcement • User plane anchoring for mobility between 3GPP access networks
E-UTRAN	eNodeB Evolved Node B	• Provides user plane protocol layers: PDCP, RLC, MAC, physical and control plane (RRC) with the user • Radio resource management • E-UTRAN synchronization and interface control • MME selection

whilst operating in FDD and/or TDD. The UE may also support advanced antenna features such as MIMO with different numbers of antenna configurations.

The physical layer and radio capabilities of the UE are advertised to the EPS at the initiation of the connection with the eNB in order to adjust the radio resources accordingly. An LTE-capable device advertises one of the categories listed in Table 1.10 according to its software and hardware capabilities. UE categories > 5 are considered part of LTE-A capabilities, with CA starting from Category 6 and above.

It should also be noted that different combinations of 3G and LTE categories can be supported in the same device, depending on the price and the market in which the device is being sold.

The UE categories are summarized in Table 1.10.

Table 1.11 illustrates how to estimate the peak throughput for sample LTE UE categories.

Table 1.11 shows the maximum capability of each category and which factor will limit it. The maximum data rates in each category are presented as the MAX_Throughput {MIMO configuration, higher-order modulation, carrier aggregation combination} that provides whatever maximum throughput is possible. Therefore, using advanced MIMO with advanced higher-order modulation and advanced carrier aggregation bandwidth techniques may or may not be possible simultaneously in every category on one carrier. For 600

Mbps, for example, it can be achieved within Category 11 by:

$2 \times CC \ 4 \times 4MIMO$, or
$1 \times CC \ 4 \times 4MIMO + 2CC \ 2 \times 2MIMO$, or
$4 \times CC \ 2 \times 2MIMO$
$1 \times CC \ 4 \times 4MIMO \ 256QAM + 1 \times CC \ 2 \times 2MIMO \ 256QAM$, or
$3 \times CC \ 2 \times 2MIMO \ 256QAM$

As mentioned earlier, some operators around the world are aiming to use the current LTE spectrum and deployment conditions and aggregate it with the Wi-Fi network to achieve Giga LTE. This can happen directly at the LTE protocol stack (LWA), or through application layer aggregations (MP-TCP), especially if the public Wi-Fi network is the operator's own and the core network can be integrated with the EPS network. If 4×4 MIMO or 256QAM are not currently supported in the LTE commercial network, the alternative means to reach a 1 Gbps network is to utilize Wi-Fi aggregation with the minimum LTE capabilities possible (2×2 MIMO + $2 \times CC$ 64QAM). Figure 1.12 illustrates how to achieve 1 Gbps using LTE evolution either by using LTE-U or LWA.

1.7 EPS Bearer Procedures

The EPS bearer service layered architecture may be described as follows:

Table 1.10 LTE UE categories.

UE Category	3GPP Release	Downlink			Uplink		General Description
		Max. Data Rate [Mbps]	Max. Number of Layers	Support for 256 QAM	Max. Data Rate [Mbps]	Support for 64QAM	
Category 0	Rel 12	**1**	**1**	No	**1**	No	Internet of Things
Category 1	Rel 8/9	10	1	No	5	No	
Category 2	Rel 8/9	51	2	No	25	No	Non CA
Category 3	Rel 8/9	102	2	No	51	No	Non CA
Category 4	Rel 8/9	150	2	No	51	No	Non CA
Category 5	Rel 8/9	300	4	No	75	Yes	Non CA
Category 6	Rel 10	301	2 or 4	No	51	No	DL 2 × CC
Category 7	Rel 10	301	2 or 4	No	102	No	UL 2 × CC
Category 8	Rel 10	3000	8	No	1500	Yes	Non CA
Category 9	Rel 11/12	452	2 or 4	No	51	No	DL 3 × CC
Category 10	Rel 11/12	452	2 or 4	No	102	No	DL 3 × CC + UL 2 × CC
Category 11	Rel 11/12	603	2 or 4	Optional	51	No	DL 4 × CC
Category 12	Rel 11/12	603	2 or 4	Optional	102	No	UL 2 × CC
Category 13	Rel 12	391	2 or 4	Yes	150	Yes	DL 2 × CC

$2 \times CC$, $3 \times CC$, and $4 \times CC$ indicate carrier aggregation with 40, 60, and 80 MHz maximum bandwidth, respectively.
The maximum data rates in each category are presented as the MAX_Throughput {MIMO Configuration, Higher Order Modulation, Carrier Aggregation Combination} that provides whatever maximum throughput possible.

Table 1.11 LTE UE categories throughput estimation.

	Downlink		
UE Category	**Max. Data Rate [Mbps]**	**Maximum Number of Bits of a DL-SCH Transport Block Received Within a TTI**	**Maximum DL Through put Calculation**
Category 6	301	75376 (2 layers, 64QAM) 149776 (4 layers, 64QAM)	• $2 \times$ CC with 2×2 MIMO and **64**QAM = (75376 * 2 Codewords * 2 Carriers)/10^3 = 301 Mbps • $1 \times$ CC with 4×4 MIMO and **64**QAM = (149776 * 2 Codewords)/10^3 = 300 Mbps Then, 4×4 can't be used with > 20 MHz because it would exceed the category rate.
Category 9	452	75376 (2 layers, 64QAM) 149776 (4 layers, 64QAM)	• $3 \times$ CC with 2×2 MIMO and 64QAM = (75376 * 2 Codewords * 3 Carriers)/10^3 = 452 Mbps • $1 \times$ CC with 4×4 MIMO and 64QAM = (149776 * 2 Codewords)/HT^3 = 300 Mbps Then, 4×4 can't be used with > 20 MHz because it would exceed the category rate. So, if 4×4 MIMO is used, this category acts like Category **6.**
Category 11	603	75376 (2 layers, 64QAM) 97896 (2 layers, 256QAM) 149776 (4 layers, 64QAM) 195816 (4 layers, 256QAM)	• $4 \times$ CC with 2×2 MIMO and 64QAM = (75376 * 2 Codewords * 4 Carriers)/10^3 = 603 Mbps • $3 \times$ CC with 2×2 MIMO and 256QAM = (97896 * 2 Codewords * 3 Carriers)/10^3 = 587 Mbps • $2 \times$ CC with 4×4 MIMO and 64QAM = (149776 * 2 Codewords * 2 Carriers)/10^3 = 599 Mbps Then, 256QAM can't be used with > 60 MHz, and 4×4 can't be used with > 40 MHz because it would exceed the category rate, and maximum CA is 80 MHz but with 2×2 MIMO and 64QAM. *A combination is also possible (e.g. $2 \times$ CC with 2×2MIMO + Third carrier with 4×4 MIMO)*
Category 13	391	97896 (2 layers, 256QAM) 195816 (4 layers, 256QAM)	• $2 \times$ CC with 2×2 MIMO and 256QAM – (97896 * 2 Codewords * 2 Carriers)/10^3 = 391 Mbps • $1 \times$ CC with 4×4 MIMO and 256QAM = (195816 * 2 Codewords)/10^3 – 391 Mbps Then, 256QAM can't be used with > 40 MHz, and 4x4 can't be used with > 20 MHz because it would exceed the category rate.

MIMO 2×2 and 4×4 always use two codewords (i.e. two DL-SCH transport blocks received within a TTI) that are spread over two or four layers. See the MIMO section later in this chapter.
Each carrier can transmit the entire maximum number of bits of a DL-SCH transport block received within a TTI.

A radio bearer transports the packets of an EPS bearer between a UE and an eNB. There is a one-to-one mapping between an EPS bearer and a radio bearer.

An S1 bearer transports the packets of an EPS bearer between an eNB and the S-GW.

An S5/S8 bearer transports the packets of an EPS bearer between the S-GW and the P-GW.

UE stores a mapping between an uplink packet filter and a radio bearer to create the binding between an SDF (Service Data Flow) and a radio bearer in the uplink, described later.

A P-GW stores a mapping between a downlink packet filter and an S5/S8 bearer to create the binding between an SDF and an S5/S8 bearer in the downlink.

An eNB stores a one-to-one mapping between a radio bearer and an S1 to create the binding between a radio bearer and an S1 bearer in both the uplink and downlink.

An S-GW stores a one-to-one mapping between an S1 bearer and an S5/S8 bearer to create the binding between an S1 bearer and an S5/S8 bearer in both the uplink and downlink.

Figure 1.13 illustrates EPS bearers. For an EPS bearer to be established:

Initial attach between the UE and the EPC is required to establish signaling and user plane bearers.

EPS security is negotiated at attach and subsequent calls.

Quality of Service for the bearer is negotiated.

1.7.1 EPS Registration and Attach Procedures

The attach procedure usually starts when the UE initiates the request. After establishing an RRC connection, the UE can send an attach request message to the MME. The

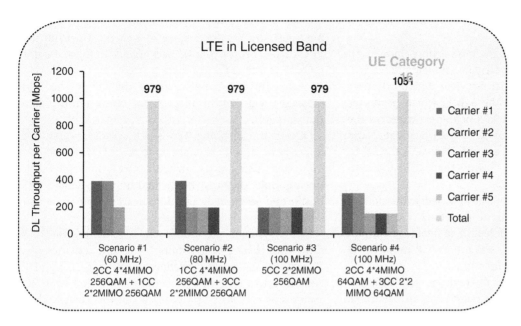

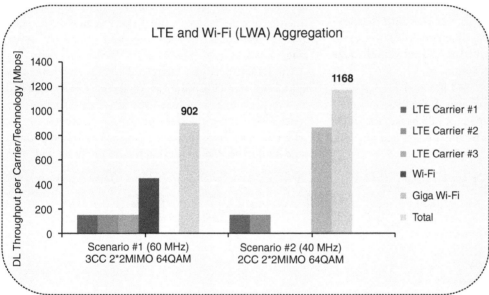

Figure 1.12 Giga LTE roadmap in licensed and unlicensed bands.

Figure 1.13 EPS bearers.

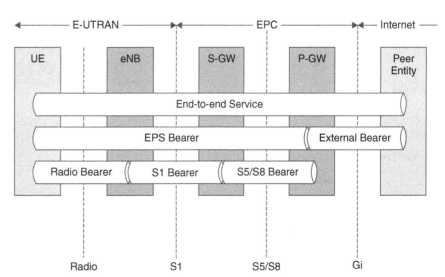

UE also requests PDN connectivity along with the attach request.

After all necessary signaling connections have been established, the EPC may trigger security functions. The HSS (Home Subscriber Service) downloads user subscriber information to the MME, which processes the UE request for default EPS bearer setup. After the default EPS bearer and QoS have been negotiated and agreed to among the MME and S-GW/P-GW, the MME forwards the default bearer setup request to the eNB and UE.

The eNB and the UE then acknowledge the default bearer setup and communicate the attach accept messages to the EPC. The EPS bearer is finally active and data can flow between the UE and the IP network, in both uplink and downlink directions.

At this point, the UE typically registers with a default APN, as per the subscription policies. If an additional APN is available, the process needs to continue setting up another EPS bearer.

Figure 1.14 illustrates the registration and attach procedure in the EPS [2]. When the UE enters LTE coverage or powers up, it first registers with the EPS network through the "Initial EPS Attach" procedure:

Register the UE for packet services in the EPS.
Establish (at a minimum) a default EPS bearer that a UE
   could use to send and receive the user application data.
IPv4 and/or IPv6 address allocation.

Once the SRBs have been established, control plane messages and parameters are sent to the UE from the EPC or E-UTRAN. The UE will adhere to these parameters to continue the protocol procedures on the access stratum. The parameters sent to the UE in the SRB messages will control all protocol layers for the data transmission.

Due to the mapping between the radio bearer and lower-layer logical channels, up to eight DRBs can be set up to carry user plane data connected to multiple PDNs (for example, they may be divided into one default EPS bearer and seven dedicated EPS bearers, or two default bearers and six dedicated bearers).

In 3G systems, the mobile registers on the network first. Then, based on downlink or uplink activities, the IP address allocation procedure starts as part of "PDP context activation". This procedure is referred to in 3G systems as establishing a PS data call (packet-switched data call). The procedure for a PS data call setup follows that for a circuit-switched call setup. When the user initiates or receives a call, the CS or PS call is established and all resources are then allocated at the call-setup stage.

The LTE procedure, compared to 3G, can provide a significant signaling reduction on the protocol layers and also improves the end user experience in terms of data reactivation after a certain period of inactivity. In 3G, when the user disconnects the data call and then initiates a new one, the PDP context activation may start all over

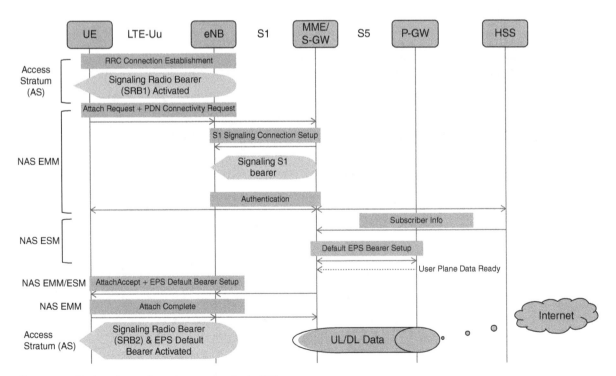

Figure 1.14 Registration and attach procedure in the EPS system.

again. However, in LTE, if the same procedure is done by the user, the call-setup time for a data call is reduced because the default DRB has already been assigned to the user when it first attached to the EPS system.

The dedicated DRB does not necessarily require an extra IP address. The protocol stack uses the Traffic Flow Template (TFT) information to decide what to do with each IP packet. Uplink and downlink traffic are mapped onto proper bearers based on TFT filters configured at the UE and P-GW.

This concept makes the dedicated bearer activation similar to the secondary PDP context activation in 3G that can be used by the IMS, for example, to ensure real-time data are delivered promptly. LTE radio bearers are summarized in Figure 1.15.

1.7.2 EPS Security Basics

Authentication and Key Agreement (AKA) involves interworking with the subscriber's HSS in order to obtain Access Authorization and Accounting (AAA) information to authenticate the subscriber. During AKA, keys are created for AS and NAS integrity protection and ciphering.

The integrity and ciphering procedures involve both NAS and AS:

NAS security context activation provides both integrity protection and ciphering for NAS signaling. The procedure takes place between the UE and the MME.

AS security context activation provides integrity and ciphering protection for RRC signaling in addition to ciphering for user plane data to be sent over the air

interface. The procedure takes place between the UE and the eNB.

The MME selects a NAS integrity algorithm and a NAS ciphering algorithm for the UE. The MME is expected to select the NAS algorithms that have the highest priority according to the ordered lists. The selected algorithm is indicated in the NAS Security Mode Command message to the UE and that message also includes the UE security capabilities. The integrity of this message is protected by the MME with the selected algorithm.

The UE verifies that the message from the MME contains the correct UE security capabilities. This enables detection of attacks in the case where an attacker has modified the UE security capabilities in the initial NAS message.

The UE then generates NAS security keys based on the algorithms indicated in the NAS Security Mode Command and replies with an integrity-protected and ciphered NAS Security Mode Complete message. NAS security is activated at this point.

After this point, the eNB creates the AS security context when it receives the keys from the MME. The eNB generates the integrity and encryption keys and selects the highest priority ciphering and integrity protection algorithms from its configured list that are also present in the UE's EPS security capabilities.

Upon receipt of the AS Security Mode Command, the UE generates integrity and encryption keys and sends an AS Security Mode Complete message to the eNB.

Figure 1.16 demonstrates key security basics for the EPS. We can summarize the key security basics as follows:

SRB	Default EPS DRB	Dedicated EPS DRB

• **SRB ID 0**
  • used to establish the RRC connection request when the UE has transitioned into **connected mode. SRB0 carries** commom control information required to establish the RRC connection
 	
• **SRB ID 1**
  • **used for RRC messages, as well as RRC messages carrying high** priority NAS signaling
 	
• **SRB ID 2**
  • used for RRC carrying low **priority NAS signaling. Prior to** its establishment, low priority signaling is sent on SRB1
	• One of the significant changes introduced in LTE is that when the mobile device connects to the network it also implicitly gets an IP address. This is called "Default EPS Bearer Activation"
	
	• With the default bearer activation, the packet call is established the same time when the UE attaches to the EPS. This is the concept that makes the LTE's connectivity to be known as "always-on"
	
	• Even though the default DRB is enough for the downlink and uplink data transfer, it comes without any quality of Service guarantees
		• For real time streaming applications, QoS may be needed especially on the air interface
		
		• Such IP packets associated with these types of applications may need to be assigned with a higher priority than other packets, especially when the bandwidth is limited
		
		• The dedicated bearer becomes important in order to support different types of applications in EPS network. Dedicated DRB can be set up right after Default DRB

Figure 1.15 LTE radio bearer description.

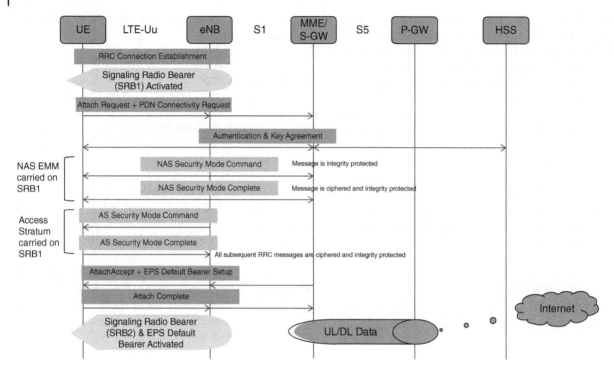

Figure 1.16 EPS security basics.

AKA: To prevent fraud that occurs when a third party obtains a copy of a subscriber's network identification information and uses it to access the system fraudulently.

Ciphering: Used to protect all user data and signaling from being overheard by an unauthorized entity.

Integrity: Protects signaling information from being corrupted. It is a message authentication function that prevents a signaling message from being intercepted and altered by an unauthorized device.

1.7.3   EPS QoS

In order to support a mixture of non-real-time and real-time applications such as voice and multimedia, delay and jitter may become excessive if the flows of traffic are not coordinated. Packet switches should be able to classify, schedule, and forward traffic based on the destination address, as well as the type of media being transported. This becomes possible with QoS-aware systems.

The QoS for data radio bearers is provided to the eNB by the MME using the standardized QoS attributes. Based on these attributes configured by the EPS, the protocol layers between the UE and the eNB can manage the ongoing scheduling of uplink and downlink traffic.

Figure 1.17 summarizes the QoS aspects of EPS. Different QoS Class Identifier (QCI) values are provided in Table 1.12.

1.8   Access and Non-access Stratum Procedures

Figure 1.18 illustrates an overview of the LTE protocol stack. The LTE air interface provides connectivity between the user equipment and the eNB. It is split into a control plane and a user plane. Of the two types of control plane signaling, the first is provided by the Access Stratum (AS) and carries signaling between the UE and the eNB; the second carries Non-Access Stratum (NAS) signaling messages between the UE and the MME, which is piggybacked onto an RRC message. The user plane delivers the IP packets to and from the EPC, S-GW, and PDN-GW.

The structure of the lower-layer protocols for the control and user planes in the AS is the same. Both planes utilize the protocols of PDCP (Packet Data Convergence Protocol), RLC (Radio Link Control), and MAC (Medium Access Control) as well as the PHY (Physical Layer) for the transmission of signaling and data packets.

The NAS is the layer above the AS layers. There are also two planes in the NAS: the higher-layer signaling related to the control plane and the IP data packets of the user plane. NAS signaling exists in two protocol layers, EMM and ESM. The NAS user plane is IP-based. The IP data packets pass directly into the PDCP layer for processing and transmission to or from the user.

Figure 1.17 QoS aspects in EPS.

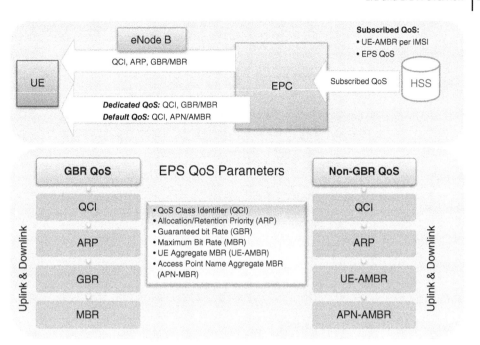

Table 1.12 Different QCI values and corresponding service requirements.

QCI	Resource Type	Priority	Packet Delay Budget (PDB)	Packet Error Loss Rate (PELR)	Examples of Services
1		2	100 ms	10^{-2}	Conversational voice
2		4	150 ms	10^{-3}	Conversational video (live streaming)
3	GBR	3	50 ms	10^{-3}	Real-time gaming
4		5	300 ms	10^{-6}	Non-conversational video (buffered streaming)
5		1	100 ms	10^{-6}	IMS signaling
6		6	300 ms	10^{-6}	Video (buffered streaming, TCP-based (www, e-mail, ftp, p2p file sharing)
7	Non-GBR	7	100 ms	10^{-3}	Voice, Video, Interactive gaming
8		8	300 ms	10^{-6}	Same as QCI 6 but used for further differentiation
9		9	300 ms	10^{-6}	

1.8.1 EMM Procedures and Description

1.8.1.1 Definitions

Attach: Used by the UE to attach to the EPC for packet services in the EPS. It can also be used to attach to non-EPS services, for example, CSFB/SMS.

Detach: Used by the UE to detach from EPS services. It can also be used for other procedures such as disconnecting from non-EPS services.

Tracking Area Updating: Initiated by the UE and used for identifying the UE location at eNB level for paging purposes in Idle mode.

Service Request (PS call): Used by the UE to get connected and establish the radio and S1 bearers when uplink user data or signaling is to be sent.

Extended Service Request: Used by the UE to initiate a circuit-switched fallback call or respond to a mobile terminated circuit-switched fallback request from the network, i.e. non-EPS services (CSFB).

GUTI Allocation: Allocate a GUTI (Globally Unique Temporary Identifier) and optionally to provide a new TAI (Tracking Area Identity) list to a particular UE.

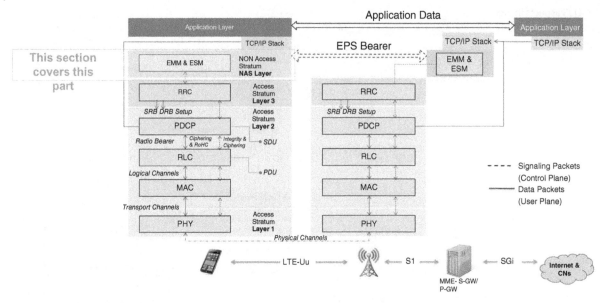

Figure 1.18 LTE protocol stack overview.

Authentication: Used for AKA (Authentication and Key Agreement) between the user and the network.

Identification: Used by the network to request a particular UE to provide specific identification parameters, for example, the IMSI (International Mobile Subscriber Identity) or the IMEI (International Mobile Equipment Identity).

Security Mode Control: Used to take an EPS security context into use and initialize NAS signaling security between the UE and the MME with the corresponding NAS keys and security algorithms.

EMM Status: Sent by the UE or by the network at any time to report certain error conditions.

EMM Information: Allows the network to provide information to the UE.

NAS Transport: Carries SMS (Short Message Service) messages in an encapsulated form between the MME and the UE.

Paging: Used by the network to request the establishment of a NAS signaling connection to the UE. Is also includes the circuit-switched service notification.

1.8.1.2 ESM Procedures and Description

Default EPS Bearer Context Activation: Used to establish a default EPS bearer context between the UE and the EPC.

Dedicated EPS Bearer Context Activation: Establishes an EPS bearer context with specific QoS (Quality of Service) between the UE and the EPC. The dedicated EPS bearer context activation procedure is initiated by the network, but may be requested by the UE by means of the UE-requested bearer resource allocation procedure.

EPS Bearer Context Modification: Modifies an EPS bearer context with a specific QoS.

EPS Bearer Context Deactivation: Deactivates an EPS bearer context or disconnects from a PDN by deactivating all EPS bearer contexts.

UE-Requested PDN Connectivity: Used by the UE to request the setup of a default EPS bearer to a PDN.

UE-Requested PDN Disconnect: Used by the UE to request disconnection from one PDN. The UE can initiate this procedure to disconnect from any PDN as long as it is connected to at least one other PDN.

UE-Requested Bearer Resource Allocation: Used by the UE to request an allocation of bearer resources for a traffic flow aggregate.

UE-Requested Bearer Resource Modification: Used by the UE to request a modification or release of bearer resources for a traffic flow aggregate or modification of a traffic flow aggregate by replacing a packet filter.

ESM Information Request: Used by the network to retrieve ESM information, i.e. protocol configuration options, APN (Access Point Name), or both from the UE during the attach procedure.

ESM Status: Report, at any time, certain error conditions detected upon receipt of ESM protocol data.

Figure 1.19 summarizes the EPS access and non-access strata and we can summarize them as follows:

Access Stratum (AS) resides between the UE and the E-UTRAN.

Consists of multiple protocol layers: RRC, PDCP, RLC, MAC, and the PHY (physical) layer.

The AS signaling provides a mechanism to deliver NAS signaling messages intended for control plane procedures as well as the lower-layer signaling and parameters required to set up, maintain, and manage the connections with the UE.

Figure 1.19 EPS access and non-access strata.

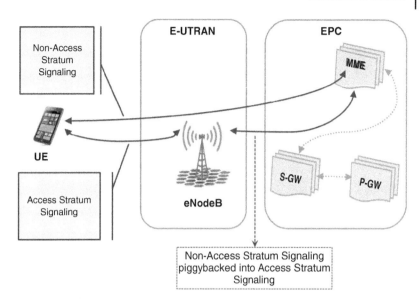

Non-Access Stratum (NAS) layer between the UE and the EPC.

Responsible for handling control plane messaging related to the core network.

NAS includes two main protocols: EMM and ESM.

The LTE system is designed to simplify the procedures carried on the EPS. This is possible by designing and assigning the required identifiers at different interfaces within the EPS system. The different identities defined in the EPS system are shown in Figure 1.20. Different types of identifiers are needed between the eNB and the UE as part of the RNTI. These RNTIs are used for different procedures such as paging, random access, and system information on the air interface. A list of EPS identifiers with descriptions and assignments is provided in Table 1.13.

1.8.2 EPS Mobility Management (EMM)

EMM is a control plane entity which manages the mobility states of the UE:

The main EMM function is similar to that in the PS domain of UMTS/GERAN – to provide attach, detach, and TAU (Tracking Area Updates).

NAS security is an additional function of the NAS, providing services to the NAS protocols, for example, integrity protection and ciphering of NAS signaling messages.

LTE has been designed with "ready-to-use" IP connectivity and an "always-on" experience, so there is a linkage between mobility management and session management procedures during the attach procedure.

The success of the attach procedure is dependent on the success of the default EPS bearer context activation procedure.

EMM procedures supported by the NAS protocol in the UE and NW are summarized in Table 1.14. The EMM common procedure can always be initiated whilst a NAS signaling connection exists. Only one UE-initiated, EMM-specific procedure can be running at any time.

1.8.3 Session Management (ESM)

The basic EPS session management (ESM) function is a control plane activity which manages the activation, modification, and deactivation of EPS bearer contexts. These can either be default or dedicated EPS bearer contexts. Table 1.15 summarizes the EMS procedures. The transaction procedure enables the UE to request resources (IP connectivity to a PDN or dedicated bearer resources). The EPS bearer context procedure is always initiated by the NW. The transaction-related procedures are initiated by the UE, with the exception of the ESM information request procedure. The ESM status and notification procedure can be related to an EPS bearer context or to a transaction procedure.

1.8.3.1 Notification Procedure

The network can use the notification procedure to inform the UE about events which are relevant to the upper layer which is using an EPS bearer context or has requested a transaction procedure. If the UE has indicated that it supports the notification procedure, the network may initiate the procedure at any time while a PDN connection exists or a transaction procedure is ongoing.

1.8.3.2 ESM Status Procedure

The purpose of sending the ESM Status message is to report, at any time, certain error conditions detected upon receipt of ESM protocol data. The ESM Status message can be sent by either the MME or the UE.

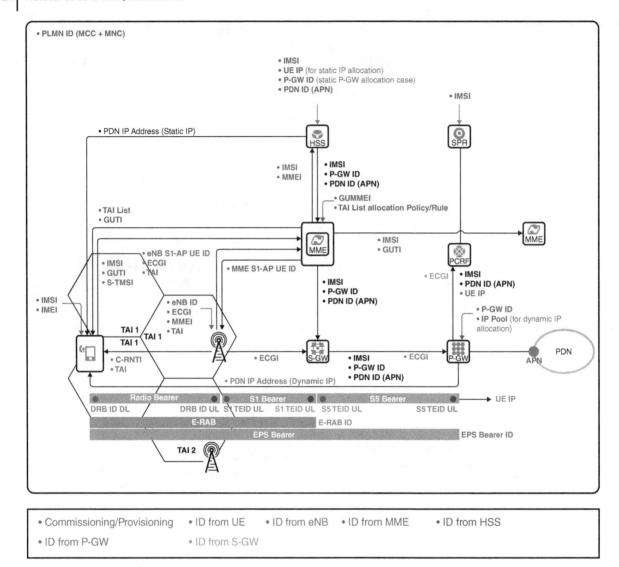

Figure 1.20 EPS identifiers.

1.8.4 EPS Idle and Active States

On the air interface, the UE typically transitions into the RRC Idle state after successfully attaching to the LTE system [3]. The UE remains in this state as long as there are no radio interface downlink or uplink packet activities with the eNB. When a data activity is initiated by a user or an application installed on the device, the UE immediately transits into RRC Connected state and remains in this state until the packet connectivity timer, known as the "User Inactivity timer," expires. The timer is configured in the eNB and is used to monitor the data activity for a user within a timed window. When the timer expires, the eNB releases the RRC connection and immediately triggers the UE's transition to the RRC Idle state.

The concepts of NAS and AS states are also available in 3G systems. In the UMTS air interface, the RRC states can be in either Connected or Idle mode. In Connected mode, the UE can be served in four different states: Cell_DCH, Cell_FACH, Cell_PCH, or URA_PCH. The state transitions in the LTE air interface are simplified to only Idle and Connected modes, avoiding all the timers and optimizations. The RRC-level state transition from Connected to Idle mode targets an improved battery lifetime of the device. The battery consumption is expected to be more efficient in Idle state when there is no connectivity or dedicated resource between the device and the eNB. The UE states are shown in Figure 1.21.

The LTE concept is a little different from the UMTS, as the main target is to keep the LTE system "always on". The user plane data can only flow when all the AS and NAS signaling connections and bearers are in Active/Connected states.

The ECM state and the EMM state are independent of each other. Table 1.16 provides a comparison between EMM and ECM states.

Table 1.13 EPS identifiers.

Identifier	Description	Assignment
IMSI	International Mobile Subscriber Identity	Unique identification of mobile (LTE) subscriber.
		Network (MME) gets the PLMN of the subscriber.
PLMN ID	Public Land Mobile Network Identifier	Unique identification of PLMN.
MCC	Mobile Country Code	Assigned by regulator.
MNC	Mobile Network Code	Assigned by regulator.
MSIN	Mobile Subscriber Identification Number	Assigned by operator.
GUTI	Globally Unique Temporary UE Identity	Identifies a UE between the UE and the MME on behalf of the IMSI for security reasons.
TIN	Temporary Identity used in Next Update	GUTI is stored in the TIN parameter of the UE's MM context. The TIN indicates which temporary ID to use in the next update.
S-TMSI	SAE Temporary Mobile Subscriber Identity	Locally identifies a UE in short within an MME group (unique within an MME pool).
M-TMSI	MME Mobile Subscriber Identity	Unique within an MME.
GUMMEI	Globally Unique MME Identity	Identifies an MME uniquely in global terms.
		GUTI contains GUMMEI.
MMEI	MME Identifier	Identifies an MME uniquely within a PLMN.
		Operator commissions at eNB.
MMEGI	MME Group Identifier	Unique within PLMN.
MMEC	MME Code	Identifies an MME uniquely within an MME group.
		S-TMSI contains MMEC.
C-RNTI	Cell-Radio Network Temporary Identifier	Identifies a UE uniquely in a cell.
eNB S1AP UE ID	eNB S1 Application Protocol UE ID	Uniquely identifies a UE on the S1-MME interface in the eNB.
MME S1AP UE ID	MME S1 Application Protocol UE ID	Uniquely identifies a UE on the S1-MME interface in the MME.
IMEI	International Mobile Equipment Identity	Identifies an ME (Mobile Equipment) uniquely.
IMEI/SV	IMEI/Software Version	Identifies an ME uniquely.
ECGI	E-UTRAN Cell Global Identifier	Identifies a cell globally.
		EPC knows the UE location based on the ECGI.
ECI	E-UTRAN Cell Identifier	Identifies a cell within a PLMN.
Global eNB ID	Global eNodeB Identifier	Identifies an eNB globally in the network.
eNB ID	eNodeB Identifier	Identifies an eNB within a PLMN.
P-GW ID	PDN GW Identifier	Identifies a specific PDN-GW.
		HSS assigns P-GW for PDN connection of each UE.

1.8.5 EPS Network Topology for Mobility Procedures

In the example shown in Figure 1.22, if the UE performs EPS registration from TAI_A, the MMEs send TAC_1 and TAC_2 in the TAI list, implying that the UE can roam around in the eNBs with the TACs belonging to this TAI list without having to re-register with the EPS network. This procedure saves on the signaling load. The UE re-registers with a tracking area update procedure if the UE enters into the coverage areas of eNBs that are part of TAC_3 (in TAI_B) and TAC_4 (in TAI_C).

The TA dimensioning and planning in the network are performed in the optimization stage. TA planning can prevent the ping-pong effect of tracking area updating to achieve optimization between paging load, registration overhead, the UE battery, and an improved paging success rate. In the same example, the paging area for the UE served in TAI_A will be for all cells belonging to TAC_1 and TAC_2, but the registration area will be limited to

Table 1.14 EMM procedures supported by the NAS protocol in both the UE and NW.

EMM Procedure Type	EMM Message
EMM common procedure	GUTI reallocation
	Authentication
	Security mode control
	Identification
	EMM information
EMM specific procedure	Attach
	Tracking area updating
	Detach
EMM connection management procedure	Service request
	Paging procedure
	Transport of NAS messages

Table 1.15 EPS session management (ESM).

ESM Procedure Type	ESM Message
Procedures related to EPS bearer contexts	Default EPS bearer context activation
	Dedicated EPS bearer context activation
	EPS bearer context modification
	EPS bearer context deactivation
Transaction-related procedures	ESM status procedure
	Notification procedure
	PDN connectivity procedure
	PDN disconnect procedure
	Bearer resource allocation procedure
	Bearer resource modification procedure
	ESM information request procedure

TAI_A only. TA updating can be either periodical or based on the mobility conditions of the device. An MME area consists of one or more tracking areas. All cells served by an eNB are included in an MME area. There is no one-to-one relationship between an MME area and an MSC/VLR area. Multiple MMEs may have the same MME area (pool area).

A tracking area corresponds to the concept of the Routing Area (RA) used in UMTS.

The TA consists of a cluster of eNBs having the same Tracking Area Code (TAC).

The TAC provides a means of tracking a UE's location in Idle mode.

TAC information is used by the MME when paging an idle UE to notify it of incoming data connections.

The MME sends the tracking area identity list, abbreviated to the TAI list, to the UE during the TA update procedure or the attach procedure.

TA updates occur periodically or when a UE enters a cell with a TAC not in the current TAI list.

The TAI list makes it possible to avoid frequent TA updates due to ping-pong effects along TA borders.

This is achieved by including the old TAC in the new TAI list received at TA update. When the MME pages a UE, a paging message is sent to all cells in the TAI list.

1.9   LTE Air Interface

1.9.1   Multiple Access in 3GPP Systems

UMTS, cdmaOne, and CDMA2000 all use the CDMA air interface. The implementation of the codes and the bandwidths used is different among these systems. UMTS utilizes a 5 MHz channel bandwidth, whereas cdmaOne uses only 1.25 MHz.

The LTE air interface utilizes two different multiple access techniques, both based on OFDM (Orthogonal Frequency Division Multiplexing):

OFDMA (Orthogonal Frequency Division Multiple Access) – used on the downlink.

SC-FDMA (Single Carrier-Frequency Division Multiple Access) – used on the uplink.

CDMA and OFDM are summarized in Figure 1.23.

OFDMA on the downlink has the following advantages:

OFDM is almost completely resistant to multi-path interference due to very long symbol duration.

Higher spectral efficiency for wideband channels.

Flexible spectrum utilization.

However, an OFDM system can suffer from high PAPR (Peak-to-Average Power Ratio) when compared to typical single-carrier systems.

OFDMA generates multiple frequencies, used to transmit useful information simultaneously. It increases spectral efficiency by reducing the spacing between the subcarriers that carry different information. Theoretically, this ensures that each subcarrier does not interfere with the adjacent subcarrier. The downlink can use resource blocks freely from different parts of the spectrum.

SC-FDMA was specified for the uplink because of its PA (Power Amplifier) characteristics. An SC-FDMA signal will operate with a lower PAPR, thus increasing the

Figure 1.21 EPS Idle and Active states.

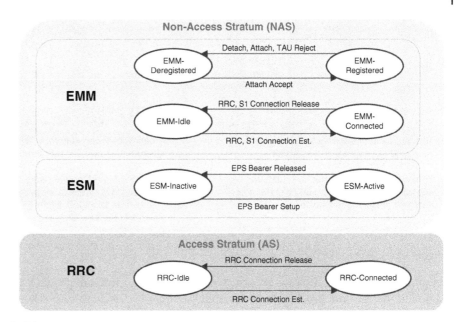

Table 1.16 Comparison between EMM and ECM states.

Layer	State	Description
EMM	EMM-Deregistered	The UE is not successfully attached to the LTE network, so the location of the UE is unknown to an MME and hence it is unreachable by an MME. The UE needs to initiate a (combined) attach procedure to establish an EMM context between the UE and the MME.
	EMM-Registered	The EMM context has been established and a default EPS bearer context has been activated in the UE. The UE position is known to the MME with an accuracy of cell level (ECM-Connected) or tracking area level (ECM-Idle).
ECM	ECM-Idle	No NAS signaling connection has been established. The UE will perform a cell selection/reselection procedure.
	ECM-Connected	A NAS signaling connection has been established and the UE has been assigned radio resource. The mobility of the UE is handled by the NW-controlled handover procedure.

Figure 1.22 EPS network topology.

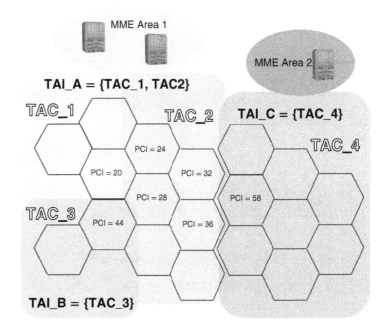

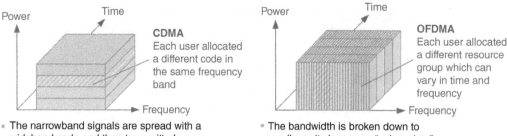

CDMA
Each user allocated a different code in the same frequency band

- The narrowband signals are spread with a wideband code and then transmitted

- The receivers are designed to extract the encoded signal (with the specific code) and treat everything else as noise

OFDMA
Each user allocated a different resource group which can vary in time and frequency

- The bandwidth is broken down to smaller units known as "subcarriers"

- These are grouped together and allocated as a resource to a device. The device can be allocated different resources in both the time and frequency domain

Figure 1.23 CDMA and OFDM.

battery life for users, allowing efficient terminal power amplifier design, and providing a better uplink cell coverage.

SC-FDMA produces a waveform associated with a single-carrier system:

Each symbol is sent one at a time, in a similar manner to Time Division Multiple Access (TDMA).
The uplink user-specific allocation is continuous to enable single-carrier transmission.

Figure 1.24 illustrates the difference between OFDMA and SC-FDMA.

1.9.2 Time–Frequency Domain Resources

The LTE air interface has procedures similar to those in HSPA. The main difference is that LTE uses OFDMA instead of WCDMA. This requires changes in the physical layer as well as enhancements in some of the MAC and RLC functionality.

Figure 1.25 shows a possible allocation of the PHY layer downlink channels into the OVSF code tree. In this figure, each channel is assigned a separate OVSF code. For example, the HSDPA channel is assigned spreading factor (SF) 16. All SFs below the used codes of SF 16 will be blocked, as they would not maintain channel orthogonality. Consequently, SF allocation between the channels is important to ensure all channels and users are allocated a separate code when a call is initiated in the cell.

The LTE-FDD frame structure is demonstrated in Figure 1.26. The E-UTRA air interface is based on OFDMA. The LTE-FDD frame structure has the following characteristics:

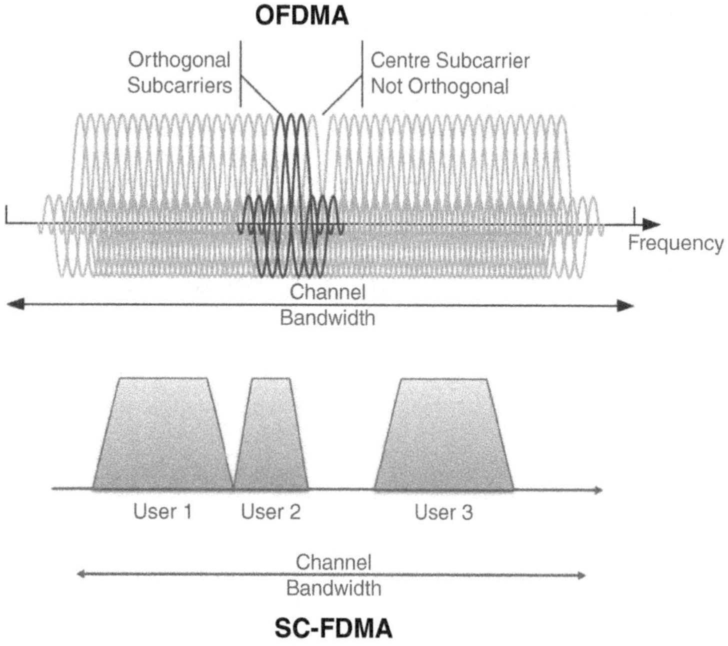

Figure 1.24 OFDMA versus SC-FDMA.

OFDMA

Orthogonal Subcarriers

Centre Subcarrier Not Orthogonal

Frequency

Channel Bandwidth

User 1   User 2   User 3

Channel Bandwidth

SC-FDMA

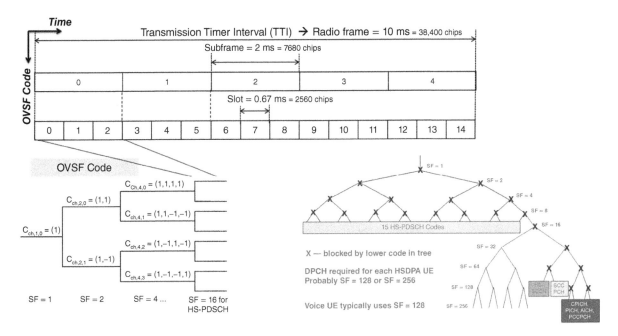

Figure 1.25 WCDMA frame structure.

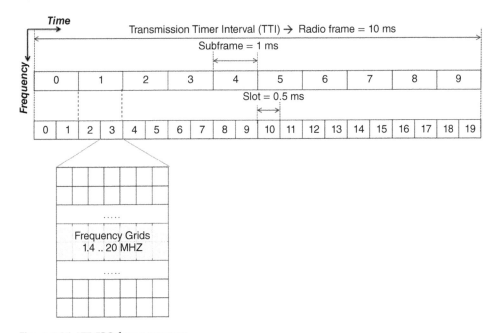

Figure 1.26 LTE-FDD frame structure.

It enables multiple devices to receive information at the same time but on different parts of the radio channel.

In most OFDMA systems, this is referred to as a "subchannel", i.e. a collection of subcarriers.

However, in E-UTRA, the term subchannel is replaced by the term PRB (Physical Resource Block).

A PRB is used in LTE to describe the physical resource in the time/frequency grid.

E-UTRA uses a variable channel bandwidth of 1.4, 3, 5, 10, 15, or 20 MHz with an OFDMA downlink and an SC-FDMA uplink.

Figure 1.27 illustrates physical resource blocks and resource elements (REs) in LTE-FDD. A PRB consists of 12 consecutive subcarriers and lasts for one slot, 0.5 ms. Each subcarrier is spaced by 15 kHz. The N_{RB}^{DL} parameter is used to define the number of RBs (resource blocks) used in the downlink. This is dependent on the channel bandwidth. In contrast, N_{RB}^{UL} is used to identify the number of resource blocks in the uplink. Each RB consists of N_{SC}^{RB} subcarriers, which, for standard operation, is set to 12 or a total of 180 kHz lasting in the 0.5 ms slot. The PRB is used to identify an allocation. It typically

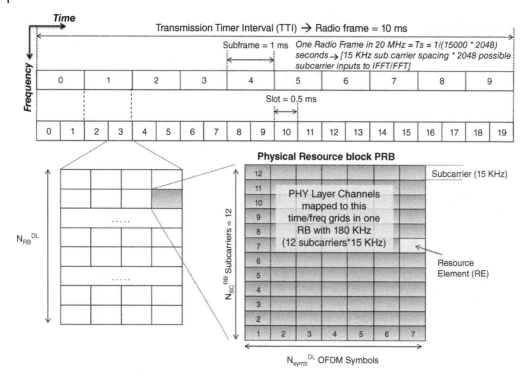

Figure 1.27 Physical resource block and resource element in LTE-FDD.

includes six or seven symbols, depending on whether an extended or normal cyclic prefix is configured [4].

The term RE (resource element) is used to describe one subcarrier lasting one symbol. This can then be assigned to carry modulated information, reference information, or nothing. The RB resources assigned to a specific UE by the eNodeB scheduler can be contiguous, where subcarriers from one RB to another are sequential; this is referred to as a "Localized Virtual Resource Block". Alternatively, the RB resources scheduled to a UE can be distributed in such a way that some resources are continuous and some are assigned a pre-defined distance away; this is referred to as a "Distributed Virtual Resource Block".

For the control information channel, such as the PDCCH, the time/frequency grid shares parts of the bandwidth with the data channels, such as PDSCH. Therefore, the control region is limited up to the first three symbols of the subframe for large bandwidths and up to four symbols for small bandwidths. This allocation is defined by the Control Channel Element (CCE) and the Resource Element Group (REG). Each CCE consists of 36 usable REs derived from 9 REGs * 4 usable REs per REG [5].

Time domain frame arrangements are illustrated in Table 1.17. The type 2 radio frame structure is used for TDD. One key addition to the TDD frame structure is the concept of "special subframes". This includes a DwPTS (Downlink Pilot Time Slot), GP (Guard Period),

and UpPTS (Uplink Pilot Time Slot). These have configurable individual lengths and a combined total length of 1 ms. There are various frame configuration options supported for TDD. Configuration options 0, 1, 2, and 6 have a 5 ms switching point and therefore require two special subframes, whereas the rest are based on a 10 ms switching point. In Table 1.17, the letter "D" is reserved for downlink transmissions, "U" denotes subframes reserved for uplink transmissions, and "S" denotes a special subframe with the three fields: DwPTS, GP, and UpPTS.

Frequency domain frame arrangements are shown in Table 1.18. A DC (direct current) subcarrier is located at the center of the frequency band. It can be used by the UE to locate the center of the frequency band. There are two guard bands at the edges of the band to avoid interference with adjacent bands.

The LTE transmission scheme provides a time resolution of 12 or 14 OFDM symbols for each subframe of 1 ms, depending on the length of the OFDM cyclic prefix. The frequency resolution provides for a number of resource blocks ranging from 6 to 100, depending on the bandwidth, each containing 12 subcarriers with 15 kHz spacing. Different types of data occupy the resource elements that make up the resource grid. The various physical channels and signals that constitute the content of the resource grid are described in the next section.

Table 1.17 Time domain frame arrangements.

Component	Configuration		
Frame types	FS-1: Used for LTE-FDD UL and DL FS-2: Used for LTE-TDD where DL and UL are sharing the same frame		
Frame	10 ms		
Subframe	1 ms (= the "TTI", Time Transmission Interval)		
Slot	0.5 ms		
	Normal Cyclic Prefix	Extended Cyclic Prefix	MBSFN [a]
# OFDM symbols/slot	7	6	3
Cyclic prefix length (µs)	5.2 for first symbol	16.67	33.3
	4.69 for other symbols		

a) MBSFN is used for multimedia broadcast over a single-frequency network (eMBMS).

Table 1.18 Frequency domain frame arrangements for different BW.

Component	1.4 MHz	3 MHz	5 MHz	10 MHz	15 MHz	20 MHz
Guard band	~22%	10%	10%	10%	10%	10%
Subcarrier spacing (kHz)	15	15	15	15	15	15
DC subcarrier	1	1	1	1	1	1
# of resource blocks	6	15	25	50	75	100
# of subcarriers (without DC)	72	180	300	600	900	1200
Total bandwidth available (MHz)	1.1	2.7	4.5	9	13.5	18

The RG for channels greater than 1.4 MHz is illustrated in Figure 1.28. Each resource element contains the modulated symbol for three different types of information contained in the physical resource grid:

User data: Physical layer channels carrying data for the users.

Variable allocation and bandwidth depending on user data demand and available cell capacity.

Figure 1.28 Resource grid arrangements.

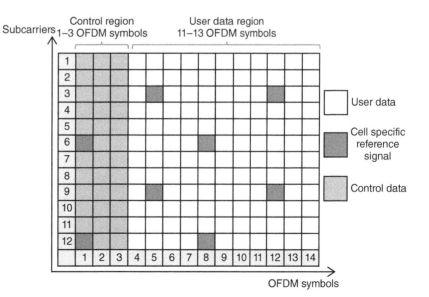

Reference or synchronization signals: Physical layer channels carrying the pilot or sync channels.

Fixed allocation and bandwidth depending on the number of antennas configured.

Control information/region: Physical layer channels carrying control-related information needed to decode the user data channels.

Variable bandwidth depending on channel conditions and cell capacity.

Total user throughput depends on the overhead that the control information and reference signals occupy from the total bandwidth available.

In an OFDM time–frequency grid, up to four symbols of the frame are used for the control channels. These channels are PCFICH, PHICH, and PDCCH. The standard allows dynamic or static symbol assignment. The PDCCH capacity depends on the channel BW (the number of REs) and the required RE for the scheduled UE. The PDCCH consists of Control Channel Elements (CCEs), each CCE is 36 REs. The CCEs form the control region per subframe and can occupy up to four symbols. The number of supported users will depend on the available number of CCEs that can serve the control channel assigned to schedule user data channels for each user. Note the following:

The control region is arranged by:

Control Channel Elements (CCEs) and the Resource Elements Group (REG).

Each CCE consists of 36 usable REs derived from 9 REGs * 4 usable REs per REG.

The total number of CCEs available in the cell depends on the system bandwidth and the number of OFDM symbols allocated for control information in a subframe.

This bandwidth is signaled dynamically in each subframe:

1, 2, or 3 OFDM symbols/subframe for bandwidths above 1.4 MHz.

2, 3, or 4 OFDM symbols/subframe for bandwidths of 1.4 MHz.

A dedicated physical channel called the PCFICH is used to indicate to the UE this information.

1.10 OFDM Signal Generation

Figure 1.29 illustrates downlink physical layer processing [5]. MIMO builds on Single Input, Multiple Output (SIMO), also called Receive Diversity (RxD), as well as Multiple Input, Single Output (MISO), also called Transmit Diversity (TxD). Both of these techniques seek to boost the SNR in order to compensate for signal degradation. As a signal passes from Tx to Rx, it gradually weakens, while interference from other RF signals also reduces the SNR. In addition, in dense urban environments, the RF signal frequently encounters objects which alter its path or degrade the signal. Multiple-antenna systems can compensate for some of the loss of SNR due to multi-path conditions by combining signals that have different fading characteristics, as the path from each antenna will be slightly different.

However, SIMO or MISO systems may not be fully suitable for the high-speed data rates promised in 3GPP's next generation of cellular systems. Therefore, the full version of MIMO can achieve benefits in terms of both increased SNR and throughput gains. MIMO in 3GPP exploits several concepts such as spatial multiplexing, transmit diversity, beamforming, and multi-user MIMO. All these techniques fall into two categories: open loop or closed loop.

A MIMO system utilizes the space and time diversity in a multi-path rich environment and creates multiple parallel data transmission pipes on which data can be carried:

The data pipes are realized with proper digital signal processing.

A transmission pipe does not correspond to an antenna transmission chain or any one particular signal path.

The rank of the MIMO system is limited by the number of transmitting or receiving antennas, whichever is lower.

Codewords, layers, antenna ports, and pre-coding are described below.

The *pre-coding* stage performs the mapping of the complex-valued modulation symbols onto each layer for transmission. Pre-coding allows the adjustment of the

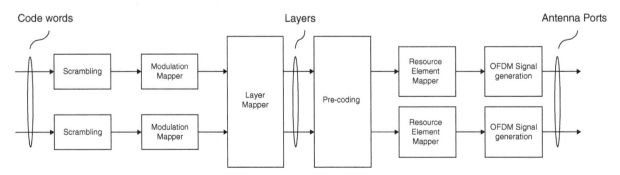

Figure 1.29 Downlink physical layer processing.

phase shift of the signal on each layer. *Resource element mapping* transforms the complex-valued symbols to the allocated resources. For each of the antenna ports used for transmission of the physical channel, the block of complex-valued symbols is mapped in sequence to *resource elements*. At this stage, the resource grids, which allocate the required physical resource blocks, are ready for transmission. The final physical layer processing stage is the actual *OFDM signal generation*. This generates the time-domain signals for each antenna. This is a signal processing procedure.

1.10.1 Main Definitions for MIMO

A MIMO antenna port, codeword, layer, and channel rank are illustrated in Figure 1.30 and we will define them below.

Antenna port

    Antennas are logical ports used for transmission. They have no one-to-one relationship with physical antennas.

    Signals on an antenna port can be transmitted over one or more physical antennas.

    Different antenna ports are used to transmit different reference signals.

Codeword

    Codewords are data blocks formed after channel coding. Different codewords represent different data blocks.

    By transmitting different data blocks, MIMO implements spatial multiplexing.

    To reduce the overhead on CQI and ACK/NACK reporting, LTE supports a maximum of two codewords. In transmit diversity, the number of codewords is one.

        When there is only one antenna at the transmit or receive end, the number of codewords can only be one.

    When there are two or more antennas at both transmit and receive ends, the number of codewords depends on the radio channel conditions and

the UE category. Dual-codeword transmission is mainly used in scenarios with high SINR, low channel correlation, and a UE category of 2 or above.

Layer

    The number of codewords may be different from the number of transmit antenna ports. Therefore, codewords need to be mapped to transmit antenna ports. This is implemented through layer mapping and pre-coding.

    In transmit diversity, the number of layers is equal to the number of antenna ports for transmitting cell-specific reference signals.

    In spatial multiplexing, the number of layers is equal to the number of independent data blocks.

    Downlink 2×2 MIMO and 4×2 MIMO support a maximum of two layers, and downlink 4×4 MIMO supports a maximum of four layers.

Channel rank

    The rank of transmit diversity is 1, and the rank of spatial multiplexing is equal to the number of layers.

    Downlink 2×2 MIMO and 4×2 MIMO support ranks 1 or 2, and downlink 4×4 MIMO supports ranks 1, 2, 3, or 4.

The DL or UL codewords are generated prior to insertion into the scrambler by applying the following steps:

Transport-block CRC (Cyclic Redundancy Check) attachment.

Code-block segmentation and code-block CRC attachment.

Turbo coding.

Rate matching to handle any requested coding rates.

Code-block concatenation to generate codewords.

1.10.2 Scrambling

In LTE downlink processing, the codeword bits generated as the outputs of the channel coding operation are scrambled using different scrambling sequences. The initial stage of the physical layer processing is scrambling.

Figure 1.30 MIMO rank, codeword, layer, and antenna port.

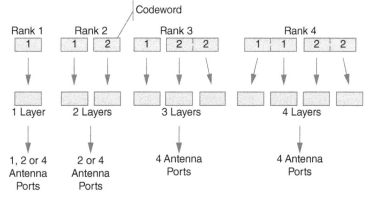

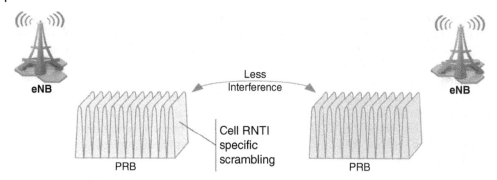

Figure 1.31 Scrambling operation.

This stage is applied to the signal in order to provide interference rejection properties. Scrambling effectively randomizes interfering signals using a pseudo-random scrambling process. Different scrambling sequences are used in neighboring cells to ensure that the interference is randomized and that transmissions from different cells are separated prior to decoding. In order to achieve this, data bits are scrambled with a sequence that is unique to each cell by initializing the sequence generators in the cell based on the PHY cell identity. Different channels are transmitted with different associated RNTIs (Radio Network Temporary Identifiers) obtained for each procedure: paging, system information blocks, data information, RACH, SPS, etc. Different RNTI assignments for each cell scramble the channels with a sequence that is unique to each cell (RNTIs are described in the next section). The scrambling improves the interference by scrambling the user data information with a scrambling code based on the physical cell ID and RNTI. Figure 1.31 highlights the scrambling operation.

1.10.3 Higher-order Modulation

For the downlink, up to 3GPP Release 11, only 64QAM was used. Later, 256QAM was introduced by 3GPP. UE Categories 11, 12, and 13 utilize 256QAM to achieve higher downlink throughput. 64QAM is used in the UL as the maximum modulation technique. The modulation stage converts the scrambled bits into complex-valued modulated symbols using one of: BPSK, QPSK, 16QAM, 64QAM, or 256QAM. Table 1.19 presents different modulation schemes.

The layer-mapping stage effectively maps the complex-valued modulated symbols onto one or several transmission layers. It splits the data into a number of layers that are configured depending on the transmission mode of the MIMO used. There are several options for layer mapping:

Single antenna mapping: For transmission on a single antenna port.

Table 1.19 Modulation schemes.

Modulation	Bits per Symbol	Typical Channel Usage in LTE
BPSK	1	Uplink or downlink control channels
QPSK	2	Uplink or downlink control and data channels
16QAM	4	Uplink and downlink data channels
64QAM	6	Uplink and downlink data channels
256QAM	8	Uplink and downlink data channels (introduced in LTE-A)

Spatial multiplexing: Mapping multiple codewords onto multiple antennas to improve the data throughput of the channel.

Transmit diversity: There is only one codeword and the number of layers is equal to the number of antenna ports used for transmission of the physical channel.

1.11 LTE Channels and Procedures

The LTE air interface provides connectivity between the user equipment and the eNB. It is split into a control plane and a user plane. Among the two types of control plane signaling, the first is provided by the access stratum and carries signaling between the UE and the eNB. The second carries non-access stratum signaling messages between the UE and the MME, which are piggybacked onto an RRC message. The user plane delivers the IP packets to and from the EPC, S-GW, and PDN-GW.

The structure of the lower-layer protocols for the control and user planes in the AS is the same. Both planes utilize the protocols of PDCP (Packet Data Convergence Protocol), RLC (Radio Link Control), and MAC (Medium Access Control) as well as the PHY (Physical Layer) for the transmission of the signaling and data packets.

The NAS is the layer above the AS layers. There are also two planes in the NAS: the higher-layer signaling related to the control plane and the IP data packets of the user plane. NAS signaling exists in two protocol layers, the EMM and the ESM. The NAS user plane is IP-based. The IP data packets pass directly into the PDCP layer for processing and transmission to or from the user.

The concept of "channels" is not new. Both GSM and UMTS define various channel categories; however, LTE terminology is closer to UMTS. There are three main categories of channels in addition to the radio bearer configured by the RRC and NAS, described in detail in the previous section. Figure 1.32 illustrates the LTE channel mapping of protocol layers.

Three layers of channels are defined in the downlink, as follows:

Logical channels: The interface between the MAC and the RLC provides the logical channels.

Control logical channels – The various forms of these channels include:

BCCH (Broadcast Control Channel) – A downlink channel used to send SI (System Information) messages from the eNB. These are defined by the RRC.

PCCH (Paging Control Channel) – A downlink channel used by the eNB to send paging information.

CCCH (Common Control Channel) – Used to establish an RRC (Radio Resource Control) connection, SRB. The SRB is also used for re-establishment procedures after any call drop. SRB0 maps to the CCCH. These are for both uplink and downlink.

DCCH (Dedicated Control Channel) – Provides a bidirectional channel for signaling. Two DCCHs

are activated. One is used for SRB1 carrying RRC messages, as well as high-priority NAS signaling. The other is used for SRB2 carrying low-priority NAS signaling piggybacked onto RRC messages. Prior to its establishment, low-priority signaling is sent on SRB1.

Traffic logical channels:

DTCH (Dedicated Traffic Channel) – Used to carry DRB (Dedicated Radio Bearer) IP packets. The DTCH is configured for uplink and downlink.

Transport channels: In UMTS, transport channels (TrCh) were split between common and dedicated channels. However, LTE has moved away from dedicated channels in favor of common/shared channels.

BCH (Broadcast Channel) – A fixed-format channel which occurs once per frame and carries the MIB (Master Information Block). Note that the majority of system information messages are carried on the DL-SCH.

PCH (Paging Channel) – Used to carry the PCCH, i.e. paging messages.

DL-SCH (Downlink-Shared Channel) – This is the main downlink channel for data and signaling. It supports dynamic scheduling as well as dynamic link adaptation. In addition, it supports HARQ operation to improve performance. As previously mentioned, it also facilitates the system information messages.

RACH (Random Access Channel) – Carries limited information and is used in conjunction with physical channels and preambles to provide contention resolution procedures.

UL-SCH (Uplink-Shared Channel) – Similar to the DL-SCH, this channel supports dynamic scheduling (eNB controlled) and dynamic link adaptation

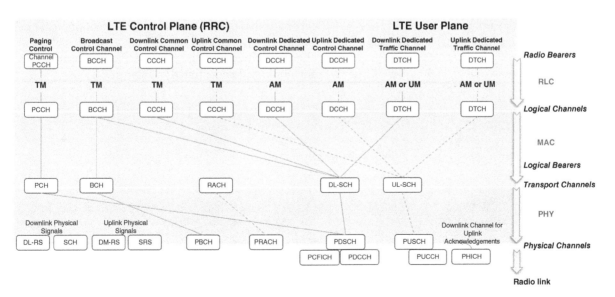

Figure 1.32 LTE channel mapping of protocol layers.

by varying the modulation and coding. In addition, it supports HARQ operation to improve the link performance.

Physical channels: The physical layer facilitates transportation of MAC TrCh as well as providing scheduling, formatting, and control indicators. The TB coming from the MAC layer is mapped onto the corresponding PHY channel to be sent over the air. The TB size is tied to the channel quality feedback from the UE such as CQI and RI. CRC (Cyclic Redundancy Check) bits are added to the TB. The purpose of the CRC is to detect errors which may have occurred when the data were being sent. The UE uses the CRC bits to detect errors on the PDSCH for HARQ retransmissions. The physical layer performs other functions on the TB such as channel coding and rate matching to ensure reliable transmission of the TB over the air.

1.11.1 LTE Physical Layer Channels

The LTE PHY layer, referred to as L1, provides a new channel structure. The main functions provided by the PHY layer in LTE are described below:

Services with higher layers
  Error detection on the transport channel and indication to higher layers.
  FEC encoding/decoding of the transport channel.
  Hybrid ARQ soft-combining.
  Rate matching of the coded transport channel to physical channels.
  Mapping of the coded transport channel onto physical channels.
Power control
  Power weighting of physical channels.
Radio link
  Modulation and demodulation of physical channels.
  Frequency and time synchronization.
  Radio characteristic measurements and indication to higher layers.
  RF signal processing.

Multiple Input, Multiple Output (MIMO)
  MIMO antenna processing.
  Transmit diversity (Tx diversity).
  Beamforming.

The main L1 physical layer channels in the DL and UL are summarized in Figure 1.33, and other channels are summarized in Table 1.20.

1.11.2 Downlink Synchronization Channels

The SCH comprises the PSS (Primary Synchronization Signal) and the SSS (Secondary Synchronization Signal). Together they enable the UE to identify the Physical Cell Identity (PCI) and then synchronize any further transmissions. There are 504 unique PCIs, divided into 168 cell identity groups, each containing three cell identities (sectors). Once a PCI is identified and both slot and frame synchronization have been done through the PSS and SSS, the UE acquires the strongest cell measured during this cell search stage, known as the acquisition stage. The same PCI should be avoided within the same site and as neighbors in order to avoid interference and degradation of the system performance. The PCI, PSS, and SSS are illustrated in Tables 1.21 and 1.22.

In UMTS, the cells are identified by primary scrambling codes (a total of 512 PSCs). During the eNB planning and deployment, the PCI and PSC planning of cells in adjacent clusters is an important topic to avoid any mismatch given the limited number of PCI/PSCs. A mismatch in PCI within two nearby cells can typically lead to system acquisition failures, low throughput, or, eventually, call drops.

1.11.3 Downlink Reference Signals

LTE utilizes different RSs (Reference Signals) to facilitate coherent demodulation, channel estimation, channel quality measurements, and timing synchronization. There are three possible reference signals used on the downlink, as demonstrated in Table 1.23.

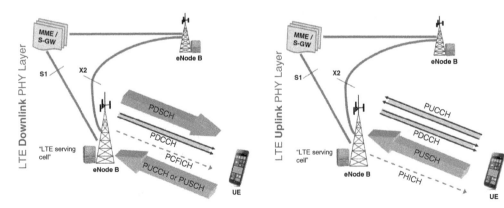

Figure 1.33 Main LTE physical channels in the DL and UL.

Table 1.20 LTE physical layer channels.

Physical Layer Channel	Direction	Main Functions	Similar Channels in UMTS
PBCH Physical Broadcast Channel	DL	• Carries RRC broadcast messages such as SIBs or MIB. • Carries SFN (System Frame Number) used for timing.	PCCPCH
SCH Synchronization Channel	DL	Used to identify the cell ID, frame and slot timing.	SCH
PCFICH Physical Control Format Indicator Channel	DL	Informs the UE of the number of OFDM symbols used for the PDCCHs. UMTS does not have an equivalent channel.	None
PHICH Physical HARQ Indicator Channel	DL	Informs the UE of the acknowledgment response (ACK/NACK) for packets in the uplink.	HS-DPCCH
DL-RS Downlink Reference Signal	DL	Used for cell signal quality estimation.	CPICH
DM-RS Demodulation Reference Signal	UL	Channel estimation for uplink coherent demodulation/detection of the uplink control and data channels.	DPCCH
SRS Sounding Reference Signal	UL	Used to provide uplink channel quality estimation feedback to uplink scheduler for channel-dependent scheduling at the eNB.	None
PRACH Physical Random Access Channel	UL	Carries the RACH preambles.	PRACH

Table 1.21 Physical cell identity.

	Physical Cell Identity (PCI)		
SSS	PSS = 0	PSS = 1	PSS = 2
0	0	1	2
1	3	4	5
2	6	7	8
3	9	10	11
....			
165	495	496	497
166	498	499	500
167	501	502	503

Table 1.22 PSS and SSS.

Information	Usage
Primary synchronization signal (PSS)	Provides downlink subframe timing for the device and unique cell ID (0, 1, or 2).
Secondary synchronization signal (SSS)	Provides downlink frame timing for the device and unique cell ID group (total of 168).
Resource grid (for LTE-FDD)	Time: Sent in subframes 0 and 5 of every frame.
	Freq: Occupies the middle bandwidth with 72 Res.

The cell-specific reference signals are arranged in time and frequency. The spacing in *time* between the reference signals is important for channel estimation and relates to the maximum Doppler spread supported. The spacing in the *frequency domain* relates to the expected coherent bandwidth and delay spread of the channel. For example, it uses one symbol in every third subcarrier (in the 12 subcarriers), resulting in four REs per RB. The *position* of the reference signals in time–frequency is dependent on the value of the Physical Cell ID (PCI). The location

Table 1.23 Downlink reference signals.

Information	Usage
Cell-specific RS (non-MBSFN)	Facilitates coherent demodulation, channel estimation, channel quality measurements.
MBSFN RS	Reference signal for MBSFN.
UE-specific RS	Typically used for beamforming. Therefore, single-antenna port transmission on the PDSCH and transmitted on antenna port 5.

of the reference signals is offset based on the PCI (Physical Cell ID mod 6). This means that there are six possible frequency shifts of RSs.

Other reference signals include Positioning Reference Signals (PRSs) and CSI (Channel State Information) reference signals (CSI-RSs). Positioning reference signals are used for the OTDOA (Observed Time Difference of Arrival) feature in LTE. The positioning reference signals have been introduced to facilitate the determination of the position of the UE – referred to as a UE-assisted positioning technique. CSI-RSs were introduced in LTE Release 10. They perform a complementary function to the DM-RS in LTE Transmission Mode 9. They can also support the Coordinated Multi-Point (CoMP) feature. The overhead of DL-RS in terms of resource elements per resource block can cause suboptimal performance when more antenna ports are added. Hence, LTE-A introduces the Channel State Information (CSI) concept. In principle, decoupling the RS for channel state information and RS for demodulation generates a new downlink reference signal known as a CSI-RS. A CSI-RS is transmitted on each physical antenna port with less overhead. It is used for measurement purposes. CSI-RSs are used with Release 10 Transmission Mode 9. Mode 9 supports both SU-MIMO and MU-MIMO with seamless switching between both.

LTE operates with multiple transmit antennas for MIMO or transmit diversity. The reference signals are defined with different patterns for multiple antenna ports. The RS pattern corresponding to a given antenna port enables the device to derive a channel estimation. The reference signals for normal CP are illustrated in Figure 1.34 and the RS overhead is calculated in Table 1.24.

RSRP and RSRQ are derived from the power level of DL-RS signals. Assume the shortest measurement bandwidth of 6 RBs (i.e. 72 REs) transmitting with 43 dBm (i.e. total downlink Tx power per cell). This means that the RSRP is 1/72 of the total power. Assuming all REs are going through a similar path loss of –100 dB, then the

RSRP can be derived as follows:

$$RSRP = 43 - 100 - 10 * \log(72) = -75.6 \text{ dBm} \tag{1.3}$$

RSRQ is the ratio between the RSRP and the RSSI, depending on the measurement bandwidth, i.e. resource blocks. Consider an ideal interference and noise-free cell where reference signals and subcarriers carrying data are of equal power over one RB (i.e. 12 REs). Then, over the 100 RBs in a 20 MHz system, for one OFDM symbol with R0, then RSRQ is estimated as inter-cell interference, which, in practice, would decrease the results. Inter-cell interference would appear as a wideband RSSI increment impacting the denominator in the RSRQ calculations.

$$RSRQ = 10 * \log((100 * 1RE)/(100 * 12RE))$$
$$= -10.79 \text{ dB} \tag{1.4}$$

1.11.4 Physical Broadcast Channel (PBCH)

Once the device has decoded the PSS and SSS it is able to:

Decode cell-specific reference signals (since their location is based on the physical cell ID).

Perform channel estimation procedures for cell selection on the searched PCI.

Decode the PBCH which carries the MIB (Master Information Block).

Based on the MIB, the UE is able to decode the PCFICH. This identifies the number of OFDM symbols assigned to the downlink control region in the subframe.

The PBCH is used to schedule the MIB while other SIBs (System Information Blocks) are sent using the PDSCH. The MIB repeats every 40 ms and uses a 40 ms TTI. It carries system configuration parameters:

The downlink bandwidth (6, 15, 25, 50, 75, or 100) resource blocks.

PHICH configuration parameter and cyclic prefix information.

SFN (System Frame Number) that enables the UE to know the subframe number for synchronization of all PHY channels transmitted.

The frame structure of the LTE downlink along with the location of the SSS, PSS, and PBCH are illustrated in Figure 1.35. Table 1.25 provides the locations of the PBCH.

1.11.5 Physical Hybrid ARQ Indicator Channel (PHICH)

The PHICH carries HARQ (Hybrid ARQ) ACK/NAKs for every uplink data transmission. It is transmitted in PHICH groups. A PHICH group consists of up

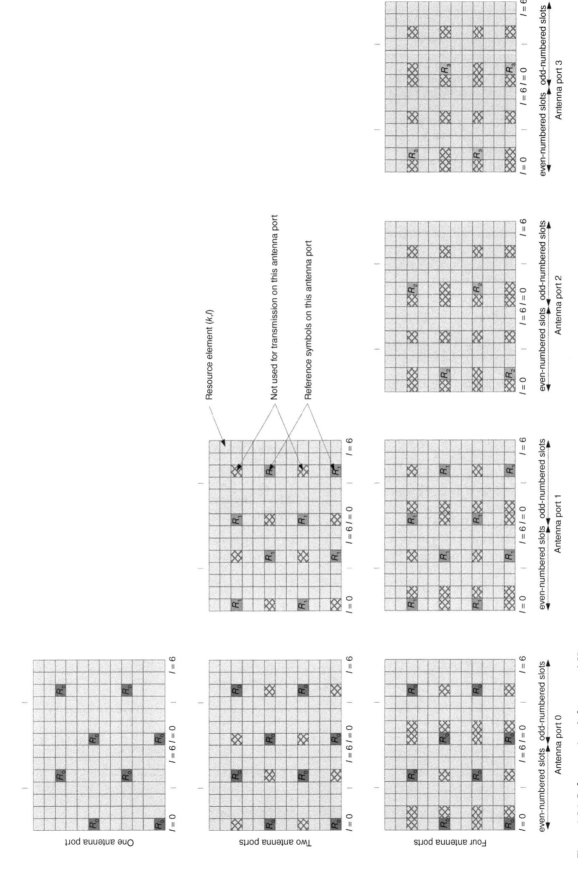

Figure 1.34 Reference signals for normal CP.

Table 1.24 Reference Signals total overhead for normal CP.

Antenna Configuration	Total Overhead (Normal CP)
1 Transmit antenna	Total REs in a subframe = 12 subcarriers × 7 OFDMA symbols (normal CP) × 2 slots = 168
	Total DL-RS REs in subframe = 4 symbols + 4 subcarriers = 8 for R_0
	Overhead = 8 / 168 = 4.76 %
2 Transmit antennas	$8 \times R_0 + 8 \times R_1 = 16$ for two antennas
	Overhead = 16 / 168 = 9.52 %
4 Transmit antennas	$8 \times R_0 + 8 \times R_1 + 4 \times R_1 + 4 \times R_1 = 24$ for four antennas
	Overhead = 24 / 168 = 14.29 %

Table 1.25 Locations of the PBCH channel.

Information	Usage
Time-domain location	Located in four symbols of second slot only (symbols 0, 1, 2, and 3).
Resource grid (for LTE-FDD)	Time: Sent in subframe 0 in every frame.
	Freq: Occupies the middle bandwidth with 72 REs.

to eight ACK/NACK processes and requires three REGs for transmission. Each PHICH within the same PHICH group is separated through different orthogonal sequences. The amount of PHICH resources (N_g) is signaled on the PBCH as part of the MIB. Table 1.26 provides PHICH-relevant information. A PHICH can be

Table 1.26 PHICH information.

Information	Usage
Configuration	N_g is equal to 1/6, 1/2, 1, or 2 and depends on whether "Normal" or "Extended" PHICH mode is being used.
Resource grid (for LTE-FDD)	Time (normal CP): Sent in first OFDM symbol of a subframe.
	Time (extended CP): Sent in the first three subframes.
	Freq: Each PHICH group consists of 3 REGs (i.e. 3*4 REs) in one frame (i.e. $N_g = 1$ in 10 MHz would generate 7 groups with a total of $7 * 3 * 4 = 84$ REs)

defined as follows:

$$\text{\# of PHICH } groups$$
$$= \begin{cases} \lceil N_g * (\# \text{ RBs}/8) \rceil & \text{for normal CP} \\ 2 * \lceil N_g * (\# \text{ RBs}/8) \rceil & \text{for extended CP} \end{cases}$$
$$(1.5)$$

Different REGs belonging to a PHICH group may be transmitted on different symbols. PHICH allocation is critical when it comes to system capacity, as will be shown later. An example of the number of PHICH symbols in a frame for normal CP at different N_g is shown in Table 1.27.

1.11.6 Physical Control Format Indicator Channel (PCFICH)

The PCFICH is used to inform the UE of the number of OFDM symbols used for the PDCCH in a subframe.

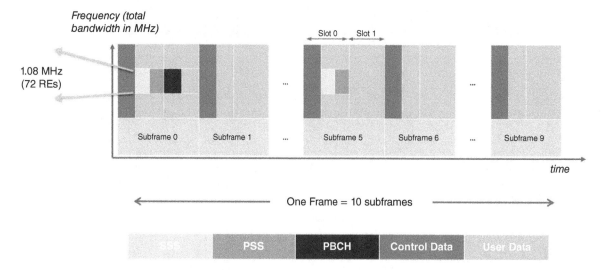

Figure 1.35 Frame structure for the LTE downlink.

Table 1.27 Example of number of PHICH symbols in a frame (normal CP).

N_g	Example of Number of PHICH Symbols in a Frame (Normal CP)	
	10 MHz bandwidth (total overhead)	20 MHz bandwidth (total overhead)
1/6	240 (0.29%)	360 (0.21%)
1/2	480 (0.57%)	840 (0.5%)
1	840 (1%)	1560 (0.93%)
2	1560 (1.86%)	3000 (1.79%)

Since the PDCCH time domain allocation (depending on capacity) is not fixed, this channel indicates how the UE is scheduled. It indicates the UE CFI (Control Format Indicator), which determines the number of OFDM symbols assigned to the PDCCH in a subframe. Specifically, it defines the number of OFDM symbols that the DCI occupies in a subframe. PCFICH-relevant information is defined in Table 1.28. Table 1.29 provides the number of OFDM symbols assigned for the PDCCH in a subframe for different CFI values.

1.11.7 Physical Downlink Control Channel (PDCCH)

The PDCCH carries scheduling assignments and other control information. It is the main channel used to schedule the UE for user data, signaling, paging, system information blocks, and Random Access (RACH) responses.

Table 1.28 PCFICH information.

Information	Usage
Frequency-domain location	The location of these varies depending on the system bandwidth and cell ID.
Resource grid (for LTE-FDD)	Time: Sent in first OFDM symbol in every subframe.
	Freq: Requires four REGs, i.e. 16 resource elements, which are distributed over the channel bandwidth.

Table 1.29 Number of OFDM symbols assigned for the PDCCH in a subframe.

CFI Value	Number of OFDM Symbols Assigned to PDCCH in a Subframe	
	System bandwidth > 10 RBs	System bandwidth ≤ 10 RBs
1	1	2
2	2	3
3	3	4

In the *frequency domain*, the PDCCH is transmitted on an aggregation of one or several consecutive CCEs (Control Channel Elements). One CCE corresponds to nine REGs (36 REs). In the *time domain*, the PDCCH occupies one to four symbols per subframe. The PCFICH is used to inform the UE of the PDCCH control region.

The PDCCH supports multiple formats (called *aggregation levels*), which vary depending on the information sent in the channel, as follows:

PDCCH Format 0 → This consists of one CCE (9 REGs).
PDCCH Format 1 → This consists of two CCEs (18 REGs).
PDCCH Format 2 → This consists of four CCEs (36 REGs).
PDCCH Format 3 → This consists of eight CCEs (72 REGs).

The information sent over the PDCCH is sent as an indicator called a Downlink Control Indicator (DCI). Each DCI conveys control and/or scheduling information to the UE. Once the UE knows the DCI, it knows how to use the data channel (PDSCH). The size of the DCI format depends on its function as well as the system bandwidth. The different DCI values and a relevant description for each value are provided in Table 1.30.

The eNB first encodes the PDCCH message by scrambling the CRC bits with the configured RNTI(s). For each DCI and based on the purpose of the PDSCH transmission, different RNTI configurations are used by the eNB. On the UE side, multiple RNTIs can be associated with one PDCCH decoding, and hence, multiple CRC checks are performed to retrieve the PDSCH data associated with the purpose of the transmission. Different RNTI format values and their usage are provided in Table 1.31.

The common search space corresponds to CCEs 0–15 at two levels:

4-CCE: CCEs 0–3, 4–7, 8–11, 12–15
8-CCE: CCEs 0–7, 8–15

These are monitored by all UEs in the cell and can be used for any PDCCH signaling. In addition, a UE must monitor one UE-specific search space at each of the aggregation levels 1, 2, 4, and 8. This may overlap with the common control search space. The location of the UE-specific search space is based on the C-RNTI, as illustrated in Table 1.32.

Putting together the explanations of DCI and RNTI thus far, we can now derive the number of PDCCH decoding hypotheses. For the *user-specific search*, a maximum of two payload sizes can be available for decoding, with the combination of aggregation level decoding candidates creating a search space up to [2 payloads $*$ (6 + 6 + 2 + 2 PDCCH candidates) = 32]. On the *common search* side, any RNTI combination can only use up to two payload sizes, and

Table 1.30 Different DCI values.

DCI Format	Description
0	PUSCH resource assignment/Aperiodic CQI/RI request.
1	PDSCH resources with no spatial multiplexing. Scheduling of one PDSCH codeword.
1A	Compact scheduling of one PDSCH codeword and random access procedure initiated by a PDCCH order.
1B	PDSCH with one codeword transmission and closed-loop SU-MIMO.
1C	Special purpose (paging, random access, system information). Very compact scheduling of one PDSCH codeword.
1D	Compact scheduling of one PDSCH codeword in MU-MIMO.
2	PDSCH with two-codeword transmission and closed-loop spatial multiplexing in SU-MIMO.
2A	PDSCH with two-codeword transmission and open-loop spatial multiplexing in SU-MIMO.
3	Transmission of TPC (Transmit Power Control) commands for PUCCH and PUSCH with 2-bit power adjustments.
3A	Transmission of TPC (Transmit Power Control) commands for PUCCH and PUSCH with 1-bit power adjustments.

Table 1.31 Different RNTI format values and their usage.

RNTI Format		Usage
SI-RNTI	System Information-Radio Network Temporary Identifier	Used when System Information Blocks (SIBs) are carried on DL-SCH.
P-RNTI	Paging-Radio Network Temporary Identifier	Used when a paging message is carried on DL-SCH.
RA-RNTI	Random Access-Radio Network Temporary Identifier	Used when a random access response is carried on DL-SCH.
C-RNTI	Cell-Radio Network Temporary Identifier	Uniquely used for identifying an RRC connection and user plane scheduling on DL-SCH.
TC-RNTI	Temporary Cell-Radio Network Temporary Identifier	Used for the random access procedure.
TPC-PUCCH-RNTI	Transmit Power Control for PUCCH-Radio Network Temporary Identifier	Used for the uplink power control of PUCCH.
TPC-PUSCH-RNTI	Transmit Power Control for PUSCH-Radio Network Temporary Identifier	Used for the uplink power control of PUSCH.

Table 1.32 The location of the UE-specific search space.

Downlink or Uplink	Common		UE-Specific		MIMO transmission mode [a]
	SI-RNTI, P-RNTI, RA-RNTI	C-RNTI, TC-RNTI	TC-RNTI	C-RNTI	
				DCI 1A, 1	1
Downlink				DCI 1A, 1	2
	DCI 1A, 1C	DCI 1A	DCI 1A, 1	DCI 1A, 2A	3
				DCI 1A, 2	4
				DCI 1A, 1D	5
				DCI 1A, 1B	6
				DCI 1A, 1	7
	C-RNTI, TC-RNTI	TPC-PUCCH-RNTI TPC-PUSCH-RNTI	C-RNTI		
Uplink	DCI 0	DCI 3, 3A	DCI 0		

a) MIMO transmission modes are discussed in the following sections.

with the combination of aggregation level decoding candidates, the common search space can be up to [2 payloads * (4 + 2 PDCCH candidates) = 12]. As a result, there is a total of up to *44 PDCCH decoding hypotheses* exercised for each subframe for the entire control region. Unlike HSDPA, where H-RNTI is configured by the RRC, LTE's RNTIs are mostly derived from the MAC layer based on computed or pre-configured values.

1.11.8 Physical Downlink Shared Channel (PDSCH)

After decoding the PDCCH, it is now easy to read the PDSCH. The PDSCH is used for various transport channels, such as paging, user data (i.e. VoLTE or PS data traffic), and signaling messages. The PDSCH will utilize the remaining available resource blocks in a cell after assigning the control regions. The PDSCH is shared among users, and the eNB scheduler decides how much bandwidth is given to the PDSCH. The PDSCH is scheduled based on MIMO modulation and can be sent from different carriers if carrier aggregation is enabled, based on UE-reported channel conditions (channel quality indicators and rank indicators).

1.12 Uplink Physical Channels

1.12.1 Uplink Reference Signals

There are two types of RSs in the uplink of the LTE. The first type is Demodulation Reference Signals (DM-RSs), which are used to enable coherent signal demodulation at the eNodeB side, similar to the UE-specific reference signals in the DL. These signals are time multiplexed with uplink data and are transmitted on the fourth or third SC-FDMA symbol of an uplink slot for normal or extended CP, respectively, using the same bandwidth as the data.

The second type is the Sounding Reference Signal (SRS), which is used to allow channel-dependent (i.e. frequency-selective) uplink scheduling. The DM-RSs cannot be used for this purpose since they are assigned over the assigned bandwidth to a UE. The SRS is introduced as a wider-band reference signal typically transmitted in the last SC-FDMA symbol of a 1 ms subframe. User data transmission is not allowed in this block, which results in about a 7% reduction in uplink capacity. The SRS is an optional feature and is highly

configurable to control overhead – it can be turned off in a cell. Users with different transmission bandwidth share this sounding channel in the frequency domain.

LTE utilizes different RSs (reference signals) on the uplink to facilitate coherent demodulation at the eNodeB and channel estimation for channel-dependent scheduling by the eNodeB. There are two possible reference signals used on the uplink, as shown in Table 1.33 and Figure 1.36.

The SRS provides the eNB with uplink channel quality information which can be used for scheduling. The UE sends an SRS in different parts of the allocated bandwidth where no uplink data transmission is available. Two modes for transmitting an SRS are supported, as follows:

Wideband mode: The SRS occupies the bandwidth required. This could, however, lead to poor channel quality estimates.

Frequency hopping mode: Sends multiple SRS signals using a narrowband transmission. This will, over time, cover the same bandwidth.

The configuration of the sounding signal (bandwidth, duration, and periodicity) is signaled to the UE by higher layers (RRC). The SRS is transmitted in the last symbol of the subframe. Since the SRS can be sent when the UE has no current PUSCH or PUCCH assignment, mechanisms must exist to stop the UE interfering with other users' PUSCHs. This is done by making sure all UEs know when the SRSs are transmitted, such that the last symbol of the subframe where an SRS is transmitted is not used by any UEs for their PUSCH. The SRS may need to interact with ACK/NACK, CQI, or SR information. If interacting with ACK/NACK, the SRS may be dropped or the ACK/NACK punctured. When interacting with CQI and SR information, the SRS is dropped.

Table 1.33 Uplink Reference Signal.

Information	Usage
DM-RS (Demodulation Reference Signal)	Associated with transmission of PUSCH or PUCCH and used by the eNodeB for coherent signal demodulation.
SRS (Sounding Reference Signal)	Not associated with transmission of PUSCH or PUCCH. This channel is power controlled and used by eNodeB for uplink channel quality information which can be used for scheduling (i.e. UL SINR estimation).

Figure 1.36 LTE subframe with SRS and DM-RS signals.

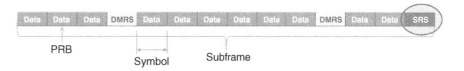

The advantages of SRS usage can be summarized as follows:

The LTE carrier typically occupies much wider spectrum.
    The RF channel response in different frequency portions of the carrier bandwidth can be noticeably different.
    This channel frequency selectivity is particularly pronounced in stationary or slow-mobility cases.
SRS transmissions help the eNB to analyze the real-time UL channel frequency selectivity and allow data transmissions to be allocated in the frequency portion that has the best channel response.
    RF spectrum efficiency is hence improved, especially under multi-user traffic conditions.

The disadvantages of SRS are:

The reservation of the SRS resource leads to 1/12 loss of UL PUSCH symbols.
The reduction of PUSCH resource leads to ~10% average UL throughput loss.
A reduction in the UL peak throughput.
    With 16QAM, up to MCS-22 may be supported with certain SRS configurations, while MCS-24 may be supported without SRS configuration.
    Higher SRS bandwidth would reduce resources available for PUSCH.
    This can cause ~ 16% peak throughput reduction (single user).

1.12.2 Physical Random Access Channel (PRACH)

The random access procedure is used in various scenarios, including initial access, handover, and call re-establishment. It is used primarily in times when the UE needs to initiate a connection with a cell to perform open-loop power control. The PRACH channel is used for the uplink to transmit preambles with certain power that is adjusted based on the path loss and number of retransmissions needed for each preamble. To minimize the collision of preambles between users, LTE defined both contention-based and contention-free RACH procedures.

1.12.3 Physical Uplink Control Channel (PUCCH)

The PUCCH carries information needed by the eNodeB to schedule the UE on the uplink and downlink. The information sent on the PUCCH is in the form of UCI (Uplink Control Information), including:

ACK/NAKs in response to downlink transmission.
DL channel conditions represented by CQI (Channel Quality Indicator) reports.
MIMO feedback such as PMI (Pre-coding Matrix Indicator) and RI (Rank Indicator).
UL scheduling information such as SRs (Scheduling Requests).

The PUCCH is transmitted on a reserved frequency region. This is configured by the higher layer and is shown in Figure 1.37.

Similar to the downlink PDCCH, the uplink PUCCH supports multiple formats. Each format, defined as an *Uplink Control Indicator (UCI)*, is used for a certain type of operation, as shown in Table 1.34.

1.12.4 Physical Uplink Shared Channel (PUSCH)

If a UE has data or signaling to be sent on the uplink, the UE requests a grant from the network and receives it on the PDCCH. The PUSCH is used for various transport channels, such as user data (i.e. VoLTE or PS data traffic) and signaling messages. Like the downlink, the uplink also has resource elements reserved for reference signals and control, and the remaining portion of the grid is used for the PUSCH. The UE is not allowed to transmit the PUCCH and PUSCH in the same subframe, thus multiplexing of different control/data information in one PUSCH is possible.

Figure 1.37 PUCCH control regions.

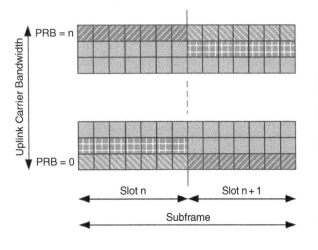

Control Region 0

Control Region 1

Control Region 2

Table 1.34 PUCCH formats.

UCI Format	Description
1	Scheduling request (SR)
1a	ACK/NACK
	$\mathrm{ACK/NACK} + \mathrm{SR}$
	With 1 bit per subframe
1b	ACK/NACK
	$\mathrm{ACK/NACK} + \mathrm{SR}$
	With 2 bits per subframe
2	CQI/PMI or RI
	$(\mathrm{CQI/PMI}$ or $\mathrm{RI}) + \mathrm{ACK/NACK}$ (extended CP only)
2a	$(\mathrm{CQI/PMI}$ or $\mathrm{RI}) + \mathrm{ACK/NACK}$ (normal CP only)
2b	$(\mathrm{CQI/PMI}$ or $\mathrm{RI}) + \mathrm{ACK/NACK}$ (normal CP only)

Figure 1.38 illustrates a combination of control signals and user data on the PUSCH. In this example, three additional types of signaling are added:

ACK/NACK – These are part of the HARQ process and are located next to the RS. This ensures that they benefit from the best possible channel estimation. The information is punctured to make way for the ACK/NACK information.

CQI/PMI – The CQI (Channel Quality Indicator) and PMI (Pre-coding Matrix Indicator) can also be multiplexed onto the PUSCH. These are rate matched with the UL-SCH. The mapping of these is sequential on one subcarrier before continuing on the next.

RI (Rank Indicator) – These are placed next to the ACK/NACK.

Various rules on the mapping and coding of control information exist. In addition, it is possible to send control information on the PUSCH without data, i.e. not the UL-SCH.

1.13 Physical Layer Procedures

The network scheduler used for the uplink and downlink is the main differentiator in terms of the performance of one infra-vendor over another. Hence, each vendor tends to utilize different mechanisms when assigning the data for a user or group of users. There are many policies to ensure fair scheduling among users, taking into account the available resources. The most common schedulers are Round Robin (RR) and Proportional Fair

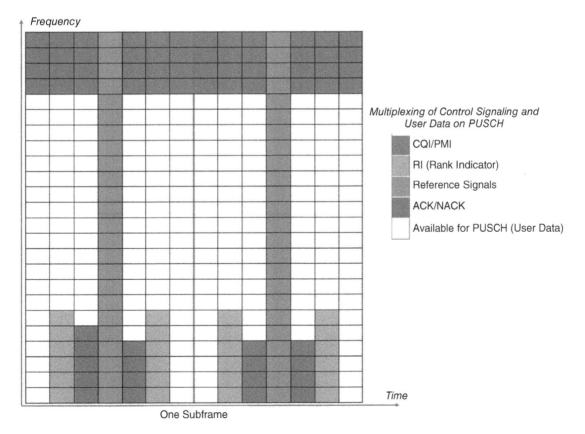

Figure 1.38 A combination of control signals and user data on the PUSCH.

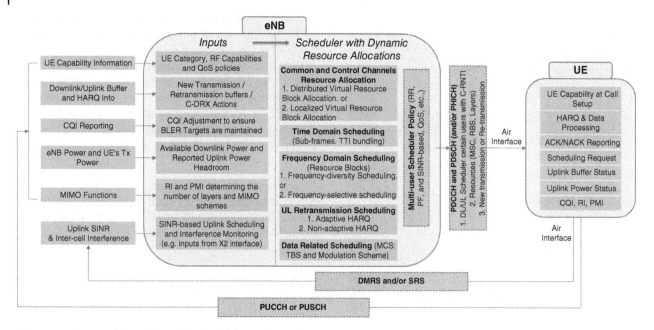

Figure 1.39 Basic uplink and downlink scheduling implementations.

(PF). Round robin typically ensures both time/frequency domain scheduling and allocation in sequential patterns. RR typically aims to ensure user fairness with limited inputs to the algorithm in terms of considering the CQI reported by different UEs. Hence, the main disadvantage is that RR may lead to lower system capacity with more unutilized resources, while fairness is guaranteed at best levels. On the other hand, proportional fair typically aims to make a tradeoff between system capacity and user fairness. This is achieved by taking the overall data available for the UE in the buffer proportionally to the actual channel quality (i.e. CQI). There are other ways of scheduling depending on the available cell power and QoS-aware schedulers. However, in most cases, the inputs to the scheduler for each user remain the same regardless of vendor implementations and scheduler policies. Figure 1.39 illustrates basic uplink and downlink scheduling implementations.

1.13.1   CQI (Channel Quality Indicator)

The CQI provides an indication of the downlink channel quality and effectively identifies an optimal modulation and coding scheme for the eNB to use. The CQI has a direct relation to:

DL channel conditions: A high CQI indicates good UE conditions and a low CQI indicates bad UE conditions.
Cell loading: As the CQI is derived from the PDSCH SINR.

The CQI mainly allows the cell to decide on the modulation technique to use, and this is periodically reported based on higher-layer (RRC) configuration.

The period can be, for example, 5 ms, 20 ms, etc. Based on the reported CQI, the eNB scheduler decides on the Modulation and Coding Scheme (MCS) to be assigned to the user. MCS indices of 0 to 31 are allowed according to the 3GPP standard, as shown in Table 1.35. 3GPP Release 12 introduced 256QAM, as shown in Table 1.36, which is the update of Table 1.35.

LTE defines multiple types of CQI. *Wideband CQI* relates to the entire system bandwidth. In contrast, *sub-band CQI* relates to a value per sub-band. This is defined and configured by the higher layers and relates

Table 1.35 CQI versus MCS and modulation.

CQI Index	MCS Range	Modulation
1		QPSK
2		QPSK
3	0–9	QPSK
4		QPSK
5		QPSK
6		QPSK
7		16QAM
8	10–16	16QAM
9		16QAM
10		64QAM
11	17–28	64QAM
12	*(MCS 29, 30, and 31*	64QAM
13	*are special MCSs used*	64QAM
14	*for retransmissions)*	64QAM
15		64QAM

Table 1.36 CQI versus MCS and modulation for 256QAM.

CQI Index	MCS Range	Modulation
0	0–4	QPSK
1		QPSK
2		QPSK
3		QPSK
4	5–10	16QAM
5		16QAM
6		16QAM
7	11–19	64QAM
8		64QAM
9		64QAM
10		64QAM
11		64QAM
12	20–27	256QAM
13		256QAM
14		256QAM
15		256QAM

to the number of resource blocks. One CQI per codeword is reported for MIMO spatial multiplexing and depends on MIMO transmission modes. Also, depending on the scheduling mode, *periodic* and *aperiodic CQI* reporting can be used. In "Frequency non-selective" and "Frequency-selective" mode, the PUCCH is used to carry periodic CQI reports. For "Frequency-selective" mode, the PUSCH is used to carry aperiodic CQI reports.

1.13.2 DL Scheduling

The DL scheduling implementation follows the following steps:

Determining MCS:

The eNB selects an MCS based on the reported CQI as follows:

If frequency-diverse scheduling is used, the eNB selects an MCS based on the wideband CQI reported by the UE.

If frequency-selective scheduling is used, the eNB selects an MCS based on the sub-band CQIs reported by the UE.

The eNB determines whether to adjust the CQI reported by the UE based on the setting of the CQI adjustment algorithm and parameters:

If this algorithm is turned on, the eNB adjusts the CQI reported by the UE and selects an MCS based on the adjusted CQI.

If this algorithm is turned off, the eNB does not adjust the CQI reported by the UE and selects an MCS based on the reported CQI.

The eNB maps the CQI reported by the UE to the Transport Block Size Index (ITBS) and then maps the ITBS to the IMCS.

The IMCS is the MCS used to schedule the UE.

Determining the number of RBs:

The amount of data to be scheduled in a TTI determines the number of RBs to be scheduled for UEs. Before determining the user data to be scheduled, the scheduler estimates the RLC overhead and MAC header overhead. This enables the scheduler to allocate scheduling resources as precisely as possible, maximizing the resource utilization efficiency.

The scheduler then obtains the amount of data to be scheduled and the IMCS, and estimates the number of RBs to schedule based on 3GPP 36.213. Based on the remaining power of the cell, the scheduler determines the number of RBs to schedule for the UE. The scheduler selects the positions of the RBs to be scheduled based on the scheduling mode: frequency-diverse scheduling or frequency-selective scheduling.

There are two types of DL scheduling that are related to CQI, and these are defined as follows:

Frequency-diverse scheduling (FDS)

Does not consider the differences in the frequency-domain channel quality for UEs.

The eNodeB calculates the scheduling priorities based on the *wideband CQIs* reported by UEs.

Based on the priority calculation results, the eNodeB allocates DL resources to the UEs from a low frequency band to a high frequency band.

Frequency-selective scheduling (FSS)

Considers the differences in the frequency-domain channel quality for UEs.

The eNodeB calculates the scheduling priorities based on *sub-band CQIs* reported by UEs.

The UEs are scheduled to the sub-bands with the optimal channel quality.

1.13.3 UL Scheduling

Data-related inputs from a UE to the eNodeB are as follows:

SR/BSR (UL)

A Scheduling Request (SR) is a 1-bit message sent by a UE to the eNodeB to request UL resources for data transmission.

A Buffer Status Report (BSR) is sent by a UE to the eNodeB to show the data amount in the UL buffer of the UE.

The procedure for triggering UL scheduling is as follows:

Before transmitting data, a UE sends the eNodeB a scheduling request (SR) on the PUCCH to request UL resources for data transmission.

Upon receiving the SR, the eNodeB schedules the UE.

The UE transmits MAC protocol data units (PDUs), including the Buffer Status Report (BSR), using the UL resources allocated by the eNodeB.

If the BSR received at the eNodeB is greater than zero, the eNodeB continues scheduling the SR UE and data transmission on the SR UE proceeds.

The UL scheduler determines the RB resources required for the user based on the buffer status reported by UEs, the QoS requirements, the power headroom of the UE, and the maximum number of RBs supported by the cell.

The processes of determining MCSs for the UL and the positions of RBs to be allocated are both based on the UL SINR measured by the eNodeB on the SRS channel (no CQI for UL), and hence SRS becomes important.

1.13.4 Multiple Input, Multiple Output (MIMO)

MIMO builds on Single Input, Multiple Output (SIMO), also called Receive Diversity (RxD), as well as Multiple Input, Single Output (MISO), also called Transmit Diversity (TxD). Both of these techniques seek to boost SNR in order to compensate for signal degradation. As a signal passes from Tx to Rx, it gradually weakens, while interference from other RF signals also reduces the SNR. In addition, in dense urban environments, the RF signal frequently encounters objects which alter its path or degrade the signal. Multiple-antenna systems can compensate for some of the loss of SNR due to multi-path conditions by combining signals that have different fading characteristics, as the path from each antenna will be slightly different.

However, SIMO and MISO systems may not be fully suitable for the high-speed data rates promised in 3GPP's next-generation cellular systems. Therefore, the full version of MIMO can achieve benefits in terms of both an increase in SNR and throughput gains. MIMO in 3GPP exploits several concepts such as spatial multiplexing, transmit diversity, beamforming, and multi-user MIMO. All these techniques fall into two main MIMO categories: open loop and closed loop.

Therefore, the MIMO system utilizes the space and time diversity in a rich multi-path environment and creates multiple parallel data transmission pipes on which data can be carried, as shown in Figure 1.40. We can summarize the MIMO operation as follows:

The data pipes are realized with proper digital signal processing by combining signals on the NxM paths.

A transmission pipe does not correspond to an antenna transmission chain or any one particular signal path.

The rank of the MIMO system is limited by the number of transmitting or receiving antennas, whichever is lower.

Codewords, layers, antenna ports, and pre-coding are described in Section 1.10.1.

In MIMO 2×2 systems, the Rank 2 multiplexing gain of increasing the throughput is mainly achieved with good channel conditions experiencing little interference in the rich, multi-path environment. With good channel conditions, the data streams are transmitted simultaneously where the total transmission power is shared among multiple data streams. Hence, the total SNR is also shared among multiple data streams, resulting in a lower SNR on each individual data stream. If the total SNR is low, the SNR on each individual stream will be small and the throughput on each data stream will suffer. This indicates that the spatial multiplexing gain for MIMO is mostly achieved in the high-SNR region, where good throughput can be achieved on each of the independent data streams.

In open-loop spatial multiplexing operations, the network receives minimal information from the UE: a Rank Indicator (RI) and a Channel Quality Indicator (CQI). The RI indicates the number of streams (the term "stream" is used for now, but in later subsections it will be changed to "layers") that can be supported under

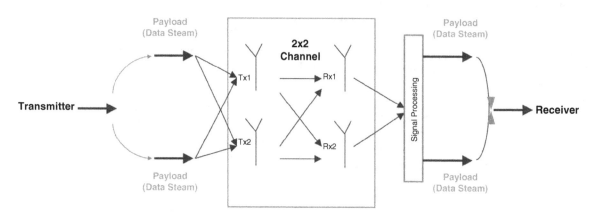

Figure 1.40 MIMO 2×2 operation.

the current channel conditions and modulation scheme. The CQI indicates the channel conditions under the current transmission scheme, roughly indicative of the corresponding SNR. Hence, only one CQI is reported by the UE, which is the spatial average of all the streams. The network scheduler then uses the CQI to select the corresponding modulation and coding scheme for the channel conditions. The network adjusts its transmission scheme and resources for the UE to match the reported CQI and RI with an acceptable block error rate.

On the other hand, at a cell edge or in other low-SNR or poor multi-path conditions, instead of increasing the data rate or capacity, MIMO is used to exploit diversity and increase the robustness of data transmission. In transmit diversity mode, MIMO functions much like a MISO system. Each antenna transmits essentially the same stream of data, so the receiver gets replicas of the same signal. This increases the SNR at the receiver side and thus the robustness of data transmission, especially in fading scenarios. Typically, additional antenna-specific coding is applied to the signals before transmission to increase the diversity effect and to minimize co-channel interference. The UE receives the signals from both Tx at both Rx and reconstructs a single data stream from all multi-path signals.

The most popular open-loop transmit diversity scheme is space/time coding, where a code known to the receiver is applied at the transmitter. Of the many types of space/time codes, the most popular are orthogonal space/time block codes (OSTBCs) or Alamouti codes. This type of code has become the most popular transmit diversity scheme, with its ease of implementation and linearity at both the transmitter and the receiver.

Dual-layer beamforming combines beamforming with 2×2 MIMO spatial multiplexing capabilities and appears in more advanced 3GPP versions (starting in Release 9, as shown in the next section) for both MU-MIMO and SU-MIMO. These beamforming techniques require the deployment of beamforming antenna arrays as well as special configurations of the network and UEs. The four types of MIMO in LTE are summarized in Figure 1.41.

A *codeword* represents an output from the channel coder. With multiple-layer transmissions, data arrive from higher-level processes in one or more codewords. In Release 8/9, one codeword is used for Rank 1 transmission, and two codewords for Rank 2/3/4 transmissions. Each codeword is then mapped onto one or more *layers*. The number of layers depends on the number of transmit antenna ports and the channel rank report by the UE (RI). There is a fixed mapping scheme of codewords to layers depending on the transmission mode used. Each layer is then mapped onto one or more antennas using a *pre-coding matrix*. In Release 8/9, there are a maximum of four antenna ports which potentially form up to four layers. Pre-coding is used to support spatial multiplexing. When the UE detects a similar SNR from both Tx, the pre-coding matrix will map each layer onto a single antenna. However, when one Tx has a high SNR and another has a low SNR, the pre-coding matrix will divide the layers between the Tx antennas in an effort to equalize the SNR between the layers. In Release 10, a non-codebook pre-coding with seamless switching between SU-MIMO and MU-MIMO with up to Rank 8 is defined. The MIMO transmission modes are provided in Table 1.37.

MIMO 2×2 is commonly used with TM3 (open-loop (OL) spatial multiplexing) or TM4 (closed-loop (CL) spatial multiplexing). Table 1.38 provides a comparison between TM3 and TM4. CL-MIMO is better suited for

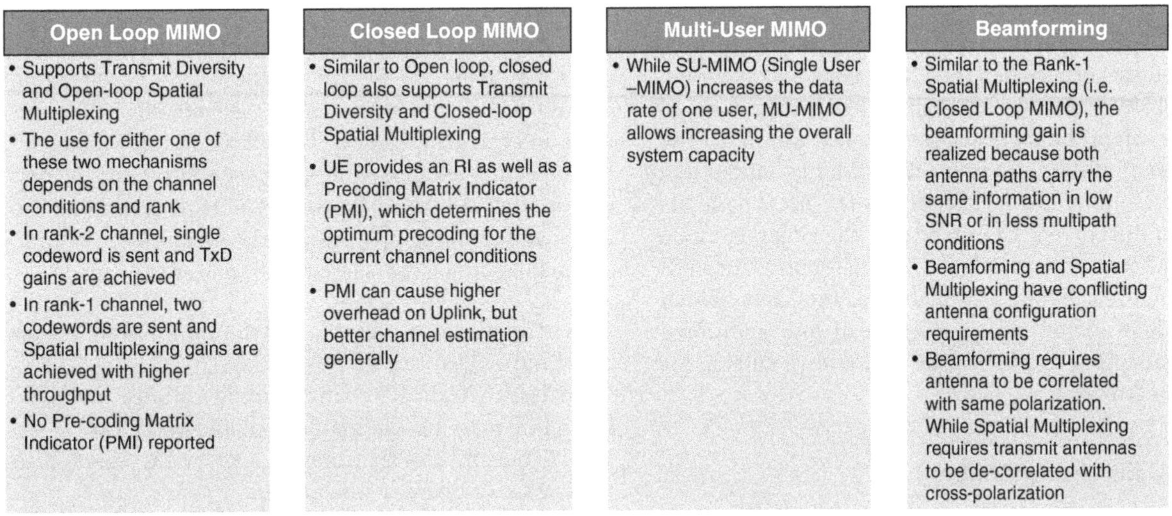

Open Loop MIMO	Closed Loop MIMO	Multi-User MIMO	Beamforming
• Supports Transmit Diversity and Open-loop Spatial Multiplexing • The use for either one of these two mechanisms depends on the channel conditions and rank • In rank-2 channel, single codeword is sent and TxD gains are achieved • In rank-1 channel, two codewords are sent and Spatial multiplexing gains are achieved with higher throughput • No Pre-coding Matrix Indicator (PMI) reported	• Similar to Open loop, closed loop also supports Transmit Diversity and Closed-loop Spatial Multiplexing • UE provides an RI as well as a Precoding Matrix Indicator (PMI), which determines the optimum precoding for the current channel conditions • PMI can cause higher overhead on Uplink, but better channel estimation generally	• While SU-MIMO (Single User –MIMO) increases the data rate of one user, MU-MIMO allows increasing the overall system capacity	• Similar to the Rank-1 Spatial Multiplexing (i.e. Closed Loop MIMO), the beamforming gain is realized because both antenna paths carry the same information in low SNR or in less multipath conditions • Beamforming and Spatial Multiplexing have conflicting antenna configuration requirements • Beamforming requires antenna to be correlated with same polarization. While Spatial Multiplexing requires transmit antennas to be de-correlated with cross-polarization

Figure 1.41 MIMO types in LTE.

Table 1.37 MIMO DL TxM mode.

Downlink Transmission Mode	PDSCH Transmission On	UE Feedback	3GPP Release
Mode 1	Single antenna port (SISO, SIMO)	CQI	
Mode 2	Transmit diversity	CQI	
Mode 3	Open-loop (OL) spatial multiplexing	CQI, RI	
Mode 4	Closed-loop (CL) spatial multiplexing	CQI, RI, PMI	Release 8
Mode 5	MU-MIMO (Rank-1 to the UE)	CQI, PMI	
Mode 6	CL, with Rank 1 spatial multiplexing (pre-coding)	CQI, PMI	
Mode 7	Beamforming with single antenna port (non-codebook based)	CQI	
Mode 8	Dual-layer beamforming	CQI, RI, PMI	Release 9
Mode 9	Non-codebook pre-coding with seamless switching between SU-MIMO and MU-MIMO up to Rank 8	CQI, RI, PMI	Release 10

Table 1.38 Comparison between TM3 and TM4.

Transmission Mode 3			Transmission Mode 4		
# of Antenna Ports	2		**# of Antenna Ports**	2	
	2A			2	
PDCCH DCI Format	Pre-coding information not sent		**PDCCH DCI Format**	Pre-coding information decides on the # of layers	
# Codewords	**# Layers**	**MIMO Technique**	**# Codewords**	**# Layers**	**MIMO Technique**
One codeword when UE reports RI = 1 and no PMI (fixed pre-coding)	2 layers	Transmit diversity	One codeword when UE reports RI = 1 with unreliable PMI	2 layers	Transmit diversity
Two codewords when UE reports RI = 2 and no PMI (fixed pre-coding)	2 layers	OL spatial multiplexing	Two codewords when UE reports RI = 2 with reliable PMI (3 pre-coding choices)	2 layers	Rank 2 CL spatial multiplexing
			One codeword when UE reports RI = 1 with reliable PMI (4 pre-coding choices)	1 layer	Rank 1 CL spatial multiplexing

low-speed scenarios when the PMI feedback is accurate, while OL-MIMO provides robustness in high-speed scenarios when the feedback may be less accurate. The advantage of CL-MIMO over OL-MIMO is limited due to the small number of PMI choices for 2 × 2 configurations in the current LTE standard. The adaptive mode selection between modes 2, 3, 4, and 6 requires the eNB to reconfigure the mode through RRC messages, which can produce increasing signaling load and additional delay in adapting to the best RF conditions suitable for the selected mode.

1.13.5 Uplink Power Control

Uplink power control reduces interference and enables it to be managed/optimized by the eNB. Downlink power control is achieved by varying the MCS and changing MIMO techniques adaptively based on the RF conditions and cell load. The PUSCH, PUCCH, and SRS are all channels that use closed-loop power control. Open-loop power control is used for the PRACH to determine the initial and ramping up power of the preambles.

The following factors impact UL power control:

System bandwidth and the maximum allowed power (by the eNB or the UE power class).

PUSCH/PUCCH/SRS channel configurations.

Downlink path loss estimate calculated in the UE.

TPC (Transmit Power Control) sent by the eNodeB to increase or decrease the power.

Other higher-layer parameters configured by the eNodeB based on the network configurations.

Table 1.39 LTE DL quality KPIs.

KPI	Definition
Serving cell RSRP, RSRQ, RSSI, SINR, CQI, path loss, uplink transmit power	Reflect RF and channel conditions observed by the UE.
Rank 2 request by UE	Distribution of the UE's samples requesting Rank 2 MIMO (two codewords).
Rank 2 served by eNB	Distribution of the eNB's samples requesting Rank 2 MIMO (two codewords).
Physical layer throughput	Throughput at PHY including all new transmissions and retransmissions.
MAC layer throughput	Throughput at MAC including only new transmissions and excluding MAC headers.
Upper layer throughput	Throughput at RLC, PDCP, and application layers.
Scheduling rate in time domain	PDCCH scheduling rate in the time domain (samples of TTIs for how frequently the UE is receiving PDCCH with C-RNTI for data scheduling) for either uplink or downlink.
Normalized physical throughput	PHY layer throughput/scheduling rate
BLER on all transmissions	Distributions of BLER on all transmissions over all packets with C-RNTI (PDSCH on downlink and PUSCH on uplink).
Number of RBs scheduled	Distribution of the samples of resource blocks assigned by the eNB scheduler for uplink or downlink.
MCS	Modulation scheme assigned by the eNB for uplink or downlink.

1.13.6 Techniques for Data Retransmission on the UL

Two techniques are defined, as follows:

UL Non-adaptive HARQ
  The RB positions and MCS for retransmissions are *identical* to those for the initial transmission.
  If the RB positions conflict with positions of the PRACH and PUCCH resources, the retransmission is suspended, affecting UL throughput.
  The eNodeB uses the PHICH channel for ACK/NACK.
UL Adaptive HARQ
  If data to be retransmitted are allocated resources that conflict with other UL resources, the eNodeB adaptively adjusts the number of RBs, their positions, and the MCS for retransmission.
  This way, UL resources are scheduled in a timely manner to reduce UL transmission delay and increase UL throughput.
  With adaptive HARQ retransmission, the eNodeB uses PDCCH scheduling to act as NACK for the UE, as the retransmission will have different resource from the retransmitted data.

Therefore, adaptive retransmission utilizes the special MCS of {29 "QPSK", 30 "16QAM", 31 "64QAM"}.

During the assessment of downlink and uplink throughput performance, several quality and system KPIs become important in order to benchmark the performance or troubleshoot certain issues raised during field testing. Table 1.39 lists several of these quality KPIs. Different tools can be used to collect these KPIs from either the UE or the network tracing point of view.

All layers above the physical layer are discussed in the following sections, so turn to those for an overview of the overheads in each one.

When multiple users are present, they share either the time or frequency domain scheduling, which reduces the peak throughput. BLER is always expected to be at 10% to improve the cell capacity. MCS typically depends on the CQI and channel conditions. MIMO TM3 above assumes near-cell conditions where Rank 2 is reported and two codewords are used for scheduling. An example to estimate the DL TP for Category 4 terminals is shown in Figure 1.42 and the factors impacting the DL TP are summarized in Table 1.40. Figure 1.43 and Table 1.41 illustrate the same for an LTE-A Category 6 terminal with carrier aggregation.

1.14 RRC Layer and Mobility Procedures

The LTE air interface provides connectivity between the user equipment and the eNB. It is split into a control plane and a user plane. Among the two types of control

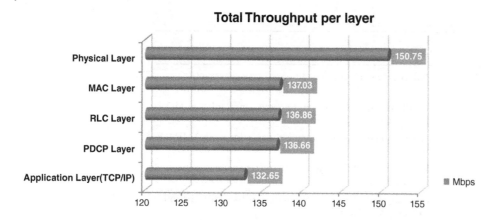

Figure 1.42 DL peak TP at differing layers for Category 4 terminals.

Table 1.40 DL peak TP factors for Category 4 terminals.

All Factors Impacting DL Peak Throughput	
LTE bandwidth (MHz)	20
UE category	4
2 × 2 MIMO transmission mode	3
Maximum MCS – DL	28
Maximum resource blocks	100
DL PHY layer PDSCH BLER (%)	10
DL PDCCH time domain scheduling rate (%)	100

the transmission of signaling and data packets. The NAS is the layer above the AS layers. There are also two planes in the NAS: the higher-layer signaling related to the control plane and the IP data packets of the user plane. NAS signaling exists in two protocol layers, the EMM and the ESM. The NAS user plane is IP-based. The IP data packets pass directly into the PDCP layer for processing and transmission to or from the user.

Table 1.41 DL peak TP factors for a Category 6 terminal.

All Factors Impacting DL Peak Throughput		
	Carrier 1	Carrier 2
LTE bandwidth (MHz)	20	20
UE category	6	N/A
MIMO transmission mode	3	3
Maximum MCS – DL	28	28
Maximum resource blocks	100	100
DL PHY layer PDSCH BLER (%)	10	10
DL PDCCH time domain scheduling rate (%)	100	100

plane signaling, the first is provided by the access stratum and carries signaling between the UE and the eNB. The second carries non-access stratum signaling messages between the UE and the MME, which are piggybacked onto an RRC message. The user plane delivers the IP packets to and from the EPC, S-GW, and PDN-GW.

The structure of the lower-layer protocols for the control and user planes in AS is the same. Both planes utilize the protocols of PDCP (Packet Data Convergence Protocol), RLC (Radio Link Control), and MAC (Medium Access Control) as well as the PHY (Physical Layer) for

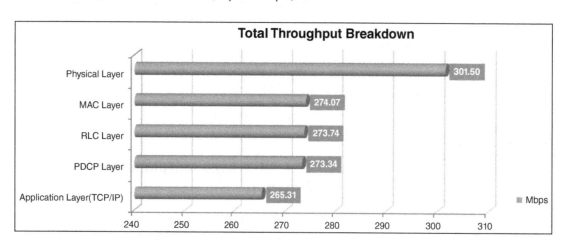

Figure 1.43 DL peak TP at differing layers for a Category 6 terminal.

The RRC constitutes the main air interface protocol for the control plane signaling messages. The RRC functions are similar to those in UMTS. Each signaling message the UE sends to the EPS, and vice versa, comprises a set of system parameters. In order for the messages to be transferred between the UE and the eNB, the RRC layer uses the services of PDCP, RLC, MAC, and PHY. During the course of this mapping, the packets are directed on a Signaling Radio Bearer (SRB). The RRC handles all the signaling between the UE and the E-UTRAN. Additionally, the core network NAS signaling is carried by a dedicated RRC message (piggybacked). When carrying NAS signaling, the RRC does not alter the information but instead provides the delivery mechanism.

The state transitions including inter-RAT are demonstrated in Figure 1.44. Since CELL_FACH in UTRAN is considered a very short period, a direct transition from UTRAN CELL_FACH to E-UTRAN RRC state is not supported. The main states for the RRC layer are provided in Figure 1.45. The upper layers can define specific DRX in RRC Idle mode, then the UE will monitor the paging DRX cycle depending on the shortest of the UE-specific DRX values, if allocated by the upper

layers, and a default DRX value broadcast in the system information.

The RRC constitutes the main air interface protocol for the control plane signaling messages. In general, signaling messages are needed to regulate the UE's behavior in order to comply with the network procedures. Each signaling message the UE sends to the EPS, or vice versa, comprises a set of system parameters. For example, the eNB needs to communicate the parameters related to mobility procedures as and when the UE needs to hand over from one cell to another. These parameters will be sent to the UE in a specific RRC message.

The RRC handles all the signaling between the UE and the E-UTRAN. Additionally, the core network NAS signaling is also carried by a dedicated RRC message. When carrying NAS signaling, the RRC does not alter the information but instead provides the delivery mechanism. SRB1 is used for RRC messages as well as for NAS messages prior to the establishment of SRB2. After SRB2 has been established, the NAS message will transfer: a UL NAS message will always transfer to SRB2; a DL NAS message will transfer to SRB2 or be piggybacked on an RRC Connection Reconfiguration message (depending

Figure 1.44 State transitions LTE, 3G, 2G.

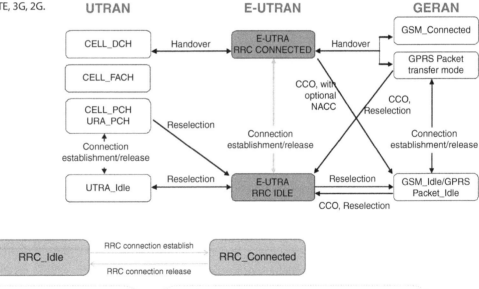

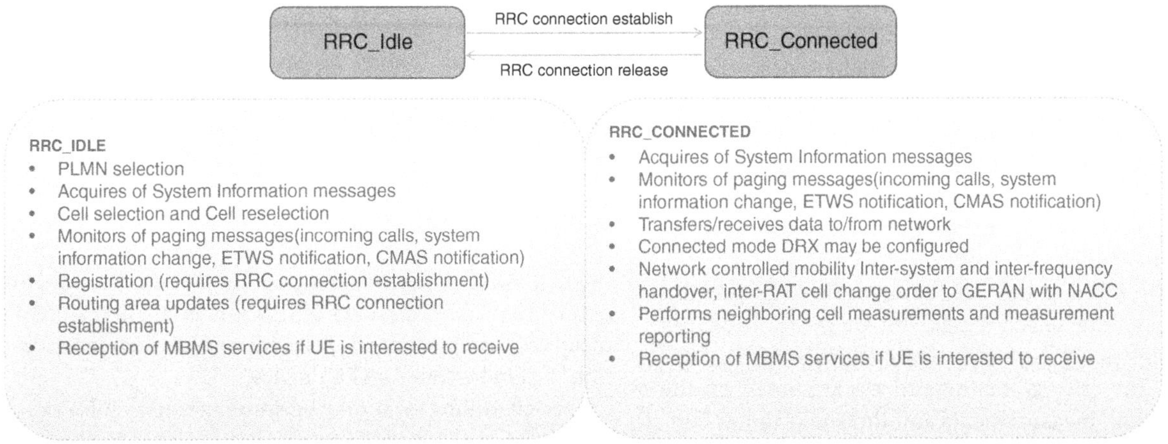

Figure 1.45 Comparison between RRC Idle and RRC Connected.

Table 1.42 Signaling bearers (SRBs) and data bearers (DRBs).

Direction	Function	SRB0	SRB1	SRB2	DRB
General	Transfer NAS message	No	Yes	Yes	No
	High priority	Yes	Yes	No	No
Downlink	PDCP	N/A	Yes	Yes	Yes
	RLC	TM	AM	AM	AM/UM
	Logical channel	CCCH	DCCH	DCCH	DTCH
	Transport channel	DL-SCH	DL-SCH	DL-SCH	DL-SCH
	Physical channel	PDSCH	PDSCH	PDSCH	PDSCH
Uplink	PDCP	N/A	Yes	Yes	Yes
	RLC	TM	AM	AM	AM/UM
	Logical channel	CCCH	DCCH	DCCH	DCCH
	Transport channel	UL-SCH	UL-SCH	UL-SCH	UL-SCH
	Physical channel	PUSCH	PUSCH	PUSCH	PUSCH

on the NW). SRB2 and DRBs are only configured after security activation. An initial UL NAS message is piggybacked on an RRC Connection Setup Complete message. The relevant signaling bearers (SRBs) and data bearers (DRBs) are provided in Table 1.42.

SRB1 is used for RRC messages (which may include a piggybacked NAS message) or NAS messages prior to the establishment of SRB2 (DCCH). In the downlink, piggybacking of NAS messages is used only for one dependent (i.e. with joint success/ failure) procedure: bearer establishment/modification/ release. In the uplink, NAS message piggybacking is used only for transferring the initial NAS message during connection setup. The RRC constitutes the main air interface protocol for the control plane signaling messages. In general, signaling messages are needed to regulate the UE's behavior in order to comply with the network procedures. Each signaling message the UE sends to the EPS, or vice versa, comprises a set of system parameters. For example, the eNB needs to communicate the parameters related to mobility procedures as and when the UE needs to hand over from one cell to another. These parameters will be sent to the UE in a specific RRC message. The RRC handles all the signaling between the UE and the E-UTRAN. Additionally, the core network NAS signaling is carried by a dedicated RRC message. When carrying NAS signaling, the RRC does not alter the information but instead provides the delivery mechanism.

After the UE decodes the required SIBs successfully, it camps on the selected cell. But to verify measured cell levels, the UE performs a procedure known as *cell selection*. The criteria for this process are based on the UE's downlink measurements and threshold values configured in the SIBs. Cell selection is needed to ensure that a UE that passes the acquisition stage (UE-dependent implementation) can only camp on the cell within a coverage threshold conveyed in the SIBs. A cell measured with levels lower than the threshold is not suitable for selection and the UE may then try another LTE acquisition or another PLMN search on other technologies. The main System Information Blocks (SIBs) are summarized in Table 1.43.

The PBCH carries the master information block. The MIB repeats every 40 ms and uses a 40 ms TTI in subframe 0 of each radio frame. It carries system configuration parameters:

The downlink bandwidth: 6, 15, 25, 50, 75, or 100 resource blocks.

PHICH configuration parameter and cyclic prefix information.

SFN (System Frame Number): The SFN is used by the UE to know the subframe number for synchronization of all PHY channels transmitted.

SIBs are carried in system information (SI) messages which are then transmitted on the DL-SCH based on various system parameters. Other than SIB1, the UE uses the SI-RNTI to decode the SIBs carried on the PDSCH. SIB1 is broadcast with 80 ms periodicity in subframe 5 of every radio frame. Scheduling of all other SIBs is specified in SIB1, except for SIB2 which is always contained in the first SI. System information transmission is illustrated in Figure 1.46. SIB1 has a fixed broadcast period (80 ms). SIB1 configures the periodicity and windows of other SIBs. The UE knows which one is the new one by means of the SI-windows. Scheduling of system information is demonstrated in Figure 1.47.

The scheduling of SI may be summarized as follows:

Determine the entry order number, *n*, of the considered SI configured by *schedulingInfoList* in SIB1.

Table 1.43 Main system information blocks.

Main LTE SIBs	Common Information Broadcast	Equivalent SIB in UMTS
MIB	Dl bandwidth, PHICH configuration parameter and cyclic prefix information. SFN (System Frame Number).	N/A
SIB1	PLMN list, tracking area code, intra-frequency reselection, closed subscriber group, frequency band indicator, SI periodicity and mapping information.	MIB and SIB 1/2
SIB2	RACH information, reference signals information, paging channel information, uplink PHY channel information, access timers and constants.	SIB 1/5/7
SIB3	Cell reselection information.	SIB 3/4
SIB4	Neighboring cell related information only for intra-frequency cell reselection. It includes cells with specific reselection parameters and blacklisted cells.	SIB 11/12
SIB5	Contains information relevant only for inter-frequency cell reselection.	SIB 11/12
SIB6	Contains information relevant only for inter-RAT cell reselection to UTRAN.	SIB 19 for UTRAN to E-UTRAN cell reselection
SIB7	Contains information relevant only for inter-RAT cell reselection to GERAN.	SIB 3/11/12
SIB8	Contains information relevant only for inter-RAT cell reselection to CDMA 2000.	SIB 3/11/12
SIB9	Contains home eNB identifier (HNBID).	SIB 20
SIB10	Contains an ETWS primary notification.	N/A
SIB11	Contains an ETWS secondary notification.	N/A
SIB12	Contains a CMAS warning notification.	N/A
SIB13	Contains MBMS-related information.	N/A

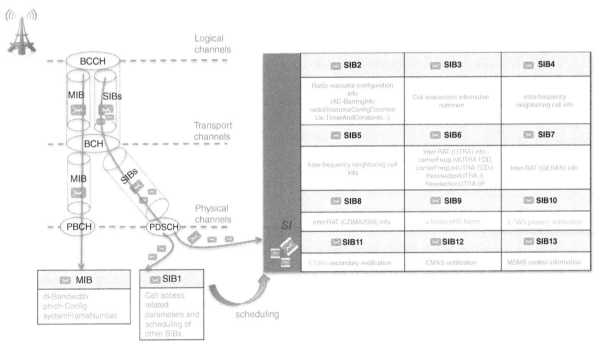

Figure 1.46 System information (SI) transmission.

Determine the integer value $x = (n - 1) *$ (*si-WindowLength*).

The start SI-window of the considered SI is:

Radio frame: SFN mod (*si-Periodicity*) = FLOOR($x / 10$).

Subframe: x mod 10.

Duration of SI-window: *si-WindowLength*.

Based on the SI message order and periodicity, this will make sure that different SI messages never overlap.

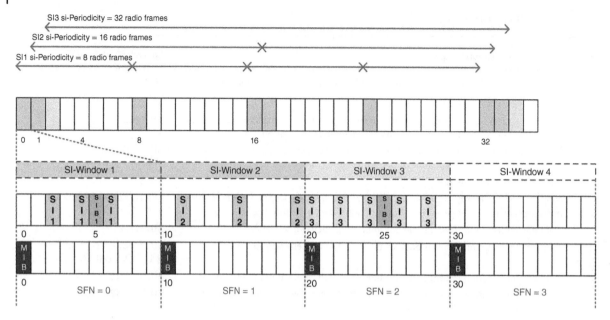

Figure 1.47 Scheduling of system information (SI).

MIB and SIB1 use fixed schedules as follows:

MIB:

First transmission: Subframe #0, SFN mod 4 = 0.
Repetitions: Subframe #0, all other radio frames.
Periodicity 40 ms.

SIB1:

First transmission: Subframe #5, SFN mod 8 = 0.
Repetitions: Subframe #5, all other radio frames for SFN mod 2 = 0.
Periodicity 80 ms.

The *system information acquisition* procedure applies to UEs in RRC Idle and UEs in RRC Connected. The system information acquisition procedure applies upon:

Selecting (for example, upon power on) and reselecting a cell.
After handover completion.
After entering E-UTRA from another RAT.
Return from out of coverage.
Receiving a notification that the system information has changed.
Receiving an indication about the presence of an ETWS notification.

Receiving an indication about the presence of a CMAS notification.
Receiving a request from CDMA2000 upper layers.
Exceeding the maximum valid duration (3 hours).

1.14.1 Paging

The UE will wake up for monitor paging every discontinuous reception (DRX) in Idle mode. When DRX is used, the UE needs only to monitor one PO per DRX cycle. A Paging Occasion (PO), as demonstrated in Figure 1.48, is a subframe where the NW may transmit a paging message. (P-RNTI transmitted on the PDCCH). A Paging Frame (PF) is one radio frame which may contain one or multiple paging occasion(s). The PO index in one PF is based on the UE identity (IMSI/IMEI), the DRX cycle length, and cell-specific parameter (nB).

Using DRX will reduce power consumption.

The PF is given by the following equation:

$$\text{SFN } mod \text{ T} = (\text{T div N}) * (\text{UE\_ID mod N})$$

$$(1.6)$$

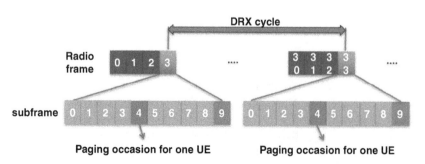

Figure 1.48 Paging occasion with DRX enabled.

Index i_s points to the PO from the subframe pattern and is derived from the following calculation:

$$i\_s = floor(UE\_ID/N) \bmod Ns \tag{1.7}$$

T is the DRX cycle of the UE and is determined by the shortest of the UE-specific DRX values, if allocated by the upper layers, and a default DRX value broadcast in SIB2. If a UE-specific DRX is not configured by the upper layers, the default value is applied.

The cell-specific parameter, nB, may take the following values:

$$nB: 4T, 2T, T, T/2, T/4, T/8, T/16, T/32 \tag{1.8}$$

This parameter indicates the number of paging occasions in T.

The value N in the above equation is given by:

$$N: min(T, nB) \tag{1.9}$$

It indicates the number of paging frames within T[1T,1/32T].

The value Ns is given by:

$$Ns: max(1, nB/T) \tag{1.10}$$

It indicates the number of paging subframes used for paging within a paging frame [1, 2, 4].

The parameter UE_ID is:

$$UE\_ID: IMSI \bmod 1024. \tag{1.11}$$

IMSI is given as a sequence of integer digits (0…9).

The network may initiate paging for EPS services using IMSI with the CN domain indicator set to "PS" if the S-TMSI is not available due to a network failure; the UE will locally detach and re-attach.

Paging informs the UE of system information changes, ETWS notifications, and CMAS notifications in RRC Idle and RRC Connected, as well as transmitting paging information to a UE in RRC Idle. A UE may initiate RRC connection establishment if paging information is provided to the upper layers. The paging message procedure is illustrated in Figure 1.49.

Upon receiving a *Paging* message, if the *ue-Identity* included in the *PagingRecord* matches one of the UE identities allocated by the upper layers for the UE in RRC Idle mode, the UE needs to forward the *ue-Identity* and the *cn-Domain* to the upper layers. The upper layers will make a decision as to whether to respond to this paging or not; if yes, while in RRC Idle mode, an RRC connection also needs to be established. The paging record UE ID will be S-TMSI/IMSI with CS/PS domain.

1.14.2 Initial Security Activation

The initial security activation (illustrated in Figure 1.50) is used to activate AS security, which includes integrity protection of the SRB and ciphering of the SRB and DRBs. SRB2 and DRB can be established after security has been activated. A UE can accept a handover message

Figure 1.49 Paging message procedure.

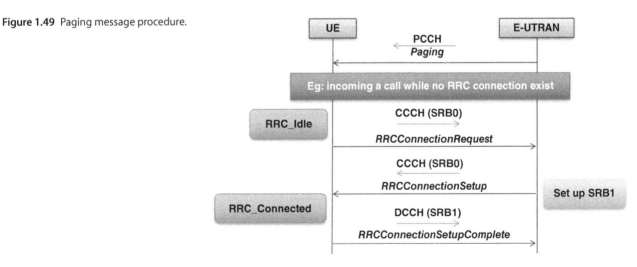

Figure 1.50 Initial security activation.

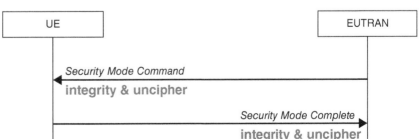

when security has been activated. The AS applies three different security keys derived from the KeNB key, as follows:

Integrity protection of RRC signaling (KRRCint)
Ciphering of RRC signaling (KRRCenc)
Ciphering of user data (KUPenc)

Integrity protection starts from *SecurityModeComplete*, while ciphering starts after the completion of *SecurityModeComplete*. Neither integrity protection nor ciphering applies for SRB0. The integrity and ciphering algorithms can only be changed upon handover. The "NULL" integrity protection algorithm (eia0) is used only for the UE in limited service mode. When the "NULL" integrity protection algorithm is used, the "NULL" ciphering algorithm is also used.

1.14.3 RRC Connection Reconfiguration

The RRC connection reconfiguration procedure (illustrated in Figure 1.51) has several functions:

Modifying an RRC connection, for example, to establish/modify/release RBs.
Performing handover (including a change in integrity and ciphering algorithms).

Set up/modify/release measurements.
May transfer NAS-dedicated information from the E-UTRAN to the UE.

During handover, if the target cell cannot inherit settings from the source cell, the NW can invoke the *full* configuration option to let the UE release and initialize with a newly configured radio configuration. Full configuration can be used to release and initialize the radio configuration if the configuration cannot continue during handover.

1.14.4 RRC Connection Release

The purpose of this procedure (illustrated in Figure 1.52) is to release the RRC connection, which includes the release of the established radio bearers as well as all radio resources. Only the E-UTRAN can initiate the RRC connection release procedure for a UE. The UE may release the RRC connection locally if requested by the upper layers. Access to the current cell may be barred for 300 s as a result of this procedure. The UE may have failed an authentication check, for example, see TS24.301. In the NAS procedure, most local RRC connection release scenarios will not cause the current cell to be barred. RRC release scenarios are outlined in Table 1.44.

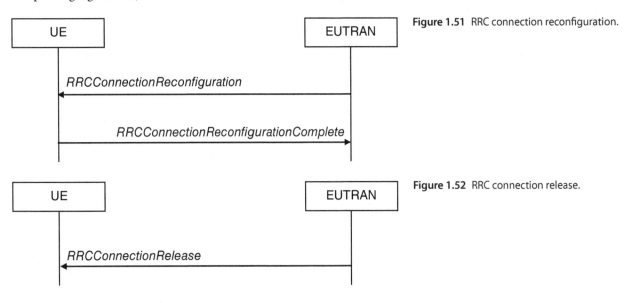

Figure 1.51 RRC connection reconfiguration.

Figure 1.52 RRC connection release.

Table 1.44 RCC release scenarios.

Possible Release Scenario	UE Behavior
With redirectedCarrierInfo IE	UE shall attempt to camp on a suitable cell according to redirectedCarrierInfo.
With idleModeMobilityControlInfo IE	Dedicated priority contained in idleModeMobilityControlInfo should take precedence over the SIB before T320 expires.
Cause: loadBalancingTAUrequired	UE will make tracking area updates.
Cause: Other	UE just moves to RRC Idle and attempts to select a suitable cell. No other specific procedure.

T320 starts upon receiving *t320* or upon cell (re)selection to E-UTRAN from another RAT with validity time configured for dedicated priorities (in which case, the remaining validity time is applied). T320 stops upon entering RRC Connected, when PLMN selection is performed on request by the NAS, or upon cell (re)selection to another RAT (in which case, the timer is carried on to the other RAT). At T320 expiration, the cell reselection priority information provided by dedicated signaling is discarded.

1.14.5 DL Information Transfer

The purpose of the DL information transfer procedure (as illustrated in Figure 1.53) from the E-UTRAN to a UE in RRC Connected mode is to transfer the following:

A NAS message
(Tunneled) non-3GPP dedicated information.

This can be transferred in SRB1 only if SRB2 has not been established yet. Otherwise, even if SRB2 is suspended, the E-UTRAN does not send this message until SRB2 is resumed. Sometimes a downlink NAS message may be piggybacked onto an RRC Connection Reconfiguration message instead of this message when there is one dependent (i.e. with joint success/failure) procedure, even if SRB2 has already been established.

1.14.6 UL Information Transfer

The purpose of the UL information transfer procedure (as illustrated in Figure 1.54) from the UE to the E-UTRAN in RRC Connected mode is to transfer:

A NAS message
(Tunneled) non-3GPP dedicated information.

During the RRC connection establishment, the NAS information is piggybacked onto the RRC Connection Setup Complete message instead of this message.

The information will be transferred in SRB1 if SRB2 has not been established; otherwise, it will always transfer in SRB2.

1.14.7 UE Capability Transfer

The purpose of the UE capability transfer procedure (as illustrated in Figure 1.55) is to transfer UE radio access capability information from the UE to the E-UTRAN. Based on the capability of the UE, the NW can make decisions about handover, measurement, and so on.

It is important to know that the RRC Connection Setup Complete does not carry any UE capability, and the network can query the capability of a specific RAT in a ue-CapabilityRequest; also, a change in E-UTRAN radio access capabilities will trigger TAU with IE "UE radio capability information update needed".

1.14.8 UE Information

The purpose of the UE information procedure (as illustrated in Figure 1.56) from the E-UTRAN is to request the UE to report information about:

The random access procedure:
  The number of preambles sent by the MAC for the last successfully completed RA procedure.
  Whether contention has been detected by the MAC for at least one of the transmitted preambles for the last successfully completed RA procedure.
Measurement information when a Radio Link Failure (RLF) occurred:
  The measurement result of the last serving cell.
  The measurement result of neighboring cells with decreasing order based on triggerQuantity (LTE) or quantityConfig (other RAT).

If the UE faces a radio link failure, the UE will report Rlf-Info_Available in the RRC Connection Reestablishment Complete message after RRC connection

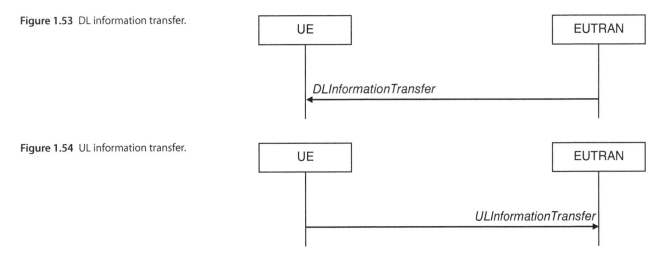

Figure 1.53 DL information transfer.

Figure 1.54 UL information transfer.

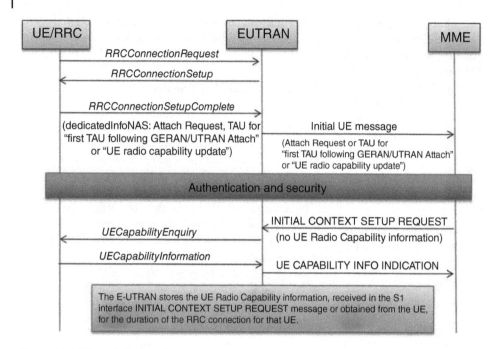

Figure 1.55 UE capability transfer.

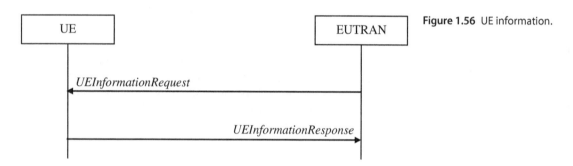

Figure 1.56 UE information.

re-establishment. Then, the network will retrieve the RLF information via UE information. The network can also retrieve an RA-related report via rach-ReportReq or a radio link failure report via Rlf-ReportReq, or both, by indicating the relevant IE in UE information.

1.15 LTE Idle Mode Mobility Procedures

1.15.1 General Mobility Procedure

The procedure for general LTE mobility is illustrated in Figure 1.57.

1.15.2 Public Land Mobile Network (PLMN) Selection

Public Land Mobile Network (PLMN) selection can be completed using either manual mode or automatic mode. The NAS will request the AS to select a cell belonging to this PLMN and then the AS will try to search the cell with the following procedure:

Initial cell selection
Stored information cell selection.

Elementary files within the USIM can support PLMN selection. These files can specify the PLMN search order, forbidden PLMNs, higher-priority PLMN timers, and previously used cell information.

Manual mode will also include PLMNs in the "forbidden PLMNs" list. The UE should use one of the following two cell selection procedures:

Initial cell selection: This procedure requires no prior knowledge of which RF channels are E-UTRA carriers. The UE will scan all RF channels in the E-UTRA bands according to its capabilities to find a suitable cell. On each carrier frequency, the UE need only search for the strongest cell. Once a suitable cell is found, this cell is selected.

Stored information cell selection: This procedure requires stored information on carrier frequencies and, optionally, also information on cell parameters

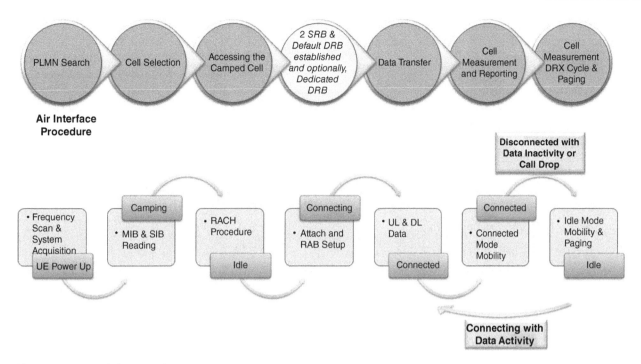

Figure 1.57 LTE mobility procedure.

from previously received measurement control information elements or from previously detected cells. Once the UE has found a suitable cell, the UE will select it. If no suitable cell is found, the initial cell selection procedure is started. The periodic attempts will only be performed in automatic mode when the MS is roaming and not while the MS is attached for emergency bearer services or has a PDN connection for emergency bearer services.

1.15.3 PLMN Selection Order

The PLMN selection order is classified into two modes: automatic mode and manual mode.

There are five PLMN selection orders from 1 to 5; each of them has its own characteristics, as described in Table 1.45.

The equivalent HPLMN list is as follows. To allow provision for multiple HPLMN codes, PLMN codes that are present within this list shall replace the HPLMN code derived from the IMSI for PLMN selection purposes. This list is stored on the USIM and is known as the *EHPLMN list*. The EHPLMN list may also contain the HPLMN code derived from the IMSI. If the HPLMN code derived from the IMSI is not present in the EHPLMN list, then it will be treated as a visited PLMN for PLMN selection purposes. HPLMN/EPLMN determination is carried out as follows: Either the

Table 1.45 PLMN selection order characteristics.

Order	PLMN Type	From
1	HPLMN/EHPLMN	Elementary file in USIM (EFIMSI and EFEHPLMN)
2	User-controlled PLMN selector with access technology (in priority order)	Elementary file in USIM (EFPLMNwACT)
3	Operator-controlled PLMN selector with access technology (in priority order)	Elementary file in USIM (EFOPLMNwACT)
4	PLMN/access technology combinations with received high-quality signal (in random order)	AS measured
5	PLMN/access technology combinations (in order of decreasing signal quality (RSRP))	AS measured

HPLMN (if the EHPLMN list is not present or is empty) or the highest priority EHPLMN that is available (if the EHPLMN list is present).

1.15.3.1 High Quality Criterion

For an E-UTRAN cell, the measured RSRP value should be greater than or equal to −110 dBm. The UE needs to read related SIBs such as MIB and SIB1 to obtain the PLMN ID of the cell and the related S-criteria , as well as SIB2. For more details, refer back to the RRC SIB description.

1.15.4 Service Type and Cell Categories

Different service types have different cell categories, as described in Table 1.46. The levels of services defined for a UE are as follows:

Limited service (emergency calls, ETWS, and CMAS on an acceptable cell).
Normal service (for public use on a suitable cell).
Operator service (for operators only on a reserved cell).

1.15.4.1 Acceptable Cell, Exception

If a UE has an ongoing emergency call, all acceptable cells of that PLMN are treated as suitable for cell reselection for the duration of the emergency call.

1.15.4.2 Suitable Cell, Exception

A cell that belongs to an RA that is forbidden for regional provision service is suitable but provides only limited service.

Access classes are applicable, as follows:

Classes 0–9, for home and visited PLMNs.
Classes 11 and 15, for the home PLMN only if the EHPLMN list is not present, or any EHPLMN.
Classes 12, 13, and 14, for home PLMN and visited PLMNs of the home country only. For this purpose, the home country is defined as the country of the MCC part of the IMSI.

1.15.5 Cell Selection – S-criteria

In LTE, cell selection criteria can be calculated from the equations below:

$$S_{rxlev} > 0 \ \text{AND} \ S_{qual} > 0 \qquad (1.12)$$

$$S_{rxlev} = Q_{rxlevmeas} - (Q_{rxlevmin} + Q_{rxlevminoffset})$$
$$ - P_{compensation} \qquad (1.13)$$

$$S_{qual} = Q_{qualmeas} - (Q_{qualmin} + Q_{qualminoffset}) \quad (1.14)$$

Descriptions and sources for each term are given in Table 1.47.

The values $Q_{rxlevminoffset}$ and $Q_{qualminoffset}$ are only signaled when a cell is evaluated for cell selection as a result of a periodic search for a higher priority PLMN while camped normally in a VPLMN. The combination of $Q_{rxlevminoffset}$ and $Q_{qualminoffset}$ is an offset to avoid a ping-pong effect in a higher priority PLMN search.

1.15.6 Camping on a Suitable Cell

When a UE is camping on a suitable cell, the UE performs the following tasks:

Register on the PLMN (done by the NAS).

Table 1.46 Service type vs cell categories.

Service Type	Cell Categories
Limited service	Acceptable cell The cell is not barred The cell selection criteria are fulfilled
Normal service	Suitable cell The cell is not barred The cell selection criteria are fulfilled In SPLMN or RPLMN or EPLMN Not in "forbidden tracking areas for roaming" For a CSG cell, the CSG ID broadcast by the cell is present in the CSG white list associated with the PLMN for which the above condition is satisfied
Operator service	Reserved cell (SIB1 -> cellReservedForOperatorUse) A cell on which camping is not allowed, except for particular UE
X	Barred cell (SIB1 → cellBarred) A cell a UE is not allowed to camp on

Table 1.47 Parameter descriptions for cell selection criteria.

Parameter	Description	Source
S_{rxlev}	Cell selection Rx level value (dB)	Based on S-criteria calculation
S_{qual}	Cell selection quality value (dB)	Based on S-criteria calculation
$Q_{rxlevmeas}$	Measured cell Rx level value (RSRP)	Based on L1 measurement
$Q_{qualmeas}$	Measured cell quality value (RSRQ)	Based on L1 measurement
$Q_{rxlevmin}$	Minimum required Rx level in the cell (dBm)	q-RxLevMin in SIB1
$Q_{qualmin}$	Minimum required quality level in the cell (dB)	q-QualMin in SIB1
$Q_{rxlevminoffset}$	Offset to the signaled $Q_{rxlevmin}$ taken into account in the S_{rxlev} evaluation as a result of a periodic search for a higher priority PLMN while camped normally in a VPLMN	q-RxLevMinOffset in SIB1
$Q_{qualminoffset}$	Offset to the signaled $Q_{qualmin}$ taken into account in the S_{qual} evaluation as a result of a periodic search for a higher priority PLMN while camped normally in a VPLMN	q-QualMinOffset in SIB1
$P_{compensation}$	$\max(P_{EMAX} - P_{PowerClass}, 0)$ (dB)	Based on $P_{PowerClass}$ and P_{EMAX}
P_{EMAX}	Maximum Tx power level a UE may use when transmitting on the uplink in the cell (dBm)	P-Max in SIB1
$P_{PowerClass}$	Maximum RF output power of the UE (dBm) according to the UE power class	For example, 23 dBm (Power class 3)

Select and monitor the indicated paging channels of the cell in the registered routing area.

Monitor relevant system information.

Perform measurements necessary for the cell reselection evaluation procedure:

Using UE internal triggers; evaluate the cell selection criterion, S, for the serving cell at least every DRX cycle.

When information on the BCCH used for the cell reselection evaluation procedure has been modified.

Execute the cell reselection evaluation process.

Reselection can be intra-frequency, inter-frequency, or inter-rat cells.

Camping on a cell in Idle mode has several purposes:

It enables the UE to receive system information from the PLMN.

When registered and if the UE wishes to establish an RRC connection, it can do this by initially accessing the network on the control channel of the cell on which it is camped.

If the PLMN receives a call for the registered UE, it knows (in most cases) the set of tracking areas in which the UE is camped. It can then send a "paging" message for the UE on the control channels of all the cells in this set of tracking areas. The UE will then receive the paging message because it is tuned to the control channel of a cell in one of the registered tracking areas and the UE can respond on that control channel.

It enables the UE to receive ETWS and CMAS notifications.

It enables the UE to receive MBMS services.

Registration on the PLMN is performed by the Attach procedure of the EMM.

1.15.7 Cell Reselection in Idle Mode

The cell reselection evaluation process in Idle mode involves the following:

Handling reselection priorities

Knowing the priorities of different E-UTRAN frequencies or inter-RAT frequencies.

Measurement rules for cell reselection

Knowing which kind of measurements should be performed, for example: intra-frequency, inter-frequency, inter-RAT.

Mobility states of a UE

If the UE is in a high- or medium-mobility state, the UE applies the speed-dependent scaling rules.

Cell reselection criteria

E-UTRAN inter-frequency and inter-RAT cell reselection criteria:

Choose the highest priority frequency among those fulfilling the cell reselection criteria.

Intra-frequency and equal priority inter-frequency cell reselection criteria:

Choose the best-ranked cell in the selected frequency.

1.15.8 Handling Reselection Priorities

Handling reselection priorities is illustrated in Figure 1.58.

If priorities are provided in dedicated signaling, the UE ignores all the priorities provided in the system information. Otherwise, the UE needs to use the priorities provided in the system information.

The UE only performs cell reselection evaluation for E-UTRAN frequencies and inter-RAT frequencies that are given in the system information and for which the UE has a priority provided.

In the system information, an E-UTRAN frequency or inter-RAT frequency may be listed without providing a priority (i.e. the field Cell_Reselection_Priority is absent for that frequency). If the UE is inthe camped on any cell state, the UE shall only apply the priorities provided by the system information from the current cell, and the UE preserves priorities provided by dedicated signaling unless specified otherwise. When the UE is in the camped normally state, it has only dedicated priorities other than for the current frequency, so the UE shall consider the current frequency to be the lowest priority frequency (i.e. lower than the eight network-configured values). While the UE is camped on a suitable CSG cell, the UE shall always consider the current frequency to be the highest priority frequency (i.e. higher than the eight values configured in the network), irrespective of any other priority value allocated to this frequency.

1.15.9 Measurement Rules for Cell Reselection

The measurement rules for cell reselection are illustrated in Figure 1.59.

If Sintrasearch is not sent, intra-frequency measurement is always performed.

If Snonintrasearch is not sent, equal or lower priority inter-frequency and inter-RAT measurements are always performed.

1.15.10 Speed-dependent Scaling of Reselection Parameters

As shown in Figure 1.60, speed-dependent scaling for reselection parameters is the process used to reselect to other cells quickly in order to avoid out-of-service periods during high/medium-speed movement. The UE shall not count consecutive reselections between the same two cells in the mobility state detection criteria if the same cell is reselected just after one other reselection. Connected-mode speed-dependent scaling will count handovers instead of cell reselections.

The criteria defining medium- and high-mobility states are as follows:**Medium-mobility state criteria:** If the number of cell reselections during time period T_{CRmax} exceeds N_{CR_M} but does not exceed N_{CR_H}.

High-mobility state criteria: If the number of cell reselections during time period T_{CRmax} exceeds N_{CR_H}.

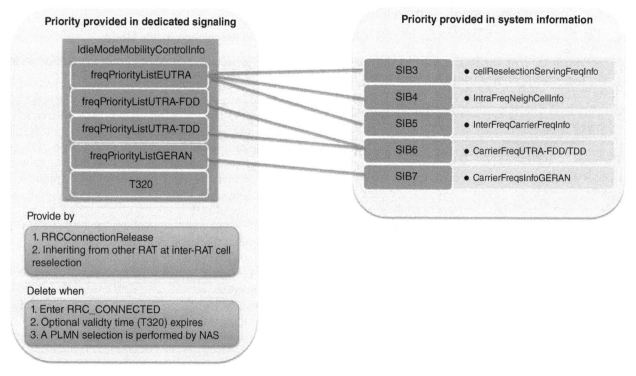

Figure 1.58 Handling reselection priorities.

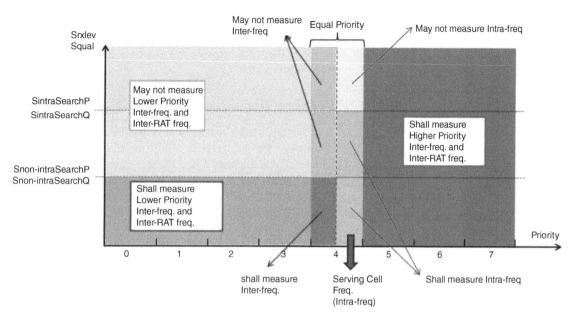

Figure 1.59 Measurement rules for cell reselection.

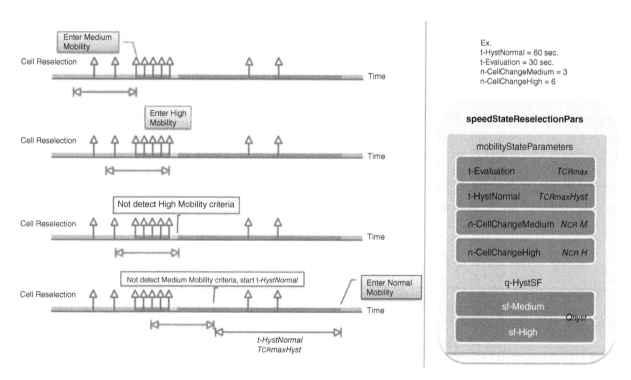

Figure 1.60 Speed-dependent scaling of reselection parameters.

If the criteria for either a medium- or high-mobility state are not detected during time period $T_{CRmaxHyst}$, then a normal-mobility state has been entered.

If a high-mobility state is detected:

Add the sf-High of "Speed dependent ScalingFactor for Qhyst" to Qhyst if sent in the system information.

For E-UTRAN cells, multiply $T_{reselectionEUTRA}$ by the sf-High of "Speed dependent ScalingFactor for TreselectionEUTRA" if sent in the system information.

For UTRAN cells, multiply $T_{reselectionUTRA}$ by the sf-High of "Speed dependent ScalingFactor for TreselectionUTRA" if sent in the system information.

For GERAN cells, multiply $T_{reselectionGERA}$ by the sf-High of "Speed dependent ScalingFactor for TreselectionGERA state" if sent in the system information.

If a medium-mobility state is detected:

Add the sf-Medium of "Speed dependent ScalingFactor for Qhyst for medium mobility state" to Qhyst if sent in the system information.

For E-UTRAN cells, multiply $T_{reselectionEUTRA}$ by the sf-Medium of "Speed dependent ScalingFactor for TreselectionEUTRA" if sent in the system information.

For UTRAN cells, multiply $T_{reselectionUTRA}$ by the sf-Medium of "Speed dependent ScalingFactor for TreselectionUTRA" if sent in the system Information.

For GERAN cells, multiply $T_{reselectionGERA}$ by the sf-Medium of "Speed dependent ScalingFactor for TreselectionGERA" if sent in the system information.

1.15.11 LTE Intra-frequency Cell Reselection

Figure 1.61 illustrates an example of LTE intra-frequency reselection parameters based on ranking due to equal priority:

Intra-frequency cell measurement criteria: $S_{rxlev} < S_{IntraSearchP}$ or $S_{Qual} > S_{IntraSearchQ}$.
Ranking based on RSRP.

The UE initially camps on the LTE F1 cell (first frequency) with physical cell ID 1. The only other neighbor defined is the F1 (same frequency) cell with physical cell ID 12. Before point 1, the serving cell fulfills $S_{rxlev} > S_{IntraSearchP}$ and $S_{qual} > S_{IntraSearchQ}$, here, the UE may choose not to perform intra-frequency measurements. At point 1, $S_{rxlev} < S_{IntraSearchP}$ (if $S_{rxlev} < S_{IntraSearchP}$ or $S_{qual} > S_{IntraSearchQ}$) and thus the UE begins the search for an intra-frequency neighbor.

The PCI = 12 cell is ranked the highest at point 2 due to its received level ($Q_{hyst} + Q_{offset}$) being better than the serving F1 cell. The UE reselects to the PCI = 12 cell when the timer expires (set at two seconds – at point 3 in this example).

1.15.12 LTE Inter-frequency Cell Reselection Rules

Inter-frequency reselection is based on absolute priorities. The UE tries to camp on the highest priority frequency available while the priorities are provided in LTE SIB5 and are valid for all UEs in the serving cell. The specific priorities per UE can be signaled in the RRC Connection Release message, and this is known as dedicated priority.

Only the frequencies listed in SIB5 are considered for inter-frequency reselection. This list can contain a maximum of eight frequencies which the UE may be allowed to monitor within the E-UTRAN. The parameters provided in SIB3 are also considered for ranking evaluations.

For inter-frequency neighboring cells, it is possible to indicate the cell-specific offset to be considered during reselection. These parameters are common to all cells on a different frequency. Blacklists can be provided to prevent the UE from reselecting to specific intra- and inter-frequencies. Cell reselection can also be speed-dependent.

1.15.13 LTE Inter-frequency Cell Reselection with Equal Priority

Figure 1.62 illustrates an example of LTE inter-frequency reselection parameters for equal priority:

Inter-frequency cell measurement criteria: $S_{rxlev} < S_{nonIntraSearchP}$ or $S_{qual} < S_{nonIntraSearchQ}$.
Ranking based on RSRP.

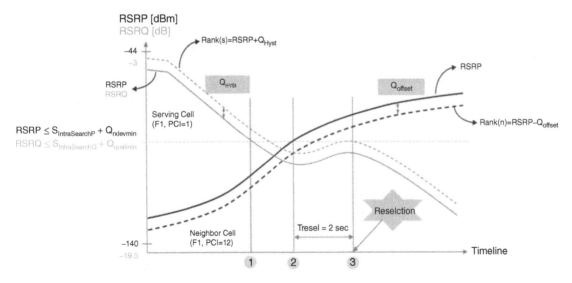

Figure 1.61 LTE intra-frequency cell reselection.

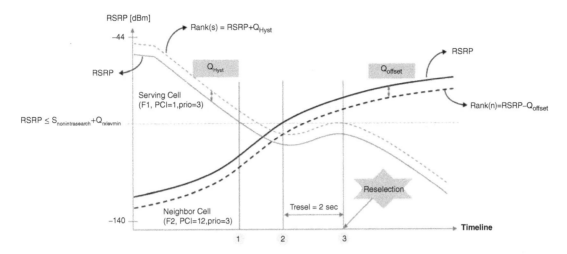

Figure 1.62 LTE inter-frequency cell reselection with equal priority.

The UE initially camps on the LTE F1 cell (first frequency) with physical cell ID 1. The only other neighbor defined is the F2 (second frequency) cell with physical cell ID 12. The priorities of the two frequencies F1 and F2 are defined as the same. At point 1, $S_{Servingcell} < S_{nonIntraSearch}$ and thus the UE begins the search for an inter-frequency neighbor. The F2 cell is ranked the highest at point 2 due to the fact that its received level ($Q_{hyst} + Q_{offset}$) is better than the serving F1 cell. The UE reselects to the F2 cell when the timer expires (set at two seconds – at point 3 in this example).

The fact that $S_{nonIntraSearch}$ is common between LTE inter-frequency measurements and inter-RAT requires additional planning. One of the ways to handle this situation is by treating sites within core LTE coverage – inter-frequency in particular – differently from edge LTE sites. Setting "$S_{nonIntraSearch}$" and "LTE SIB3 $Q_{rxlevmin}$" on a per-cell basis avoids situations where the UE is performing concurrent inter-RAT and

inter-frequency measurements. This per-cell setting also helps in performing inter-frequency LTE reselection before inter-RAT, which allows the UE to remain on LTE coverage longer. In addition, "$Q_{offsetFreq}$" can be introduced on the inter-frequency LTE layer to treat intra- and inter-frequency cell reselection differently for equal priority inter-frequency reselection. This offset is especially important if no changes to "$S_{nonIntraSearch}$" and "LTE SIB3 $Q_{rxlevmin}$" are made on the serving LTE cell. It can be concluded that inter-frequency with equal priority cell reselection is similar to that for intra-frequency.

1.15.14 LTE Inter-frequency Cell Reselection with Low Priority

Figure 1.63 illustrates an example of LTE inter-frequency reselection parameters based on low priority, where the inter-frequency cell measurement criteria are:

$S_{rxlev} < S_{nonIntraSearchP}$ or $S_{qual} < S_{nonIntraSearchQ}$

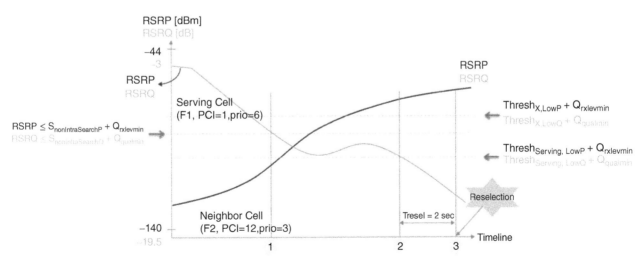

Figure 1.63 LTE inter-frequency cell reselection with low priority.

Table 1.48 Serving and neighboring cell threshold definitions.

Parameter	Serving Cell	Neighboring Cell
threshServingLowQ provided in SIB3	$S_{qual} < Thresh_{Serving, LowQ}$	$S_{qual} > Thresh_{X, LowQ}$
threshServingLowQ not provided in SIB3	$S_{rxlev} < Thresh_{Serving, LowP}$	$S_{rxlev} > Thresh_{X, LowP}$

The UE initially camps on the LTE F1 cell (first frequency) with physical cell ID 1 with high priority (such as priority 6). The only other neighbor defined is the F2 (different frequency) cell with physical cell ID 12 with a lower priority than the serving cell (such as priority 3). Before point 1, the serving cell fulfills $S_{rxlev} > S_{nonIntraSearchP}$ and $S_{qual} > S_{nonIntraSearchQ}$, so the UE may choose not to perform inter-frequency measurements. At point 1, $S_{rxlev} < S_{nonIntraSearchP}$ or $S_{qual} < S_{nonIntraSearchQ}$ and thus the UE begins the search for an inter-frequency neighbor. The quality of the serving cell is lower than a threshold ($S_{qual} < Thresh_{Serving, LowQ}$ or $S_{rxlev} < Thresh_{Serving, LowP}$), meanwhile, the quality of the neighboring inter-frequency cell is better than a threshold ($S_{qual} > Thresh_{X, LowQ}$ or $S_{rxlev} > Thresh_{X, LowP}$) at point 2 to meet the lower priority cell reselection criteria. The UE reselects to the PCI = 12 cell when the timer expires (set at two seconds – at point 3 in this example). Descriptions of threshold parameters for serving and neighboring cells are given in Table 1.48.

1.15.15 LTE Inter-frequency Cell Reselection with High Priority

Figure 1.64 illustrates an example of LTE inter-frequency reselection parameters based on high priority. The UE initially camps on the LTE F1 cell (first frequency) with physical cell ID 1 with lower priority (such as priority 3).

Table 1.49 Neighboring cell threshold definitions.

Parameter	Neighboring Cell
threshServingLowQ provided in SIB3	$S_{qual} > Thresh_{X, HighQ}$
threshServingLowQ not provided in SIB3	$S_{rxlev} > Thresh_{X, HighP}$

The only other neighbor defined is the F2 (different frequency) cell with physical cell ID 12 with a higher priority than the serving cell (such as priority 6). Before point 1, the UE always performs high-priority measurements. At point 1, the quality of the neighboring inter-frequency cell is better than a threshold ($S_{qual} > Thresh_{X, HighQ}$ or $S_{rxlev} > Thresh_{X, HighP}$ or $S_{qual} > Thresh_{X, HighQ}$). The UE reselects to the PCI = 12 cell when the timer expires (set at two seconds – at point 2 in this example).

Descriptions of the threshold parameters for neighboring cells are given in Table 1.49.

1.16 LTE Connected Mode Mobility Procedures

1.16.1 Measurement Parameters

The Connected mode measurement objects, reporting configurations, quantity configurations including

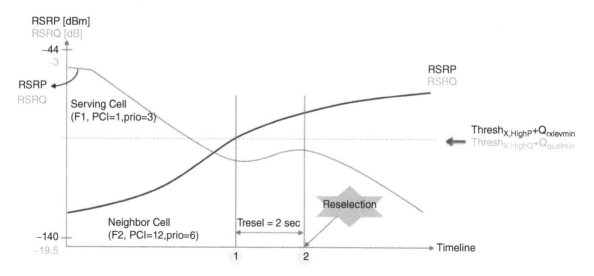

Figure 1.64 LTE inter-frequency cell reselection with high priority.

different RATs (E-UTRAN, UTRAN, GERAN), and measurement gaps are all illustrated in Figure 1.65. The measurement parameters consist of set up/modify/release via the RRC Connection Reconfiguration message in RRC Connected mode, while intra-frequency measurements are the measurements at the downlink carrier frequency of the serving cell and inter-frequency measurements are the measurements at frequencies that differ from the downlink carrier frequency of the serving cell. The network must configure *measObject* for the serving frequencies (earfcn) during measurement setup.

1.16.2 Measurement Procedure in RRC_Connected Mode

LTE events are shown in detail in Table 1.50.

The eNodeB executes the handover based on the UE measurements. The eNB's RRC layer requests the UE to measure intra-frequency cells, inter-frequency cells, or inter-RAT cells belonging to another 3GPP or non-3GPP system. The reporting by the UE is controlled by the eNB through periodic or event-based measurements. The list of events configured by the eNB's RRC layer is summarized in Table 1.50.

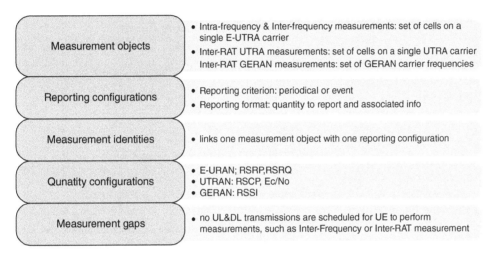

Figure 1.65 Measurement parameters.

Table 1.50 Serving and neighboring cell threshold definitions.

Handover Event	Definition
A1	Serving cell becomes better than a threshold. UE measurements can be based on RSRP and/or RSRQ.
A2	Serving cell becomes worse than a threshold. UE measurements can be based on RSRP and/or RSRQ.
A3	Neighboring cell offset becomes better than the serving cell. UE measurements can be based on RSRP and/or RSRQ.
A4	Neighboring cell becomes better than a threshold. UE measurements can be based on RSRP and/or RSRQ.
A5	Serving cell becomes worse than threshold 1 and neighboring cell becomes better than threshold 2. UE measurements can be based on RSRP and/or RSRQ.
B1	Inter-RAT neighbor becomes better than a threshold. UE measurements can be based on other RAT measurement values (i.e. WCDMA CPICH RSCP and/or Ec/No).
B2	Serving cell becomes worse than threshold 1 and inter-RAT neighbor becomes better than threshold 2. UE measurements can be based on LTE and other RAT measurements.

RSRP (Reference Signal Received Power): Calculated as the linear average over the power from the first antenna port, indicated as R0. Within the measurement cycle (or bandwidth), RSRP is estimated over all REs where R0 is transmitted. If the UE can reliably detect that R1 is available, it may use R1 in addition to R0 to determine the RSRP in each transmission symbol over the measurement cycle.
RSRQ (Reference Signal Received Quality): The ratio between the RSRP and the RSSI, depending on the measurement bandwidth, i.e. resource blocks.

In UMTS, the same concept of event-based measurement applies. However, because UMTS supports handover between multiple cells, the events have different definition scope: for example, event 1A is used to add a cell into the active set, and event 1B is used to remove a cell from the active set.

Assume the shortest measurement bandwidth of 6 RBs (i.e. 72 REs) transmitting with 43 dBm (i.e. total downlink Tx power per cell). This means that the RSRP is 1/72 of the total power. Assuming all REs are going through a similar path loss of −100 dB, the RSRP can be derived as follows:

$$RSRP = 43 - 100 - 10 * \log(72) = -75.6 \text{ dBm.}$$
$$(1.15)$$

RSRQ is the ratio between the RSRP and the RSSI, depending on the measurement bandwidth, i.e. the resource blocks. Consider an ideal interference and noise-free cell where reference signals and subcarriers carrying data are of equal power over one RB (i.e. 12 REs). Then, over the 100 RBs in a 20 MHz system, for one OFDM symbol with R0, RSRQ is estimated as inter-cell interference, which, in practice, would decrease the results. Inter-cell interference would appear as a wideband RSSI increment impacting the denominator in the RSRQ calculation.

$$RSRQ = 10 * \log \quad ((100 * 1RE)/(100 * 12RE))$$
$$= -10.79 \text{ dB} \qquad (1.16)$$

1.16.3 DRB Establishment During Initial Attach

DRB establishment during initial attach is illustrated in the signaling flow in Figure 1.66. During the initial attach, the network will assign a default EPS bearer and related DRB.

The NAS requests the registration procedure and it then requires the AS to establish the RRC connection. The PRACH procedure in the MAC and L1 is needed before the RRC connection can be established.

In Step 7, SRB2, DRB, and the default EPS bearer can be set up at the same time after security has been activated:

The link between the EPS bearer and the DRB is via eps-BearerIdentity and drb-Identity.
The NW will also indicate QoS-related parameters in the ESM message.
A dedicated EPS bearer is optional during this procedure, and if included it will be contained in another dedicatedInfoNASList.
A measurement-related configuration can be set up during this procedure or later.
It will also contain dedicated radio resource configuration.

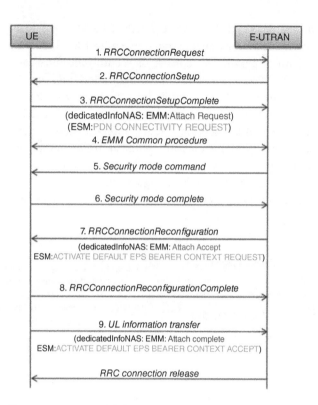

Figure 1.66 DRB establishment during initial attach.

After the related DRBs have been established, the UE can send and receive user data with the corresponding DRB and EPS bearer. The number of DRBs that a UE of Categories 1–5 can support is eight.

1.16.4 DRB Establishment After Initial Attach

DRB establishment after initial attach is illustrated in the signaling flow shown in Figure 1.67. During the initial attach and due to the application package not always being continuous (the NW will release the RRC connection due to the inactivity of a user), then, upon receipt of the application package or paging for an MT package, the UE will trigger the establishment of the DRB.

In Step 6, SRB2 and DRB can be set up after security has been activated:

The link between the EPS bearer and the DRB is via eps-BearerIdentity and drb-Identity.
A measurement-related configuration can be set up during this procedure or later.
It will also contain dedicated radio resource configuration.

1.16.5 Connected Mode Mobility

In RRC Connected mode, the network controls UE mobility and decides when the UE should move to another cell, which may be on another frequency or

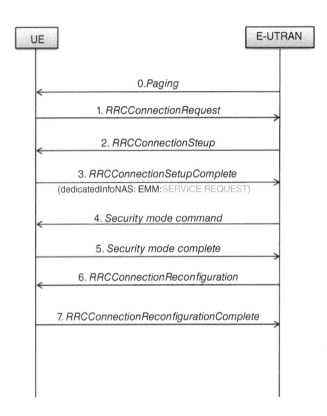

Figure 1.67 DRB establishment after initial attach.

RAT. The network triggers a handover procedure as a result of, for example, radio conditions or loading. This may be based on a measurement report from the UE or it may be a blind handover.

The handover type may be:

An intra-frequency handover
An inter-frequency handover
An inter-RAT handover to UTRAN, GERAN, CDMA2000, or a cell change order to GERAN.

Compared with UMTS, which can have several active set cells, LTE only has one serving cell, so every handover is a hard handover instead of a soft handover.

1.16.6 LTE Intra-frequency Handover

An LTE cell change always involves a hard handover. LTE introduces handovers that can be supported through S1 or X2 interfaces, as follows:

In an S1-type handover, the source and target eNBs communicate via the MME through the S1 interface to exchange handover-related signaling messages.
X2-based handovers are the most commonly used handovers because they involve fewer delays, and the eNBs communicate directly.
X2 is a logical interface that needs to be set up between neighboring cells.

If an X2 handover fails, typically the handover is re-tried over the S1 interface.

The messages between the UE and the source/target eNB are transmitted on the RRC layer of the control plane. For LTE's X2-based handover, when the UE sends a measurement report message over the RRC layer, the source eNB sends a handover request to the target eNB including a list of the bearers to be transferred, and whether downlink data forwarding is being proposed. Then, the source eNB sends an RRC Connection Reconfiguration message to the UE over the RRC layer. At the same time, the downlink packets received at the source eNB from the S-GW are forwarded to the target eNB. Then, the UE synchronizes with the target eNB using the RA procedure, after which, the UE sends an RRC Connection Reconfiguration Complete message over the newly established RRC with the target eNB. The UE then starts collecting the SIBs from the target eNB carrying the required information about the cell parameters. The target eNB sends a UE Context Release message to the source eNB confirming the successful handover and enabling source eNB resources to be released. Finally, the data start flowing directly from the new serving eNB to the UE.

1.16.7 Delay Assessment During Handover

Delay assessment during handover is illustrated in Figure 1.68. Refer to [1] for more details on handover delay and total interruption times.

1.16.8 Event A3 Measurement Report Triggering

An A3 measurement report is used to compare the serving cell with an E-UTRAN neighboring cell. It can be used by the NW for intra-LTE handover.

Figure 1.69 illustrates an example of A3 measurement reporting. The NW has already configured the intra/inter-frequency neighboring cell with an A3 measurement event. At point 1, the UE satisfies the A3 measurement entering condition until time to trigger (TTT) seconds. The UE will trigger and send the A3 measurement report at point 2 (TTT = 320 ms in this example). An A3 measurement report can be configured by the eNB to trigger A3 measurement reports when the leaving condition for this event is met. At point 3, the UE satisfies the A3 measurement leaving condition until time to trigger (TTT) seconds. The UE will trigger and send the A3 measurement report at point 4 (TTT = 320 ms in this example). The details of A3 measurement are available at TS36.331 section 5.5.4.4 EventA3 (Neighbor becomes offset better than serving).

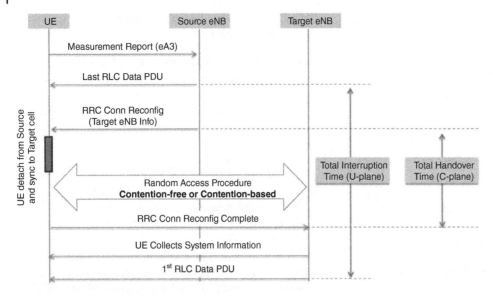

Figure 1.68 Delay assessment during handover.

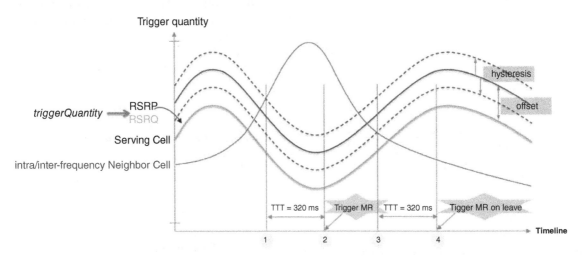

Figure 1.69 Event A3 measurement report triggering.

1.16.9 Intra-frequency Handover Call Flow

The intra-frequency handover call flow is illustrated in Figure 1.70. Refer to [1] for more details on intra-frequency handover.

In Step 2, handover can only occur after security has been activated. In Step 3, the NW configures the measurement of the neighboring cell (intra-frequency and/or inter-frequency) and in Step 9, the target eNB generates an RRC Connection Reconfiguration message including the mobilityControlInformation to perform the handover. This is sent by the source eNB to the UE and it contains the necessary parameters (i.e. new C-RNTI, target eNB security algorithm identifiers, and optionally dedicated RACH preamble, target eNB SIBs, etc.) for handover. Meanwhile, the UE will start the T304 safeguard for the handover execution phase.

In Steps 11 and 12, the random access procedure can be contention-free or contention-based depending on the RA configuration in Step 9. After a successful RA procedure, T304 will be stopped.

1.16.10 Intra-frequency Parameter Tradeoffs

Intra-frequency parameter tradeoffs are described in Table 1.51.

The main areas to consider while optimizing event A3 are:

Reducing pilot pollution areas that can impact SINR and therefore the throughput.
Reducing the amount of handover ping-pongs.
Reducing call drops.
Data interruption time during handover.

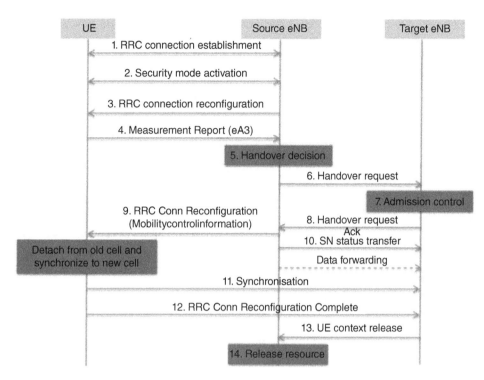

Figure 1.70 Intra-frequency handover call flow.

Table 1.51 Intra-frequency parameter tradeoffs.

Parameter	Example of Settings	Discussion
a3-Triggerquantity	RSRP	RSRP is more robust than RSRQ for handovers, especially in unloaded conditions. RSRQ can be used as an additional A3 trigger.
a3-Hysteresis	4 (2 dB)	Small values can cause cases of going out of triggering conditions earlier and TTT reset.
a3-offset	6 (3 dB)	Low values would cause more handovers and possible ping-pong effects, causing data interruption.
a3-timeToTrigger	240 ms	Higher values delay handover.
a3-ReportAmount	Infinity	If the handover is not triggered (due to eNB issues/failures), it is better to keep reporting event A3.
a3-ReportInterval	240 ms	Re-trigger of event A3 when the handover command is not sent.
filterCoefficient	8 to 11	If this value is set too low, measurement reports could be triggered by rapid, temporary, or short-term fluctuations in RSRP.

The impact of these phenomena is low SINR while the RSRP of the serving cell is good, especially around handover regions where a sharp fall in SINR is observed:

Impacting overall throughput performance where RLF occurs mainly due to SIB read failure or RACH failure (both are DL-related failures).

Fixing RF pilot pollution and cell overshooting can be a first step to improving the serving cell SINR and overall CQI and throughput.

1.16.11 Radio Link Failures and Re-establishment

Radio link failure can be caused by:

Normal radio link failure, for example, due to poor quality:

No N311 consecutive "in-sync" indications from the lower layers before T310 expires.

An RA problem being indicated from the MAC while T300, T301, T304, and T311 are not running.

For example, RA fails when the UL synchronization status is "non-synchronized" or there are no PUCCH resources for SR available.

Reaching the maximum number of retransmissions in the RLC layer.

Intra-LTE handover failure (T304 expiry).

Mobility from E-UTRA failure (inter-RAT handover failure).

An integrity check failure indication from the PDCP.

RRC connection reconfiguration failure (inability to comply with the configuration).

While RRC connection re-establishment can be made for the following reasons:

The UE triggers an RRC connection re-establishment procedure to resume SRB1 operation and the reactivation of security.

Only when AS security has not been activated. Otherwise, the UE will move directly to RRC Idle.

During a radio link failure, the UE will start T311 to find a suitable cell to recover the connection. If it cannot find a cell after the expiry of T311, the UE will move directly to Idle mode. If the UE can find a suitable cell, it will trigger the re-establishment procedure and start T301 to guard the re-establishment. If T301 expires, the UE will also move to Idle mode and the re-establishment also fails, as illustrated in Figure 1.71.

There are several reasons leading to LTE call drops. They vary from PHY layer issues all the way to RRC-related problems. Some of these factors are handover failures, RACH failures, RLC unrecoverable errors, or misconfigured RRC parameters. In LTE, any of these air interface failures leads to a loss of the radio

link between the UE and the eNB, and this is known as Radio Link Failure (RLF). RLF does not necessarily cause a call drop, as there are methods to restore the connection through a re-establishment procedure. If this procedure subsequently fails, the call drops and a new RRC connection is then required. Figure 1.71 illustrates the factors involved in RLF and call drops. The figure shows a summary of procedures in each layer (timers and constants used to detect RLF) and the call re-establishment procedures. It also lists some of the common reasons for observing such failure in any of the layers (i.e. coverage, parameters, RF issues, etc.). Besides a weak RF condition or coverage issues causing RLF, the other common reason is handover failure. Troubleshooting and optimizing handover success rate is essential in ensuring a satisfactory end user experience and stabilizing the network KPIs.

1.16.12 RLF Timers in Idle and Connected Modes

In 3GPP, RLF parameters can be set differently in order to distinguish RLF timers for services like QCI = 1 and QCI = 9.

According to the 3GPP (36.331), RLF parameters T301, T310, N310, T311, and N311 can be set in such a way as to distinguish between the SIB2 RLF timers and the Connected mode RLF timer, which can be set per QCI:

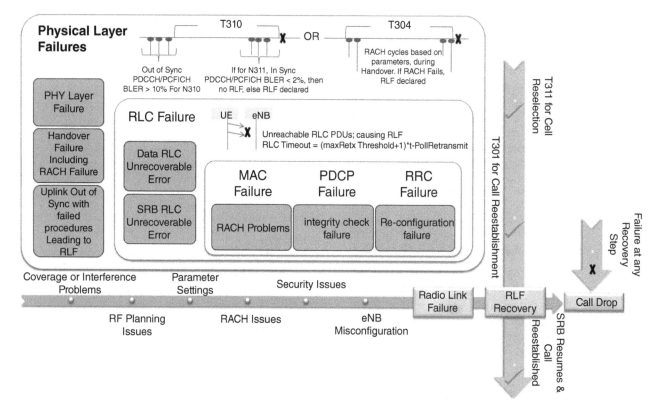

Figure 1.71 Radio link failure and re-establishment procedure.

In addition to SIB2, an RRC Reconfiguration message in Connected mode can be re-signaled when certain services are established (i.e. VoLTE).

If used, the values will override those in SIB2.

New IE rlf-TimersAndConstants including RLF-related timers and counters is introduced, allowing for these to be set per UE.

The IE is to be signaled within RadioResourceConfig Dedicated.

1.16.13   RLF Parameter Tradeoffs

The RLF parameters in SIB2 applied to PS data QCIs are explained in Table 1.52. RLF parameters in Connected mode applied to VoIP QCIs are explained in Table 1.53.

1.16.13.1   Timers/Counters Definitions

T311: This timer is started when the UE starts the RRC connection re-establishment procedure. The timer is stopped if, before the timer expires, the UE selects an E-UTRAN or inter-RAT cell to camp on. After the timer expires, the UE enters the RRC Idle mode.

T301: This timer is started when the UE sends an RRC Connection Re-establishment Request message. The timer is stopped if, before it expires, the UE receives an RRC Connection Re-establishment or RRC Connection Re-establishment Reject message. The timer is also stopped if the selected cell becomes an unsuitable cell.

T310: This timer is started when the UE detects any fault at the physical layer. The timer is stopped if the UE detects one of the following before the timer expires: (1) The physical-layer fault is rectified; (2) a handover is triggered; (3) the UE initiates an RRC connection re-establishment procedure. After the timer expires, the UE enters RRC Idle mode if the security mode is not activated. If the security mode is activated, the UE initiates an RRC connection re-establishment procedure.

N310: Indicates the maximum number of successive "out-of-sync" indications received from L1.

N311: Indicates the maximum number of successive "in-sync" indications received from L1.

A UE detects an RLF when any of the following conditions is met:

The timer T310 expires.

An RA failure occurs and the timer T300, T301, T304, or T311 is not running.

Upon receiving consecutive N310 "out-of-sync" indications from lower layers while timers T300, T301, T304, and T311 are not running, the UE starts T310.

Upon receiving consecutive N311 "in-sync" indications from lower layers while T310 is running, the UE stops T310.

Table 1.52 RLF parameters in SIB2 applying to PS data QCIs (i.e. QCI = 9).

Parameter	Sample Guideline	Discussion
T301	1000 ms	• The current parameters are set to delay RLF in LTE for QCI = 9.
T310	1000 ms	• However, for VoLTE-related services, this may cause a high RTP interruption and hence RTP timeout or longer mute time (one-way audio).
N310	10	• These current QCI = 9 timers aim to delay the RLF as long as possible to improve call drop rate KPIs, but may not be suitable for VoLTE traffic.
T311	3000 ms	
N311	1	

Table 1.53 RLF parameters in Connected mode applying to VoIP QCIs (i.e. QCI = 1).

Parameter	Sample Guideline	Discussion
T301	600 to 1000 ms	• The target of these settings is to drop quickly and re-establish quickly.
T310	1000 ms (600 ms if the call re-establishment is stable)	• Can be set to a lower value for VoLTE to do quicker re-establishment rather than higher mute time, depending on how stable the re-establishment process is.
N310	5	• If re-establishment is unstable, then set N311 = 1 to delay RLF as RTP interruption may be higher due to the UE going to Idle if re-establishment does not succeed.
T311	1000 ms	
N311	2	

1.17 Interworking with Other 3GPP Radio Access

1.17.1 E-UTRA States and Inter-RAT Mobility Procedures

The EPS system is designed to be able to interwork with other RATs. When making a mobility decision between cells in the same RAT, different RATs or frequencies, many criteria (e.g. coverage, load balancing, terminal capabilities, access restrictions, and subscription attributes) are considered at the network design stage. The optimization process can be extremely complex when multiple systems are deployed in the same network, for example, multiple UTRAN frequencies/bands, GERAN, multiple LTE-FDD frequencies/bands, and/or LTE-TDD. The legacy UTRAN to GERAN inter-RAT must be revisited to adopt the new frequency(s) deployed for the LTE system. State transition for different RATs is shown in Figure 1.72.

1.17.2 Inter-RAT Cell Reselection

3GPP introduces a priority layer concept for cell reselection. Any IRAT or inter-frequency cell reselection between cells or frequencies (or even bands) is controlled by the assigned priority. The layer priority is not applied to cells from the same frequency as that of the serving cell, as illustrated in Table 1.54.

As the complexity of the deployed system topologies increases, such priority-based reselection becomes important. With the diversity of cells deployed in the network (femto, micro, or macro) within the same or different RATs, priority reselection can assist the operator in enforcing the targeted camping strategy. In this situation, cell reselection can be layered up by assigning the cells into high, low, or equal priorities. Priorities are typically provided to the UE via system information or RRC Release messages.

Table 1.54 summarizes the concept of layer priority and the measurements requirements by the UE for

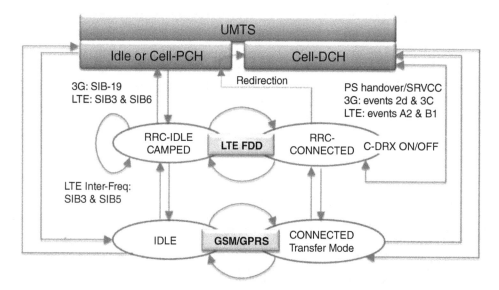

Figure 1.72 E-UTRA states and inter-RAT mobility procedures.

Table 1.54 Measurement configurations for different priorities.

Priority Configured by Network	No Priority	Lower Serving cell signal power > threshold	Lower Serving cell signal power < threshold	Equal Serving cell signal power > threshold	Equal Serving cell signal power < threshold	Higher Thresholds to regulate the measurements
LTE inter-frequency	No reselection allowed	Measurements not mandatory	Measurements mandatory	Measurements not mandatory	Measurements mandatory	Measurements mandatory
Inter-RAT from LTE to WCDMA	No reselection allowed	Measurements not mandatory	Measurements mandatory	Not allowed in standard		Measurements mandatory but regulated for battery saving
Inter-RAT from WCDMA to LTE	No reselection allowed	Measurements not mandatory	Measurements mandatory			

inter-frequency and inter-RAT. In LTE deployment, it is expected that the priority of LTE will be higher than other RATs, especially as LTE provides a higher data rate than other RATs (for example, UMTS). In other scenarios, the LTE priority might be lower than non-LTE femto cells (i.e. home NodeB) in the deployed areas. Hence, the priority setting is an optimization choice that depends on the designed camping strategy, LTE deployment coverage, the targeted performance, and the end user's perceived experience.

1.17.3 E-UTRAN to UTRAN Cell Reselection Flow

The E-UTRAN to UTRAN cell reselection flow is illustrated in Figure 1.73.

SIB type 6 defines the reselection parameters for UTRAN. Reselection from LTE is always based on the comparative priority of the other RAT. LTE and any other RAT cannot be defined to have the same priority, and other RATs will not be considered for reselection unless a priority (cell reselection priority) is specified in SIB6. Based on the priority level defined in SIB6, the UE will use the RAT-specific value of either $\text{Thresh}_{x,high}$ or $\text{Thresh}_{x,low}$ (and $\text{Thresh}_{serving,low}$) to decide if reselection takes place. Additionally, each RAT will have an associated value for $T_{reselection}$ as well as the optional mobility state-related parameters (discussed previously) and maximum allowed transmit power. Moreover, the

maximum UE transmit power for UTRAN is specified along with suitably related values of $Q_{rxlevmin}$ and $Q_{qualmin}$ for RSCP and Ec/No, respectively.

1.17.4 E-UTRAN to UTRAN Measurement Rules

3GPP 36.133 provides the following *minimum* requirements for UTRAN cell measurements while camped in LTE Idle mode. The eNB is only required to provide a list of frequencies on which UTRAN measurements need to be conducted in SIB6. This means that detailed neighbor list information (via PSCs) is not available to the UE.

Table 1.55 shows, in detail, the timer values for different DRX cycle lengths. There are three different timers: detection of UTRAN frequency, measurements of UTRAN frequency, and evaluation of UTRAN frequency. The UTRAN neighboring cells need to be

Table 1.55 UTRAN FDD timers based on DRX cycle length.

DRX Cycle Length (s)	$T_{DetectUTRA,FDD}$ (s)	$T_{MeasureUTRA,FDD}$ (s)	$T_{EvaluateUTRA,FDD}$ (s)
0.320	30	5.12	15.36
0.640	30	5.12	15.36
1.28	30	6.4	19.20
2.560	60	7.68	23.04

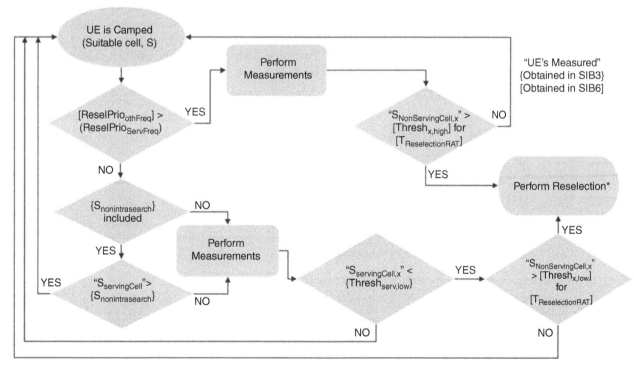

*Only if at least 1 sec has elapsed since camping on serving cell

Figure 1.73 E-UTRAN to UTRAN cell reselection flow.

measured according to the following rules from 3GPP 36.133:

Inter-RAT cell reselection evaluation will be performed only on those cells for which priority has been assigned.

In Idle mode, in 3GPP Release 8, the eNB is only required to provide a list of frequencies on which UTRAN measurements need to be conducted in SIB6. This means that detailed neighbor list information (via PSCs) is not available to the UE.

A new UMTS cell needs to be detected by $N_{UTRA}^* \cdot T_{DetectUTRA,FDD}$ seconds.

An already-detected lower-priority UMTS cell needs to be measured at least once every $N_{UTRA}^* T_{MeasureUTRA,FDD}$ seconds.

An already-detected higher-priority UMTS cell needs to be measured at least once every $T_{MeasureUTRA,FDD}$ seconds.

The measurements need to be filtered using at least two measurements.

Measurements should be spaced by at least $(N_{UTRA}^* T_{MeasureUTRA,FDD})/2$.

The filtering should be such that the time constant of the filter is $< N_{UTRA}^* T_{EvaluateUTRA,FDD}$.

N_{UTRA} is the number of carriers which are measured for UTRAN cells.

1.17.5 E-UTRAN to UTRAN Measurement Stages

A UTRAN cell search by a UE camped in LTE is categorized into two phases:

Cell-detection phase: The UE is trying to find the best cell among the 512 primary scrambling codes (PSCs). The cell is considered detectable when any UTRAN cell CPICH Ec/Io $\geq$ −20 dB:

Once every N_{UTRA}^*94 DRX cycles (for 0.32 second DRX).
Once every 47 DRX cycles (for 0.64 second DRX).
Once every 47 DRX cycles (for 1.28 second DRX).
Once every 24 DRX cycles (for 2.56 second DRX).

Cell-measurement phase: When any cell is considered detectable in the previous phase, the UE is required to keep measuring that particular cell for a possible inter-RAT reselection:

Once every 8 DRX cycles (for 0.32 second DRX).
Once every 4 DRX cycles (for 0.64 second DRX).
Once every 2 DRX cycles (for 1.28 second DRX).
Once every 2 DRX cycles (for 2.56 second DRX).

Both phases – the cell-detection phase and the cell-measurement phase – have different configurations for different DRX cycles, as illustrated in Figure 1.74.

1.17.6 LTE Inter-RAT to UTRAN Cell Reselection with Low Priority

Inter-RAT UTRAN-FDD cell measurement criteria are as follows:

$$S_{rxlev} < S_{nonIntraSearchP} \text{ or } S_{qual} < S_{nonIntraSearchQ}$$

A description of the threshold parameters for serving and neighboring cells is provided in Table 1.56.

Figure 1.75 illustrates an example of LTE inter-RAT UTRAN-FDD reselection parameters based on low priority. The UE initially camps on the LTE F1 cell (first frequency) with physical cell ID 1 with high priority (such as priority 6). The only other neighbor defined is the F2 (UTRAN-FDD frequency) cell with PSC = 12 with a lower priority than the serving cell (such as priority 3). Before point 1, the serving cell fulfills $S_{rxlev} > S_{nonIntraSearchP}$ and $S_{qual} > S_{nonIntraSearchQ}$, so the UE may choose not to perform inter-RAT measurements. At

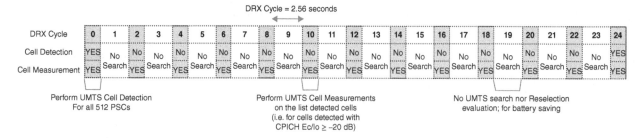

Figure 1.74 Cell detection and cell measurements for different DRX cycles.

Table 1.56 Serving and neighboring cell thresholds.

Parameter	Serving Cell	Neighboring Cell
threshServingLowQ provided in SIB3	$S_{qual} < Thresh_{Serving,LowQ}$	$S_{qual} > Thresh_{X,LowQ}$
threshServingLowQ not provided in SIB3	$S_{rxlev} < Thresh_{Serving,LowP}$	$S_{rxlev} > Thresh_{X,LowP}$

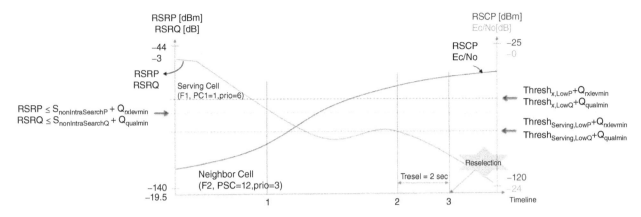

Figure 1.75 LTE inter-RAT to UTRAN cell reselection with low priority.

point 1, $S_{rxlev} < S_{nonIntraSearchP}$ or $S_{qual} < S_{nonIntraSearchQ}$ and thus the UE begins the search for an inter-RAT neighbor. The quality of the serving cell is lower than a threshold ($S_{qual} < Thresh_{Serving, LowQ}$ or $S_{rxlev} < Thresh_{Serving, LowP}$), meanwhile the quality of the neighboring inter-RAT cell is better than a threshold ($S_{qual} > Thresh_{X, LowQ}$ or $S_{rxlev} > Thresh_{X, LowP}$) at point 2 to meet lower priority cell reselection criteria. The UE reselects to the inter-RAT UTRAN-FDD PSC = 12 cell when the timer expires (set at two seconds – at point 3 in this example).

1.17.7 LTE Inter-RAT to UTRAN Cell Reselection with High Priority

Measurements of higher priority are always performed, but are regulated with the threshold parameters described in Table 1.57 for serving and neighboring cells.

Figure 1.76 illustrates an example of LTE inter-RAT reselection parameters based on high priority. The UE initially camps on the LTE F1 cell (first frequency) with physical cell ID 1 with lower priority (such as priority 3). The only other neighbor defined is the F2

Table 1.57 Serving and neighboring cell thresholds.

Parameter	Neighboring Cell
threshServingLowQ provided in SIB3	$S_{qual} > Thresh_{X,HighQ}$
threshServingLowQ not provided in SIB3	$S_{rxlev} > Thresh_{X,HighP}$

(UTRAN-FDD frequency) cell with PCS = 12 with a higher priority than the serving cell (such as priority 6). Before point 1, the UE always performs high-priority measurements. At point 1, the quality of the neighboring inter-RAT UTRAN-FDD cell is better than a threshold ($S_{rxlev} > Thresh_{X, HighP}$ or $S_{qual} > Thresh_{X, HighQ}$). The UE reselects to the inter-RAT UTRAN-FDD PSC = 12 cell when the timer expires (set at two seconds – at point 2 in this example).

1.17.8 UTRAN to E-UTRAN Cell Reselection Flow

3GPP defines a new SIB type (SIB19) for broadcasting LTE-related information. If SIB19 is broadcast with a list

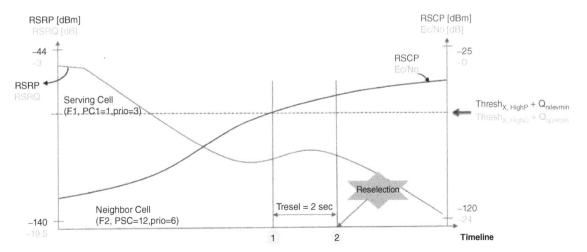

Figure 1.76 LTE inter-RAT to UTRAN cell reselection with high priority.

of E-UTRAN frequencies and priorities, the UE will measure and reselect to the LTE E-UTRAN cells according to the rules explained below.

The UE follows the Absolute Priority reselection according to the criteria defined by 3GPP:

Criterion 1: The $S_{rxlev,nonServingCell,x}$ of a cell on an evaluated higher absolute priority layer is greater than $Thresh_{x,high}$ during a time interval $T_{reselection}$.

Criterion 3: $S_{rxlev,ServingCell} < Thresh_{serving,low}$ or $S_{qual,ServingCell} < 0$ and the $S_{rxlev,nonServingCell,x}$ of a cell on an evaluated lower absolute priority layer is greater than $Thresh_{x,low}$ during a time interval $T_{reselection}$.

As described in Figure 1.77, cell reselection to a cell on a higher absolute priority layer than the camped frequency is performed if Criterion 1 is fulfilled. Cell reselection to another UTRAN inter-frequency cell on an equal absolute priority layer to the camped frequency is performed if Criterion 2 is fulfilled (this is not shown above since it is not related to inter-RAT).

Reselection to a cell on a lower absolute priority layer than the camped frequency is performed if Criterion 3 is fulfilled. If more than one cell meets the criterion, the UE shall reselect the cell with the highest $S_{rxlev,nonServingCell,x}$. The UE shall not perform cell reselection until more than 1 second has elapsed since the UE camped on the current serving cell. For UE in RRC Connected mode state CELL_PCH or URA_PCH, the interval

$T_{reselections,PCH}$ applies if provided in SIB4. The reselection rules discussed here are specified in 3GPP Release 8.

3GPP Release 9 provides slightly different criteria for introducing cell reselection based on RSRQ measurements. The same concepts discussed here also apply, but with different parameters.

1.17.9 UTRAN to E-UTRAN Measurement Rules

The UE must be able to identify new E-UTRA cells and perform RSRP measurements on identified E-UTRA cells if carrier frequency information is provided by the serving cell, even if no explicit neighbor list with physical layer cell identities is provided. Figure 1.78 shows the UTRAN to E-UTRAN measurement rules flow chart.

There are mainly two different rates of E-UTRAN cell search defined by 3GPP to prevent high consumption of UE battery life. If $S_{rxlev,ServingCell} > S_{prioritysearch1}$ and $S_{qual,ServingCell} > S_{prioritysearch2}$ then the UE shall search for E-UTRA layers of higher priority at least every $T_{higher\_priority\_search}$ where $T_{higher\_priority\_search}$ is "60*N_layer" seconds, where N_layer is the number of high-priority frequencies across all RATs. Hence, the purpose of this procedure is to save battery consumption by reducing the amount of UE measurements.

If $S_{rxlev,ServingCell} \leq S_{prioritysearch1}$ or $S_{qual,ServingCell} \leq S_{prioritysearch2}$ then the UE must search for and measure E-UTRA frequency layers of higher or lower priority in preparation for possible reselection. In this

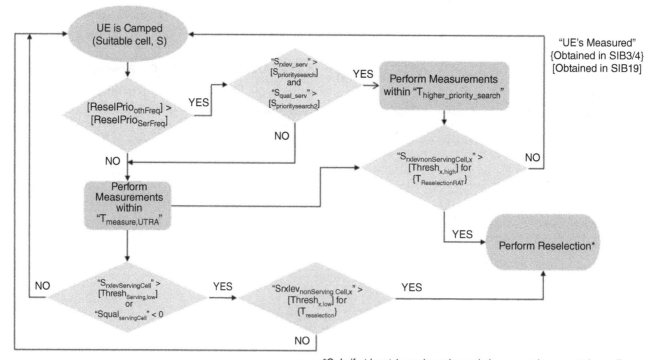

Only if at least 1 sec has elapsed since camping on serving cell

Figure 1.77 UTRAN to E-UTRAN cell reselection flow.

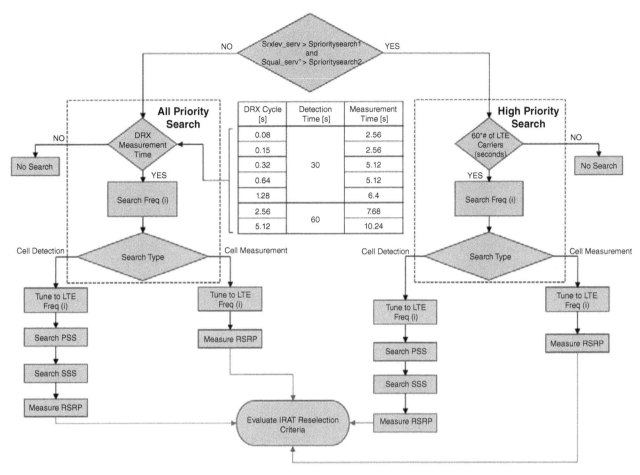

Figure 1.78 UTRAN to E-UTRAN measurement rules.

scenario, the minimum rate at which the UE is required to search/measure higher-priority layers is the same as that defined for lower-priority layers.

1.17.10   UTRAN to E-UTRAN Measurement Stages

An E-UTRAN cell search by a UE camped in 3G is categorized into two phases:

Cell-detection phase: The UE is trying to find the best cell that is considered detectable when any E-UTRAN cell RSRP ≥ -123 dBm. The UE is required to detect a new LTE cell within K_carrier * 30 seconds (K_carrier is the number of LTE frequencies).

Cell-measurement phase: When any cell is considered detectable in the previous phase, the UE is required to keep measuring the cell for a possible inter-RAT reselection.

Figure 1.79 describes the characteristics of cell detection and cell measurements for different DRX cycles.

Similar to the search phases described earlier for UTRAN, 3GPP also allows different searches for

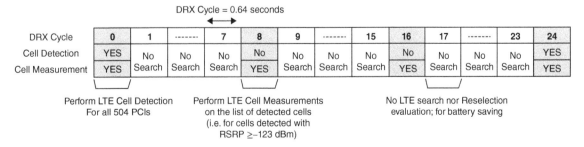

Figure 1.79 UTRAN to E-UTRAN measurement stages.

E-UTRAN cells to reduce the amount of full-space (504 PCI) searches.

Cell-detection phase: (K_carrier is the number of LTE frequencies configured in SIB19):
0.08 s DRX: Once every 96*K_carrier DRX cycles.
0.16 s DRX: Once every 48*K_carrier DRX cycles.
0.32 s DRX: Once every 24*K_carrier DRX cycles.
0.64 s DRX: Once every 24*K_carrier DRX cycles.
1.28 s DRX: Once every 12*K_carrier DRX cycles.
2.56 s DRX: Once every 6*K_carrier DRX cycles.
5.12 s DRX: Once every 3*K_carrier DRX cycles.

Cell-measurement phase:
0.08 s DRX: Once every 32*K_carrier DRX cycles.
0.16 s DRX: Once every 16*K_carrier DRX cycles.
0.32 s DRX: Once every 16*K_carrier DRX cycles.
0.64 ms DRX: Once every 8*K_carrier DRX cycles.
1.28 ms DRX: Once every 4*K_carrier DRX cycles.
2.56 ms DRX: Once every 2*K_carrier DRX cycles.
5.12 ms DRX: Once every 1*K_carrier DRX cycles.

1.17.11 LTE Inter-RAT to GERAN Cell Reselection with Low Priority

The inter-RAT UTRAN-FDD cell measurement criteria are as follows:

$$S_{rxlev} < S_{nonIntraSearchP} \text{ or } S_{qual} < S_{nonIntraSearchQ}$$

A description of the threshold parameters for serving and neighboring cells is provided in Table 1.58.

Figure 1.80 illustrates an example of LTE inter-RAT GERAN reselection parameters based on low priority. The UE initially camps on the LTE F1 cell (first frequency) with physical cell ID 1 with a high priority

(such as priority 6). The only other neighbor defined is the F2 (GERAN frequency) cell with BSIC = 12 with a lower priority than the serving cell (such as priority 3). Before point 1, the serving cell fulfills $S_{rxlev} > S_{nonIntraSearchP}$ and $S_{qual} > S_{nonIntraSearchQ}$, so the UE may choose not to perform inter-RAT measurements. At point 1, $S_{rxlev} < S_{nonIntraSearchP}$ or $S_{qual} < S_{nonIntraSearchQ}$ and thus the UE begins the search for an inter-RAT neighbor. The quality of the serving cell is lower than a threshold ($S_{qual} < \text{Thresh}_{Serving, LowQ}$ or $S_{rxlev} < \text{Thresh}_{Serving, LowP}$), meanwhile the quality of the neighboring inter-RAT cell is better than a threshold ($S_{rxlev} > \text{Thresh}_{X, LowP}$) at point 2 to meet lower-priority cell reselection criteria. The UE reselects to the inter-RAT GERAN BSIC = 12 cell when the timer expires (set at two seconds – at point 3 in this example).

1.17.12 LTE Inter-RAT to GERAN Cell Reselection with High Priority

The UE always performs measurements of higher priority, but these are regulated by the following parameters; the neighboring cell fulfills $S_{rxlev} > \text{Thresh}_{X, HighP}$.

Figure 1.81 illustrates an example of LTE inter-RAT reselection parameters based on high priority. The UE initially camps on the LTE F1 cell (first frequency) with physical cell ID 1 with a lower priority (such as priority 3). The only other neighbor defined is the F2 (GERAN frequency) cell with BSIC = 12 with a higher priority than the serving cell (such as priority 6). Before point 1, the UE always performs high-priority measurements. At point 1, the quality of the neighboring inter-RAT GERAN cell is better than a threshold ($S_{rxlev} > \text{Thresh}_{X, HighP}$). The

Table 1.58 Serving and neighboring cell thresholds.

Parameter	Serving Cell	Neighboring Cell
threshServingLowQ provided in SIB3	$S_{qual} < \text{Thresh}_{Serving,LowQ}$	$S_{rxlev} > \text{Thresh}_{X,LowP}$
threshServingLowQ not provided in SIB3	$S_{rxlev} < \text{Thresh}_{Serving, LowP}$	

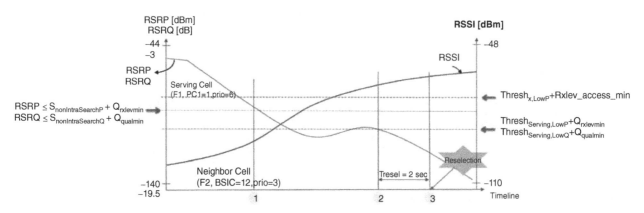

Figure 1.80 LTE inter-RAT to GERAN cell reselection with low priority.

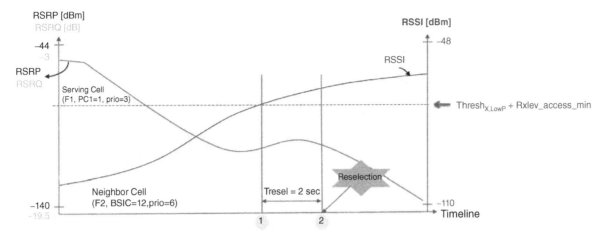

Figure 1.81 LTE inter-RAT to GERAN cell reselection with high priority.

UE reselects to the inter-RAT GERAN BSIC = 12 cell when the timer expires (set at two seconds – at point 2 in this example).

1.17.13   Inter-RAT Handover

Inter-RAT measurement procedures in RRC Connected mode include the following characteristics:

The start of inter-RAT measurements, based on event A2.

The stopping of inter-RAT measurements, based on event A1.

The triggering of an inter-RAT handover. For inter-RAT handover, either event B1 or event B2 can be used. The difference is that event B2 considers the current LTE serving cell before entering into triggering conditions. Usage is typically dependent on the deployment strategy. Inter-RAT handover events were listed in Table 1.50.

1.17.14   Event B1 Measurement Report Triggering

The procedure for triggering event B1 measurements is shown in Figure 1.82.

Inter-RAT handover (or SRVCC) may be blind or non-blind. In the blind scenario, the handover takes place without knowledge of the radio conditions of the other technology. The handover may be triggered by the radio conditions of the serving cell or by other considerations such as cell or infrastructure loading. The disadvantage of this approach is the unknown coverage quality of the target cell, while the advantage is that the handover can be executed quickly without additional measurements. This is particularly helpful in circuit-switch fallback (CSFB) voice calls from LTE.

For non-blind inter-RAT handover, the eNB configures the UE with one of the following events:

Event B1: The inter RAT neighboring cell becomes better than a threshold.

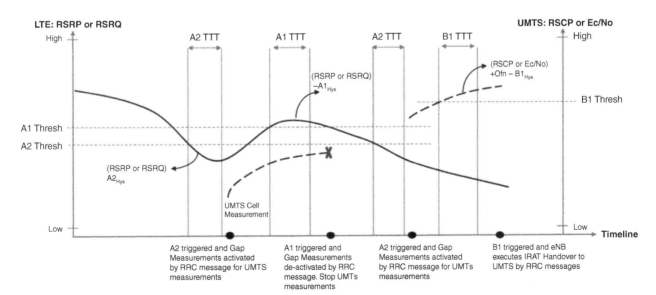

Figure 1.82 Event B1 measurement report triggering.

Event B2: The serving cell becomes worse than threshold 1 and the inter-RAT neighboring cell becomes better than threshold 2.

Other events like A2 + B2 can be configured, and this is a vendor-related implementation.

1.17.15 Inter-RAT Handover Gap Measurements

For the UE to report an inter-RAT or inter-frequency handover, the eNB needs to configure the UE with a measurement gap:

The gap duration is always 6 ms with a repetition period of either 40 or 80 ms, corresponding to 15% or 7.5% of the available subframes respectively, and referred to as gp0 or gp1.

The specific starting SFN (System Frame Number) and subframe for each gap are defined for the UE via the RRC, enabling diversity of gap location throughout the available time slots.

The UE reports when gaps are necessary for either inter-frequency and/or inter-RAT during its capability reporting based on the FGI (Feature Group Indicator).

For the duration of the 6 ms gap, no downlink transmissions are scheduled for the UE and the UE does not transmit on the uplink.

An additional restriction is that the UE cannot transmit in the first subframe following a measurement gap. This corresponds to a 7 ms silence period for the UE in the uplink.

With carrier aggregation devices, gapless inter-frequency handover is possible for certain bands, depending on the UE capabilities.

The measurement gap is a time period during which the UE performs measurements on a neighboring frequency of the serving frequency. Measurement gaps apply to inter-frequency and inter-RAT measurements. Each UE typically has only one receiver, and consequently, one UE can receive signals on only one frequency at a time. After receiving measurement gap configurations from the eNodeB, the UE starts gap-assisted inter-frequency or inter-RAT measurement based on the configurations.

Gap-assisted measurement may conflict with discontinuous reception (DRX), semi-persistent scheduling (SPS), or both. In the case of such a conflict, the gap-assisted measurement takes precedence. This, however, affects data transmission quality. To solve this problem, the design of measurement gaps takes DRX and SPS into consideration. The start time of the measurement gaps can be adjusted in such a way that the conflicts with DRX and/or SPS are minimized whereas gap-assisted measurements are performed on time.

1.17.16 Gap Measurement Example

When gap-assisted measurements for various handover types co-exist, the eNodeB records the measurements based on these handover types. Different gap-assisted measurements can share the same measurement gap configuration. A UE releases measurement gaps only after all gap-assisted measurements have stopped. Figure 1.83 shows an example of gap measurements; two

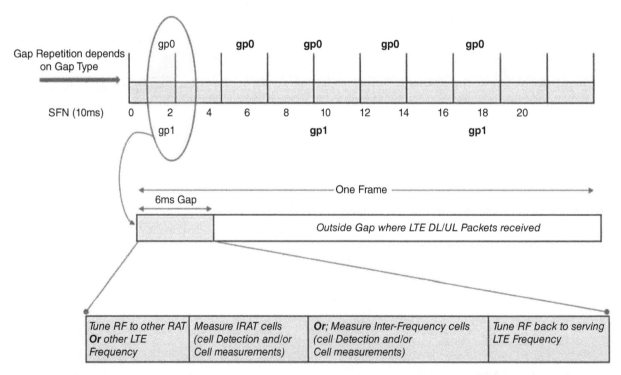

Figure 1.83 Gap measurement example.

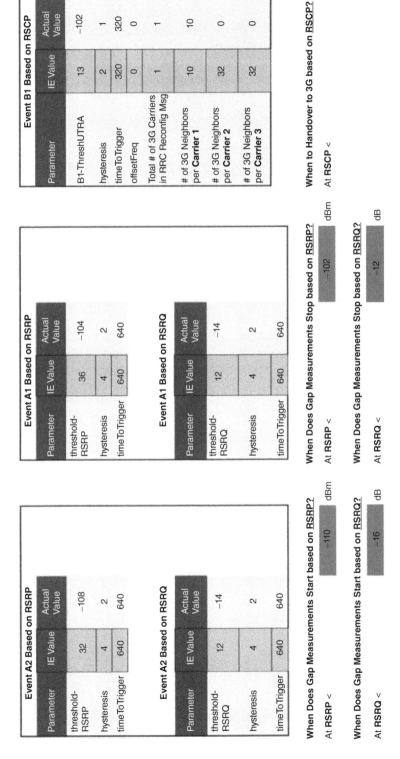

Event A2 Based on RSRP

Parameter	IE Value	Actual Value
threshold-RSRP	32	−108
hysteresis	4	2
timeToTrigger	640	640

Event A2 Based on RSRQ

Parameter	IE Value	Actual Value
threshold-RSRQ	12	−14
hysteresis	4	2
timeToTrigger	640	640

When Does Gap Measurements Start based on RSRP?

At **RSRP** < ⬚ −110 dBm

When Does Gap Measurements Start based on RSRQ?

At **RSRQ** < ⬚ −16 dB

Event A1 Based on RSRP

Parameter	IE Value	Actual Value
threshold-RSRP	36	−104
hysteresis	4	2
timeToTrigger	640	640

Event A1 Based on RSRQ

Parameter	IE Value	Actual Value
threshold-RSRQ	12	−14
hysteresis	4	2
timeToTrigger	640	640

When Does Gap Measurements Stop based on RSRP?

At **RSRP** < ⬚ −102 dBm

When Does Gap Measurements Stop based on RSRQ?

At **RSRQ** < ⬚ −12 dB

Event B1 Based on RSCP

Parameter	IE Value	Actual Value
B1-ThreshUTRA	13	−102
hysteresis	2	1
timeToTrigger	320	320
offsetFreq	0	0
Total # of 3G Carriers in RRC Reconfig Msg	1	1
# of 3G Neighbors per **Carrier 1**	10	10
# of 3G Neighbors per **Carrier 2**	32	0
# of 3G Neighbors per **Carrier 3**	32	0

When to Handover to 3G based on RSCP?

At **RSCP** < ⬚ −101 dBm

Figure 1.84 Example of inter-RAT handover parameters.

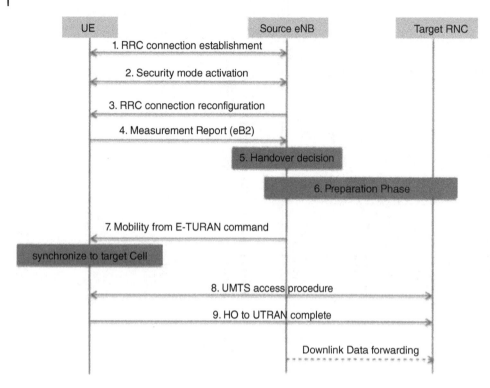

Figure 1.85 Inter-RAT handover to UTRAN call flow.

measurement gap patterns are available: pattern 1 and pattern 2. In pattern 1, the gap duration is 6 ms and the repetition of the gap is 40 ms. In pattern 2, there is a 6 ms gap every 80 ms period.

1.17.17 Example of Inter-RAT Handover Parameters

An example of inter-RAT handover parameters is shown in Figure 1.84.

1.17.18 Inter-RAT Handover to UTRAN Call Flow

The call flow for inter-RAT handover is illustrated in Figure 1.85. It is clear that:

The inter-RAT handover procedure is initiated only when AS security has been activated and SRB2 with at least one DRB has been set up and is not suspended.

Step 3: The NW will configure the measurement of the neighboring cell (inter-RAT).

Step 7: After the preparation phase, the source eNB prepares and sends the Mobility from E-UTRAN Command message to the UE for the handover to the target network. This will contain the parameters (i.e. mobility purpose, target RAT type, the handover message of target RAT, security parameter, etc.) for the handover.

Step 9: After accessing the target UMTS cell, the UE sends a Handover to UTRAN Complete message to the target RNC to indicate a successful handover.

References

1 Elnashar, A., El-saidny, M. A. *et.al.* (2014) *Design, Deployment, and Performance of 4G-LTE Networks: A Practical Approach*, John Wiley & Sons.

2 3GPP (2010) TS 36.300 V8.12.0, "3rd Generation Partnership Project; Technical Specification Group Radio Access Network; Evolved Universal Terrestrial Radio Access (E-UTRA) and Evolved Universal Terrestrial Radio Access Network (E-UTRAN); Overall Description."

3 3GPP (2010) TS 36.331 V8.9.0, "3rd Generation Partnership Project; Technical Specification Group Radio

Access Network; Evolved Universal Terrestrial Radio Access (E-UTRA); Radio Resource Control (RRC)."

4 3GPP (2009) TS 36.201 V8.3.0, "3rd Generation Partnership Project; Technical Specification Group Radio Access Network; Evolved Universal Terrestrial Radio Access (E-UTRA); LTE Physical Layer – General Description."

5 3GPP (2009) TS 36.212 V8.8.0, "3rd Generation Partnership Project; Technical Specification Group Radio Access Network; Evolved Universal Terrestrial Radio Access (E-UTRA); Multiplexing and Channel Coding."

2

Introduction to the IP Multimedia Subsystem (IMS)

2.1 Introduction

In this chapter, we will introduce the IP Multimedia Subsystem (IMS), which is the core network of Voice over LTE (VoLTE). We will focus on IMS architecture, IMS core network elements, and the IMS core network call and signaling flow between different network elements. In addition, the chapter will cover practical deployment scenarios for IMS-network-based real use cases from a live network deployment. Moreover, the high-level design (HLD) of a practical IMS system will be presented, addressing key design aspects. Although, the IMS target is to be the core for all IP multimedia communication traffic, we will focus only on VoLTE in this chapter. Other IMS services such as Voice over Wi-Fi (VoWiFi), Video over LTE (ViLTE), and Web Real-Time Communication (WebRTC) are not discussed in detail. Our aim in this chapter is to provide the reader with the foundations of IMS, which will be necessary when we reach the VoLTE practical performance analysis provided in Chapters 3 and 4. IMS can also support access modes such as Wideband CDMA (WCDMA), Code Division Multiple Access 2000 (CDMA2000), Wi-Fi, xDSL, local area network, PSTN/integrated services digital network (ISDN), and worldwide interoperability for microwave access (WiMAX).

2.1.1 Voice over LTE Overview

As mobile broadband networks continue to evolve, mobile broadband has become a development trend for mobile speech services that rely heavily on soft switch. The radio access network (RAN) is evolving from GERAN/UTRAN towards LTE, and the voice core network is evolving towards the IP Multimedia Subsystem (IMS). An IMS-based VoLTE solution is the most popular Voice over IP (VoIP) solution in the telecom industry. With IMS-based VoLTE, both speech and data services are transmitted in the packet-switched (PS) domain. IP multimedia applications are supported by IP multimedia sessions in the IMS. Therefore, VoLTE will eliminate

the need for legacy circuit-switched (CS) infrastructure. However, the CS domain will be phased out gradually with the increase in VoLTE terminals. The majority of service providers are currently providing VoLTE services or testing them. In addition to this, the IMS will be a unified IP core for voice and video services over the PS domain. Therefore, a voice and data session can co-exist at the same time and the user can enjoy high-speed Internet services in addition to high-definition (HD) voice and video services. Other, supplementary services will also be supported.

IMS-based VoLTE introduces an IMS network to the LTE-EPC architecture. The IMS network implements a number of signaling functions, including session setup, number analysis, and session release. Speech data are carried over the EPC core instead of the legacy media gateway (MGW) of 3G. The IMS entity takes advantage of bearer services provided by the PS domain and UTRAN to transmit IMS signaling and subscriber data. A handover function is supported when the location of a UE changes. This means that the IMS bearer service provided by a PS domain will not be interrupted. Figure 2.1 illustrates the overall IMS topology. The IMS is a service control network deployed on top of the LTE PS domain. The same concept is applied, for example, to critical communication services when deployed on top of the LTE network. The IMS co-exists with core networks in the CS domain which support non-VoLTE devices and fixed services. Therefore, the IMS allows for fixed-mobile convergence.

The following improvements led to the adoption of IMS in a 3G access domain for CS networks instead of the legacy CS MGWs:

- Air interfaces in the PS domain are continuously optimized based on a QoS mechanism applied to real-time IP services.
- The IP-based processing capabilities of UE are enhanced.
- In addition, the end-to-end QoS mechanism applied to session-based IMS bearer networks has become more mature. However, improvement is expected

Practical Guide to LTE-A, VoLTE and IoT: Paving the Way Towards 5G, First Edition. Ayman Elnashar and Mohamed El-saidny.
© 2018 John Wiley & Sons Ltd. Published 2018 by John Wiley & Sons Ltd.

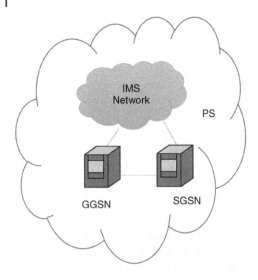

Figure 2.1 IMS position in the PS domain.

with enhanced VoLTE (eVoLTE) which is part of 3GPP Release 14. The eVoLTE solution is an enhanced version of VoLTE that provides HD voice, video, rich media communication, VoWiFi, WebRTC, and rich multimedia services, which helps operators to stabilize and expand voice revenues and compete with over-the-top (OTT) service providers.

2.1.2 SIP Protocol

In an IMS network, Session Initiation Protocol (SIP) messages are exchanged between UEs using signaling PDP contexts through the PS domain of the LTE network. Similarly, media streams (voice/video) are transmitted using media PDP contexts. Like common interactive PDP contexts, media PDP contexts do not require special processing. However, signaling PDP contexts do require special processing.

The SIP is an application-layer control protocol for carrying out multimedia communications on the IP network. The SIP is used to create, change, and terminate session processes with one or more than one participants. The SIP originated in the multicast backbone M-bone experiment in 1996. By 1999, the SIP was defined by the IETF-MMUSIC in RFC2543 [1]. In 1999, the IETF-MMUSIC became the independent SIP task force. Later, two SIP-centered task forces, namely SIPPING and SIMPLE, were added. In July, 2002, the SIP was defined in the new RFC3261 [2] and most content was rewritten and cleaned up. In addition, certain new features were added and most content was backward-compatible with RFC2543 [1]. The SIP cooperates with the Session Description Protocol (SDP), Real-time Transport Protocol (RTP)/RTP Control Protocol (RTCP), Real Time Streaming Protocol (RTSP), and the Domain Name System (DNS) to set up sessions

in IMS and carry out the media negotiation. The SIP is a simple, flexible protocol for creating, modifying, and terminating sessions. It works independently of underlying transport protocols.

The main features of the SIP are as follows:

- Simplicity: Text style is similar to the HTTP and SMTP protocols.
- Extensibility: Can cooperate with other protocols, such as SDP/MSML.
- Flexibility: Special headers can be added by vendors.

The SIP is an application-layer control protocol for creating, modifying, and terminating sessions such as Internet multimedia conferences, Internet telephone calls, and multimedia distribution. The SIP messages used to create sessions carry session descriptions that allow participants to agree on a set of compatible media types. These session descriptions are commonly formatted using SDP. When used with SIP, the offer/answer model [3] provides a limited framework for negotiation using SDP. The SIP is an IETF protocol for multimedia sessions including text, video, voice, and so on. The main functions of the SIP are to set up, control, and tear down sessions. Protocols used in conjunction with the SIP, such as SDP, RTP/RTCP, and RTSP can be summarized as follows:

Session Description Protocol (SDP): A control protocol of the application layer. It is used for the media negotiation during session establishment.
Real-time Transport Protocol/RTP Control Protocol (RTP/RTCP): Bearer plane protocols of the application layer. After session establishment, the RTP ensures the real-time transmission of a media stream. The RTCP monitors the media stream.
Real Time Streaming Protocol (RTSP): The control protocol of the application layer. It is used to control the change of some parameters such as QoS after session establishment.

The high-level protocol stack including the SIP and other related protocols is described in Figure 2.2.

The SIP can be based on the User Datagram Protocol (UDP), the Transmission Control Protocol (TCP), or the Stream Control Transmission Protocol (SCTP). However, most cases now use the UDP.

A simplified SIP network communication between two UEs is shown in Figure 2.3.

2.1.3 SIP Basic Flow

SIP messages are classified into two main types: request messages (which initiate a session) and response messages (which respond to a request). The request message includes basic request message contents and extended request message contents. Table 2.1 illustrates a SIP

Figure 2.2 The protocol stack including the SIP and other related protocols.

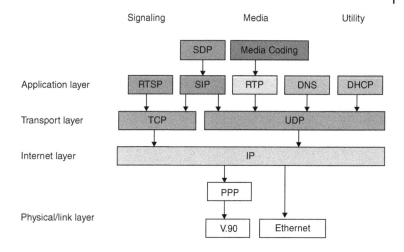

Figure 2.3 Simplified SIP network communication.

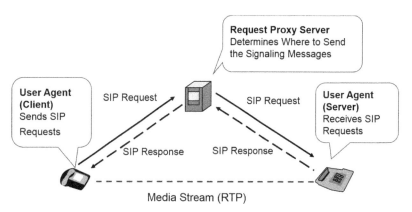

Table 2.1 Message content of SIP basic request and SIP extended request.

Basic Request	Extended Request
INVITE: To initiate a session	MESSAGE: Is applied to IM
ACK: The response to INVITE	SUBSCRIBE: To subscribe to a notify event
CANCEL: To cancel a session	NOTIFY: To send a notify event
BYE: To terminate a session	UPDATE: To modify the session attributes at the establishment stage of a call
REGISTER: To register on a server	PUBLISH: To distribute its event state to the status server
OPTIONS: For querying servers about their capabilities	PRACK: To indicate the reliability of a temporary response

Table 2.2 Message content of a SIP response message.

Item	Description
Provisional	Request received, continuing to process the request.
Success	The action was successfully received, understood, and accepted.
Redirection	Further action needs to be taken in order to complete the request.
Client error	The request contains bad syntax or cannot be fulfilled at this server.
Server error	The server failed to fulfill an apparently valid request.
Global failure	The request cannot be fulfilled at any server.

session for a basic request and an extended request. Table 2.2 demonstrates th response message contents. The SIP basic flow for session setup and registration are illustrated in Figure 2.4.

2.1.4 SIP Session Flow in IMS

The logical SIP entities, as described in [4], consist of User Agents (UA) and Network Servers. The SIP uses

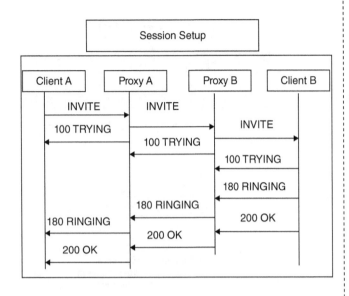

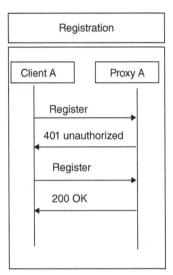

Figure 2.4 SIP basic flow.

client/server architecture in which the UAs are classified into User Agent Clients (UACs) and User Agent Servers (UASs); the UAC initiates SIP requests, while the UAS returns SIP responses. Both UACs and UASs can terminate a call. On the other side, Network Servers consist of three main servers:

Register: Maintains the location of SIP users; the SIP client needs to update the location using a register request.
Proxy: Decides the next hop and forwards a request.
Redirect: Accepts the SIP request and translates to the new address.

The registration server on the IMS network is called the Serving Call Session Control Function (S-CSCF), while proxy servers on the IMS network are the Proxy-CSCF (P-CSCF), Interrogating-CSCF (I-CSCF), and Serving-CSCF (S-CSCF). The initial session request flow sent by the UE is described in Figure 2.5. The update request flow sent by the UE is described in Figure 2.6 [5].

An UPDATE message can only be sent when resources have been reserved successfully and the called party carries out the 200 (OK) response. 180 (RING) occurs only when the called party receives the UPDATE message sent by the calling party (the calling party has successfully reserved resources) and after the called party has successfully reserved resources.

The initial response flow sent by the UE is described in Figure 2.7 [5].

In Figure 2.7, the first 200 (OK) is for PRACK, while the second 200 (OK) is for an INVITE request. ACK

is for INVITE. Only an INVITE request has the ACK response.

2.1.5 SDP Protocol

As defined in [6], when initiating multimedia teleconferences, Voice over IP (VoIP) calls, streaming video, or other sessions, there is a requirement to convey media details, transport addresses, and other session description metadata to the participants. SDP protocols provide a standard representation for such information, irrespective of how that information is transported. The SDP is purely a format for session description; it does not incorporate a transport protocol, and it is intended to use different transport protocols as appropriate, including the Session Announcement Protocol (SAP) [7], Session Initiation Protocol (SIP) [2], Real Time Streaming Protocol (RTSP) [8], electronic mail using the MIME extensions, and the Hypertext Transport Protocol (HTP). The SDP is intended to be a general-purpose protocol, so it can be used in a wide range of network environments and applications. Examples of SDP usage include session initiation, streaming media, email and the World Wide Web (www), and multicast session announcement. An example of media negotiation is described in Figure 2.8.

The SDP is intended for describing multimedia sessions for the purposes of session announcement, session invitation, and other forms of multimedia session initiation. The position of the SDP in the protocol stack is shown in Figure 2.2. By default, the UDP protocol is used with the SDP. If a message contains more than 1300 bytes, the TCP protocol is used.

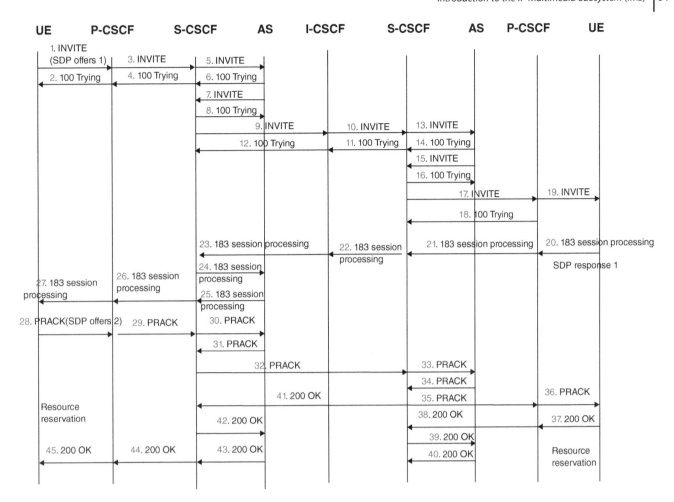

Figure 2.5 Initial session request.

2.2 IMS Network Description

2.2.1 IMS Network Architecture

Figure 2.9 represents the IMS reference architecture including interfaces towards legacy networks and other IP-based multimedia systems [5]. Details of the roles of these nodes are described later in this chapter.

To simplify the IMS topology, a practical layer-based IMS architecture is classified into four layers, as follows:

- Bearer control and access layer
- Session control layer
- Service capability layer
- Application layer.

Figure 2.10 illustrates the layer-based IMS architecture including the above four layers. The four layers are described below (only three paras below - nothing for application layer from above list - combined with service capability layer?).

The *service capability layer* consists of all types of ASs and resource servers. It provides services such as gaming, conferencing, and IM. It also provides service capabilities such as presence, group, and media resources. Operators can combine the service capabilities into different services based on actual demands to launch services faster.

To support the product development of third parties, operators can open certain service capabilities and provide the API interface for third parties, which can be developed independently at the *application layer*.

The functions of the service capability layer and the application layer are described briefly in Figure 2.11.

The *session control and media resources layer* consists of functional network elements such as the P-CSCF, I-CSCF, S-CSCF, BGCF, MRFC/MRFP, HSS, and AGCF. This layer performs functions such as registration, authentication, session route control, service triggering, topology hiding of the inter-network gateway (THIG), routing, resource control, and interworking.

The *access/bearer control layer* consists of functional network elements such as the PDF, PCRF, SPDF, A-RACF, NACF, and CLF. It provides QoS resource control, depending on Service-Based Local Policy (SBLP)/Service-Based Bearer Control (SBBC). It controls

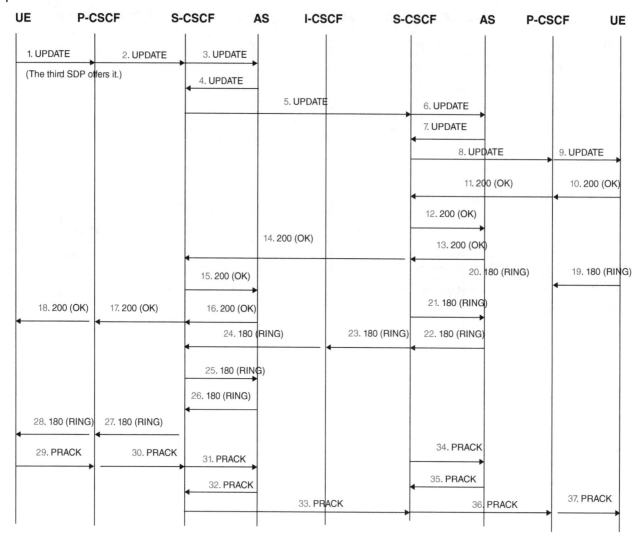

Figure 2.6 Update request flow.

the resources of the access network through the PDF, PCRF, SPDF, and A-RACF.

The functions of the session control and media resources layer and the access/bearer layer are outlined briefly in Figure 2.12.

If fixed networks access the IMS, the access and bearer control layer can include the Network Attachment Subsystem (NASS). The NASS consists of the CLF and NACF, and performs the functions of location management and network access configuration to access fixed networks. This layer also performs the NAT control function to support transversal NAT in enterprise networks.

The main IMS network elements in the session control layer are summarized in Table 2.3.

2.2.2 Call Session Control Function – CSCF

As described in [9], the Call Session Control Function (CSCF) establishes, monitors, supports, and releases

multimedia sessions and manages the user's service interactions. For more details, see clause 4a.7.1 in [10]. The CSCF can act as Proxy CSCF (P-CSCF), Serving CSCF (S-CSCF), or Interrogating CSCF (I-CSCF). The P-CSCF is the first contact point for the UE within the IMS system; the S-CSCF handles the session states in the network; the I-CSCF is mainly the contact point within an operator's network for all IMS connections destined for a subscriber of that network operator, or a roaming subscriber currently located within that network operator's service area. This functional entity is identical to the CSCF defined in [10], except for when acting as a P-CSCF. The P-CSCF behavior differs from the behavior in TS 123 002 [10] by the following main points:

- The e2 reference point is supported to enable the P-CSCF to retrieve information (for example, the physical location of the user equipment) from the NASS.
- The Gq and Iq reference points are not supported but are replaced by the Gq' reference point.

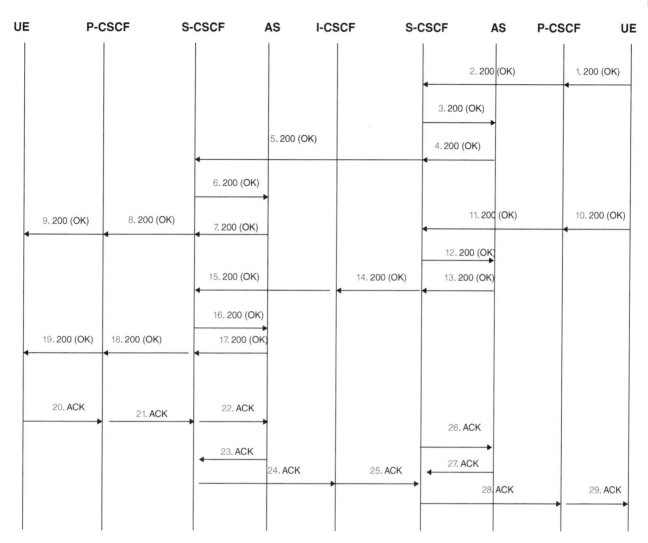

Figure 2.7 Initial response flow.

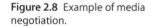

Figure 2.8 Example of media negotiation.

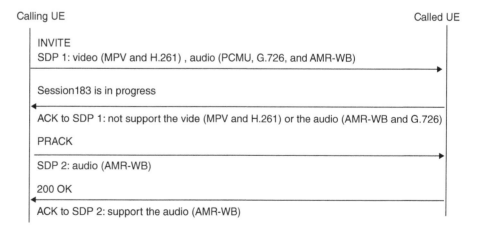

Based on the local configuration, the I-CSCF may perform transit routing functions [11]. Further specifications of the P-, S-, and I-CSCF are provided in [11]. Support for emergency communications requires the use of an E-CSCF, as described in [12]. Call session control function entities are shown in Figure 2.13.

The key functions of CSCF network elements are summarized in Table 2.4.

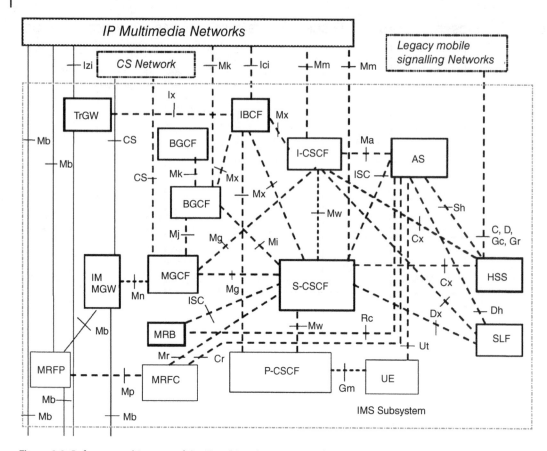

Figure 2.9 Reference architecture of the IP multimedia core network subsystem.

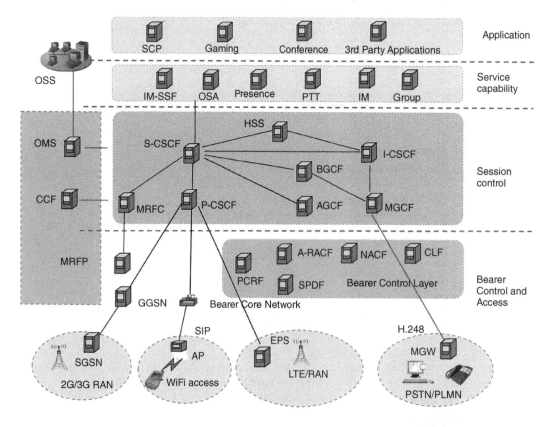

Figure 2.10 IMS architecture.

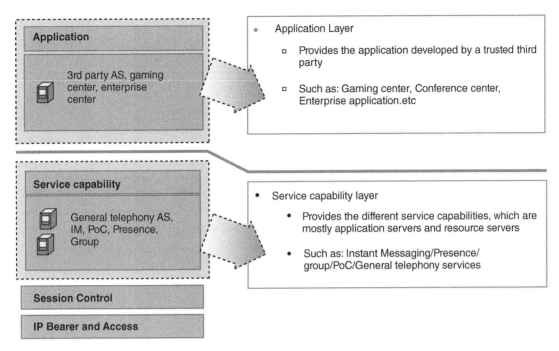

Figure 2.11 Functions of the service capability layer and the application layer.

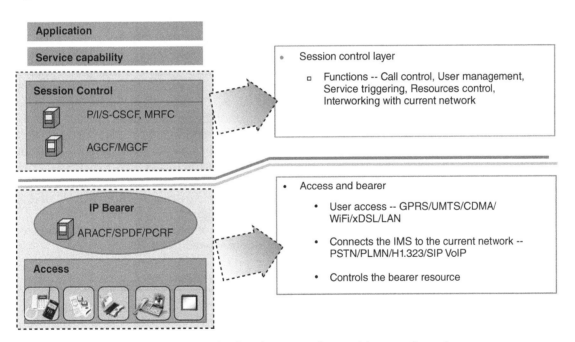

Figure 2.12 Functions of the session control and media resources layer and the access/bearer layer.

A simple model for the IMS call procedure is illustrated in Figure 2.14. While two IMS users are involved in a call, the P/I/S-CSCF can handle the whole signal-routing procedure.

2.2.3 Interworking Nodes

IMS interworking nodes consist of the Media Gateway Controller Function (MGCF), the Breakout Gateway Control Function (BGCF), and the IM-MGW. The MGCF is described in [9] and provides the ability to control a Trunking Media Gateway Function (T-MGF) through a standardized interface. Such control includes allocation and de-allocation of resources of the media gateway as well as modification of the usage of these resources. The MGCF communicates with the CSCF, the BGCF, and circuit-switched networks. The MGCF performs protocol conversion between the ISUP and the

Table 2.3 IMS network elements (NEs) in the session control layer.

Function	NE	Function	NE
Call control	P-CSCF	Network interworking	MGCF
	I-CSCF		IM-MGW
	S-CSCF		BGCF
User management	HSS	Media resource	MRFC
	SLF		MRFP

Table 2.4 Key functions of CSCF network elements (NEs).

CSCF	Key Functions
P-CSCF (Proxy)	The first contact point to the IMS network in the visiting domain or home domain.
	Access network control.
	QoS control, NAT control, and security control.
I-CSCF (Interrogating)	First entry to the IMS network of a carrier.
	S-CSCF assignment and session routing.
	Topology hiding.
S-CSCF (Serving)	User registration authentication.
	Session route control (normal, interworking, emergency call).
	Service trigger.

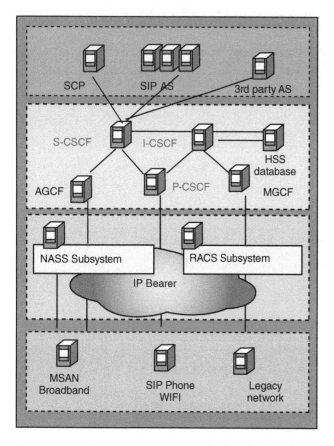

Figure 2.13 Call session control function network elements.

SIP. It also supports interworking between the SIP and non-call-related SS7 signaling (i.e. TCAP-based signaling for supplementary services such as CCBS). In the case of incoming calls from legacy networks, the MGCF determines the next hop in IP routing depending on received signaling information. Based on the local configuration, the MGCF may perform transit-routing functions [11]. This functional entity is identical to the MGCF defined in [10], except that, in addition, it supports TCAP interworking. A node implementing this functional entity in an NGN network and a node implementing it in a 3GPP network may differ in terms of supported resources (for example, codecs) and configuration.

The BGCF as described in [9] is the next hop in SIP routing. This determination may be based on information received in the protocol, administrative information, and/or database access. For PSTN terminations, the BGCF selects the network in which PSTN breakout is to occur and – within the network where the breakout is to occur – selects the MGCF. This functional entity is identical to the BGCF defined in [10], although a node implementing this functional entity in an NGN network and a node implementing it in a 3GPP network may differ in terms of configuration (for example, breakout criteria). In cases of transit, the BGCF may have extra functionality for routing transit traffic [11]. Interworking nodes' network elements consist of the BGCF, MGCF, and MGW, as highlighted in Figure 2.15.

The key functions of interworking nodes' network elements are summarized in Table 2.5.

2.2.4 Connection to Legacy Networks

There are three scenarios for connection to legacy networks (i.e., PLMN/PSTN). The three scenarios are summarized as follows:

Scenario 1: An IMS user calls a PSTN/CS user. As shown in Figure 2.16, the S-CSCF connects to PLMN/PSTN via the BGCF and the MGCF.

Scenario 2: An IMS user calls a PSTN/CS user who belongs to another operator. As shown in Figure 2.17, the local S-CSCF connects to the local BGCF1, which connects to the BGCF2 of the other operator.

Scenario 3: An IMS user calls a PSTN/CS user for whom topology hiding is required. As shown in Figure 2.18, the local S-CSCF connects to the local I-CSCF1, which connects to the BGCF2 of the other operator.

Figure 2.14 Simple model for the IMS call procedure.

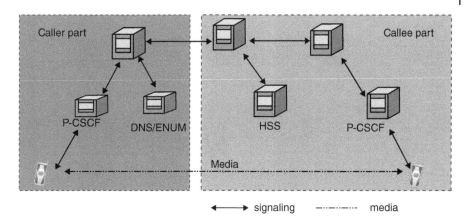

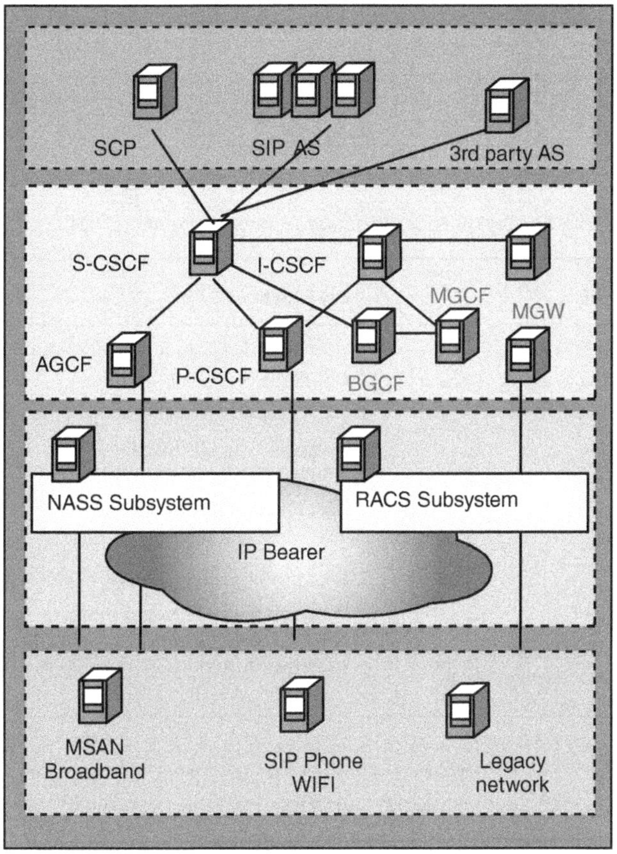

◄──────► signaling   ─·──·──·─ media

Table 2.5 Key functions of interworking nodes' network elements (NEs).

NE	Key Functions
MGCF	Controls the IMS-MGW for establish/modify/delete media channels.
	Selects the I-CSCF for incoming calls from PSTN/CS.
	Performs protocol conversion between ISUP and SIP.
BGCF	Selects a proper MGCF for interworking with the PSTN/CS domain.
IM-MGW	Terminates bearer channels from a circuit-switched network and media streams from a packet network.

Figure 2.15 Interworking nodes' network elements (NEs).

2.2.5 Database Function

The database function of the IMS consists of the Home Subscriber Server (HSS) and the Subscription Locator Function (SLF). Database function network elements are shown in Figure 2.19.

The SLF appears in the User Identity to HSS Resolution mechanism. As described in [13], the User Identity to HSS Resolution mechanism enables the I-CSCF and the S-CSCF to find the address of the HSS that holds the subscriber data for a given user identity when multiple and separately addressable HSSs have been deployed by the network operator. The resolution mechanism is not required in networks that utilize a single HSS. An example of a single-HSS solution is server farm architecture. The resolution mechanism described in [5] is based on the Subscription Locator Function (SLF). The subscription locator is accessed via the Dx interface. The Dx interface is always used in conjunction with the Cx interface. The Dx interface is based on Diameter. Its functionality is implemented by means of the routing mechanism provided by an enhanced Diameter redirect agent, which is able to extract the identity of the user from the received requests.

To obtain the HSS address, the I-CSCF and the S-CSCF send to the SLF the Cx requests aimed at the HSS. On receipt of the HSS address from the SLF, the I-CSCF and S-CSCF send the Cx requests to the HSS. While the I-CSCF is stateless, the S-CSCF stores the HSS address/name, as specified in [5]. Further requests associated with the same user make use of the stored HSS address. In networks where the use of the User Identity to HSS Resolution mechanism is required, each I-CSCF and S-CSCF should be configured with the

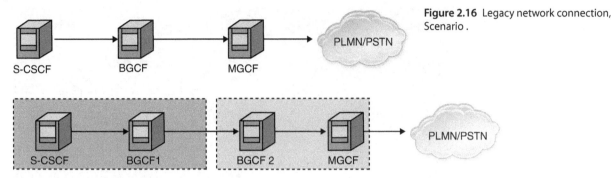

Figure 2.16 Legacy network connection, Scenario .

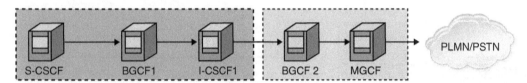

Figure 2.17 Legacy network connection, Scenario 2.

Figure 2.18 Legacy network connection, Scenario 3.

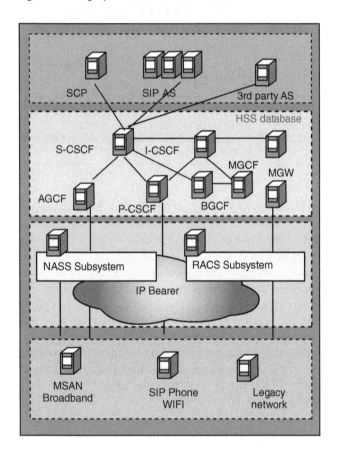

Figure 2.19 Database function network elements (NEs).

Table 2.6 Key functions of database network elements (NEs).

NE	Key Functions
HSS (Home Subscriber Server)	User identification, numbering and addressing information.
	User security information: network access control information for authentication and authorization.
	User location information at inter-system level.
	User profile information (iFC, etc).
SLF (Subscription Locator Function)	When an operator has more than one HSS, the SLF is used to select the related HSS, and usually the SLF is combined with the HSS.

address/name of the SLF implementing this resolution mechanism. The key functions of database network elements are summarized in Table 2.6.

The main functions of the database HSS are illustrated in Figure 2.20. It consists of mobility management, user security information generation, user security support, service provisioning support, session establishment support, identification handling, service authorization support, access authorization, application services support, and CAMEL service support. These functions can be summarized as follows:

Mobility management: The function that supports the user mobility through the CS domain, the PS domain, and the IM CN subsystem.

Call and/or session establishment support: The HSS supports the call and/or session establishment procedures in the CS domain, the PS domain, and the IM CN subsystem. For terminating traffic, it provides information on which call and/or session control entity currently hosts the user.

User security information generation: The HSS generates user authentication, integrity and ciphering

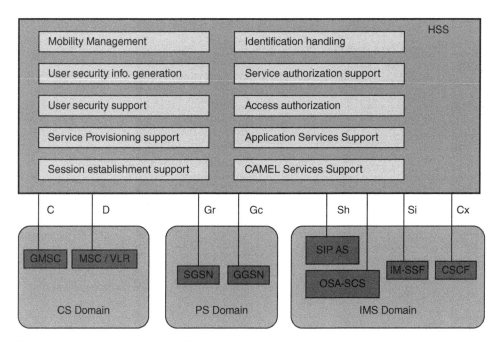

Figure 2.20 Main functions of the HSS database.

data for the CS and PS domains and for the IM CN subsystem.

User security support: The HSS supports the authentication procedures to access the CS domain, the PS domain, and the IM CN subsystem services by storing the generated data for authentication, integrity, and ciphering and by providing these data to the appropriate entity in the CN (i.e. MSC/VLR, SGSN, 3GPP AAA server, or CSCF).

User identification handling: The HSS provides the appropriate relations among all the identifiers uniquely determining the user in the system, i.e. the CS domain, the PS domain, and the IM CN subsystem (IMSI and MSISDNs for the CS domain; IMSI, MSISDNs and IP addresses for the PS domain; private identity and public identities for the IM CN subsystem).

Access authorization: The HSS authorizes the user for mobile access when requested by the MSC/VLR, SGSN, 3GPP AAA server, or CSCF, by checking that the user is allowed to roam to that visited network.

Service authorization support: The HSS provides basic authorization for MT call/session establishment and service invocation. In addition, the HSS updates the appropriate serving entities (i.e., MSC/VLR, SGSN, 3GPP AAA server, CSCF) with relevant information related to the services to be provided to the user.

Service provisioning support: The HSS provides access to the service profile data for use within the CS domain, the PS domain, and/or the IM CN subsystem.

Application services and CAMEL services support: The HSS communicates with the SIP application

server and the OSA-SCS to support application services in the IM CN subsystem. It communicates with the IM-SSF to support the CAMEL services related to the IM CN subsystem. It also communicates with the gsmSCF to support CAMEL services in the CS domain and the PS domain.

The HSS can save IMS subscription information of carriers, and can support carriers or end users in customizing or modifying the subscription information through the interface connected to the service management system as well. Moreover, the HSS provides the Cx interface based on the Diameter protocol to connect I-CSCF. The I-CSCF can provide subscribers with the S-CSCF information through the Cx interface, which is used as proof that subscribers use S-CSCF. Also, the HSS provides the service of querying the routes of called parties for IMS subscribers, such as S-CSCF domain name or address information.

The HSS can register the S-CSCF domain name and routing information through the Cx interface during the IMS registry process, and support the download of IMS subscription information to the S-CSCF through the Cx interface as well. It can also calculate authentication tuples according to subscriber security contexts, and provide authentication tuples required by subscribers or networks to the S-CSCF through the Cx interface based on the Diameter protocol as well. The HSS provides the Sh interface based on the Diameter protocol to connect the SIP AS so as to provide subscription information for value-added services. The HSS is also responsible for transparently storing the AS's value-added service data of special subscribers, but not translating semantically.

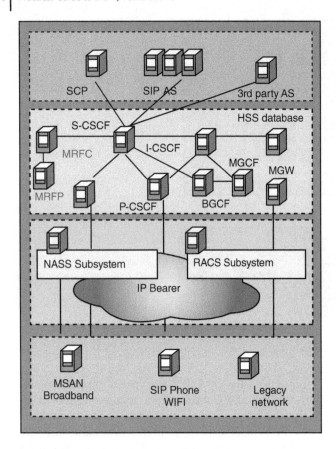

Figure 2.21 IMS multimedia resources' network elements.

2.2.6 IMS Multimedia Resources Function

As described in [9], the Media Resource Function Controller (MRFC), in conjunction with a Media Resource Function Processor (MRFP), is located in the transport layer (see [14]) and provides a set of resources within the core network for supporting services. The MRFC interprets information coming from an AS via an S-CSCF and controls the MRFP accordingly. The MRFC, in conjunction with the MRFP, provides, for example, multi-way conference bridges, announcement playback, media transcoding, and so on.

This functional entity is identical to the MRFC defined in [10], although a node implementing this functional entity in an NGN network and a node implementing it in a 3GPP network may differ in terms of supported resources and configuration. Multimedia resources' network elements are highlighted in Figure 2.21.

The key functions of the MRFC and the MRFP are summarized in Table 2.7.

2.2.7 IMS Network Elements in Access and Bearer Layer

The IMS network elements in the access and bearer layer are summarized in Table 2.8.

Table 2.7 Key functions of IMS multimedia resources.

NE	Key Functions
MRFC	Controls the media stream resources in the MRFP.
	Interprets information coming from an AS and S-CSCF (i.e. session identifier) and controls MRFP accordingly.
MRFP	Provides resources to be controlled by the MRFC.
	Mixes incoming media streams (i.e. for multiple parties).
	Processes media streams (i.e. audio transcoding, media analysis).

Table 2.8 IMS network elements (NEs) in the access and bearer layer.

Function	NE	Function	NE
Resource Control	PCRF	Charging	CCF
	SPDF+A-RACF	NE manage	OMS
Access Control	NACF	SBC	ABGF
	CLF	Naming and address	DNS/ENUM

Table 2.9 Functions of IMS entities of the access and bearer layer.

NE	Key Functions
PCRF/SPDF	Performs QoS control function in IMS network.
DNS/ENUM	DNS (Domain Name System) is responsible for translating the URL to IP address for session routing.
	ENUM (E.164 Number URI Mapping) is used to translate a Tel URI to SIP URI.
NAT/SBC	SBC can perform functions such as: NAT traversal, security, IPv4/IPv6 conversion, etc.
	Media proxy, fixed network QoS control and security.
NACF/CLF	Acts as the DHCP and AAA server function for the fixed-network access user.

The functions of the IMS entities of the access and bearer layer are summarized in Table 2.9.

2.2.8 IMS Service Capability Layers

The Application Server (AS) provides value-added services for IMS subscribers. It can be located in a home network or provided by a third party. The AS can be divided into three types: SIP AS, OSA AS, and IM-SSF. The OSA AS communicates with IMS network elements through the OSA service capability servers. The IM-SSF provides the mapping between IMS SIP and CS CAP and the SSP triggers capability so that the VoIP subscribers of the IMS domain can seamlessly inherit CS intelligent services. The AS can obtain the subscribers' service data and subscribers' status information through the interface of the HSS. SIP AS and OSA AS obtain data through the Sh interface based on the Diameter protocol. The IM-SSF obtains data through the Si interface based on the MAP.

As described in [5], the ISC interface is between the S-CSCF and the service platform(s). An AS offering value-added IM services resides either in the user's home network or in a third party location. The third party could be a network or simply a standalone AS. The S-CSCF to AS interface is used to provide services residing in an AS. Two cases are identified:

- S-CSCF to an AS in the home network.
- S-CSCF to an AS in an external network (for example, a third party or visited network).

The SIP application server may host and execute services; it can influence and impact the SIP session on behalf of the services and it uses the ISC interface to communicate with the S-CSCF. The S-CSCF is able to supply the AS with information to allow it to execute multiple services in order within a single SIP transaction.

The ISC interface is able to support subscription to event notifications between the application server and the S-CSCF to allow the application server to be notified of the implicit registered public user identities, registration state, and UE capabilities and characteristics in terms of SIP user agent capabilities and characteristics.

The S-CSCF decides whether an application server is required to receive information related to an incoming initial SIP request to ensure appropriate service handling. The decision at the S-CSCF is based on (filter) information received from the HSS. This filter information is stored and conveyed on a per-application server basis for each user. It is possible to include a service indication in the filter information, which is used to identify services and the order in which they are executed on an application server within a single SIP transaction. The name(s)/address information of the application server(s) are received from the HSS. For an incoming SIP request, the S-CSCF performs any filtering for ISC interaction before performing other routing procedures towards the terminating user, for example, forking, caller preferences, etc.

The S-CSCF does not handle service interaction issues. Once the IM SSF, the OSA SCS, or the SIP application server has been informed of a SIP session request by the S-CSCF, the IM SSF, OSA SCS, or SIP application server ensures that the S-CSCF is made aware of any resulting activity by sending messages to the S-CSCF. From the perspective of the S-CSCF, the "SIP application server", "OSA SCS", and "IM-SSF" exhibit the same interface behavior. When the name/address of more than one application server is transferred from the HSS, the S-CSCF contacts the application servers in the order supplied by the HSS. The response from the first application server is used as the input to the second application server. Note that these multiple application servers may be any combination of SIP application server, OSA service capability server, or IM-SSF types.

The S-CSCF does not provide authentication and security functionality for secure, direct third-party access to the IMS. The OSA framework provides a standardized means of third-party secure access to the IMS. If an S-CSCF receives a SIP request on the ISC interface that was originated by an application server destined for a user served by that S-CSCF, then the S-CSCF treats the request as a terminating request to that user and provides the terminating request functionality as described above. Both registered and unregistered terminating requests are supported. It is possible for an application server to generate SIP requests and dialogs on behalf of users. Request originating sessions on behalf of a user are forwarded to the S-CSCF serving the user if the AS has knowledge of the S-CSCF assigned to that user, and the S-CSCF performs regular originating procedures for these requests.

Originating requests on behalf of registered and unregistered users are supported. More specifically, the following requirements apply to the IMS Service Control interface:

1. The ISC interface is able to convey charging information as per [15] and [16].
2. The protocol on the ISC interface allows the S-CSCF to differentiate between SIP requests on Mw, Mm, and Mg interfaces and SIP requests on the ISC interface. Figure 2.22 describes Open Services Architecture (OSA) interfaces.

The application server offers value-added IM services and resides either in the user's home network or in a third party location. The main function is to process SIP messages from the IMS network and initiate SIP messages. As described previously, the IMS application server may be a Service Access-Service Capability Server (SA-SCS), an IP Multimedia-Service Switching Function (IM-SSF), or an Application Server (SIP AS). The IMS service capability layer is illustrated in Figure 2.23.

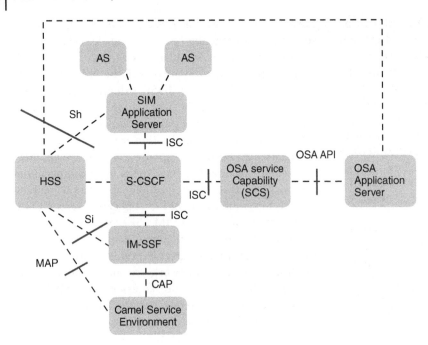

Figure 2.22 OSA services architecture.

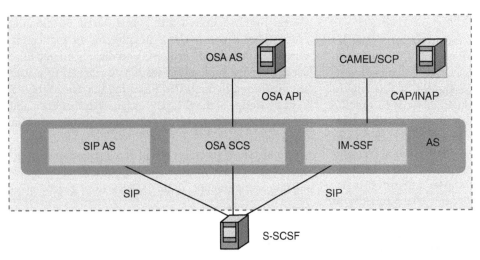

Figure 2.23 IMS service capability layer.

2.2.9 IMS Protocols and Reference Interfaces

Figure 2.24 provides an overview of the functional entities that compose the NGN IMS, and the reference points between them and with components outside the IMS.

The location and function of each interface are described in Table 2.10.

The IMS main protocols are summarized in Table 2.11.

The Diameter base protocol is intended to provide an Authentication, Authorization, and Accounting (AAA) framework for applications. It is based on the RADIUS protocol. The H.248 protocol (media gateway control protocol) for the MGC is used to control the media resource of the MG.

2.2.10 Call Session Control Function (CSCF)

The main network element in SIP signaling between the UE and the IMS is the CSCF; it controls and manages the registration of UEs on the IMS network and conducts service session management. Registration includes authentication. IMS-AKA is the primary authentication method, however other methods, such as HTTP digest and TLS-based HTTP digest, can also be used. The CSCF is defined as having three different roles that may collocate in the same node or separate nodes connected through the Mw interface, as indicated in Figure 2.24, and all are involved in UE-related SIP signaling transactions.

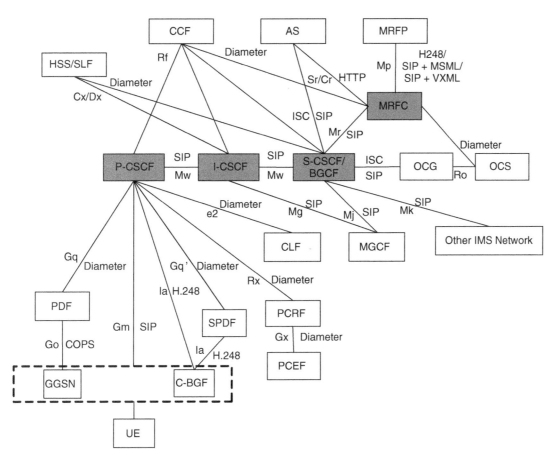

Figure 2.24 IMS protocols and reference interfaces.

The logical functions of the CSCF include:

- P-CSCF (Proxy-Call Session Control Function)
- I-CSCF (Interrogating-Call Session Control Function)
- S-CSCF (Serving-Call Session Control Function)
- MRFC (Multimedia Resource Function Controller)

The P-CSCF, I-CSCF, and S-CSCF are the core parts in the IMS network, and they are located in the session control layer, providing the call/session control function, while the MRFC provides various media control functions.

2.2.10.1 Proxy CSCF (P-CSCF)

The Proxy CSCF (P-CSCF) is the first contact point with which theUE interacts from the IMS side. The P-CSCF is responsible for all functions related to controlling the IP connectivity layer, i.e. the EPS. Therefore, the P-CSCF contains the Application Function (AF) that is a logical element for the PCC concept. The P-CSCF is typically located in the same network as the EPS, but Release 8 includes the concept of Local Breakout, which permits the P-CSCF to remain in the home network while a PCRF on a visited network may still be used. The functions of the P-CSCF are classified into: SIP proxy, access network management, security control, signaling

compression, emergency call identification and routing, SIP signaling filtering, flooding attack prevention, bearer control, and session timer maintenance.

The main functions of the P-CSCF can be summarized as follows:

Proxy function: The P-CSCF function as a SIP proxy means that the P-CSCF is the unified entry point to the IMS visited network in which the P-CSCF routes the SIP transactions from the visited access network to the S-CSCF or I-CSCF on the home network. In the scenario when the subscriber acts as the caller, the subscriber only needs to send the request to the P-CSCF. Then, the P-CSCF acts as the entry point to the IMS and initiates registration and the call for the subscriber. On the other side, and in the case of a called subscriber, the IMS network forwards the request message to the P-CSCF with which the caller is registered. Then, the P-CSCF sets up a connection between the caller and the called subscriber.

Access network management: The P-CSCF function of access network management means that the P-CSCF performs functions such as IP bearer resource authentication, NAT, and QoS management for the local access network of the IMS, and hence provides

Table 2.10 Location and function of IMS interfaces.

Interface	Location and Function	Protocol
Gm	Between IMS user terminals and the P-CSCF. It is used for the registration and session control of the IMS user.	SIP
Mw	Used for message communication and proxy forwarding between the CSCFs in the IMS registration and session flows.	SIP
Gq	Between the P-CSCF and the PDF. The P-CSCF informs the PDF of the current session and bearer-related information through the Gq interface. Therefore, the PDF can execute SBLP, interwork with PDFs in other backbone networks, and find out the QoS guarantee path to the peer access network or interworking node.	Diameter
Gq'	Between the P-CSCF and the SPDF. It performs similar functions to the Gq interface.	Diameter
Go	Used between the PDF and the GGSN to send QoS policies to the convergence node in the IP access network to execute policies.	COPS
Gx	Between the PCRF and PCEF. The PCRF sends instructions indicating the installation or removal of related charging policies. The PCEF reports related bearer events to the PCRF.	Diameter
Cx	Used to transfer the following information to and fro between the CSCF and the HSS: All the information required when the I-CSCF selects the S-CSCF. Information of the route from the CSCF to the HSS. Information related to roaming authorization that the CSCF obtains from the HSS. Security parameters that the CSCF downloads from the HSS which are required for the access authentication of IMS subscribers. Subscription data of the IMS session filter that the HSS sends to the CSCF.	Diameter
Dx	Between the CSCF and the SLF. When the IMS operator has multiple HSS subscription databases, the Dx interface is used for the CSCF to obtain from the SLF the address of the HSS that has the subscription data of the subscriber being processed. When the operator has only one HSS (server array), this interface is not required.	Diameter
Mg	Between the CSCF and the MGCF. Through the interface, the CSCF indirectly controls other non-IMS networks, such as the CS network, mobile 3G R4 network based on IP, and fixed NGN.	SIP
Mj	Between the S-CSCF and the BGCF. Through the interface, the S-CSCF forwards session control signaling to the BGCF. Then the BGCF selects the MGCF required for interworking with the PSTN and 3G/2G CS network.	SIP
Mk	Between the BGCFs. Through the interface, the BGCF in the same network with the calling S-CSCF forwards session control signaling to the BGCF in the same network with the interworking node MGCF.	SIP
Mm	Between the CSCF and other external non-IMS IP networks. Through the interface, the CSCF interworks with other IMS networks or non-SIP IMS networks.	SIP
Mr	Between the CSCF and the MRFC. Through the interface, the S-CSCF obtains related network resource services, such as announcement playing, digit collection, conference bridging, and video stream media.	SIP
Mx	Between the I-BCF and other entities such as the CSCF, AGCF, and BGCF. It implements interworking between the IMS and other packet networks.	SIP
ISC	Between the S-CSCF and the AS. According to the rules of triggering IMS subscription obtained from the HSS and the SIP service request sent from the IMS terminal, the S-CSCF decides whether to trigger the service and then directs the session to a specific AS for final processing of the value-added service.	SIP
Rf	Between the CCF and other entities such as the CSCF, MRFC, BGCF, and AS. It implements session-related offline charging.	Diameter
Ro	Between the OCS and other entities such as the AS, MRFC, and OCG to implement session-related online charging.	Diameter
Rx	Between the PCRF and the AF to exchange session information in the application layer. The information is required when the PCRF determines policies and charging control rules. The PCRF exchanges the determined policies and charging control rules with the PCEF.	Diameter

(Continued)

Table 2.10 (Continued)

Interface	Location and Function	Protocol
Sh	Between the HSS and the SIP AS or OSA-SCS. Through the interface, the AS queries the HSS to obtain the data relevant to value-added service logic and synchronizes relevant data to the HSS.	Diameter
Dh	Between the SLF and the AS or OSA-SCS. According to the given subscriber identity and home domain information, the AS confirms the address of the HSS where the user data is located.	Diameter
BOSS	Between the HSS/AS and the business hall server. It is used to transfer corresponding communication information.	SOAP
Mn	Between the MGCF and the IM-MGW. Through the interface, the MGCF controls the interworking of the media streams on the IM-MGW and the invoking of special resources.	H.248
Mp	Between the MRFC and the MRFP. Through the interface, the MRFC controls announcement playing, conferencing, and DTMF receiving and sending of the MRFP.	H.248 or SIP
Mb	Between the IMS access network and the IPv6 network. Through the interface, the data on the control plane and user plane in the IMS network can be transmitted over the IPv6 network.	IPv6
Ut	Between the UE and the AS. Terminal users can manage and customize service options on the AS through the interface.	HTTPs
e2	Between the P-CSCF and the CLF. The P-CSCF obtains location information of fixed users through the interface.	Diameter/SOAP
e4	Between the A-RACF and the CLF. The A-RACF obtains user configuration information through the interface.	Diameter
Rq	Between the SPDF and the A-RACF. The SPDF interacts with the A-RACF through the interface to control QoS resources in the access network.	Diameter
Ia	Used between the SPDF and the C-BGF to send QoS policies to the C-BGF so as to implement policies.	H.248
a1	Between the NACF and the access network. The access network obtains such information as the IP address of the UE through the interface.	DHCP
a2	Between the NACF and the CLF. The NACF registers the association between the allocated IP address and the user identity to the CLF.	Diameter

Table 2.11 IMS main protocols.

Protocol	Location and Function
SIP	Used to control calls between the CSCF and other entities such as the UE, AS, and MGCF.
COPS	Used for NAT control between the CSCF and NAT network devices. It is also used for the policies between the PDF and the GGSN to interact with each other.
Diameter	Used to exchange information between the CSCF and the HSS or PDF. It is also used to transfer charging information between the CCF/OCS and the CSCF, MRFC, BGCF, MGCF, or AS.
ENUM	Used to change an E.164 number to a SIP domain name between the CSCF and the ENUM server.
DNS	Used to change a domain name to an IP address between the CSCF/AS and the DNS server.
H.248	Used to control messages between the MGCF and the MGW, and between the MRFC and the MRFP.
SNMP	Used for the NMS in the IMS to interwork with managed elements.
IPv6	Defines the IP address scheme of the next generation.
IPsec	Used to protect network security between the UE and the CSCF and between security gateways.
SOAP	Used to exchange information between the P-CSCF and the CLF, the NACF and the CLF, and between the AS and service portal. The exchange is over the extensible mark-up language or the Hypertext Transfer Protocol.
NTP	Used to support the time synchronization between the OMS2600 and the NTP server such that all IMS equipment can synchronize with each other.

end-to-end service assurance for the IMS. Since the P-CSCF is the entry node to the IMS, it can categorize subscribers into different access types based on the access network configuration. The access types are registration and de-registration, implicit registration, and third party registration. The P-CSCF can then perform different NAT, QoS management, and security authentication policies for these different access types.

Security control function: The security control function of P-CSCF means that the P-CSCF sets up and maintains a security association between the UE and the P-CSCF, and performs the required security encryption policies for the access networks that do not provide the security protection mechanism on the link layer. The P-CSCF can compress and decompress the SIP messages from the UE that are included in the P-CSCF signaling compression function.

Emergency control: The emergency call identification and routing function of the P-CSCF means that the P-CSCF, as an entry node to the IMS, can identify emergency calls and ordinary calls and route different calls to different S-CSCFs/E-CSCFs. The P-CSCF can determine whether the current call is an emergency call based on the emergency call flag in the priority header field in the INVITE message sent by the UE. If the INVITE message does not contain the emergency call flag, the P-CSCF can determine whether the current call is an emergency call for mobile subscribers by querying the universal emergency call number table or the emergency call number table based on the Request-URI and P-Access-Network-Info header fields. For fixed subscribers, the P-CSCF can determine whether the current call is an emergency call by querying the universal emergency call number table based on the Request-URI header field. For emergency call routing when a subscriber is roaming on a visited domain, the P-CSCF routes the emergency call request to the S-CSCF of the visited domain instead of the S-CSCF in the home domain of the subscriber.

SIP filtering: SIP signal filtering means that the P-CSCF can filter SIP messages based on the source office direction of received SIP messages and the destination office direction of transmitted SIP messages. The operator can effectively prevent subscribers from using invalid SIP messages based on this feature. The P-CSCF has a role in flooding attack prevention, by means of the P-CSCF protecting SIP messages against flooding attack, which protects the IMS network against attacks resulting from external networks. In addition, the P-CSCF adds the IP addresses from which the IMS network is being attacked to a blacklist and then the P-CSCF rejects service requests directly from the IP addresses on the blacklist.

Bearer control: The bearer control function of the P-CSCF means that the P-CSCF interacts with the Policy Decision Function (PDF) to implement authorization decisions (i.e. enabling, disabling, and updating the bearer) on the media layer. The P-CSCF function of session timer maintenance means that, for a long-duration session, the P-CSCF starts a timer to perform periodical detection so that the scenario where the session is released but the resources are not released can be prevented.

2.2.10.2 Serving CSCF (S-CSCF)

The S-CSCF is located in the user's home network; it maintains the user's registration and session. At registration, it interfaces with the HSS to receive the subscription profile, including authentication information, and it authenticates the UE. For service sessions, the S-CSCF interacts with the UE through the other CSCFs, and may also interact with the application servers or the MRFCs to set up the service session properly. It also carries the main responsibility for controlling the interworking elements. The S-CSCF may also need to interact with the MGCF for interworking with CS networks, or with other multimedia networks for UE-requested services. The S-CSCF acts as the central node on the session control layer of the entire IMS core.

The S-CSCF performs the following functions:

Subscriber registration: The S-CSCF accepts the subscriber registration request forwarded by the P-CSCF. Then, the S-CSCF interacts with the HSS to authenticate IMS end users and downloads the basic subscription data of IMS end users from the HSS.

Session processing: The S-CSCF performs the basic session-routing function for originating and terminating IMS subscribers.

Service triggering: The S-CSCF downloads the subscription data of subscribers from the HSS and triggers different subscribed services to the related application servers based on the iFC policy in the subscription data.

2.2.10.3 Interrogating CSCF (I-CSCF)

The I-CSCF is located at the border of the home network; it is responsible for finding out the UE's registration status and either assigning a new S-CSCF or routing to the right existing S-CSCF. The request may come from the P-CSCF, from other multimedia networks, or from CS networks through the MGCF. In addition, the I-CSCF may need to interact with the application servers for service handling. The Ma interface is used for this when the Public Service Identity (PSI) is used to identify the service, and the I-CSCF can route the request directly to the proper AS. The I-CSCF acts as an entry point to the IMS network operator.

The I-CSCF performs the following functions:

Acts as the entry point for the operator's IMS network: The I-CSCF functions as the unified entry point for the operator's IMS home network. The I-CSCF generally functions as the entry point for the terminating IMS network, and therefore can determine whether the callee belongs to the local domain by querying the HSS. If the callee belongs to the domain, the I-CSCF routes the call to the S-CSCF. Otherwise, the I-CSCF rejects the call.

S-CSCF assignment: When a subscriber is registered, the I-CSCF assigns an S-CSCF for the registered subscriber based on the information obtained from the HSS and forwards the registration request to the S-CSCF.

Session routing: During a session, the I-CSCF obtains the address of the S-CSCF to which the the callee belongs from the HSS and routes the SIP call to the S-CSCF.

Topology hiding: The I-CSCF can hide the topology of the operator network to which the I-CSCF belongs by encrypting and decrypting the SIP address information.

2.2.10.4 CSCF Functions and Features

CSCF functions and features include, but are not limited to, subscriber registration, user authentication, roaming and roaming restriction, service triggering, media control, topology hiding, SIP compression, session control, session timer, number portability, PSI, flow control, number analysis, NAT, QoS, emergency call service,

forking, advanced user tracing, transit supported by the I-CSCF, SIP signal filtering, embedded DNS/ENUM, and network disaster tolerance.

2.2.11 IMS Subscriber Registration

The registration of CSCF subscribers is required in the following cases:

- Users use IMPUs (SIP URIs) for communication.
- The current IP addresses of users are mapped to IMPUs.
- Information about the locations and service capabilities of users can be obtained.
- Authentication and authorization during registration ensures network security.

The home network is that to which the user is subscribed and where the user's information is held, while a visited network is one accessed by a user but run by another operator.. Initial registration in home and visited networks is shown in Figure 2.25.

In an IMS network, the registration procedures are the same for user access to a home network or a visited network. A CSCF can provide the register function, including authentication and subscription download. There are three kinds of registration:

- Registration and de-registration
- Implicit registration
- Third-party registration.

The CSCF subscriber registration function is outlined in Figure 2.26.

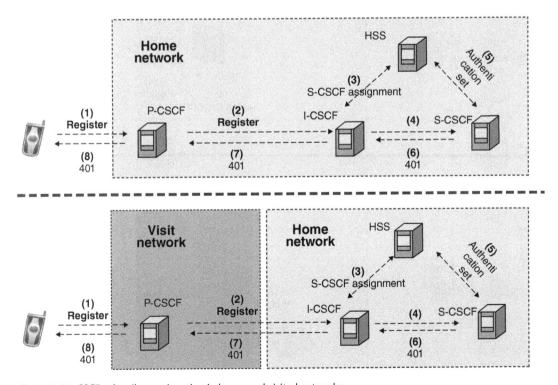

Figure 2.25 CSCF subscriber registration in home and visited networks.

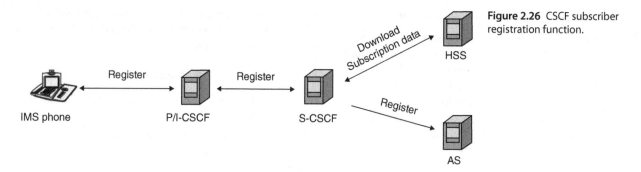

Figure 2.26 CSCF subscriber registration function.

In implicit registration [17], only one IMPI is associated with multiple IMPUs which are in the same IRS; the registration of all the IMPUs in the IRS can be implemented through the IMS registration of one IMPU. While in third-party registration, if a user has subscribed to IMS services, after the user has registered successfully, the S-CSCF checks the downloaded iFC and triggers the route to the AS which serves the user. Implicit registration and third-party registration are included in the registration procedure and are described below:

Implicit registration: If multiple identities are set as an implicit registration set during user definition, after the user registration is complete, calls can be made directly and none of the users in the implicit registration set needs to be registered again. If any of the users is de-registered, all the users are de-registered.

Third-party registration: After a user is re-registered successfully, the S-CSCF also sends a registration request to the AS of the user service. In this way, the AS knows that the user is registered and downloads the service data of the user from the HSS for use in user calls.

The purposes of the register function include:

- Verifying the validity of subscribers.
- Downloading subscription data of subscribers so that services can be triggered during a call.
- Initiating third-party registration and instructing the AS to download the subscriber data so that subscribed services can be used during a call.

The registration procedure is shown in Figure 2.27 for the initial registration message and a second registration

1 — Initial register message

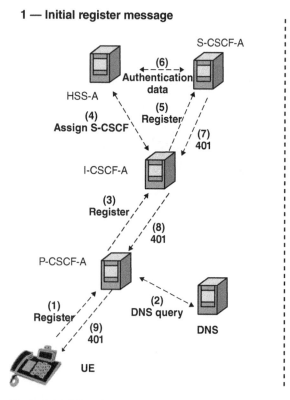

2 — 2nd register message

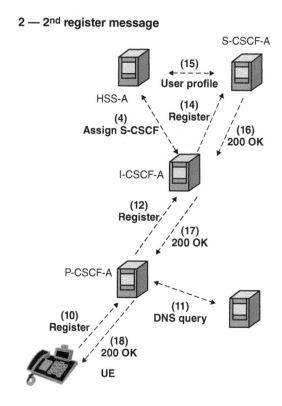

Figure 2.27 IMS registration procedure.

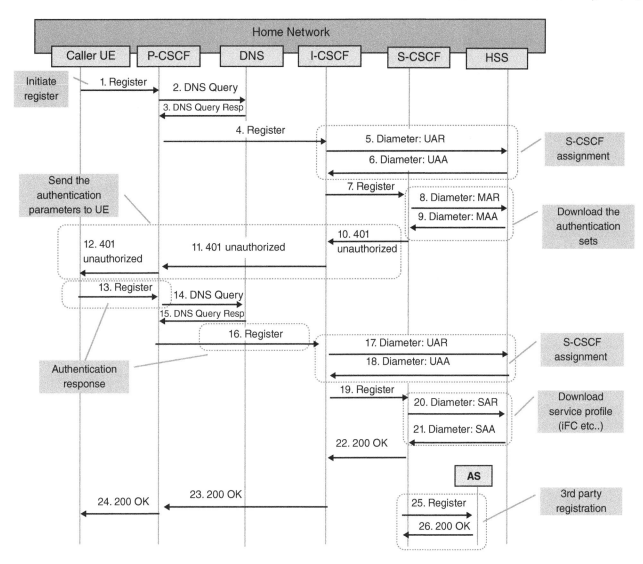

Figure 2.28 IMS registration signaling flow.

message. The registration is considered successful when the UE receives a 200 OK message from the P-CSCF.

The registration signaling flow is illustrated in Figure 2.28. In Step 1, the caller UE initiates a registration request to the P-CSCF. For Steps 2 to 4, based on the registration server domain name (configured on the caller UE) carried in the registration request, the P-CSCF routes the request to the I-CSCF. In Step 5, the I-CSCF queries the HSS by using UAR messages. In Step 6, the HSS returns a UAA message that carries the S-CSCF capability set required by the user. In Step 7, based on the subscribed requirements of the user, the I-CSCF selects the appropriate S-CSCF and sends the registration request to the S-CSCF. In Step 8, based on the user information carried in the registration request, the S-CSCF sends an MAR message to request the download of the user authentication vector from the HSS. In Step 9, the HSS returns an MAA message that

carries the user authentication vector to the S-CSCF (the HSS calculates the user authentication vector according to the password specified during user definition). In Step 10, after obtaining the authentication vector, the S-CSCF generates a random value and performs an encryption operation by using the random value and authentication vector to obtain a verification code and store the code. Then, the S-CSCF returns a 401 authentication challenge that carries the randomly generated value (the nonce parameter in the WWW-Authenticate header field in the 401 message).

In Steps 11 and 12, the P-CSCF returns the 401 message to the UE. In Step 13, the UE obtains a verification code by using the random value and its authentication vector in the 401 message and sends the second registration request carrying this verification code (the response parameter in the authorization header field in the second registration request) to the P-CSCF. Steps 14 to 19 are the same as Steps 2 to 7. In Step 20, the S-CSCF checks whether the verification code carried in the user registration request and verification code generated by the S-CSCF are the same. If yes, the authentication is successful and it indicates that the user is valid. The reason is that only a valid user knows his user definition information. After successful authentication, the S-CSCF sends an SAR message to the HSS to download the subscription data of the user for triggering ASs in calls. In Step 21, the HSS returns an SAA message carrying the user subscription data. In Steps 22 to 24, the S-CSCF saves the user subscription data and returns a 200 message to the UE, which means that registration authentication is successful. In Steps 25 to 26, after successful registration, the S-CSCF initiates third-party registration to the AS to which the user is subscribed, instructing the AS to download user data for use in calls.

The DNS queries in Steps 2, 3, 14, and 15 are the same. The UE may initiate a secondary registration by using different domain names, and the requests may be routed to different I-CSCFs. In most cases, however, the same domain names are used.

Diameter message explanations are given in Table 2.12.

The functions of the different nodes during registration are described in Table 2.13.

The information stored during the registration is detailed in Table 2.14.

After the P-CSCF registration has been completed successfully, the following information is stored:

S-CSCF IP address: After being registered successfully, the user initiates a call to the P-CSCF and the P-CSCF knows to which S-CSCF to route the call (instead of selecting any S-CSCF as it would during registration). The S-CSCF sends its IP address in the Service-Route header field in a registration 200 message to the P-CSCF.

UE information, including the IMPI and IMPU of the UE, access network information, and UE IP address: When the user serves as the called party, the P-CSCF knows how to route the call to the UE.

After the S-CSCF registration has been completed successfully, the following information is stored:

Table 2.12 Diameter message explanations.

Message	Explanation
UAR	User Authorization Request
UAA	User Authorization Answer
MAR	Multimedia Authentication Request
MAA	Multimedia Authentication Answer
SAR	Server Assignment Request
SAA	Server Assignment Answer

Table 2.13 Functions of different nodes during registration.

Node	Function
P-CSCF	Checking the IMPI, IMPU, and home domain.
	Querying the DNS to obtain the I-CSCF IP address based on the home domain and forwarding the initial registration request.
I-CSCF	Querying the HSS to select an S-CSCF and specifying the S-CSCF.
	Forwarding the registration request to the S-CSCF.
S-CSCF	Downloading authentication data from the HSS to authenticate the terminal.
	Downloading the service subscription data (service profile) of the user from the HSS.
	Performing third-party authentication based on the iFC.
HSS	Interacting with the I-CSCF (delivering the S-CSCF list and the functions supported by each S-CSCF) to determine the S-CSCF.
	Delivering authentication data and user service subscription data and recording the user registration status.

P-CSCF IP address: When the user is the called party, the S-CSCF knows the P-CSCF that serves the called party. The P-CSCF sends its IP address in the Path header field in a calling message to the S-CSCF.

UE information, including authentication information (authentication information does not need to be downloaded during user re-registration) and service-triggering data (service triggering during a call).

After the HSS registration has been completed successfully, the S-CSCF IP address is stored. During user re-registration, the HSS directly returns a UAA message carrying the S-CSCF IP address and, unlike in the initial registration, the I-CSCF does not need to select any S-CSCF.

2.2.12 IMS Subscriber De-registration

There are two types of de-registration: UE-initiated de-registration and network-initiated de-registration. When de-registration is performed:

A user is not de-registered unless the UE is powered off.

The HSS generally de-registers a user for an administration or security reason. For example, the user has modified the authentication information. In this case, the user is de-registered so that the user is re-registered to use new authentication information.

Table 2.14 The information stored during registration.

Node	Before Registration	During Registration	After Registration
UE	IMPI, IMPU, Credentials Home domain Proxy name/address	Same as before registration	IMPI, IMPU, Credentials Home domain Proxy name/address
P-CSCF (in home or visited network)	DNS address	I-CSCF address/name UE address,IMPI,IMPU	Final network entry point UE address, IMPI,IMPU S-CSCF address
I-CSCF (in home network)	HSS address	S-CSCF address/name (temporary)	No state information
S-CSCF (in home network)	HSS address	HSS address/name User profile P-CSCF address/name P-CSCF network ID UE IP address, IMPI,IMPU	May have session state information same as during registration Service profile P-CSCF address
HSS	User service profile	P-CSCF network ID	S-CSCF address/name

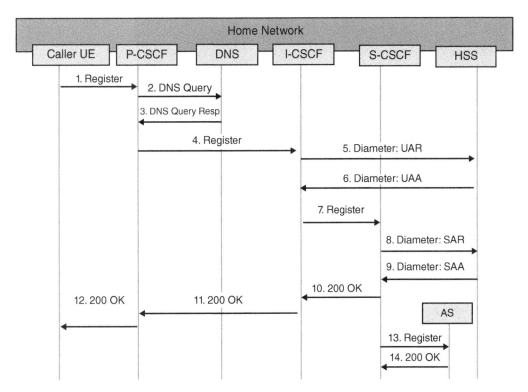

Figure 2.29 User-initiated de-register signaling flow.

The mechanism of de-registering an AS is the same as the mechanism of de-registering the S-CSCF. The user registration request carries a register timer, which is saved by both the AS and S-CSCF. When the register timer expires, the user is de-registered.

The user-initiated de-register signaling flow is shown in Figure 2.29.

The only difference between the registration procedure and the de-registration procedure is whether the value of the Expire field in the register message initiated by

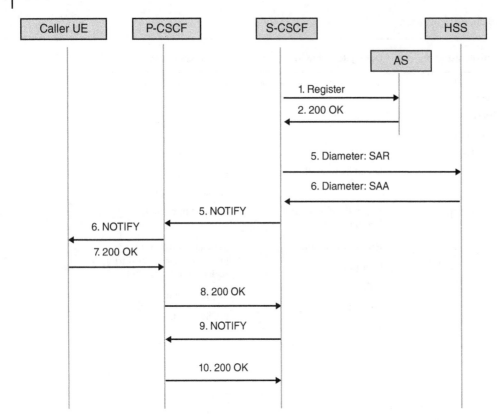

Figure 2.30 S-CSCF-initiated de-register signaling flow.

the UE is 0. If the value of this field is 0, it indicates de-registration; otherwise, it indicates registration or re-registration.

The S-CSCF-initiated de-register signaling flow is illustrated in Figure 2.30.

S-CSCF de-registration can be automatic or manual. When the user register timer expires, S-CSCF de-registration is performed automatically. Generally, manual S-CSCF de-registration is performed when the operator needs to expand or migrate users or an upgrade is needed.

2.2.13 IMS Discovery

P-CSCF discovery is the procedure through which the UE gets to know the address of the P-CSCF. In order to communicate with the IMS network, a UE must know at least one IP address of the P-CSCF. There are three ways to discover the P-CSCF: the GPRS procedure, the DHCP/DNS procedure, and setting the static configuration in the UE. In the GPRS procedure, the UE obtains the P-CSCF IP from the GGSN. In the DHCP/DNS procedure, the DHCP server can provide the P-CSCF IP/domain name while assigning dynamic IP addresses. Two options are considered for P-CSCF discovery:

- Pre-configured P-CSCF addresses in the client.
- P-CSCF discovery during default bearer attachment.

Figure 2.31 shows a generic call flow for IMS discovery during default bearer attachment.

2.2.14 CSCF Authentication

Authentication is a method used to identify a user and ensure the validity of the user. The home network authenticates users by using the user initial registration procedure. When a user terminal initiates initial registration, the S-CSCF authenticates the terminal according to the header field carried in the Register message and the authentication mode selected when the user is defined on the HSS. Currently, the HTTP digest authentication mode is adopted for fixed terminals; that is, the authentication is performed on the basis of user names and passwords. Authentication and authorization during registration ensure network security.

Without a SIM, UMTS Subscriber Identity Module (USIM), or ISIM, SIP terminals on the fixed network do not support the IMS AKA authentication mechanism or Early IMS authentication mechanism. The SIP digest and HTTP digest authentication mechanisms are adopted to connect terminals on the fixed network to the IMS network. SIP digest is the upgraded version of HTTP digest, and a user can be subscribed to either SIP digest or HTTP digest. HTTP digest authentication is

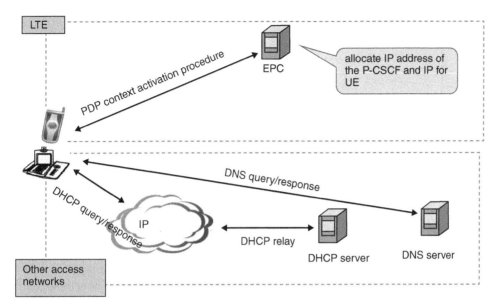

Figure 2.31 IMS discovery during default bearer attachment.

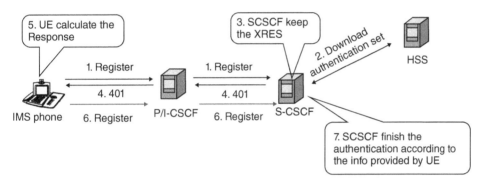

Figure 2.32 CSCF authentication function.

an HTTP-based authentication mechanism, with user names and passwords used as authentication parameters. The CSCF can provide the registration function, including authentication and subscription download; the authentication process is shown in Figure 2.32. The S-CSCF authenticates the user according to the authentication parameters downloaded from the HSS then the S-CSCF downloads the subscription from the HSS to finish the registration.

2.2.14.1 Service Profile

A service profile is the set of service- and user-related data. A basic IMS service profile consists of the charging address setting, block setting, registration authority setting, roaming authority setting, subscription media ID, iFC, and the S-CSCF capability set. The iFC is used by the S-CSCF to trigger a service request to the AS; it is stored in the HSS and downloaded by the S-CSCF during registration. Service profile functions are described in Figure 2.33.

A service profile is the service data to which the user is subscribed on the HSS. When performing user registration or unregistered services, the S-CSCF downloads service data from the HSS and performs the related service processing. A service profile consists of three parts: the public identity, core network service authorization, and the iFC. A service profile cannot be downloaded manually from the HSS.

The service profile parts are described as follows:

Public identity: A public identity consists of the user public identity associated with the service configuration. A public identity can be a SIP URI or a Tel URI. Each identity contains one call-barring flag. If the call-barring flag is set, the user can only be registered, instead of making calls.

Core network service authorization: This contains the media policy and user level of the user.

iFC: This specifies the AS that needs to be triggered when the user implements the services. An iFC consists of trigger points and ASs. A trigger point

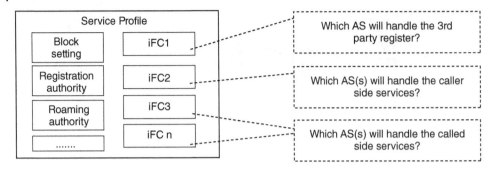

Figure 2.33 IMS service profile.

mainly describes the trigger conditions, for example, triggering during registration or triggering in the caller registration procedure. For an AS, the IP address of the server to be triggered needs to be specified.

2.2.15 CSCF Roaming and Roaming Restriction

The CSCF works in conjunction with the HSS to implement roaming and roaming restriction functions. Thus, the carrier can specify the networks on which the subscribers can roam. Figure 2.34 illustrates CSCF roaming and roaming restriction. The trust domain modes are configured in the I-CSCF – the I-CSCF checks if the registered user is from the trust domain. The HSS configuration mode is the roaming address lists, and roaming authorization profiles of the subscribers are stored in the HSS. A roaming subscriber can access services registered on the home network if the visited network falls in the trust domain list of the home P-CSCF and in the roaming address list of the HSS.

This section describes roaming implementation in the trusted domain mode, how to configure the roaming data, and how to check and compare the roaming permission. It also describes the roaming implementation in the HSS configuration mode.

Note that subscribers can roam only when roaming is allowed in both the trusted domain mode and HSS configuration mode. If roaming restriction is set in either mode, subscribers are not allowed to roam.

The procedure for judging and handling a trusted domain is:

- If the P-CSCF and I-CSCF are deployed on the same NE, the P-CSCF is a trusted domain. Otherwise, go to Step 2.
- The I-CSCF extracts the IP address of the P-CSCF from the Via header field in the registration message and checks whether the IP address falls into the IP address range of the local domain table. If yes, the P-CSCF is a trusted domain. Otherwise, go to Step 3.
- The I-CSCF checks whether the IP address falls into the IP address range of the trusted domain table. If yes, the P-CSCF is a trusted domain. Otherwise, the P-CSCF is not a trusted domain.

If the P-CSCF is a trusted domain, the I-CSCF continues to process the registration request. Otherwise, the I-CSCF returns a 403 message responding to the registration request.

2.2.16 S-CSCF Service Trigger

In the S-CSCF service trigger, the S-CSCF triggers the service to the AS according to the iFC downloaded from the HSS. Figure 2.35 illustrates the S-CSCF service trigger process.

The initial Filter Criteria (iFC) describe the application subscription information that can be provided to subscribers. The iFC are stored on the HSS as a subscriber

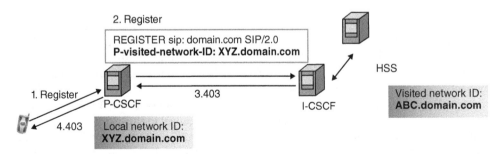

Figure 2.34 CSCF roaming and roaming restriction.

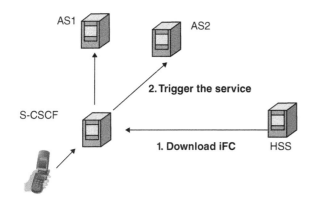

Figure 2.35 S-CSCF service trigger.

configuration and are downloaded to the S-CSCF during subscriber registration. The iFC contain the address of the AS on which the subscriber needs to subscribe to services. The S-CSCF triggers services based on the address of the AS in the iFC.

2.2.17 S-CSCF Media Control

The S-CSCF can undertake media control according to the user's subscriber data. A user may subscribe to the media capabilities in the HSS – [SUBSCRIBED-MEDIA-PROFILE-ID] is used in the service profile data. The

Figure 2.36 S-CSCF media control.

S-CSCF will permit or reject a session request according to the user's subscribed media downloaded from the HSS (the media capabilities of the subscribers are obtained by querying the SMPF, including the media type, data codec, image codec, and audio codec supported by the subscriber). When the subscribed media information does not meet the subscriber's service requirements, the S-CSCF can reject the request. The S-CSCF media control process is illustrated in Figure 2.36.

2.2.18 CSCF Emergency Call

Carriers can configure certain numbers such as the emergency numbers (for example, 911). Subscribers can set up these calls irrespective of whether they are registered or in default. Figure 2.37 describes the CSCF emergency call procedure.

In an emergency call, the E-CSCF (Emergency-CSCF) has a decisive role. It receives a request from a P-CSCF or an S-CSCF which contains emergency session establishment. For example, if additional location information is required or location information is missing, the E-CSCF will request such location information from the LRF (Location Retrieval Function). For a VoLTE emergency call to be successful, support is needed from LTE radio, EPC, PCRF (emergency bearer setup), and IMS. The basic procedure for emergency calls is described in Figure 2.38.

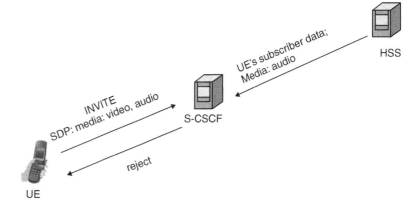

Figure 2.37 CSCF emergency call procedure.

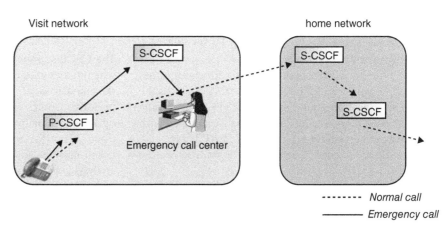

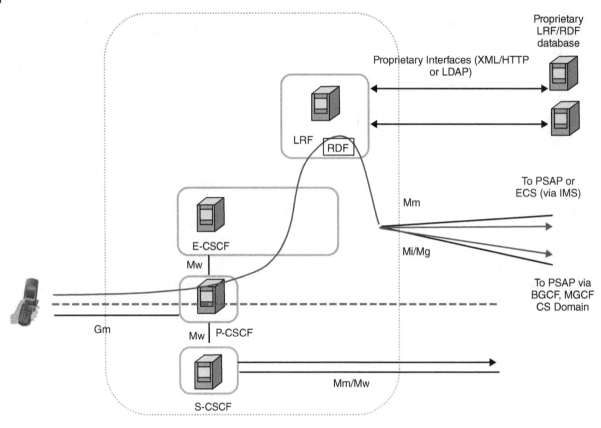

Figure 2.38 VoLTE emergency call handling.

The procedure may be described as follows:

The IMS end-user device sends a SIP INVITE request to a P-CSCF with an emergency Tel URI, for example, tel:911, or a SIP URI, for example, sip:911@ims.domain.com in the R-URI.

After the P-CSCF receives the SIP INVITE request, it performs digit analysis to check whether one of the configured emergency numbers, for example 911, matches the user part of the SIP URI or the Tel URI found in the R-URI. If one of the numbers matches, then, depending on the selected operator option, the P-CSCF forwards the request to the E-CSCF.

The E-CSCF can serve emergency sessions for registered and unregistered subscribers. If the operator has configured privacy to be disabled, then the P-CSCF also removes the Privacy header.

The E-CSCF first contacts the Location Retrieval Function (LRF) over the Ml interface. The LRF contains the Routing Determination Function (RDF) for a specific Public Safety Answering Point (PSAP) associated with the particular emergency call. If no PSAP is received, the E-CSCF inserts a default PSAP address in the R-URI.

The E-CSCF does analysis on the received R-URI to determine the next hop in the route. If the R-URI can be identified as an emergency call URI (as configured),

internationalization is not executed but the call is forwarded directly to the BGCF/PSTN. Otherwise (i.e. if the LRF inserted a PSAP number), the received number contained in the R-URI is treated as an ordinary number and, if necessary, internationalized. After an ENUM query, the emergency call is either forwarded to a PSAP in the PSTN via the BGCF/MGCF or located in the IMS.

Compared to other calls, emergency calls are handled with the highest priority. This ensures that none of the CSCF roles, including the BGCF, drops an emergency call in the case of overload.

2.2.19 Numbering Analysis

The CSCF supports the number analysis function, with which it analyzes the Tel URI and SIP URI, supplements the extensions, and performs number conversion. It also performs route selection according to the additional options such as the caller type and SDP. The basic functions of numbering analysis include call source analysis, called number analysis, and call rules analysis. Number analysis is a process used by the S-CSCF to search for the portal gateway of the network where the call is located and to send call requests to the gateway. Figure 2.39 shows a description of S-CSCF number analysis.

Figure 2.39 S-CSCF number analysis description.

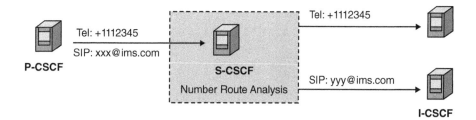

Table 2.15 The number types for fixed and mobile phones in the IMS system.

Number Type	Tel	SIP URI
Fixed phone	Tel:+ CC-AC-456781	SIP: +CC-AC-456781@ domain name
	CC - Country Code; AC - Area Code; 456781 -phone number; domain name – IMS domain name	
Mobile phone	Tel: +CC1234567890	SIP: +CC1234567890@domainname
	CC - Country Code; 1234567890 -phone number; domainname – IMS domain name	

The number format in the IMS system consists of a SIP number and a Tel number. The Tel number may be a global number or a local number. Table 2.15 illustrates the number types for fixed and mobile phones in the IMS system.

2.2.20 Route Selection Analysis

Route selection analysis is the process in which the route number is generated as a result of combining three processes. These processes are: Route Selection Source Code (RTSSC), Route Selection Code (RTSC), and Route Selection Attribute Code (RTSAC). The S-CSCF obtains the next hop address for routing the call request in one of the following ways:

The caller is an IMS subscriber and the Request-URI carried in the request message is a SIP number or a Tel number mapping with a SIP number on the ENUM server. In this case, the S-CSCF needs to query the DNS to obtain the IP address of the next hop NE according to the host name corresponding to the SIP number.

The caller is not an IMS subscriber and the Request-URI carried in the request message is a Tel number. In this case, the S-CSCF needs to implement number analysis on the Tel number to obtain the IP address of the next hop NE.

Figure 2.40 shows the IMS route selection process.

As described in Figure 2.41, the RTSAC is determined by the caller CPC priority and the media type in call requests.

The Route Selection Code (RTSC) is needed when the same user dials different numbers and the S-CSCF

needs to select different routes. A different caller prefix corresponds to a different RTSC. Figure 2.42 describes the Route Selection Code process.

The RTSSC is needed when different users dial the same number and the S-CSCF needs to select different routes. A different caller prefix corresponds to a different RTSSC in the fixed network. Figure 2.43 describes the Route Selection Source Code process.

A different caller prefix also corresponds to a different RTSSC in the mobile network. An example of an IMS user dialing an emergency call located in the IMS domain is shown in Figure 2.44.

2.2.21 Topology Hiding

The topology-hiding function, known as the Topology Hiding Inter-network Gateway (THIG), helps to improve the security of the IMS network [18]. It prevents the equipment information of the home network from being exposed to external networks. The THIG functionality is contained in the I-CSCF, which encrypts and decrypts the S-CSCF address and (or) P-CSCF address contained in the Via, Record-Route, Route, and Service-Route headers. The CSCF entities in other domains receive the encrypted route information without knowing how to decrypt it. The topology-hiding procedure in the CSCF is described in Figure 2.45.

The THIG provides the following functions:

Determines the message direction (incoming/outgoing).

Decrypts the routing header fields encrypted by the NEs in the local domain carried in incoming messages.

Encrypts the routing header fields of the NEs in the local domain carried in outgoing messages.

RTSSC: Route selection source code   RTSC: Route selection code

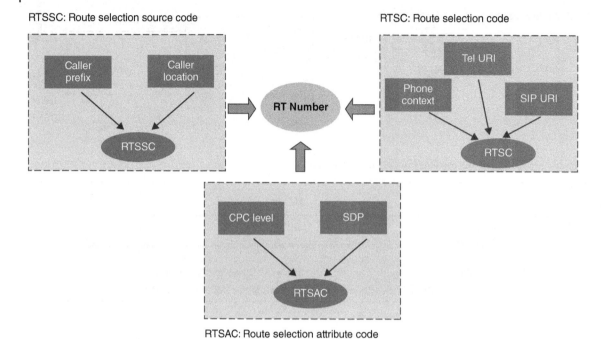

RTSAC: Route selection attribute code

Figure 2.40 IMS route selection process.

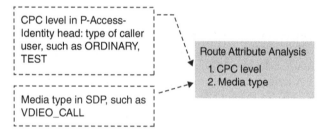

Figure 2.41 Route Selection Attribute Code (RTSAC).

2.2.22 Embedded DNS/ENUM

Before describing the functions of the DNS/ENUM in the CSCF, let us define the DNS/ENUM. The DNS is the Domain Name Server, via which we can query the IP address of a node by its host name or the opposite – query the host name by its IP address. A simple DNS query request is shown in Figure 2.46.

For instance, in response to the request to send a host name ("P-CSCF. Domain.com"), the IP address "x.x.x.x" is returned by the DNS, which is the IP of the P-CSCF. The DNS is a database with a distributed hierarchy located in the DNS servers all around the world. As shown in Figure 2.47, the whole DNS database is pictured as an inverted tree, with the "Root Node" at the top. A DNS server resolves a certain range of addresses, called a zone. Management of zones is achieved by delegation from the root server to the servers on the level immediately below, and finally to the lowest level.

The ENUM server (E.164 Number Server) performs the phone number mapping function. E.164 is the format of the common phone number, +12-34-567890, for instance. The ENUM server is used to query the corresponding SIP URI of a Tel number, and then the SIP URI will be sent to the DNS for an IP address query. The ENUM query procedure is illustrated in Figure 2.48.

An E.164 number is the only format adopted by callees in the CS/PSTN domain; an E.164 number can be IMPU for an IMS user. In the IMS, the IMPU can be in the form of a SIP URI or a Tel URI. A Tel URI uses the E.164 format. When an IMS user calls another IMS user or a CS user, the S-CSCF transforms a Tel URI into a SIP URI that can be routed by the SIP through the ENUM DNS mechanism. If the transformation is successful, the call is set up. Otherwise, it indicates that the called number is in the PSTN or the CS domain. In such a case, the S-CSCF routes the session to the CS domain through the BGCF. When a user in the PSTN or CS domain calls an IMS user, the MGCF transforms the Tel URI into a SIP URI that can be routed by the SIP through the ENUM DNS mechanism. The ENUM DNS architecture is explained in Figure 2.49.

The domain "e164.arpa" is recommended in RFC2916 [19] as the one to use for storage of ENUM numbers. In order to facilitate distributed operations, this domain is divided into subdomains.

An E.164 number contains a country code and thus distinguishes itself from phone numbers. The subdomains can also use country codes.

Figure 2.42 Route Selection Code (RTSC) process.

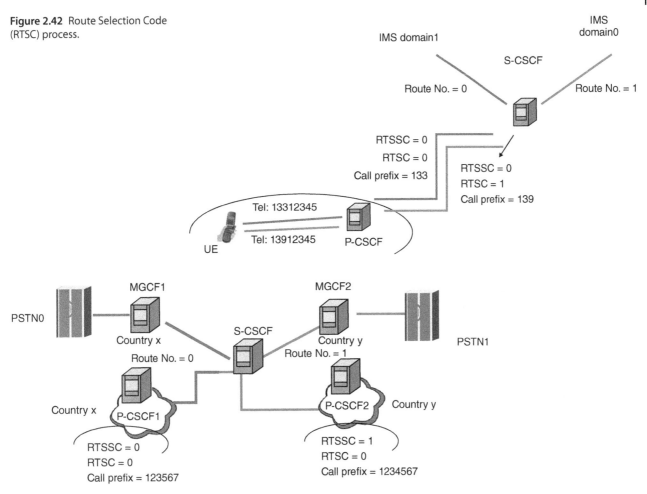

Figure 2.43 Route Selection Source Code (RTSSC) process.

Figure 2.44 Example of an IMS user dialing an emergency call.

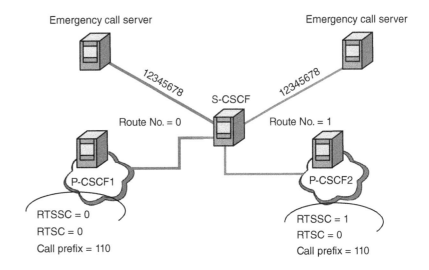

2.2.23 Functions of the DNS/ENUM in the CSCF

The CSCF is integrated with the function of the DNS/ENUM server. The data of the embedded DNS/ENUM can only be queried by the functional entities of the CSCF. When a functional entity queries the peer DNS/ENUM server, it queries the Database Management System (DBMS) first. If the required DNS/ENUM data are configured in the DBMS, the functional entity does not query the DNS/ENUM server. If the required DNS/ENUM data are not configured in the DBMS,

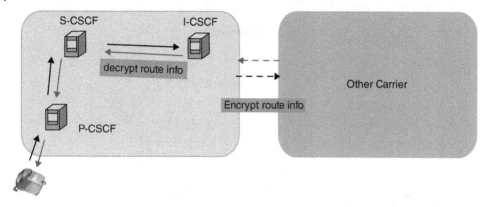

Figure 2.45 CSCF topology-hiding procedure.

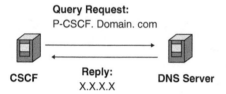

Figure 2.46 DNS query request.

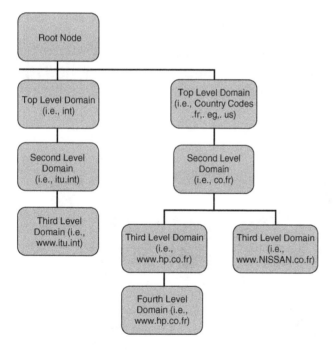

Figure 2.47 DNS server database.

the functional entity queries the DNS/ENUM server. Compared with the DNS/ENUM server, the embedded DNS/ENUM has the following features: reliability and high speed.

As illustrated in Figure 2.50, the S-CSCF interacts with the DNS and ENUM by using the Diameter protocol instead of the SIP. Before routing a call, the calling S-CSCF performs different processing according to the URI format of the called party: if the URI of the called party is a SIP URI, the S-CSCF sends a Diameter request to the DNS server by using the domain name of the called party to obtain the IP address (generally, the I-CSCF) of the IMS ingress NE corresponding to the domain name of the called party and routes the request to the called domain. If the URI of the called party is a Tel URI, the S-CSCF sends a Diameter request to the ENUM server by using the Tel number of the called party to determine whether the called party is an IMS domain user. If the called party is an IMS user, the ENUM server returns the SIP URI corresponding to the user and the S-CSCF queries the DNS server by using the domain name corresponding to the SIP URI to route the call. If the ENUM server does not return the corresponding SIP URI, this indicates that the user is not an IMS domain user. In this case, number analysis needs to be performed on the basis of the Tel URI to route the call to the ingress NEs in other domains.

Generally, a defined IMS user has one SIP number and one Tel number. A local domain IMS user is generally registered to the ENUM server. The DNS is the basis for the realization of the ENUM standard. The essential task of the ENUM server is the mapping of E.164 numbers (telephone numbers) to URIs used in VoIP systems. The requirement for ENUM was driven by the convergence of fixed-line, mobile, and IP networks for mapping telephone numbers with (SIP) Internet addresses. Every phone number can be converted or linked with one or more unique (SIP) Internet addresses or URI Web addresses. NAPTR records defined in the ENUM server provide this functionality, allowing a SIP-based VoIP telephone to forward a call to an E.164 number directly via the Internet.

2.2.24 CSCF Charging

The CSCF supports the offline charging mode. As described in [16], the offline charging functionality is based on the IMS network nodes reporting accounting

Figure 2.48 ENUM query procedure.

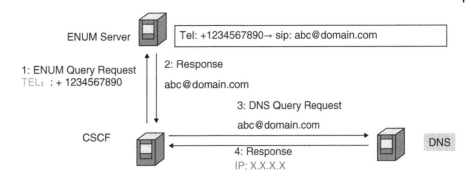

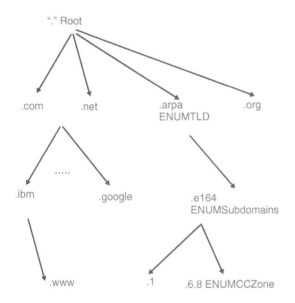

Figure 2.49 ENUM DNS architecture.

information upon receipt of various SIP methods or ISUP messages, as most of the accounting-relevant information is contained in these messages. This reporting is achieved by sending Diameter Accounting Requests (ACRs) [start, interim, stop and event] from the IMS network elements to the CDF. The Diameter client uses ACR start, interim, and stop in procedures related to successful SIP sessions. It uses ACR events for unsuccessful SIP sessions and for session-unrelated procedures. Figure 2.51 shows the IMS charging procedure. It is operator configurable in the nodes for which SIP methods or ISUP messages and accounting requests are sent. The operators may enable or disable the generation of an ACR message by the IMS node in response to a particular "triggering SIP method/ISUP message". However, the operator can enable or disable the ACR message based on whether or not the SIP (Re) INVITE message that is replied to by the "triggering SIP method/ISUP message" carried piggybacked user data.

If it detects that the AS has failed to be triggered by the INVITE message due to an exception such as a triggering timeout, the S-CSCF puts through the call. If the CCCH parameter of the SCHP command is set to AS Failure,

the S-CSCF needs to generate a charging CDR in this AS-triggering-failure scenario; the scenario is explained in Figure 2.52.

2.2.25 PSI Service

This section describes the PSI and focuses on the changes of processes and header fields. As described in RFC5002 [20], a Public Service Identifier (PSI) identifies an Application Server (AS). It can also identify the service or group in the AS. The CSCF will handle the call forwarding in both directions when the PSI is the caller and when the PSI is the called. Figure 2.53 describes the CSCF PSI service. The PSI service involves the AS initiating services such as conferencing on behalf of subscribers. The process of the PSI service is the same as that of an ordinary call. More details are given in [13].

2.2.26 SIP Forking

As detailed earlier, the SIP is the Session Initiation Protocol. SIP forking is the ability of a SIP proxy server to fork SIP request messages to multiple destinations according to 3GPP CSCFs, and ASs that behave according to this version of the specification shall not fork any request. Other networks outside the IM CN subsystem are able to perform SIP forking. Hence, 3GPP UEs should be ready to receive responses generated due to a forked request and behave according to the procedures specified in this section.

The UE may accept or reject early dialogs, for example, if the UE is only capable of supporting a limited number of simultaneous dialogs. Upon receipt of a first final 200 OK (for an INVITE), the UE acknowledges the 200 OK and cancels other early dialogs that may have been established. The UE may need to update the allocated resources according to the resources needed. If the UE receives a subsequent 200 OK, the UE acknowledges the dialog and immediately sends a BYE to drop the dialog. On the terminating side, a UE is able to receive, as specified in [15], several requests for the same dialog that were forked by a previous SIP entity.

The UE is able to include preferences in INVITEs, indicating that proxies should not fork the INVITE

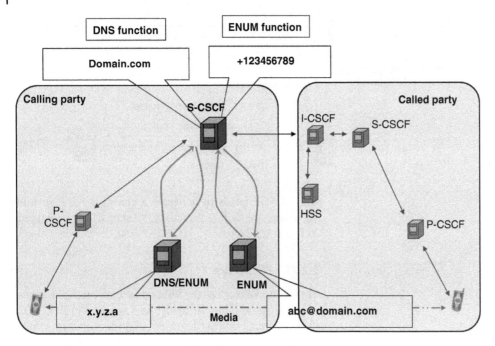

Figure 2.50 DNS and ENUM functions in the IMS.

Figure 2.51 CSCF charging.

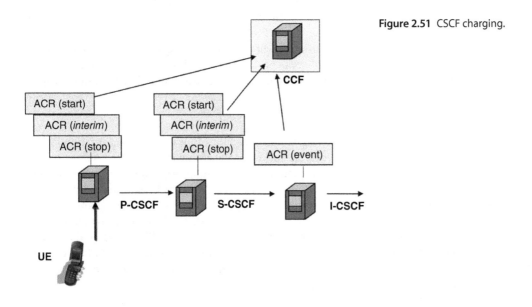

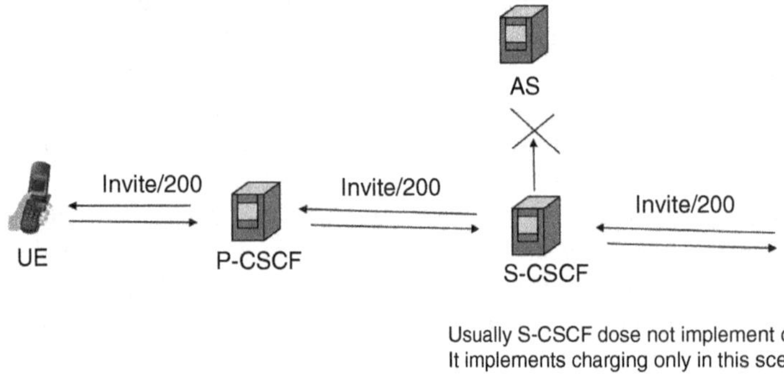

Figure 2.52 AS-triggering-failure scenario.

Usually S-CSCF dose not implement charging,
It implements charging only in this scenario

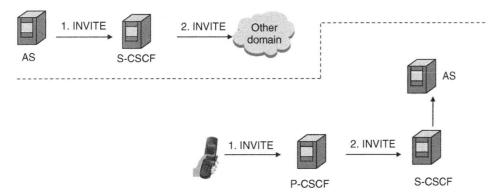

Figure 2.53 CSCF PSI service.

request. When one IMPU has more than one IMPI, and these IMPI are registered in one network at the same time, the S-CSCF can do the forking function to send the call INVITE to all the terminals. There are two search options for forking: parallel search and sequential search. Figure 2.54 shows the search options for forking.

The forking service targets the callee. Generally, a subscriber registers with only one UE. If the same subscriber registers with multiple UEs, how does the CSC connect the callee if any other subscriber initiates a call towards the subscriber? The forking concept is introduced to answer this question. The CSC connects the callee by using two methods: the first method is the parallel search, in which the CSC sends the call request to all the registered UEs at the same time. If a UE is connected successfully, other UEs are disconnected. The second method is the sequential search, in which the CSC sends the call request to the first UE. If the first UE does not respond to the request, the CSC sends the call request to the second UE, and so on. The CSC sends the call request until one UE is connected successfully. If no UE is connected successfully, the call fails.

As per [21], the S-CSCF determines whether to use a parallel search or a sequential search, as follows:

If the call request received by the S-CSCF does not contain the Request Disposition header field, the S-CSCF processes the call based on the default forking policy of the configured object.

If the call request received by the S-CSCF contains the Request Disposition header field, the S-CSCF processes the call based on the indication in the header field.

2.2.27 Session and Service Control

2.2.27.1 Session Control

The CSCF supports session setup, modification, and release. Registered subscribers initiate SIP sessions and receive SIP session invitations whether they are currently involved in other sessions or not. A CSCF is able to handle several SIP sessions for a single subscriber in parallel. A subscriber initiates a session setup by selecting an application on an end-user device. This request is sent to a P-CSCF in the IMS control network. Subscribers

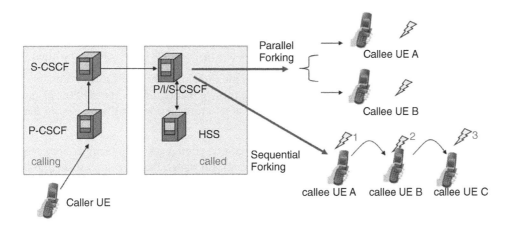

Figure 2.54 CSCF SIP forking.

are only connected to services and applications to which they have subscribed and for which they are authorized. The HSS, which stores this information, temporarily downloads it to the S-CSCF, which serves the subscriber.

Session control consists of session setup, session modification, and session release. A session setup is triggered by a SIP INVITE request originated in an end-user device of a registered subscriber, an AS acting on behalf of a subscriber, or an MGCF. If multiple contacts are registered by a terminating subscriber, the S-CSCF, if configured, could fork a SIP INVITE request as specified in RFC3841 [22], which takes caller preferences as well as fork and parallel directives into account. For session modification during an ongoing session, an IMS subscriber or an AS acting on behalf of a subscriber in either User Agent (UA) or Back-to-Back User Agent (B2BUA) mode, or an MGCF initiates a session modification by sending a SIP re-INVITE or UPDATE request.

In session release, if a user-initiated session release occurs, the session is released when the end-user device involved in the SIP session sends a SIP BYE request. Network-initiated session release is initiated for various reasons, for example, by an operator using Subscriber Administration (SA), by an outdated registration, by an empty pre-paid budget, or loss of bearer reported by the Policy Control Server (PCS). The CSCF can be configured to resume ongoing sessions for subscribers that become unregistered because of not updating their registration, or de-register themselves in the IMS. This is especially necessary for handover of a Voice Call Continuity (VCC) subscriber, if the subscriber first roams within the IMS and later within the CS domain.

2.2.27.2 Service Control
For distributing requests from and to application servers over the ISC interface, the S-CSCF acts as a SIP proxy. It hosts dedicated service control, which takes the following into account:

The subscriber's service information obtained from the user profile.
The filter criteria.

Generic operator service triggers are configured and stored locally at the S-CSCF. The S-CSCF downloads the user profile from the HSS during IMS registration. If an unregistered subscriber is involved, the user profile is downloaded during the processing of a SIP request. To simplify user profile provisioning, shared filter criteria can be configured. In this case, only a unique identifier is provisioned in a subscriber's user profile, and this is delivered through the Cx interface. The content of the shared filter criteria is configured at the S-CSCF and retrieved by means of the received identifier. Shared and normal filter criteria co-exist, and the S-CSCF uses both.

Filter criteria specify rules and priorities to invoke ASs for a particular subscriber, and are organized as follows:

Filter criteria for registered calling subscribers, which are executed at the originating S-CSCF only.
Filter criteria for registered called subscribers, which are executed at the terminating S-CSCF only.
Filter criteria for unregistered calling and called subscribers, which are executed at the default S-CSCF. By means of the Resource Management Function (RMF) of the S-CSCF, functionally equivalent ASs can be grouped into a resource pool; every time an AS resource is required by service control, the RMF is contacted to determine and assign available resources. By such an assignment, load-sharing aspects are considered. The Operator Service Profile (OSP) specifies user-independent service triggers, which are permanently stored at the S-CSCFs. Operator service triggers are valid for all requests received over the Mw and ISC interfaces.

The operator service triggers are handled in a similar way to filter criteria, and are organized as follows:

Operator service triggers executed before subscriber-specific filter criteria.
Operator service triggers executed after subscriber-specific filter criteria.
Forced operator service triggers for transit scenarios and specific services.

If the filter criteria are set appropriately, a call setup request can be passed to several ASs in series. This holds for the originating side of the call, where the ASs are selected based on the identity of the caller, as well as for the terminating side of the call, where the ASs are selected based on the identity of the called party. A specific situation arises on the terminating side when one of the ASs in the service chain forwards the call setup request to another destination, thus changing the identity of the (finally) called party. The S-CSCF stops the evaluation of any further filter criteria of the originally called party, but has two options by which to continue:

Simply forward the request to the new destination.
Start evaluating the originating filter criteria of the originally called party before forwarding the request to the new destination.

The S-CSCF can be configured to determine between the options dynamically depending on information found in the call setup request.

2.2.28 Session Timer

After a session has been established, the session timer enables the UE to initiate session update request messages (re-INVITE or UPDATE) periodically in order to keep the session in an active state. During the session establishment for an INVITE request, the CSCF negotiates the session update interval. When the session is established successfully, the negotiation about the session update interval is complete and the UE (refresher) responsible for initiating session update requests is determined. During the session process, the UE periodically initiates session update request messages (re-INVITE or UPDATE) at the negotiated intervals. The receiver returns the 2XX message to ensure that the session is in an active state. The CSCF session timer is described in Figure 2.55.

After a session has been established, the CSCF starts the session timer. If the CSCF fails to receive an UPDATE, re-INVITE, or 2XX message when the session timer expires, the CSCF will release the session-related resources. After a session has been established, if the BYE message of the UE is lost, the CSCF can send a BYE message to release the session-related resources after the session timer expires. Therefore, related resources can be released in time, preventing network resources from being suspended due to exceptions. During a session setup, the CSCF adds the negotiation of session update interval. Subsequently, the UE periodically originates the session update request, re-INVITE or UPDATE, according to the determined interval, thus ensuring the session is active. The session timer negotiation is illustrated in Figure 2.56.

During the session setup process, the callee adds the negotiated update period and update party in the

Session-Expires header field in the 2XX message based on the value of the Session-Expires header field in the initial INVITE message. The CSC uses the value of the Session-Expires header field in the 2XX message as the expiration period of the session timer. If the UE fails to send a re-INVITE or UPDATE message within the negotiated update period, the CSCF sends a BYE message after the negotiated update period expires to release related session resources.

2.2.29 Number Portability Service

Number portability (NP) means that a subscriber can keep the original number unchanged even if the subscriber changes to another carrier. For example, a mobile subscriber chooses B as a new carrier, but the subscriber keeps and uses the number that he/she applied to the original carrier, A. The number portability service procedure is explained in Figure 2.57.

Let's now describe the process of implementing the NP service based on the session process. As shown in Figure 2.57, after the subscriber registers with the network of carrier A, the subscriber wants to use the network of carrier B without changing the number, due to the low quality of the network of carrier A or for some other reason. In this case, the subscriber can subscribe to the NP service. After the subscriber subscribes to the NP service, the carrier adds a new route record on the global NP server (central NPDB) (generally, certain route parameters, such as rn = 123, are added to the

Figure 2.55 CSCF session timer.

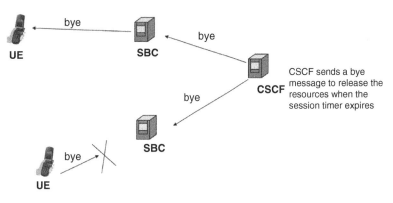

Figure 2.56 CSCF session timer negotiation.

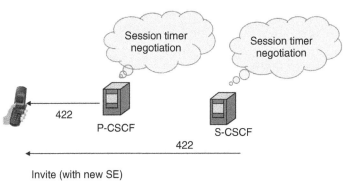

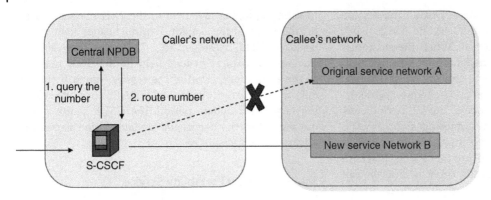

Figure 2.57 Number portability service.

end of the called number for use in the route to network B). After receiving a request to initiate a call to the subscriber, the S-CSCF queries the NP server based on the called number (the called numbers to be queried from the NP server can be configured on the S-CSCF. Generally, not all the called numbers must be queried from the NP server, depending on the planning of the carrier). If the S-CSCF obtains a route parameter from the NP server, the S-CSCF routes the call to network B based on the route parameter instead of the called number. If the call is routed based on the called number, the call is routed to network A of the original carrier.

2.2.30 Network Disaster Tolerance

When CSCF failure occurs, the IMS network automatically detects the failure and selects another entity that has the same functions in other areas to continue service processing for subscribers. This is called the *network disaster tolerance* of the CSCFs and it is explained in Figure 2.58.

2.2.31 S-CSCF Assignment

As illustrated in Figure 2.59, when a user is registered with a network for accessing the IMS service, the I-CSCF allocates an S-CSCF to the UE.

The I-CSCF communicates with the HSS by using the Cx interface to obtain the information required for selecting an S-CSCF. If the HSS returns the name of an S-CSCF, the I-CSCF checks the IP address of the S-CSCF by using the S-CSCF name returned by the HSS. If the HSS returns the capability set of an S-CSCF, the I-CSCF executes an algorithm according to the capability set of each received S-CSCF to select an appropriate S-CSCF. The I-CSCF selects the S-CSCF according to its capabilities – capability ID and meaning are defined by the operator. The S-CSCF selection process is shown in Figure 2.60.

2.2.32 Routing

This section describes the typical routing of a session setup request from one IMS subscriber to another

Figure 2.58 Network disaster tolerance.

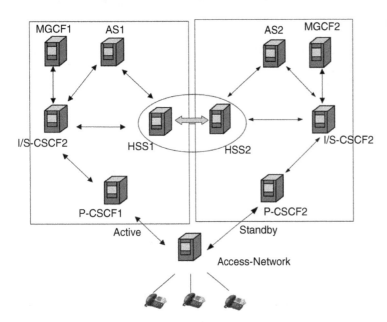

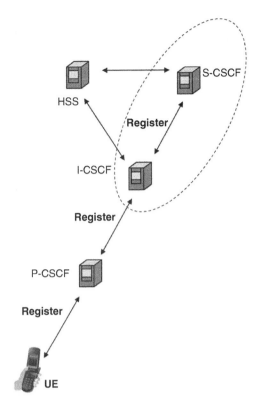

Figure 2.59 S-CSCF assignment.

an IMS originating breakout to a CS/PSTN domain. If the number resolution is successful, the S-CSCF routes the message to the terminating network; that is, to the terminating I-CSCF. The terminating I-CSCF queries the HSS for the address of the S-CSCF that serves the called subscriber. If the subscriber is registered, the HSS returns the address of the S-CSCF of the registered subscriber and the I-CSCF forwards the request to the S-CSCF. If the subscriber is not registered, no S-CSCF is currently assigned to the called subscriber, so the HSS does not provide an S-CSCF address. The I-CSCF then selects a default S-CSCF for executing services for unregistered subscribers. Here, services such as call forwarding are executed if configured.

When an originating S-CSCF receives a Request-URI with no telephone number but an alphanumeric user part, then an ENUM enquiry is not necessary and it is forwarded to an I-CSCF. During an ongoing session, both subscribers initiate a session modification by sending a SIP re-INVITE or UPDATE request. Finally, a session is released by either subscriber involved in the SIP session sending a SIP BYE request. Normally, a session is terminated when a subscriber de-registers during an ongoing session. For special scenarios, such as voice call continuity, the CSCF is configured to keep ongoing sessions for subscribers who become unregistered.

2.2.33 Telephone Number Mapping

To enable IMS subscribers to address other subscribers, the CSCF supports the use of both telephone number uniform resource identifiers (Tel URIs) and SIP URIs. To further route the ensuing SIP messages belonging to the respective connections in the IMS core network, Tel URIs must be translated to SIP URIs using the standard telephone number mapping (ENUM) method specified in RFC3761 [23]. The originating S-CSCF initiates an ENUM query if the R-URI of the received SIP request includes a Tel URI or a SIP URI, the domain name in the R-URI is configured in the S-CSCF-based ENUM

subscriber in the same IMS domain. It describes telephone number mapping and the resolution of domain names.

The SIP standard IETF RFC3261 [2] describes the session setup that includes the three-way handshake without the establishment of a secondary PDP context for the media flow. If a Request-URI, in SIP or Tel URI format and containing a telephone number, is received by the originating S-CSCF, it extracts this telephone number. If necessary, the S-CSCF then converts the telephone number into an international format before using it as the input for an ENUM enquiry. If the number resolution fails and no SIP URI is resolved, the S-CSCF initiates

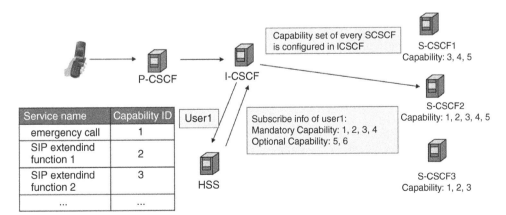

Figure 2.60 S-CSCF selection.

table, and the user part of the SIP URI begins with a "+" sign or contains only digits and visual separators, as specified in IETF RFC3966 [24]. This ensures that a SIP URI containing a dial string is also translated into a public user ID in the SIP URI format and is assigned to a subscriber in the home domain.

If the number resolution fails and no SIP URI is resolved, the S-CSCF passes the SIP INVITE request to the BGCF for an IMS originating breakout to a CS/PSTN domain. The BGCF then selects an MGCF according to an administrable number plan and routes the call to the selected MGCF. The S-CSCF supports ENUM queries to different, pre-configurable domains. For redundancy reasons, the S-CSCF also supports configurations for a primary and secondary ENUM server to be used if the first ENUM query fails. The ENUM query logic is separated from the S-CSCF core and can be executed anywhere in the service chain. This enables services to be executed after an ENUM query. Post-ENUM services are informed of the success or failure of the ENUM query, so that they consider this in their message processing.

The following can also be implemented in the CSCF:

Dial plan handling and internationalization
Service number detection
Carrier selection.

2.2.34 Domain Name Portability

The CSCF uses the Domain Name System (DNS) for the ordinary host name-to-address translation as well as to locate SIP services and perform telephone number mapping (ENUM). It therefore supports all the DNS resource record formats necessary for these activities: the naming authority pointer (NAPTR) and server (SRV) resource records according to the RFCs 3764 [25], 3761 [23], 3263 [26], and 2782 [27]; the Ipv4 address of a host (A) resource records according to the RFCs 1034 [28] and 1035 [29]; and the Ipv6 address of a host (AAAA) resource records according to RFC 3596 [30]. For SIP-based interfaces,

the CSCF provides a full implementation of RFC3263 [26]. It is thus possible to use the DNS-based fail-over and load-balancing mechanisms described therein.

The CSCF allows the configuration of distinct ENUM servers. ENUM translation requests are sent directly to these servers and are not forwarded across the ordinary DNS infrastructure that is used to resolve network nodes. Both Ipv4 and Ipv6 transport towards the DNS server and ENUM server are supported.

2.2.35 NAT Keep Alive

When a UE initiates an initial registration request message, if the P-CSCF determines that the subscriber accesses the network through the NAT and uses the fast registration mode to implement NAT keep alive, the P-CSCF will change the value of the Expire header field carried in the 200 OK message returned by the S-CSCF into a smaller value (smaller than the general NAT mapping aging duration, namely the value of the FREGP parameter of the PACN). Therefore, the UE regards the NAT mapping aging duration as the registration period and initiates re-registration requests at a frequency smaller than the registration period. The registration period saved on the S-CSCF is still the original registration period (the value of the Expire header field carried in the initial registration request message initiated by the UE).

After receiving a re-registration request message from the UE, the P-CSCF generally returns a 200 OK message directly, instead of forwarding the re-registration request message to the I-CSCF. Therefore, the system performance will not be affected by frequent re-registration request messages. The P-CSCF does not forward any re-registration request message to the I-CSCF until the registration remaining duration on the S-CSCF becomes smaller than the re-registration period of the UE. Therefore, the subscribers are prevented from being de-registered due to expiration. Figure 2.61 describes the CSCF NAT keep alive procedure.

Figure 2.61 CSCF NAT keep alive procedure.

P sends Hello

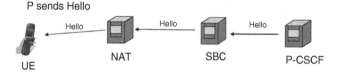

Fast Re-registration keep alive

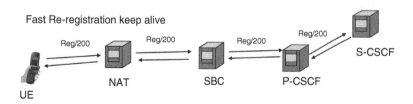

2.2.36 Fast Re-registration

The P-CSCF saves the registration duration of the core side. After receiving a 200 OK response message from the S-CSCF, the P-CSCF saves the value of the Expire header field carried in the 200 OK message as the registration duration of the core side. (For the P-CSCF, the S-CSCF is the core side.) In fast re-registration, after receiving an initial registration request message, if the P-CSCF discovers that the registration duration is too short (the value of the Expire header field carried in the registration request message is smaller than the registration duration of the core side saved on the P-CSCF), the P-CSCF will change the value of the Expire header field carried in the registration request message to another parameter that is handled by PACN before forwarding the registration request message to the S-CSCF. After receiving a re-registration request message from the UE, if the P-CSCF discovers that the subscriber registration address remains the same and that the value of the Expire header field carried in the registration request message is smaller than the saved registration duration of the core side, the P-CSCF directly returns a 200 OK message to the UE instead of forwarding the re-registration request message to the S-CSCF.

The CSCF supports the fast re-registration of subscribers. Therefore, attacks on the network resources brought by frequent subscriber registrations are reduced effectively. If a subscriber intends to keep saving his registration information on the P-CSCF when the registration duration on the P-CSCF expires, the subscriber needs to initiate a re-registration request message to the P-CSCF. If the registration duration applied by the subscriber is too short, re-registration request messages may be initiated frequently. Therefore, the network load increases. The fast re-registration function of the CSCF can reduce the inter-NE information exchange and the time spent on re-registrations. Therefore, the network load is reduced. The CSCF fast re-registration procedure is described in Figure 2.62.

2.2.37 Media Bypass

The media bypass procedure in the IMS is described in Figure 2.63.

The dotted lines in Figure 2.63 indicate signaling flows, while the gray lines indicate media flows. In a normal signaling media flow, the media flows between two UEs need to be forwarded by the SBC. For media bypass, the media flows between two UEs do not need to pass through the SBC and the media flows can be forwarded by the network connecting the UEs.

2.2.38 Media Resource Function Controller

The Media Resource Controller (MRC) performs the function of the Multimedia Resource Function Controller (MRFC) network element in the IMS network. The MRFC provides management of the media resources and works together with the MRFP to provide the mechanism of separation of the control and bearer of the media resources. The MRFC is the media resource control unit of the IMS network and converts the SIP resource control commands from the S-CSCF and AS into commands for controling the MRFP (MCU). The MRFC can manage multiple types of MRFPs and multiple MRFPs (MCUs). In addition, the MRFC can interact with the AS and MRFP (MCU) to perform functions such as announcement playback and digit collection for basic sessions, video/audio conferencing, high-definition conferencing, and TelePresence.

The position of the MRFC in the IMS network is illustrated in Figure 2.64.

The reference interfaces of the MRC are described in Figure 2.65.

Figure 2.62 CSCF fast re-registration.

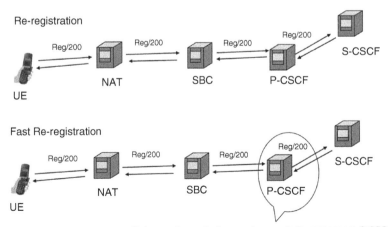

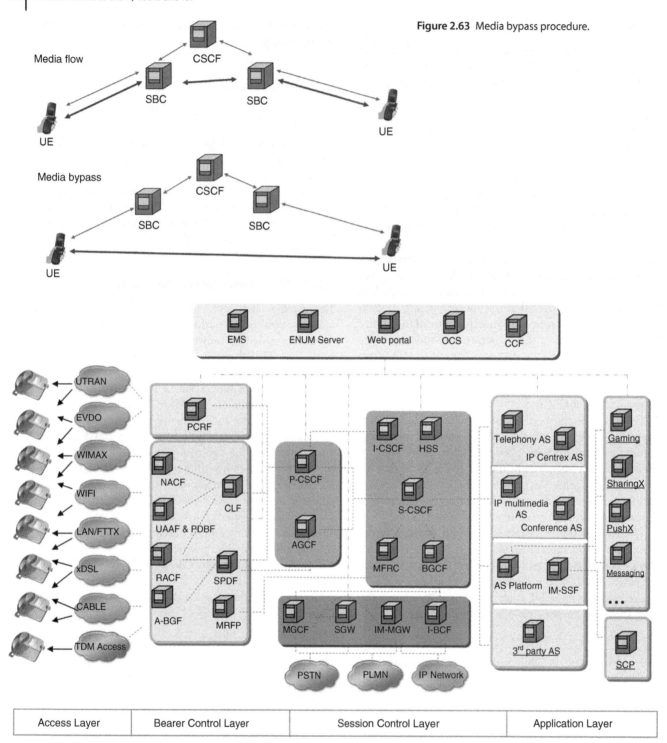

Figure 2.63 Media bypass procedure.

| Access Layer | Bearer Control Layer | Session Control Layer | Application Layer |

Figure 2.64 The position of the MRFC in the IMS network.

The Multipoint Control Unit (MCU) functions as the Media Resource Processing unit (MRP) on the IMS network. The MCU parses the control commands from the MRC and sets up video/audio RTP media streams with the conference terminal, performing functions such as audio mixing, standard-definition and high-definition video, continuous presence, transcoding, bit rate adaptation, and media playback. The MCU can enhance high-definition video conferencing, TelePresence conferencing, and hybrid video conferencing (including the voice terminal, standard-definition terminal, high-definition terminal, and TelePresence terminal).

The Sr interface is located between the MRFC and the AS. For a VXML service, the Sr interface uses the HTTP

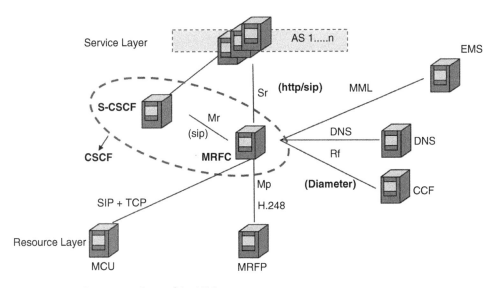

Figure 2.65 Reference interfaces of the MRC.

Table 2.16 Protocols of MRC interfaces.

Interface Name	Interconnected Entities	Interface Protocol
Sr	MRFC–AS	SIP
Mr	MRFC–S-CSCF	SIP
Mp	MRFC–MRFP	H.248
	MRFC–MCU	SIP+TCP
Rf	MRFC–CCF	Diameter
DNS	MRFC–DNS server	DNS
North-bound interface	MRFC–NMS	SNMP

protocol when the MRFC needs to obtain scripts through the HTTP address. In other scenarios, the Sr interface uses the SIP protocol.

For the interface between the MRFC and the MCU, as the MCU of the UE has been applied successfully in the H323 system, the MCU connects to the IMS system in the access mode that is consistent with the access mode used by the previous interface; that is, the interface uses the SIP protocol for signaling flow and uses the TCP protocol for media control. Protocols of MRC interfaces are described in Table 2.16.

2.3 IMS Identities and Subscription

This section describes private and public user IDs in IMS, the support of user IDs, implicit and explicit registration, and the dependencies for service subscription. The CSCF supports multiple private user IDs (IMPI) and multiple public user IDs (IMPU) for each IMS subscriber. IMS subscribers are addressed by means of symbolic names denoted user identities (user IDs). In IMS, two types of user ID are defined in [31, 32], and [5]:

IMS private user ID – The format of the Network Access Identifier (NAI) defined in [31] and RFC 4282 [33] is used for the IMPI. The IMPI belongs to the IMS operator and the IMS operator assigns the IMPI. It is associated with the IMS subscription, but is not used for the routing of SIP messages. Each IMS end-user device of an IMS subscriber has exactly one private user ID.

IMS public user ID – The IMPU has the following formats:

SIP URI: The SIP URI format is defined in [2] and [8]. Examples of SIP URI formats are sip: user-x@ operator.com and sip:+1234567@ims.domain.com.

Tel URI: The Tel URI is described in RFC 3966. An example is tel:+1234567.

The IMPU is publicly known and is used by other subscribers to set up sessions with the associated subscriber. It is used within IMS for the routing of SIP messages. IMS supports different IMS service profiles for each public user ID, but one service profile is shared by multiple public user IDs even from different implicit registration sets. Therefore, the HSS maintains special service information, such as filter criteria, for each public user ID. Figure 2.66 demonstrates the relationship between user IDs and services. Different IMS service profiles are assigned to each public user ID. The HSS assigns them using data provisioning.

2.3.1 Data Model of IMS

All subscriber-relevant data (permanent or temporary) that correspond to the individual IMS subscription are stored centrally in the HSS. These data can be retrieved and used by other network elements. To access these

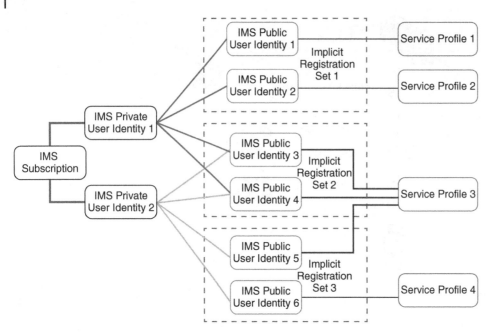

Figure 2.66 IMS public and private identities.

data, the HSS provides the Cx interface to I-/S-CSCF and the Sh interface to the application servers. The HSS stores the data in the central subscriber repository. The IMS network elements access the data using the HSS-FE. For the operator to administer the data, subscriber administration interfaces for single and bulk operations are provided. At the HSS, a dedicated provisioning gateway provisions the subscriber repository using Service Provisioning Markup Language (SPML) interfaces. The stored IMS user profile includes:

IMS subscriber authentication data (for example, the HTTP digest key).
IMS subscriber identity information (for example, IMS public user identity, IMS private user identity).
IMS subscriber administration data (for example, blocking the public user identity for the subscriber).
Service provisioning information (for example, filter criteria).
User mobility information (S-CSCF address).
Charging information (for example, CCF address, ECF address).
Display name of the subscriber.

The operations that can be performed on IMS subscriber data are "create", "modify", "delete", "display", "block", "unblock", and "de-register". Different types of identities (IMS private identities or IMS public identities) are attached to each IMS subscriber and stored in the subscriber repository. A single public user identity can be shared among multiple private user identities and a single IMS private user identity can be shared among multiple public user identities of the IMS subscription, as shown in Figure 2.66. An IMS public user identity

does not need to be part of an implicit registration set. The IMS subscriber profiles for different access types are basically the same. A subscriber can access the IMS using different access methods with the same subscription.

The HSS uses a common subscriber profile to support IMS registration for subscribers that do not have a user profile. The profile template is filled at the time of registration by communicating with the S-CSCF. These profiles are authenticated using early IMS authentication and, optionally, these profiles are authenticated using IMS AKA, digest AKA (v1or v2), or HTTP digest. The HSS also supports users with ordinary user-specific profiles stored in the HSS database.

2.3.2 Initial Filter Criteria (iFC)

The iFC are configured and stored in the HSS in the subscriber service profile with an initial registration set IRS. Shared iFC are stored outside of a subscriber profile but can be referred to. During IMS registration, the service profile including the iFC is provided to the S-CSCF. The iFC contain the following: **A priority:** The priority defines the order in which the iFC are executed. iFC with a higher priority are executed before iFC with a lower priority.

A trigger point: The trigger point defines a pattern that is compared with the SIP message the iFC are acting on. For instance, a SIP INVITE request related to a voice call must be forwarded to a different application server than a SIP MESSAGE request that is transporting an SMS. A trigger point to detect the request for a voice call could specify the pattern "(SIP method=INVITE) AND (SDP media = audio)". If all INVITE requests carry SDP for audio media in the specific IMS network, the trigger point could be even simpler.

An application server address: If the trigger point pattern matches with the SIP message, the SIP request is forwarded to the application server (AS) whose address is specified in the iFC. iFC are only applied to standalone SIP requests or initial requests for a dialog. A standalone request is a SIP request that is not related to a session (for example, a SIP MESSAGE request received outside of a session). An initial request for a dialog is a SIP request like a SIP INVITE or SIP SUBSCRIBE when they are used to set up a SIP dialog (re-INVITE or re-SUBSCRIBE requests that are sent within the dialog do not belong to this category). iFC are not applied to SIP messages that are exchanged within a session.

iFC are only applied to standalone SIP requests or initial requests for a dialog. A standalone request is a SIP request that is not related to a session (example: a SIP MESSAGE request received outside of a session). An initial request for a dialog is a SIP request like a SIP INVITE or SIP SUBSCRIBE when they are used to set up a SIP dialog (re-INVITE or re-SUBSCRIBE requests that are sent within the dialog do not belong to this category). iFC are not applied to SIP messages that are exchanged within a session.

Figure 2.67 demonstrates the iFC set, which includes an AS, FEE, and an integrated feature just as a reminder that the implementation of a service is irrelevant for iFC handling. The iFC containing pattern 1 have the highest

priority and are executed first. The pattern match and the dialog-forming INVITE request are forwarded over the ISC interface to the application server, AS1. AS1 processes the INVITE request according to its service logic and forwards the INVITE to the S-CSCF. The S-CSCF now evaluates the next iFC in its list, but pattern 2 does not match. The application of pattern 3 (the third iFC) matches again, and the INVITE is forwarded to AS3, which happens to be one of the integrated features.

iFC entail a decision on how to forward a SIP request. In the case of a dialog-forming request, each server traversed in the dialog setup phase individually decides whether it will stay in the path for mid-dialog requests and informs the other servers of its decision. Once the dialog has been established, the route is fixed and known to the involved servers. Neither iFC nor any other route-selection procedures are executed in the middle of a dialog. iFC are provisioned and stored in the HSS as part of a subscriber's user profile. Each subscriber has his/her own individual set of iFC, although usually iFC sets of large subscriber groups have the same content.

Many subscribers have iFC sets of the same content. By default, the S-CSCF separately downloads and stores each iFC set for each subscriber, even if all iFC are identical. The 3GPP has therefore designed an optimization called Shared iFC (SiFC). Shared iFC are iFC sets that are locally configured in the S-CSCF and have unique, shared iFC set identifiers. During registration, the S-CSCF does not download the full iFC sets of the user profiles, but only the shared iFC set identifiers related to these sets which refer to the iFC sets that have been locally configured in the S-CSCF. The use of SiFC instead of individual iFC saves a considerable amount

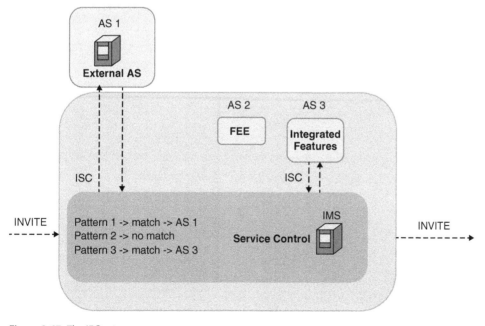

Figure 2.67 The iFC set.

of memory in the S-CSCF and reduces the network bandwidth required to download the user profiles from the HSS. The SiFC sets must be configured and maintained equally in all S-CSCFs in the network and in the HSS. When the S-CSCF requests the download of a user profile with the Cx SAR request, the S-CSCF sets a flag in that request if SiFC sets are supported. If this flag is not set, the HSS does not return the SiFC set identifiers but the complete iFC sets. Individual iFC and SiFC can be mixed in a subscriber profile. The sequence of execution is determined by iFC priority.

2.4 IMS Architecture and Interfaces

2.4.1 Session Border Control

SBC is a session-aware node that is responsible for controlling the signaling and media streams involved in communications over IP. It provides application-level connectivity and security protection across the borders and boundaries of different IP networks. SBC is typically deployed in VoIP and IMS networks at network borders such as those between an untrusted access network and a trusted core network, or between two service providers. SBC consists of Access SBC (A-SBC) and Interconnect SBC (I-SBC). Figure 2.68 demonstrates the interfaces of SBC.

The CSCF offers a decomposed border controller together with the media-handling component. With its border controller (BC) configuration, the CSCF provides an alternative solution to monolithic SBCs. The CSCF border control is based on the CSCF software functions, which constitute the access and interconnection session border controls. The signaling controller, A-BCF, of the

CSCF border control controls an Access/core Border Gateway Function (A-BGF) through an integrated Iq (H.248) interface at the edge of the access network. The signaling controller, I-BCF, of the CSCF border control controls an Interconnection Border Gateway Function (I-BGF) via an Ix (H.248) interface connecting to a peer network. The decomposed border control architecture offers clear advantages. An example of an RCS use case is explained in Figure 2.69.

The decomposed SBC architecture offers the following benefits compared to one-box SBC solutions:

The decomposed SBC architecture provides better scalability than a one-box approach. The CSCF-based SBC also takes full advantage of the yearly performance improvements obtained by using standard server blades.

Important features available with CSCF IMS development are automatically available with CSCF BC. In addition, all standard features associated with CSCF including reliability, security, quality, performance, overload protection, failover, and many more are also inherently available.

Border control included in the CSCF which also serves in core IMS roles (i.e. S-CSCF) reduces the number of products in a network and hence accounts for OPEX and CAPEX advantages for the operator.

Decomposed SBC will also be evolved to the cloud, and the following are the additional benefits:

Agile architecture with quick time to market (TTM) for new services.

CAPEX reduction thanks to separating HW from SW and using off-the-shelf HW.

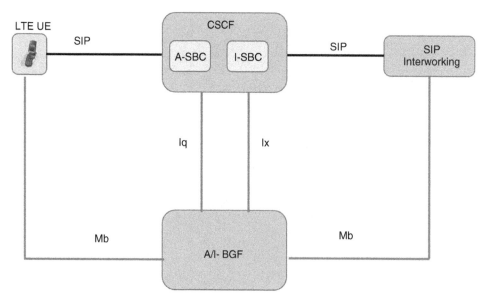

Figure 2.68 The SBC interfaces.

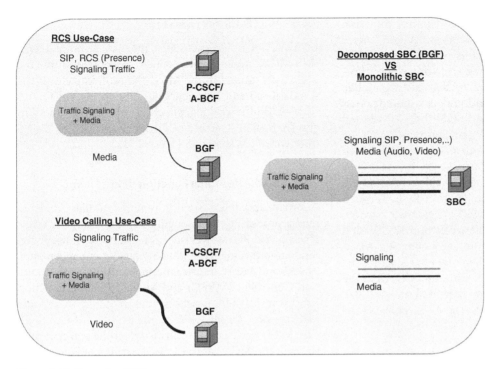

Figure 2.69 Example of RCS use case.

Separating a signaling plan from a media plan will provide dynamic resource allocation and therefore more efficient control of the available resources.

2.4.2 Access Session Border Control (A-SBC)

Access session border control provides a set of security functions for mobile, fixed, wireless, and cable access. It provides protection against complex or frequent attacks at the SIP layer. Such attacks originate from end-user devices trying to access network elements to which they are not permitted access, denial of service (DoS) attacks, message flooding, or malformed messages. The aim is to prevent single end-user devices from making IMS unavailable by denying them service and to continue serving other IMS users while under attack. The components of the A-SBC are the Access Border Control Function (A-BCF) and the Access Border Gateway Function (A-BGF). These components are distributed within IMS to offer the greatest flexibility. The signaling controller, A-BCF, is collocated with the P-CSCF and controls an A-BGF at the edge of the access network, which itself correspondingly controls traffic throughput. From a signaling viewpoint, the A-BCF operates in SIP proxy mode, although the SDP as well as several SIP headers and parameters are modified.

Among other features, the A-SBC supports Network Address Translation (NAT)/Network Address Port Translation (NAPT), far-end NAT, pin-holing, and network address hiding. Pin-holing denotes the dynamic opening and closing of gates at the A-BGF. Far-end NAT denotes network address translation when an end-user device is located behind a remote device. With NAPT, the UDP and TCP port numbers are translated additionally. The A-SBC provides lawful interception (LI) support for calls initiated over non-mobile access networks, for example over fixed and WLAN through PDGW access networks. In such cases, the A-SBC is used to deliver the required communication content.

2.4.3 Interconnection Session Border Control (I-SBC)

Interconnection session border control provides handling for border-crossing issues such as connections to other IMS, SIP networks over non-trusted interconnections, interconnections over public IP networks, and protection at the network border. The components of the I-SBC are the Interconnection Border Control Function (I-BCF) and the Interconnection Border Gateway Function (I-BGF), which consists of IP version interworking, SIP and SDP message manipulation, trunk-based routing, topology hiding, denial of service, rate limitation, and the access control list.

2.4.4 Access Transfer Control Function (ATCF)

In order to maintain the continuity of CS calls in an area where there is no proper LTE coverage, handover from LTE to 3G/2G is necessary, and this is called enhanced Single Radio Voice Call Continuity (eSRVCC); refer to Chapters 3 and 4 for more details on eSRVCC. For this purpose, the new Access Transfer Control Function (ATCF) and Access Transfer Gateway (ATGW) were introduced by 3GPP. The ATCF is part of the eSRVCC architecture to support roaming scenarios when introducing LTE access in addition to the existing mobile access networks, and voice continuity is required across these networks. The ATCF is part of the P-CSCF role and controls the ATGW. The CSCF provides for eSRVCC by implementing the ATCF and controlling the BGF which serves the ATGW. The key purpose of the ATGW is to anchor the media stream between the eSRVCC UE and the remote end point, similar to a BGF. Through this anchor, in the case of an SRVCC domain transfer, the media path is switched between the ATGW and the MSC server until the remote end point. The ATCF resides in the SIP signaling path in order to recognize the domain transfer and to interact with the SCC AS in the home network. The ATCF is implemented on the CSCF and integrated into the P-CSCF. The ATCF controls the ATGW through the IQ interface (H.248 based).

2.4.5 Media GW Functions

In a VoLTE–VoLTE call, the MGW can also be used to provide Media Resource Function Processor (MRFP) functionalities for voice traffic: for instance, transcoding, tone/announcement, conferencing, and lawful interception services. The MGW can be used to provide voice quality enhancements, comprehensive statistics, and seamless interaction for real-time quality metrics. The MGW in IMS-MGW and MRFP roles is connected to the Evolved Packet Core (EPC) via the Mb interface (the MGW VoLTE access port). In VoLTE calls, the AS/NVS controls the MGW (IMS-MGW/MRFP) through the H.248 (Mn) interface. The connectivity diagram is shown in Figure 2.70.

2.4.6 VoLTE–CS Interworking

In VoLTE–CS interworking, the MGW implements IMS-MGW functionality which is controlled by the MGCF, integrated in the MSS. The MSS introduces the standardized and interoperable solution for VoLTE–CS (IMS–CS) interworking, which is an essential functionality in the first stage of VoLTE deployments. VoLTE–CS interworking is shown in Figure 2.71.

2.4.7 Media Resource Function (MRF) for VoLTE

A broadband voice service requires similar kinds of media processing functionalities as those already available for 2G/3G networks – for example, transcoding, tone/announcements, and voice quality enhancements. The AS and MGW implement the Media Resource Function Controller (MRFC) and Media Resource Function Processor (MRFP) functionalities required to provide MRF services for HD voice traffic over LTE. In a VoLTE call, MRF services are available when the call is routed through the MGW and AS. As described in Figure 2.72, the MRFC controls the MGW's MRFP functionalities through the H.248 interface.

2.5 MMTel (Multimedia Telephony) Services

As described in [34], the IMS multimedia telephony service should allow multimedia conversational communications between two or more users. It provides real-time, bidirectional conversational transfer of speech, video, or, optionally, other types of data. IMS multimedia telephony is point-to-point communication between terminals or between a terminal and a network entity. This

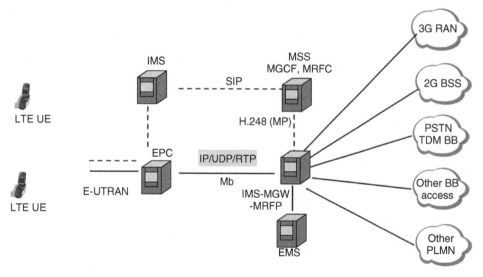

Figure 2.70 VoLTE in the MGW.

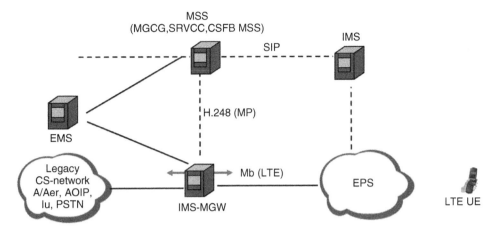

Figure 2.71 VoLTE–CS interworking function in the MGW.

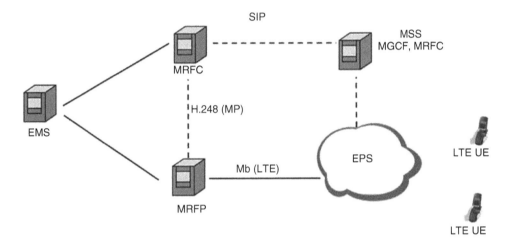

Figure 2.72 MRF connectivity for VoLTE.

communication is usually symmetrical, but in special cases the media components present in each direction may be different, or they may be the same but with different bit rates and different QoSs.(repeated below)

An IMS multimedia telephony communication can start with only one type of media, and additional types of media may or may not be added by users as the communication progresses. Therefore, a particular IMS multimedia telephony communication may consist of only one type of media, such as voice. The IMS multimedia telephony service is different from other IMS-based services, such as Push to Talk over Cellular. Its characteristics include the following:

IMS multimedia telephony is a service where speech, and speech combined with other media components, is the typical platform, but the service is not limited to including speech, it also caters for other media or combinations of media (for example, text and video).

The IMS multimedia telephony service includes supplementary services. The behavior of supplementary services is almost identical to supplementary services

for CS voice (TS 11), noting that most supplementary services are active in the setup phase. Mid-session supplementary services such as session transfer and session hold exist.

The anticipated usage model is that of traditional telephony: one user connecting to any other user, regardless of operator and access technology.

When a supplementary service is invoked, it applies to all media components of an IMS multimedia telephony communication. A supplementary service can be activated by the user for one or more types of media components if one or more of these media components is present.

IMS multimedia telephony can support many different types of media. The IMS multimedia telephony service includes the following standardized media capabilities:

Full-duplex speech.

Real-time video (simplex, full-duplex), synchronized with speech if present.

Text communication.

File transfer.

Video clip sharing, picture sharing, audio clip sharing.
Transferred files may be displayed/replayed on the receiving terminal for specified file formats.

Support for at least one common, standardized format (for example, JPEG, AMR) per media type.

Support for each of the media capabilities listed in points 1–5 above is optional for a UE.

The IMS multimedia telephony service should also at least support the addition and removal of individual media to/from an IMS multimedia telephony communication.

Within the IMS domain, one or more telephony application servers are designated for voice services. This application server can provide the controlling logic for relevant voice and video services including GSMA IR.92 and GSMA IR.94. Communication between the IMS core (i.e. the S-CSCF) and the voice application server is via the standardized IMS ISC interface. In some deployment scenarios, the IR.92 and IR.94 IMS feature set implementation role is an integrated feature in the application server. The following is a list of features and functions that are supported by the application server:

MMTEL (IR.92/IR.94)
SCC-AS/T-ADS
IP-SM-GW
IM-SSF
Ut/XCAP server.

2.5.1 Ut XCAP

XCAP is a 3GPP standards-compliant protocol. A user terminal functioning as an XCAP client sends XCAP requests to an XCAP server located on an application server.

These requests are instructions to the XCAP server to read or modify in some way supplementary service data related to the subscriber's device.

The XCAP client is a GSMA VoLTE (IR.92) or a Video over LTE (IR.94) compliant device. The XCAP server interacts with the selected repository (HLR/HSS), in accordance with the application server architecture, to gain access to the subscriber's supplementary service data. All interaction with the HLR is performed over the MAP interface and all interaction with the HSS is performed over the Sh interface. Interaction between the subscriber's device (XCAP client) and the XCAP server is performed over the Ut interface. Figure 2.73 illustrates the connectivity between the XCAP client and the XCAP server.

2.5.2 MMTel Modification from an XCAP Client

Subscribers registered on a CS network update their supplementary services using the services listed in the HLR supplementary services column, while subscribers on a PS (LTE and/or VoIP) network update their supplementary services using the services listed in the MMTEL supplementary services column (assuming they have MMTEL-capable terminals).

Regardless of the serving network on which a subscriber updates his/her supplementary services, subscribers with XCAP-capable terminals can be sure they always have access to the latest service information through header conditions and Etag comparison.

Users with an XCAP client can interrogate and modify their supplementary service data through a simple

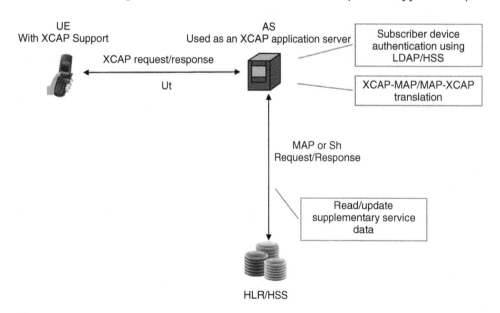

Figure 2.73 The XCAP connectivity.

user interface that issues HTTP requests, like HTTP GET, HTTP PUT, and HTTP DELETE to the XCAP server. The XCAP server then accesses and modifies the user's service data, stored in the selected repository, in accordance with the request sent from the XCAP client. This means that the XCAP server then interacts with the subscriber's supplementary service data, maintained in the Home Location Register (HLR), and either downloads or updates it depending on the nature of the request.

2.5.3 MMTel Data Synchronization Between XCAP Client and Server

The XCAP client stores a copy of the XML document that contains the user's own supplementary service data. This XML document is downloaded whenever the XCAP client issues the HTTP GET request. However, it is possible that another client might change the XML document at the source – that is, in the HLR or in the HSS – by using facility codes.

In such a case, the version cached by the XCAP client may not be up to date. It is imperative, therefore, for the service data to be synchronized between the XCAP client and the network. This synchronization is achieved by assigning an entity tag (Etag) header to the XML document. This Etag acts as a version identifier for the XML document. Each time the XML document is updated, the Etag value is updated.

The same subscriber could change their supplementary service settings using a different UE as well. In fact, the subscriber can even change their settings by submitting Man Machine Interface (MMI) instructions from their CS mobile phone, in accordance with the standard [35]. The XCAP client can use conditional headers like If-Match and If-None-Match when performing HTTP requests. For example, when using the HTTP GET request, If-None-Match is used to indicate to the XCAP server that the XML document should only be returned – downloaded – if the Etag of the client and the Etag of the XML document, cached with the subscriber's data on the XCAP server, are different. This helps to minimize network traffic. If-Match is used with HTTP PUT and HTTP DELETE requests. If the Etag value given in the request, issued by the client, matches the Etag value of the cached XML document, the XML document is modified.

2.5.4 Location Mapping

Location information is very important and plays a key role in CS routing as well as many other services including call routing, charging, LI, statistics, IN, and other location-based services. In VoLTE architecture, providing CS-comparable services – that is, service parity – is a must. E-UTRAN location information management over the external interfaces, however, is not yet fully standardized for VoLTE. Importantly, in early VoLTE deployments, it would put a heavy burden on the back-end infrastructure. The AS can provide location-based services in various roles: MMTEL, SCC-AS, and IM-SSF without the need to implement major changes to the backend infrastructure.

The AS is enabled to support two means to manage location information in IMS VoLTE. The PANI-based solution allows the AS to retrieve E-UTRAN location information from the P-Access-Network-Info header, which is then utilized for the purposes of location-based charging, location-based routing, and location-based interaction with the Intelligent Network SCP platform. The PVNI-based solution allows the AS to retrieve location identifiers from the P-Visited-Network-ID header, which is then utilized for the purposes of providing roaming-specific services, like subscriber-controlled and operator-determined barring, number formatting, and IN interworking, for example, for pre-paid charging management.

The AS is also enabled to retrieve and generate location information when an LTE user is camping on CS access. The AS supports location information retrieval over the SIP ISC interface, and allows the interworking of both solutions. This allows the integration of existing IN and AS services to VoLTE without the necessity of service adaptation to LTE location.

2.5.5 Location Parameters

Location parameters in the IMS network are described in Table 2.17.

The IMS network can connect to various access networks and thus, access-network-specific information is relayed to the AS over the SIP interface. As defined by RFC3455, over the SIP interface, the access network information and the location information are transferred in the P-Access-Network-Info (PANI) header. To include information that is specific to particular access technologies, for example, E-UTRAN, [21] extends the standard definition and the usage of the PANI header field. PANI headers are distinguished by being provided either by the network or the user (UE). From the perspective of the AS, this means that the source of the location information, embedded in the PANI header of the SIP request/response, is the network or the UE. Hence, the following two types of PANI header are differentiated:

Network-provided PANI header: This contains network-provided location information (NPLI).
UE-provided PANI header: This contains UE-provided location information (UPLI).

Table 2.17 Different location parameters in the IMS network.

Radio Network ID	Definition
LAI	The Location Area Identification (LAI) is defined with the Mobile Country Code (MCC), the Mobile Network Code (MNC), and the Location Area Code (LAC). The LAC identifies a 2G/3G location area within a PLMN. LAI = MCC + MNC + LAC
RAI	The Routing Area Identification (RAI) is defined with the LAI and the Routing Area Code (RAC). The RAC identifies a routing area within a 3G location area. RAI = LAI + RAC = MCC + MNC + LAC + RAC
SAI	The Service Area Identification (SAI) is defined with the LAI and the Service Area Code (SAC). The SAC identifies a service area of one or more cells that belong to the same 3G location area. SAI = LAI + SAC = MCC + MNC + LAC + SAC
CGI	The Cell Global Identification (CGI) is defined with the LAI and the Cell Identity (CI). The CI identifies the cell(s) of a BTS within a 2G/3G location or routing area. CGI = LAI + CI = MCC + MNC + LAC + CI
TAI	The Tracking Area Identity (TAI) is defined with the Mobile Country Code (MCC), the Mobile Network Code (MNC), and the Tracking Area Code (TAC). The TAC identifies a tracking area (LTE location area) within a PLMN. TAI = MCC + MNC + TAC
ECGI	The E-UTRAN Cell Global Identification (ECGI) is defined with the Mobile Country Code (MCC), the Mobile Network Code (MNC), and the E-UTRAN Cell Identity (ECI). The ECI identifies the cell(s) of an eNodeB within an LTE tracking area. ECGI = MCC + MNC + ECI Note: The TAC is not part of the ECGI

2.5.6 ECGI–CGI Mapping

During ECGI–CGI mapping, the AS executes a search in the ECGI–CGI mapping configuration. The input for the search is the parsed ECGI, which was retrieved from the PANI header after a successful PANI header selection; if the ECGI is found, the routine returns the corresponding 2G/3G internal cell ID. Based on the internal cell ID, the AS reads the associated 2G/3G record content. By accessing this 2G/3G record content, the AS has all the relevant information – for example, the CGI (including the LAC/CI or LAC/SAC), the location number, the routing zone, and the virtual MSC/VLR address – which can be used in the existing configuration over the existing interfaces by the existing service logic.

The TAC value of the ECGI is not used during the configuration search. It is, however, used as a parameter in various subsequent call control procedures. When the search is unsuccessful (the retrieved ECGI and the corresponding index is not found in the ECGI–CGI mapping configuration), the AS uses static location information.

The AS configuration allows the use of static location information by default. The mapping logic is disabled even if an LTE location is available in the PANI headers.

If the parameter is configured accordingly, mapping is allowed as per the above logic.

Generally, the AS uses static location information in two scenarios:

If no E-UTRAN location is retrieved during the PANI header selection.

If an E-UTRAN location is retrieved during the PANI header selection but the ECGI–CGI mapping was unsuccessful (the ECGI was not found in the ECGI–CGI mapping configuration).

The static location used depends on the AS configuration, the actual scenario, and the deployed subscriber database.

As the static location, the AS uses the Static CGI. In roaming/visited network scenarios, the AS uses the Roaming Static CGI as static location. The Roaming Static CGI is the concatenation of the MCC, the MNC, the LAC, and the SAC/CI.

2.5.6.1 Location Information in a Visited Network

The P-Visited-Network-ID (PVNI) embedded in the P-Visited-Network-ID header of the SIP REGISTER request identifies the access network where the user is camping. During a third party registration, the AS parses the PVNI header and stores it in the SPD. In order for this ID to be a meaningful input for service execution, the AS implements the visited network configuration and a logic, based on which the PVNI header is mapped to those existing parameters that can be used in the existing configuration over the existing interfaces by the existing service logic. 3GPP defines that the PVNI is inserted only in REGISTER requests. The AS also supports PVNI in INVITE/MESSAGE requests. If no PVNI is received in the INVITE/MESSAGE during call and message-related service execution, the AS uses the PVNI stored in the SPD, which was retrieved as part of the registration procedure.

2.6 Service Centralization and Continuity AS (SCC AS)

2.6.1 SCC AS Homing

Users with LTE terminals that support Voice over LTE (VoLTE), that is, VoLTE subscribers, can be registered under CS and PS networks at the same time. This simultaneous registration presents a challenge for mobility management, since the core network has to be able to determine the location of each Idle mode terminal in the network for, among other reasons, originating and terminating call delivery purposes. This is normally achieved by querying the HLR for the subscriber's location, but the HLR only contains the location information for one network registration – usually the circuit-switched network registration – since the subscriber's HLR record can contain only one VMSC/VLR address. Thus, when calls have to be routed to LTE terminals in Idle mode, the core network needs to know which access network to route them to. Likewise, when a call is performed from the CS side of a VoLTE subscriber, the network needs to associate it with the correct access networks.

2.6.2 T-ADS/Domain Selection

Terminating-Access Domain Selection, or T-ADS [36], is a functionality which determines where a call will be terminated for a VoLTE user. T-ADS ensures that the IMS can route the call to the UE even when it is in 2G/3G coverage.

In order to route MT calls appropriately, the IMS must know whether the UE is in LTE (VoIP-capable packet access network) or UTRAN/GERAN CS coverage. T-ADS is the function that provides this information.

T-ADS belongs to the functional block of the SCC-AS and is implemented within the AS product. The MMTel AS functionality of the application server is responsible for providing the ringing tone in T-ADS use cases; this includes interacting with a media gateway.

2.6.3 SRVCC

Single Radio Voice Call Continuity (SRVCC) provides the continuation of voice services when the terminal/UE is moving between a packet-switched (PS) and circuit-switched (CS) network in overlapping radio coverage areas. SRVCC requires that the voice service is anchored in an IP Multimedia Subsystem (IMS) application server enhanced with SRVCC functionality. SRVCC specifies that the terminal/UE will transmit/receive information from/to the core network using a single radio access technology at any given time.

The SRVCC procedure requires network-element-level support. Besides the MSS (SRVCC), the LTE eNodeB, the MME, the registers (HLR/HSS), the SCC AS, the ATCF and the ATGW, and also the user terminal are required to be compliant to support SRVCC.

When the subscriber moves from LTE to 2G/3G, the SRVCC-capable terminal measures the signal strength of the neighboring cells and sends measurement reports to E-UTRAN. When the inter-system access change criteria from LTE to 2G/3G are met, the eNodeB initiates a PS to CS handover request to the MME. Based on the Session Transfer Number for SRVCC (STN-SR) – that is, SRVCC-specific subscription information stored in the HSS and the HLR – the MME knows whether the SRVCC procedure can be applied to the terminal and hence to the subscriber. If the subscriber is entitled to SRVCC based on the STN-SR information, a radio access change is required based on the measurement reports, and also a real-time bearer exists, the MME triggers the SRVCC procedure for the voice component with the MSS (SRVCC) over the Sv interface.

The MSS (SRVCC) coordinates the SRVCC procedure with the CS handover procedure. It can reserve the target cell resources from 2G/3G access in the same MSS (intra-MSS handover) or in the target MSS (inter-MSS handover).

The MSS (SRVCC) also initiates the session transfer procedure to the Voice Call Continuity anchoring point. The anchoring point is responsible for the session continuity from PS to CS bearer. To facilitate session transfer of the voice component to the CS domain, depending on the deployment setup, the IMS multimedia telephony sessions need to be anchored in the IMS in either the SCC AS or the ATCF.

The MSS (SRVCC) then sends a PS to CS handover response to the MME, which includes the necessary CS handover command information for the UE to access UTRAN/GERAN.

Enhanced SRVCC (eSRVCC) or Release 10-based SRVCC has introduced two main components: the Access Transfer Control Function (ATCF) and the Access Transfer Gateway Function (ATGW). eSRVCC is intended to improve the handover between PS and CS when a subscriber is roaming. In eSRVCC, the ATCF and the ATGW are located on the visited network.

2.6.4 Policy Control

LTE is an all-IP flat network architecture. LTE provides the user with IP connectivity to a PDN for accessing the Internet, as well as for running services such as VoIP. The LTE network layout with PCS-relevant elements is shown in Figure 2.74.

The target is to establish an IMS-based voice/video call over LTE infrastructure with appropriate QoS: i.e., voice quality, delay, jitter, etc. Call signaling and payload messages are packet switched over LTE infrastructure using IP encapsulation.

A dedicated bearer is used to maintain QoS for a voice call. The call establishment procedure is similar to a standard IMS procedure with extra steps for dedicated bearer establishment. A dedicated bearer is associated with:

QoS Class Identifier (QCI)
Allocation and Retention Priority (ARP).

The ARP is used for call admission control while the QCI defines the QoS treatment.

Traffic Flow Templates (TFTs) are used to filter packets into different bearers at the end points of the EPS, i.e., at the UE or P-GW. TFTs use IP header information such as source and destination IP and TCP port. TFTs define the UE end points' IP address and port; these pieces of information are copied to the P-GW to set up filters for the VoLTE-dedicated bearer so that only packets related to the VoLTE call go to the

dedicated bearer. However, the CSCF also needs to specify Framed-IP-address AVP in the AAR so that the PCRF can match the subscriber session in the P-GW and apply the changes via a Re-authentication message to the correct subscriber session. For this reason, two AARs are needed: one for each UE end point. Each AAR contains TFTs that are applicable only to one UE end point.

The principle for mapping TFTs to LTE bearers is shown in Figure 2.75.

2.6.4.1 Bearer Establishment

As illustrated in Figure 2.76, the bearer establishment process in the IMS involves default bearer establishment and dedicated bearer establishment. The UE subscription data are stored in the HSS, while dedicated bearer parameters are pre-defined in the PCRF.

Policy and Charging Control (PCC) has a key role in the way users' services are handled in the Release LTE/SAE system. PCC determines how bearer resources are allocated for a given service, including how the service flows are partitioned to bearers, what QoS characteristics those bearers will have, and finally, what kind of accounting and charging will be applied. A PCRF allows the operator to set these parameters dynamically for each service, and even each user, separately.

The PCRF makes the decision on which policy to use for the service in question. If subscriber-specific policies are used, the PCRF may request subscription-related policies from the Subscription Profile Repository (SPR). Based on the decision, the PCRF creates the PCC rules that determine the handling in the EPS.

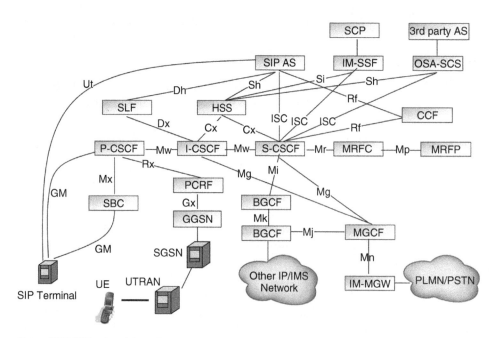

Figure 2.74 LTE network layout.

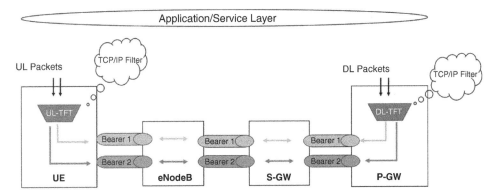

Figure 2.75 LTE TFTs.

Figure 2.76 Bearer establishment.

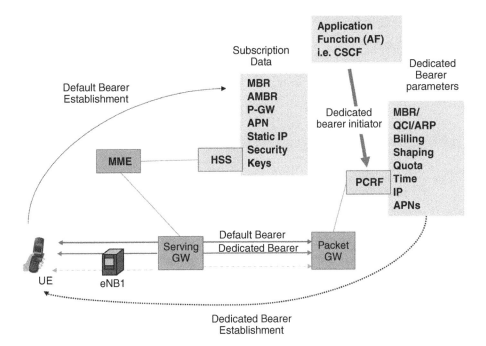

2.6.4.2 QoS

The EPC bearer model is shown in Figure 2.77. The bearer model itself is similar to the GPRS bearer model, but it has fewer layers. The EPS supports the always-on concept.

Each UE registered on the system has at least one bearer called the default bearer. The default bearer has basic QoS capabilities. Additional bearers may be set up on demand for services that need more stringent QoS; these are called dedicated bearers. The network may also map several IP flows that have matching QoS characteristics to the same EPS bearer.

The bearer setup logic works so that the UE first signals on the application layer, on top of the default bearer, to an application server in the operator service cloud, i.e. with IMS, to set up the end-to-end service. This signaling may include QoS parameters or simply indication to a known service.

The QoS parameters have been optimized for the EPC. Only a limited set of signaled QoS parameters are included in the specifications; they are:

QoS Class Identifier (QCI): This is an index that identifies a set of locally configured values for three QoS attributes – priority, delay, and loss rate. The QCI is signaled instead of the values of these parameters. Ten pre-configured classes have been specified in two categories of bearers: Guaranteed Bit Rate (GBR) and Non-Guaranteed Bit-Rate (Non-GBR) bearers. In addition, operators can create their own classes that apply within their network. The standard QCI classes and the values for the parameters within each class are shown in Table 2.18.

Allocation and Retention Priority (ARP): This indicates the priority of the bearer compared to other bearers. It provides the basis for admission control in

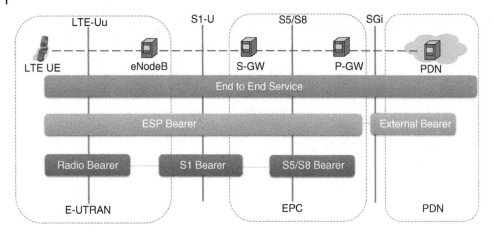

Figure 2.77 LTE bearer model.

Table 2.18 LTE QCIs.

QCI	Resource Type	Priority	Packet Delay (ms)	Packet Loss Rate	Example Service
1	GBR	2	100	10^{-2}	Conversational voice
2	GBR	4	150	10^{-3}	Conversational video
3	GBR	5	300	10^{-6}	Buffered streaming
4	GBR	3	50	10^{-3}	Real-time gaming
5	Non-GBR	1	100	10^{-6}	IMS signaling
6	Non-GBR	7	100	10^{-3}	Interactive gaming
7	Non-GBR	6	300	10^{-6}	Video buffered streaming
8	Non-GBR	8	300	10^{-6}	WWW, FTP, p2p
9	Non-GBR	9	300	10^{-6}	Progressive video

bearer setup, and further in a congestion situation if bearers need to be dropped.

Maximum Bit Rate (MBR): Identifies the maximum bit rate for the bearer. Note that a Release 8 network is not required to support differentiation between the MBR and GBR, and the MBR value is always set equal to the GBR.

Guaranteed Bit Rate (GBR): Identifies the bit rate that will be guaranteed to the bearer.

Aggregate Maximum Bit Rate (AMBR): Many IP flows may be mapped to the same bearer, and this parameter indicates the total maximum bit rate a UE may have for all bearers in the same PDN connection.

Table 2.18 represents the QoS parameters that are part of the QCI class, and the nine standardized classes. The QoS parameters are:

Resource Type: Indicates which classes will have GBR associated with them.

Priority: Used to define the priority for the packet scheduling of the radio interface.

Delay Budget: Helps the packet scheduler to maintain a scheduling rate sufficient to meet the delay requirements for the bearer.

Loss Rate: Helps to use appropriate RLC settings, for example, the number of retransmissions.

2.6.4.3 Rx Diameter Call Flow

The P-CSCF opens an Rx Diameter session with the PCRF using an AA-Request command. The P-CSCF provides the UE's IP address using either Framed-IP-Address AVP or Framed-IPv6-Prefix AVP, and the corresponding service information within Media-Component-Description AVP(s). The P-CSCF indicates to the PCRF as part of the Media-Component-Description whether the media IP flow(s) should be enabled or disabled with the Flow-Status AVP.

When the PCRF receives an initial AA-Request from the P-CSCF, the PCRF performs session binding, as described in [37]. To allow the PCRF to identify the IP-CAN session for which this request applies, the AF provides either the Framed-IP-Address or the Framed-IPv6-Prefix containing the full IP address applicable to an IP flow or IP flows towards the UE. In the case of a private IP address being used, the AF may also provide PDN information if available in the Called-Station-Id AVP for session binding; the

PCRF replies with an AA-Answer to the P-CSCF. The acknowledgment towards the P-CSCF should take place before or in parallel with any required PCC Rule provisioning towards the PCEF and should include the Access-Network-Charging-Identifier(s) and may include the Access-Network-Charging-Address AVP, if they are available. The AA-Answer message also includes the IP-CAN-Type AVP, if such information is available. In the case where the IP-CAN-Type AVP is included with a value of 3GPP (0), the AA-Answer message includes the 3GPP-RAT-Type AVP. Figure 2.78 illustrates the Rx Diameter call flow.

There is no overlapping between the VoLTE subscriber profile in the HSS and the QoS subscriber profile in the SPR, thus no specific integration is required. The HSS profile provides QoS parameters for basic attach (default bearer QCI, bandwidth); only these are needed for basic LTE access. The SPR profile provides advanced QoS parameters and policing features, for example, for voice bearer, time/location-based services and others.

2.7 Operator X IMS–VoLTE Architecture

2.7.1 Architecture Design Concepts and Principles

Some generic principles should be taken into consideration while designing a VoLTE/IMS network for an operator; these principles include, but are not limited to, the following:

- The network has to be capable of interconnecting with other networks based on 3GPP standard interfaces.
- The network should be capable of being integrated with other network nodes or other platforms based on 3GPP standard interfaces.
- The deployment of the network can be done in phases to support variable service launch needs.
- The network should be scalable and capable of accommodating both new markets and capacity demands.
- Subscriber data and application platforms such as the HLR/HSS and SMSC are centralized.
- It is highly recommended that IMS network elements support redundancy and avoid single points of failure.
- It is recommended to deploy a Border Media Gateway at each city/pop location to keep the media close to the access.
- The media gateway to be controlled by the AS provides the tones and announcements within the VoLTE/IMS network.

For operator X, the above principles have been considered. Moreover, there are several vendor dependencies, as explained below:

- Operator X's EPC core consists of an MME, a serving gateway, a PDN gateway, an HSS, and a PCRF.

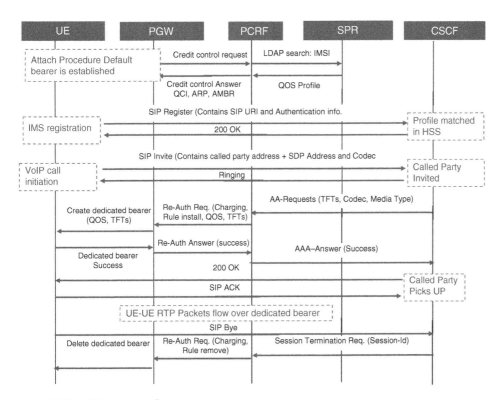

Figure 2.78 Rx Diameter call flow.

- All Diameter-related signaling exchanges are done by one vendor (Vendor A).
- The Mobility Management Entity (MME) is provided by another vendor (Vendor B).
- The serving gateway (S-GW) is provided by another vendor (Vendor B).
- The Packet Data Network Gateway (P-GW) is provided by another vendor (Vendor B).
- The Home Subscriber Server (HSS) is provided by Vendor A.
- The Policy Charging and Rules Function (PCRF) is provided by Vendor A.

Figure 2.79 represents the VoLTE/IMS high-level architecture for operator X; it can be seen that redundancy exists for core network elements of the IMS network, i.e. S/I/E−CSCF, MGCF, AGCF, MRFP, I-BGF and SBCs connect City A with City B via the IP network.

Operator X's VoLTE/IMS network was built on top of the existing wireless IP network, which includes the following:

- **Evolved Universal Terrestrial Radio Access Network (E-UTRAN):** Operator X has already deployed an E-UTRAN.
- **Evolved Packet Core (EPC):** Operator X's packet core is an Evolved Packet Core, consisting of Vendor A's MME, S-GW, HSS, and policy control provided by the PCRF.
- **3G circuit-switched network (3G CS):** Operator X's circuit-switched network is provided by Vendor C and consists of Vendor C's MSS and MGW.

2.7.2 Interconnection Topology

IMS users interconnect with the existing users of the current network, as shown in Figure 2.80. The interconnection design rules are as follows:

- IMS users interconnect with the former PSTN through the MGCF and IM-MGW. The MGCF completes the conversion of broadband/narrowband signaling; the IM-MGW completes the conversion of broadband/narrowband media and codec conversion.
- IMS users interconnect with the former PSTN through the MGCF. The MGCF implements signaling conversion and interconnection. The media IP between the IMS and NGN can connect to each other directly. The IMS interworks with the PLMN through the BICC protocol via the MGCF and IM-MGW. The IMS interworks with other IP networks through the SIP protocol via the I-BCF and I-BGF (or by the I-BCF and I-SBC). (Other SIP networks include the IMS of other carriers and the SIP networks of present carriers.)

The distribution of interconnected network elements is explained in Table 2.19.

2.7.3 Topology of Interworking with PSTN

The topology of IMS interworking with the PSTN is described in Figure 2.81.

An IMS signaling connection to the PSTN can be made in three ways:

- Through M2UA by the STP, as shown in Figure 2.82. The ISUP signaling stack between the MGCF and the PSTN through M2UA by the STP is shown in Figure 2.83.
- Through M3UA by the SG, as shown in Figure 2.84. The ISUP signaling stack between the MGCF and the PSTN through M3UA by the SG is shown in Figure 2.85.
- Through M2UA by the IM-MGW, as shown in Figure 2.86. The ISUP signaling stack between the MGCF and the PSTN through M2UA by the IM-MGW is shown in Figure 2.87.

2.7.4 Topology of Interworking with NGN

The IMS topology of interworking with the NGN is described in Figure 2.88.

2.7.5 Location-based Routing

2.7.5.1 Routing Zone and Use of LTE Location

The routing zone (RZ) is a logical parameter with range from 0 to 2999 used to identify service area coverage. The RZ parameter is defined in the service area configuration using the MMI interface of the MSS and the application server.

In order to re-use RZs with VoLTE (LTE access), the following are required:

- Current LTE radio access location information needs to be delivered to the application server. Without this information, the application server is not able to determine the RZ.
- The routing zone configuration is defined for the application server in a manner similar to the MSS today. This can be done with the same MMI (ZEPR, for example, to modify the RZ).
- The application server has the M16.2 feature 2037 implemented and activated. This feature provides mapping configuration from the received E-UTRAN cell global identification (part of the SIP P-Access-Network-Info header) to the service area, which then has the proper routing zone configured.

Figure 2.89 represents this process in the application server.

In Operator X's network, 800 series location-based services are being used on the existing network. For LTE-subscriber-originated calls, the application server will be configured for 800 series location-based service routing.

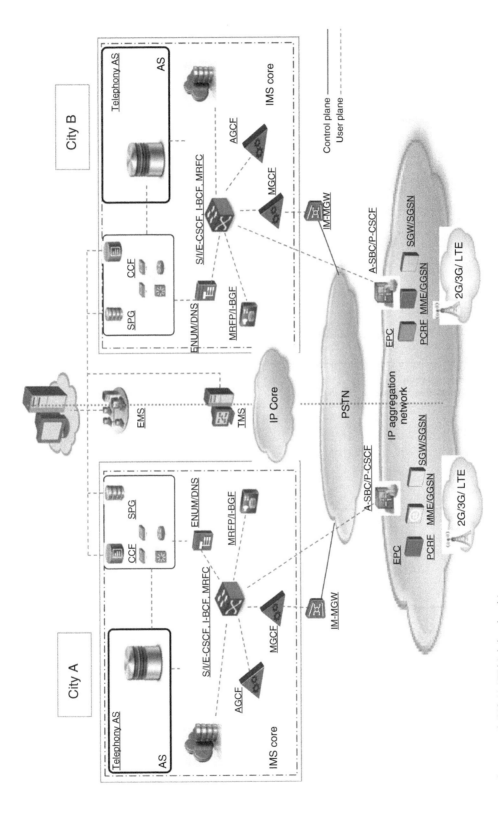

Figure 2.79 Operator X's IMS–VoLTE high-level architecture.

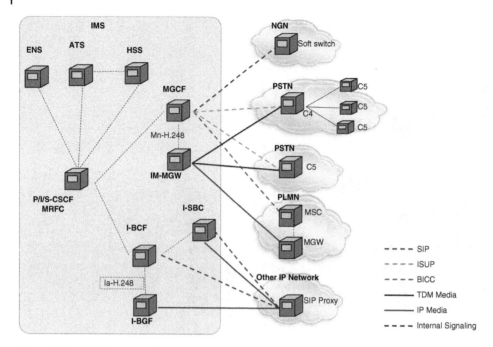

Figure 2.80 IMS interconnection.

Table 2.19 Distribution of interconnected Network Elements (NEs).

Interconnection Network	Interconnection Network Element Name	Interconnection Signaling	Interconnection Media Gateway
PSTN	MGCF	ISUP over M2UA and MTP3	IM-MGW
NGN	MGCF	SIP (or SIP-I/SIP-T) over UDP	No need for interworking media gateway, because the media IP between IMS and NGN can connect to each other directly.
PLMN	MGCF	ISUP over M3UA	IM-MGW
SIP network	I-BCF	SIP over UDP	I-BGF (or I-SBC)

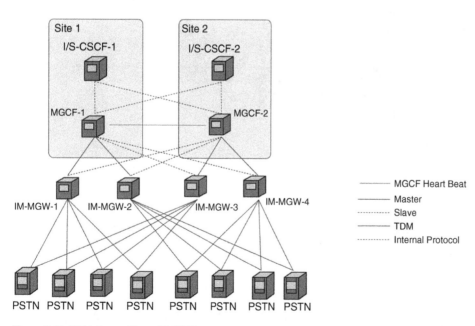

Figure 2.81 IMS interworking with PSTN.

Figure 2.82 IMS signaling connection to the PSTN through M2UAby the STP.

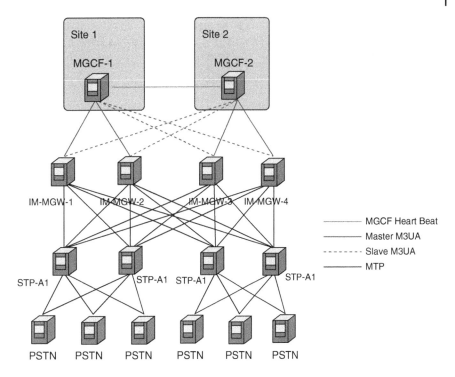

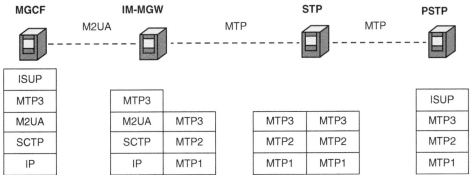

Figure 2.83 ISUP signaling stack between the MGCF and the PSTNthrough M2UA by the STP.

Figure 2.84 IMS signaling connection to the PSTN through M3UA by the SG.

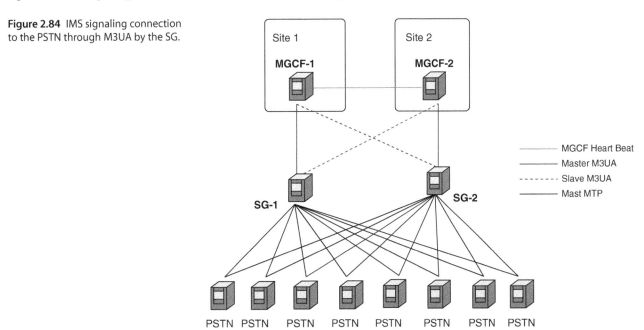

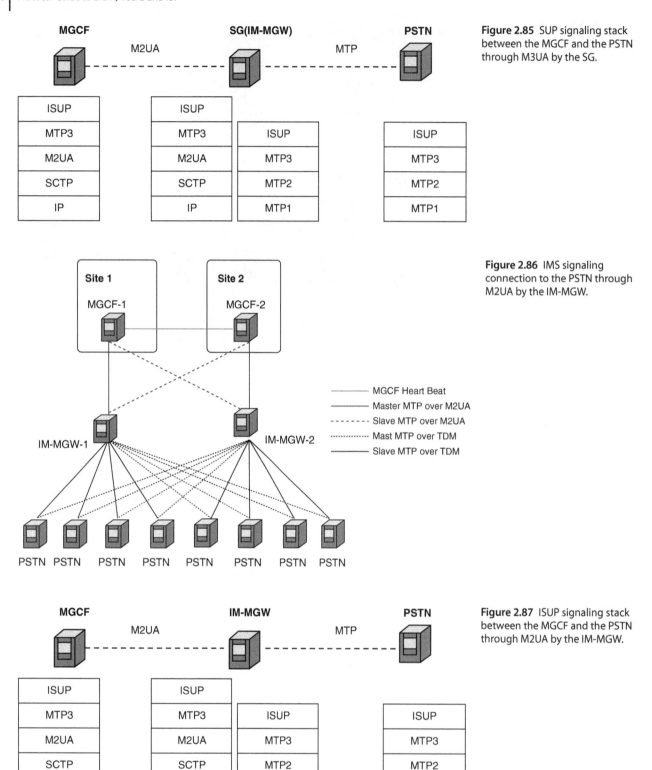

MGCF   **SG(IM-MGW)**   **PSTN**

M2UA   MTP

ISUP
MTP3
M2UA
SCTP
IP

ISUP
MTP3
M2UA
SCTP
IP

ISUP
MTP3
MTP2
MTP1

ISUP
MTP3
MTP2
MTP1

Figure 2.85 SUP signaling stack between the MGCF and the PSTN through M3UA by the SG.

Site 1   **Site 2**

MGCF-1   MGCF-2

——————— MGCF Heart Beat
——————— Master MTP over M2UA
- - - - - Slave MTP over M2UA
·········· Mast MTP over TDM
——————— Slave MTP over TDM

IM-MGW-1   IM-MGW-2

PSTN PSTN PSTN PSTN PSTN PSTN PSTN PSTN

Figure 2.86 IMS signaling connection to the PSTN through M2UA by the IM-MGW.

MGCF   **IM-MGW**   **PSTN**

M2UA   MTP

ISUP
MTP3
M2UA
SCTP
IP

ISUP
MTP3
M2UA
SCTP
IP

ISUP
MTP3
MTP2
MTP1

ISUP
MTP3
MTP2
MTP1

Figure 2.87 ISUP signaling stack between the MGCF and the PSTN through M2UA by the IM-MGW.

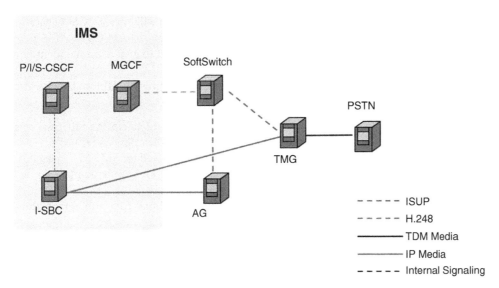

Figure 2.88 IMS interworking with the NGN.

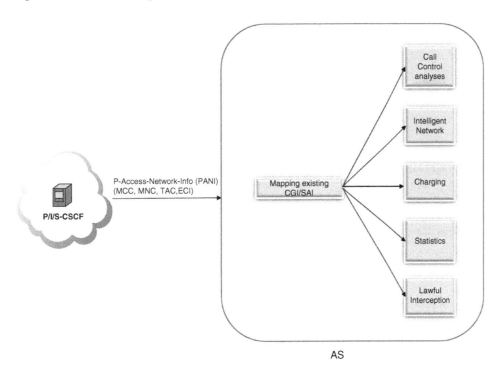

Figure 2.89 AS location-mapping design.

2.7.6 Ut/XCAP

In Operator X's network, the change in supplementary services via a subscriber's device will be handled by the CS fallback (CSFB) solution. The request is terminated at the HLR, and requested changes will be provisioned in the subscriber profile.

2.7.7 Existing Network

VoLTE nodes are connected to the existing network elements described in the following subsections.

2.7.7.1 Operator X's Network MGCF

The MGCF is needed for interworking between the SIP-based VoIP/FAX service of the IMS and voice/FAX services of the PSTN/CS domain. It performs the signaling conversion between the SIP and the ISUP and controls the MGW. The MGCF function will be collocated with the existing MSS. The MSS and GCS in the network will be connected to both ASs using the M3UA interface. The MSS will be connected to the MME over the Sv interface for SRVCC handover.

The MSS and the GCS in the network will be connected to both CSCFs using the SIP interface for calls from a PS to a CS network, a CS to a PS network, and eSRVCC handover.

The existing SG interface will remain in service for subscribers requiring CS fallback to use CS services due to handset or network configurations.

In Operator X's network, only one TCSI trigger (123930339) will be used to indicate the homing function required by the GMSC in the case of an MT call to LTE subscribers. The MSS and GCS are expected to route the GT (IDP) in load-sharing mode to both ASs.

If a subscriber is registered on a PS network, the AS will provide the Connect message as an IDP reply. The Connect message will have a prefix added before the B number to indicate the routing required towards the PS network. The MSS and GCS will use the prefix to route the call towards the CSCF using the SIP connectivity; the MSS and GCS will remove the prefix before sending on the SIP interface. The CSCF will route the call to the AS where the subscriber is registered for the T-ADS function.

If a subscriber is not registered on a PS network, the AS will check the service KEY and relay the IDP towards the SCP. The MSS and GMSS are expected to configure the IDP relay support for AS GT.

In the case of a PS to CS handover being required due to coverage limitations of the LTE, the MME is expected to provide the STN-SR number to the MSS for the handover on the Sv interface. The MSS is expected to route the STN-SR number towards the CSCF.

The following working assumptions have been taken into account for the solution design:

- A homing function is used on the network for a CS to PS MT call.
- Each MSS is connected to the MME using the Sv interface.
- Each MSS is connected to the CSCF using the SIP interface.
- Each MSS is connected to the AS using the MAP interface.

2.7.7.2 HSS

The HSS is planned to be connected to the application server using the Sh interface and to the CSCF using the Cx interface. In Operator X's network, MMTEL services are planned to be from the HSS over the Sh interface. Operator X's network will have an HSS with three HSS FEs.

2.7.7.3 HLR

The HLR in Operator X's network will be connected to the application server using the MAP interface. The LTE subscriber configuration will have a TCSI trigger in the HLR for the homing function from CS to PS. A new service key will be configured to give a unique identifier for each service.

The application server requires the HLR to provide the following data:

- The TCSI trigger and new SKEY in the SRI response for an MT call of an LTE subscriber in HPLMN.
- The IP-SM-GW function, used for SMS MT messages.
- The NSDM function, used for supplementary service updates to the AS.
- The MSRN (routing) address to route terminating VoLTE calls.
- The VLR address and IMSI to route terminating short messages.
- Intelligent network-related information for SCP service activations.
- Additional service information.
- Basic service configuration.

2.7.7.4 SMSC

SMSCs in Operator X's network will be connected to the application server using the MAP interface for the SMSMO and SMSMT service of a subscriber registered on the application server. An SMSC is expected to route application server GT and IP-SM-GW GT towards the application server.

2.7.7.5 IN

INs in Operator X's network will be connected to the application server for SCP-related services of a subscriber registered on the application server.

2.7.7.6 STP

The application server will be connected to the STP via the M3UA interface. The STP will be used to route signaling messages to existing network nodes: HLR, SMSC, IN, etc. The MSS and GMSS, with direct connectivity to the application server, will not use the STP for signaling traffic.

2.7.7.7 MME

The MME will be connected to the existing MSS using the Sv interface.

2.7.7.8 PCRF

A VoLTE solution needs to be integrated with the existing PCRF server. The integration point between the PCRF and the VoLTE solution is the Rx interface (Diameter protocol).

2.7.7.9 IP-SM-GW

The Short Message Service (SMS) can be provided either with the CS fallback solution, as specified by 3GPP [38], in an MSS and MME deployment, or, alternatively, as SMS over IP in an IMS deployment. To support the SMS over IP solution, 3GPP defined the IP-SM-GW role.

The IP-SM-GW role manages the following tasks:

- Notifying the HLR of the SMSIP user's availability (at registration) and unavailability (at de-registration).
- Homing terminating messages to the serving IP-SM-GW.
- Available domain selection (IMS/SIP, PS, CS).
- Hunting among available domains.

2.7.7.10   Messaging Profile

Every IP-SM-GW user provided with access to the services of the IP-SM-GW must be provisioned in a repository. Hence, the IP-SM-GW user is allocated to, and linked with, a set of allowed services, configured with the messaging profile parameters in the application server.

IMS subscribers using IP-SM-GW services must have iFC provisioned such that IP-SM-GW is triggered in the event of:

- Registration events, including initial registration, re-registration, and de-registration.
- A SIP message request.

The supplementary services that the IP-SM-GW uses can be provisioned in the HSS.

2.7.7.11   IP-SM-GW and User Types

The IP-SM-GW domain selection role selects among SIP/IMS, 2G/3G PS and CS domains belonging to one subscriber. Therefore, IP-SM-GW functionality is valid and can be utilized for subscribers whose MMTEL supplementary service data are either downloaded from the HLR using the MAP Restore Data (RD) operation or from the HSS using the User-Data-Request (UDR) Diameter message over the Sh interface.

2.7.7.12   IP-SM-GW Interworking with Multi-SIM/ONS

Multi-SIM/ONS sequential or parallel hunting logic is executed for members of a multi-SIM/ONS group. If a group member is an IP-SM-GW user, additional sequential hunting is performed for that member according to the IP-SM-GW domain selection and hunting logic to provide the messaging-specific terminating access domain selection functionality, so a selection is made among the available domains of the subscriber.

2.7.7.13   HLR Support for IP-SM-GW

The standard IP-SM-GW functionality requires support in the HLR. To support IP-SM-GW functionality, the HLR is enhanced with the following:

J interface (MAP) support between the IP-SM-GW and the HLR:

- Registration and de-registration from the IP-SM-GW to the HLR for SMS delivery.
- Re-routing of the SRIforSM requests from the HLR to the IP-SM-GW in order to obtain the return address to which the SM should be forwarded.
- Allowing the interrogation of the HLR using SRI-forSM to retrieve the IMSI and the current MSC and/or SGSN addresses.
- Allowing the HLR to be informed when a memory-capacity-exceeded condition ceases.
- Allowing SMS-related data to be retrieved from the HLR.

The optional subscriber-level configuration of an IP-SM-GW address from the list of allowed IP-SM-GW addresses.

2.7.7.14   IP-SM-GW Address Registration to the HLR

To register the IP-SM-GW address in the HLR, the application server uses the MAP Any Time Modification (ATM) operation. The procedure is shown in Figure 2.90.

2.7.7.15   IP-SM-GW Homing

After a successful IP-SM-GW user registration [38], the HLR stores the IP-SM-GW user subscription information and the dedicated subscriber-level IP-SM-GW address.

Figure 2.90 IP-SM-GW and MAP Any Time Modification (ATM) operation.

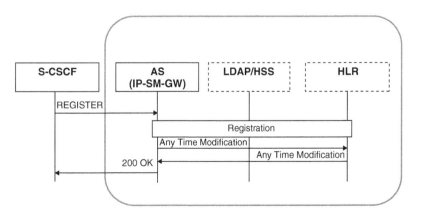

The IP-SM-GW homing logic is as follows:

- The SMSC/SMS-GMSC makes an HLR inquiry to retrieve routing information, using the MAP SRIforSM.
- The HLR checks the calling party address (CgPA = SMS-GMSC).
  - The subscriber has an IP-SM-GW address registered in the HLR so the target of the MAP operation is the subscriber's IP-SM-GW address.
- The HLR sends SRIforSM to the IP-SM-GW.
  - The IP-SM-GW accepts the SRIforSM operation from the HLR if the called party GT address is the same as its own virtual IP-SM-GW address.
  - The CdPA GT address (used for SCCP-level GT routing) is changed from the MSISDN to the registered IP-SM-GW address. Hence, the HLR relays the SRIforSM request to the retrieved IP-SM-GW address.
  - If the called party address and the own virtual IP-SM-GW address do not correspond, an error is returned with the "facilityNotSupported" reason.
- The IP-SM-GW sends back the MAP SRIforSM request to the HLR to retrieve routing information, including IMSI, VMSC, and/or SGSN addresses and MWD.
  - The CdPA GT address is exchanged back to the MSISDN (used for SCCP-level routing towards the HLR) and the CgPA GT is set to the own logical IP-SM-GW address.
- The HLR receives the SRIforSM request and checks the CgPA GT address (= IP-SM-GW). The request is from the IP-SM-GW so the HLR processes the request.
- The HLR responds with an SRIforSM to the IP-SM-GW.
  - The SRIforSM response contains subscriber and SMSC/SMS-GMSC related information. The IP-SM-GW saves this information:
    - o IMSI
    - o SGSN/VMSC address
    - o MNRG/MNRF flag.
- The IP-SM-GW generates a correlation ID.
- After saving the subscriber and SMSC/SMS-GMSC related information, the IP-SM-GW puts a unique correlation ID and its own IP-SM-GW address into the SRIforSM response instead of the real IMSI number and VMSC/SGSN addresses.
- The IP-SM-GW sends back a MAP SRIforSMRes operation to the SMSC/SMS-GMSC – that is, to the CgPA GT address of the originally received MAP SRIforSM forwarded by the HLR.
- The SRIforSM response now contains IP-SM-GW-related information:
  - Correlation ID
  - IP-SM-GW address.

- As a result of this, the SMS-GMSC sends the MTForwardSM operation to the IP-SM-GW (and only to the IP-SM-GW) instead of the MT-VMSC/SGSN.
- The SMSC/SMS-GMSC sends the MTForwardSM to the IP-SM-GW.

Figure 2.91 illustrates IP-SM-GW homing, in which, if the IP-SM-GW address in the MTForwardSM operation is its own IP-SM-GW address and the IP-SM-GW Domain Selection (FEA3855) license is active, then the processing of the received MTForwardSM is started.

The previously saved data are compared to the data in the received MT-SM. If the information corresponds, then the IP-SM-GW starts the selection logic. If there is any difference, the MT-SM is rejected by the IP-SM-GW.

2.7.8 External Network Connections

2.7.8.1 PSTN and IMS Subscription
Operator X's network will interconnect with the PSTN using the existing connectivity with the GMSS.

For the IMS subscription of Operator X's network, each subscriber will be provisioned with two IMPIs and three IMPUs. Each IMPI corresponds to a specific authentication method – AKA or digest – and whenever subscribers use any of these two IMPIs to register, the three IMPUs will be registered implicitly in the IMS domain. The two IMPIs and three IMPUs are as follows:

IMPIs:
The IMSI-derived IMPI to be used for USIM AKA authentication.
Example: 1234567890@ims.mnc:xyz.mcc:123.3gpp network.org.
The MSISDN-based IMPI to be used for HTTP digest authentication and ISIM AKA authentication.
Example: +123-1234567@ims.domain.com.

IMPUs:
The IMSI-derived IMPU to be used for USIM AKA authentication. This IMPU is only used for AKA authentication and should not be used for VoIP sessions; therefore, it has to be barred for such scenarios. For VoIP sessions, a user can use the other two IMPUs below.
Example: 1234567890@ims.mncxyz.mcc123.3gppnet work.org.
The MSISDN-based IMPU to be used for the VoIP service.
Example: sip:+123-1234567@ims.domain.com.
The Telephone URI.
Example: tel: +123-1234567.

The three IMPUs described above will be provisioned as part of one implicit registration set (IRS). The

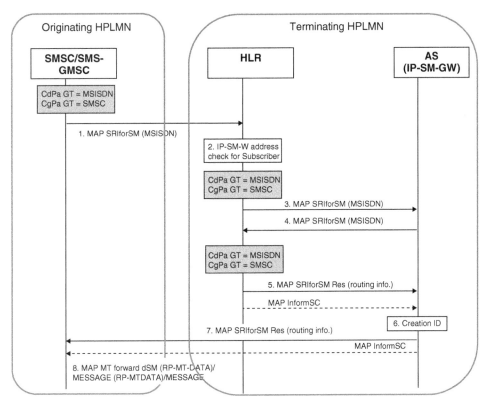

Figure 2.91 IP-SM-GW homing.

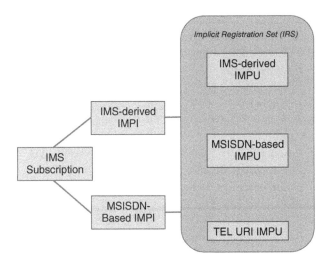

Figure 2.92 IMS subscription for a VoLTE user.

relationship between these identities and groups with an IMS subscription is illustrated in Figure 2.92.

The IMS private user, public user, private service IDs and public service IDs are stored in the IMS user profile in the HSS. The S-CSCF receives private and public user IDs from the IMS user profile that is downloaded from the HSS to the S-CSCF during IMS registration. On the IMS end-user device, the private ID is, for example, either

configured by using a configuration file or Over-The-Air Provisioning (OTAP), stored on the ISIM, or derived from the IMSI in accordance with [31]. When the public user ID is derived from the IMSI in accordance with [31], only one single public user ID is derived. However, the IMS end-user device supports multiple public user IDs, either pre-configured or retrieved from the CSCF during the initial registration.

2.7.9 Supported Explicit and Implicit Registration

In IMS, public user IDs are registered explicitly or implicitly. Explicit IMS registration is processed for the public user ID which is sent in the SIP REGISTER request. Implicit registration is processed for all other public user IDs from an implicit registration set, to which the explicitly registered public user ID belongs. The implicit registration set is provisioned at the HSS on a per-subscriber basis and includes a list of public user IDs (IMPUs). With IMS, multiple implicit registration sets are supported for each IMS subscriber. The advantage of implicit registration is that only a single IMS registration procedure is required in order to register many public user IDs. All public user IDs belonging to the same registration set are registered, re-registered, or de-registered if one of them is registered, re-registered, or de-registered. For explicit registration, no group IMPUs are used.

Group IMPUs are only registered implicitly together with an ordinary public user ID in an implicit registration set.

Registrations using the same private user ID must always use the same authentication method. Therefore, multiple explicit registrations from the same end-user device must use the same authentication mechanism as used during the last (re-)authentication procedure. When IMS AKA is applied, and if multiple explicit registrations are initiated from an end-user device, they must all undergo the security association established during the initial registration. The security association is only removed when all public user IDs of the end-user device are de-registered. Nevertheless, multiple explicit registrations are also sent from different end-user devices. In this case, diverse authentication mechanisms are used.

Each IMPU stored in the HSS is optionally associated with a display name. It is transported with the IMPU whenever present and unblocked. The feature allows the "Name" of the calling party to be presented to the called party, in addition to the calling number.

References

1 Schulzrinne, H. and Schooler, E. (1999) "Session Initiation Protocol(SIP)," RFC 2543, March.

2 Rosenberg, J., Schulzrinne, H., Camarillo, G., Johnston, A., Peterson, J., Sparks, R., Handley, M., and Schooler, E. (2002) "SIP: Session Initiation Protocol," RFC 3261, June.

3 Rosenberg, J. and Schulzrinne, H. (2002) "An Offer/Answer Model with Session Description Protocol (SDP)," RFC 3264, June.

4 Lingle, K., Mule, J.-F., Maeng, J. and Walker, D. (2007) "Management Information Base for the Session Initiation Protocol (SIP)," RFC 4780, SIP MIB Modules, April.

5 3GPP TS 23.228: "IP Multimedia Subsystem (IMS); Stage 2; (Release 8)."

6 Handley, M., Jacobson, V. and Perkins, C. (2006) "Session Description Protocol (SDP)," RFC 4566, July.

7 Handley, M., Perkins, C. and Whelan, E. (2000) "Session Announcement Protocol (SAP)," RFC 2974, October.

8 Schulzrinne, H., Rao, A. and Lanphier, R. (1998) "Real Time Streaming Protocol (RTSP)," RFC 2326, April.

9 3GPP TS 23.517: V8.0.0 (2007-12).

10 3GPP TS 123.002 V12.5.0: "LTE; Network architecture", (Release 12).

11 3GPP TS 182.006: "IP Multimedia Subsystem (IMS); Stage 2 description," (Release 7, modified).

12 3GPP TS 182.009: "Protocols for Advanced Networking (TISPAN); NGN architecture to support emergency communication from citizen to authority."

13 3GPP TS 29.228: V5.22.0.

14 ETSI ES 282 001: "Telecommunications and Internet converged Services and Protocols for Advanced Networking (TISPAN); NGN Functional Architecture," (Release 1).

15 3GPP TS 32.240: "Charging management, Charging architecture and principles," (Release 9).

16 3GPP TS 32.260: "Charging management; IP Multimedia Subsystem (IMS) charging," (Release 9).

17 3GPP TR 33.cde: V0.0.2 (2004-07) "Security Aspects of Early IMS."

18 3GPP TS 33.203 V10.2.0: "Access security for IP-based services," (Release 10).

19 Faltstrom, P. (2000) *"RFC 2916, E.164 number and DNS,"* September.

20 Camarillo, G. and Blanco, G. (2007) *"RFC 5002, P-Profile-Key P-Header,"* August.

21 3GPP TS 24.229 V12.6.0: "Multimedia call control protocol based on Session Initiation Protocol (SIP) and Session Description Protocol (SDP); Stage 3," (Release 12).

22 Rosenberg, J., Schulzrinne, H. and Kyzivat, P. (2004) *"RFC 3841, Caller Preferences for SIP,"* August.

23 Faltstrom, P. and Mealling, M. (2004) *"RFC 3761, ENUM,"* April.

24 Schulzrinne, H. (2004) *"RFC 3966, The Tel URI,"* December.

25 Peterson, J. (2004) *"RFC 3764, SIP enum service,"* April.

26 Rosenberg, J. and Schulzrinne, H. (2002) *"RFC 3263, SIP: Locating SIP Servers,"* June.

27 Gulbrandsen, A., Vixie, P. and Esibov, L. (2000) *"RFC 2782, DNS SRV RR,"* February.

28 Mockapetris, P. (1987) *"RFC 1034, Domain Concepts and Facilities,"* November.

29 Mockapetris, P. (1987) *"RFC 1035, Domain Implementation and Specification,"* November.

30 Thomson, S., Huitema, C., Ksinant, V. and Souissi, M. (2003) *"RFC 3596, DNS Extensions to Support IPv6,"* October.

31 3GPP TS 23.003 V10.5.0: "Numbering, addressing and identification," (Release 10).

32 3GPP TS 23.008: "Organization of subscriber data" (Release 8).

33 Aboba, B., Beadles, M., Arkko, J. and Eronen, P. (2005) *"RFC 4282, The Network Access Identifier,"* December.

34 3GPP TS 22.173 V10.3.0: "Multimedia Core Network Subsystem (IMS) Multimedia Telephony Service and supplementary services; Stage 1," (Release 10).

35 3GPP TS 22.030 V7.0.1: "Man–Machine Interface (MMI) of the User Equipment (UE)," (Release 7).

36 3GPP TS 23.292: "IP Multimedia Subsystem (IMS) Centralized Services; Stage 2" (Release 8).

37 3GPP TS 29.213 V11.4.0: "Policy and charging control signalling flows and Quality of Service (QoS) parameter mapping," (Release 11).

38 3GPP TS 23.204: "Support of Short Message Service (SMS) over generic 3GPP Internet Protocol (IP) access; Stage 2," (Release 7).

3

VoLTE/CSFB Call Setup Delay and Handover Analysis

3.1 Overview

The IP Multimedia Subsystem (IMS) is a framework for delivering all-IP-based services. Voice via IMS has been defined as a possible solution from 3GPP Release 5, prior to LTE. However, the cost and immediate need for IMS services prevented carriers from migration to IMS-based services, especially with the widespread presence of circuit-switched (CS) services offered in GSM and UMTS networks. CS fallback (CSFB), introduced in 3GPP Release 8, enables the support of a voice service without IMS. This is the commonly deployed scenario for many existing LTE networks. Voice over LTE (VoLTE) is compulsory in terms of offering rich communication services (RCS) via IMS in addition to improving the performance of the already-deployed CSFB voice solution. However, CSFB and VoLTE are deployed concurrently, so carriers can gradually roll out an LTE/IMS system while still supporting a 2G/3G fallback mechanism. It is therefore important to benchmark the performance of both solutions, highlight the deployment challenges, and study the impacts on the end-user experience. This chapter provides performance analysis, deployment challenges, and comparisons between these voice solutions. The chapter presents practical performance analysis including end-to-end assessment of call setup delay in different radio conditions, main challenges impacting the in-call performance, and performance aspects of Single Radio Voice Call Continuity (SRVCC) and its evolution releases. Therefore, the chapter provides comprehensive analysis for call setup delay including CSFB and VoLTE with different scenarios including stationary and mobility scenarios, and handover analysis including SRVCC in terms of data interruption time. Finally, recent topics in handover performance and data interruption reduction during handover are presented. Synchronized handover is introduced as a potential solution to reduce data interruption time during handover.

3.2 Introduction

The IP Multimedia Subsystem (IMS) is an all-IP system designed to assist mobile carriers in delivering next-generation interactive and interoperable services cost-effectively and over an architecture providing the flexibility of the Internet [1, 2]. Although the IMS was initiated in 3GPP in the early releases (Release 5), it has not yet been widely implemented. The main reason is the use of the well-established 2G/3G circuit-switched (CS) networks offering voice and other CS services. Network carriers need to invest to deploy a complete IMS-based network consisting of multiple core network elements. An IMS client also needs to be incorporated in the UE to interact with the network. Various key telephony functions should be supported by the IMS core network and the IMS client on the UE to ensure a satisfactory customer telephony experience. Therefore, IMS deployment requires major investment from carriers, network vendors, and device manufacturers. Both the Long Term Evolution (LTE) device and the network need to support various features across the LTE protocol stack to ensure satisfactory voice over LTE (VoLTE) performance. Some features are also needed to ensure optimal LTE system capacity for large numbers of VoLTE users. Moreover, Single Radio Voice Call Continuity (SRVCC) provides an interim solution for handing over a VoLTE call to legacy 2G/3G networks.

On the other hand, circuit-switched fallback (CSFB) to 2G/3G networks has been widely deployed as an interim voice solution within the deployed LTE PS networks. As soon as the UE originates or receives a voice call, the eNodeB redirects the UE to UMTS or a GSM network, depending on the configuration and underlying coverage [3]. Therefore, the CSFB architecture requires interworking between the evolved packet core (EPC) and the 2G/3G CS core network. Specifically, it requires the SGi interface between the Mobility Management Entity

(MME) in the EPC and the Mobile Switching Center (MSC) in the 2G/3G CS core network [4]. The requirements to minimize the number of interfaces between the core networks and to re-use the air interface of the existing 3G/2G network for voice calls have accelerated CSFB deployment, and therefore it has been adopted as the first choice for voice with LTE. However, with the rapid spread of LTE deployment worldwide, there is a need for a framework that enables subscribers to access a range of multimedia services without fallback to the legacy 2G/3G systems. There is also a growing motivation to compete with Over-The-Top (OTT) voice over IP (VoIP) services, which have fragmented the value-added services using cross platforms and technologies. This has also restricted the ability of mobile carriers to provide just access and dump pipe while OTT services are indirectly being subsidized.

Figure 3.1 summarizes the voice evolution in 3GPP after the introduction of the LTE system. In the first phase, all voice traffic initiated in LTE is handed over to legacy CS networks (3G/2G). Single radio solutions use CSFB to switch between LTE and 2G/3G access modes, ensuring data continuity on the redirected radio access technology (RAT). CSFB has become the predominant global solution for voice, primarily due to cost, size, and power advantages of the UE with single radio solutions. The second phase is the VoLTE, in which, single radio voice call continuity (SRVCC) has been introduced. SRVCC enables the handover to 3G/2G while the user is outside the VoLTE coverage in order to ensure voice call continuity. The third phase introduces enhanced capacity and services of all-IP networks (voice and video

over IP) for continuous coverage across the broader range of network access methods, including LTE, and 3G/HSPA+, with interoperability across carriers (i.e. roaming) and legacy telephony domains. In addition, it is expected that 3G voice will evolve towards better capacity and voice quality, referred to as WCDMA+ voice evolution.

Voice is a real-time service with tight delay requirements, thus it requires a robust underlying radio network to ensure an optimal user experience. VoLTE requires end-to-end Quality of Service (QoS) to be supported in all layers from the device through the radio network up to the core network and including interaction with the IMS core. This is necessary to ensure robust signaling performance and optimal voice quality in the presence of other data traffic (especially in loaded scenarios). These factors add strict requirements to the LTE network for enabling VoLTE as a voice solution to the end user [3]. 3GPP has adopted the GSMA IR.92 IMS profile for voice and SMS [5] and the GSMA IR.94 IMS profile for conversational video [6] to provide high-quality, IMS-based telephony services over LTE radio access. The profiles define optimal sets of existing 3GPP-specified functionalities that all industry stakeholders, including network vendors, service providers, and handset manufacturers, can use to offer compatible LTE voice/video solutions.

As explained, a mix of VoLTE and CSFB devices can co-exist in the same LTE network. 3GPP allows mechanisms for the UE's preferences, whether it is capable of, or prefers, CS voice and/or IMS PS voice [7]. Four settings are possible: "CS Voice only," "IMS PS Voice only," "Prefer CS Voice with IMS PS Voice secondary,"

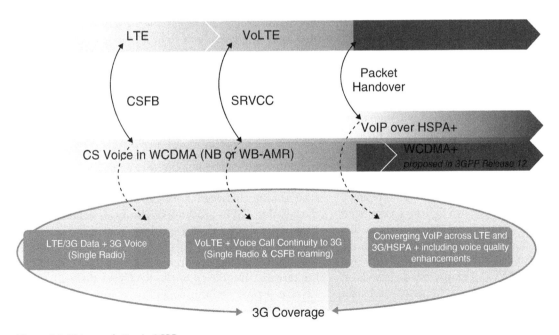

Figure 3.1 Voice evolution in 3GPP.

and "Prefer IMS PS Voice with CS Voice Secondary" [7]. Specific combinations of these UE settings and the Evolved Packet System (EPS) attach results from the core network to make sure that the UE either remains in E-UTRAN or reselects to another RAT, depending on the network support and device capabilities. Hence, 3GPP generally covers all possible combinations of UE and network capabilities to ensure voice continuity to the best level of support. Such alignment between device capability and network support is negotiated at the time the device is initially attached to the EPS.

The performance of CSFB, VoLTE, and SRVCC are key elements in guaranteeing a good voice experience within the LTE network [7–11]; in particular, it is important that a mix of VoLTE and CSFB devices can co-exist in the same LTE network. CSFB performance based on redirection has been analyzed in [8] for 3GPP Release 8 and Release 9 and compared with UMTS. The analysis in [8] shows that, on average, the mobile-terminated (MT) or mobile-originated (MO) call setup delay for CSFB from LTE to UMTS is around 1 second greater than legacy UMTS CS calls. A delayed return (DR) approach was proposed in [9] to minimize the effect of CSFB by 60%; using this approach, the UE is kept in UMTS without returning to LTE if there is no active data session. A limited access transfer algorithm that reduces the number of ping-pong effects between LTE and UMTS in an SRVCC call was proposed in [10], by which the large access transfer traffic delay can be improved. VoLTE in call performance has been evaluated in commercial networks in terms of audio quality and call reliability [11].

The focus of this chapter is the call setup delay for CSFB and VoLTE and SRVCC performance in terms of voice and signaling interruption. Firstly, the CSFB call flow and relevant call setup KPIs are established and analyzed in detail for the redirection and PS handover strategies. The call setup KPIs for different strategies (basic Release 9 redirection, enhanced Release 9 redirection, and PS handover with/without gap measurements) are evaluated along with the associated performance and challenges. The second contribution is the analysis of the VoLTE call setup delay in a comparative manner with CSFB. The relevant VoLTE KPIs are established and analyzed for different scenarios including near-cell, far-cell, and mobility. The third contribution is the analysis of the enhanced SRVCC (eSRVCC) performance and its relevant KPIs in comparison to delays associated with LTE intra-frequency handover. All analysis in this chapter is based on commercial LTE/HSPA+ networks. The best practices and optimization techniques for CSFB, VoLTE, and eSRVCC are provided, based on live network performance and the targeted end-user experience. In this chapter, we use the terms LTE and E-UTRAN interchangeably, as well as UMTS and UTRAN.

3.3 CSFB Call Flow and Relevant KPIs

Various methods are specified by 3GPP for handling CSFB from LTE to UMTS once a voice call is initiated [3, 7, 12]. CSFB to UTRAN can be executed by two methods. The first method is "redirection" based, in which, upon initiation of a voice call, the E-UTRAN Radio Resource Control (RRC) layer releases the UE from LTE and redirects it by the same message to the other RAT to handle the voice call while the PS data session is interrupted until the data session is re-initiated on the target RAT. The second method is "PS handover" based, where a packet-switched handover from E-UTRAN to the target RAT (UTRAN in this chapter) is initiated by the eNB to handle the voice call while the PS data session can partially remain uninterrupted until the handover is completed. Both mechanisms are applicable for MT or MO calls while the UE is in RRC Idle or RRC Connected mode. There are a number of improvements and flavors in the "redirection" mechanism, targeting enhanced voice call setup delay and PS interruption time [12].

The most common CSFB mechanisms are summarized in Table 3.1. Each type is typically deployed based on the network and device capabilities. Several methods can be deployed together to address device capabilities.

Figure 3.2 illustrates the call signaling flow for two CSFB mechanisms: RRC redirection and PS handover [3]. For CSFB with RRC redirection, the call is initiated and the UE sends an Extended Service Request (ESR). The redirection method works by releasing the RRC connection while the UE is camped on an LTE cell. The RRC release message indicates the specific frequency and RAT information for the UE to be redirected after the release. The device can then search for any cell on the redirection frequency and RAT, acquire the targeted RAT and frequency, and then initiate a normal CS call setup procedure.

Step 8 in Figure 3.2 indicates the concept of CSFB redirection occurrence as part of the RRC connection release in LTE. Typically, CSFB redirection is performed without any prior inter-RAT (IRAT) measurements on the targeted system. Therefore, Steps 4–7 are optional. Initiating a CSFB call with or without IRAT measurements depends on the deployment strategy. Measurement-based redirection may help in achieving load sharing between multiple 3G frequency carriers during the CSFB procedure. However, achieving this kind of load sharing at a cell level within the same 3G frequency carrier is subject to the CSFB mechanism. Release 8 redirection is restricted to deliver the UE to the targeted frequency of the RAT (not the cell), while Release 9 provides the flexibility to indicate both cell and frequency carrier information. Hence, load sharing at cell level cannot be achieved with Release 8 even if the UE is able to measure and report the best cell

Table 3.1 CSFB deployment strategies.

#	CSFB Type	Description	RRC Mechanism
S1	3GPP Release 8 with basic functions	Redirection based, with SIB skipping feature: Deferred Measurement Control Reading (DMCR)	RRC redirection with RRC release
S2	3GPP Release 8 with optimized functions	Redirection based, with DMCR and optimized SIB scheduling	RRC redirection with RRC release
S3	3GPP Release 9	Redirection based, with DMCR and RRC release including a list of 3G cells and tunneled SIB information	RRC redirection with RRC release
S4	CSFB LTE to UMTS through PS handover *without* measurement	Target cell is prepared in advance and device can camp on the cell directly in Connected mode once the voice call is initiated	PS handover with UTRAN mobility from E-UTRAN
S5	CSFB LTE to UMTS through PS handover *with* measurement		

Figure 3.2 MO CSFB from E-UTRAN Connected mode to UTRAN with redirection.

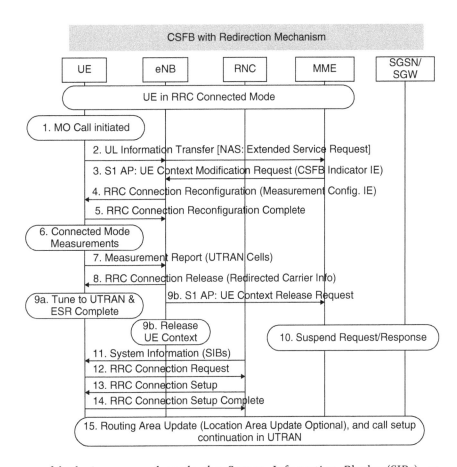

within one carrier. This limits the use case of deploying a measurement-based Release 8 CSFB redirection strategy.

For Release 8 CSFB, the UE performs a full search to acquire any cell on the redirected RAT (for example, if the redirection is to UMTS, the UE will search the full 512 primary scrambling codes (PSCs) on the redirected frequency). After the RRC release message has been received, the UE is moved directly to UMTS Idle mode

and reads the System Information Blocks (SIBs), as shown in Step 11. The SIBs are broadcast continuously by the cell, allowing the UE to know the initial configurations of the cell's parameters. Compared to UMTS-only CS calls, the entire SIB reading prior to call setup is not required, as the UE would have previously decoded them when initially camped on the cell. Therefore, the main challenge of the redirection method is the call

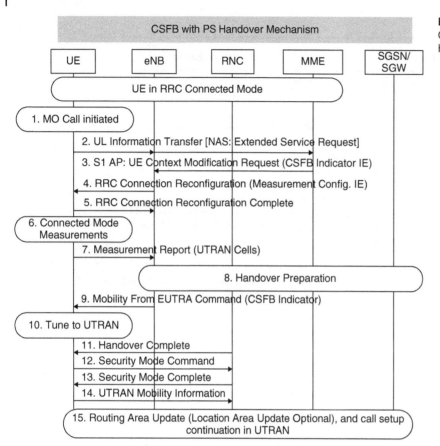

Figure 3.3 MO CSFB from E-UTRAN Connected mode to UTRAN with PS handover.

setup delay due to SIB reading. To overcome this burden, Deferred Measurement Control Reading (DMCR) is introduced. This feature allows the UE to skip reading non-mandatory SIBs (such as SIBs 11/12/19, or any of their extensions) during the CS call setup. Therefore, the UE will read the mandatory SIBs such as 1, 3, 5, and 7 in Step 11. This further reduces the delays coming from reading all of the SIBs. However, for DMCR to work efficiently, the UE needs to read SIB3 to understand whether DMCR is supported in the cell or not. Therefore, SIB scheduling optimization in a deployed network is still required, in which the UMTS Radio Network Controller (RNC) ensures that SIB3 is scheduled to the UE as soon as possible. On the other side, UMTS SIB tunneling is introduced during the redirection process from LTE to UMTS [12]. In this mechanism, a list of PSCs is defined in the LTE RRC release message, with a container that includes the associated SIBs for each cell.

The second mechanism is CSFB with PS handover, denoted PSHO and shown in Figure 3.3. In the PSHO procedure, the target cell is prepared before the fallback to UTRAN is triggered. The device can camp on the target cell directly in UTRAN Connected mode (i.e., Cell_DCH). During the execution of the handover procedure, the eNB typically configures the UE to perform IRAT measurements within a specified gap duration

(i.e. a tune-away mechanism for the UE to measure another RAT while camped on LTE, specified by a gap repetition every 40 or 80 ms and a duration of 6 ms). Alternatively, the PSHO can be triggered blindly, where a neighbor relationship is defined between the UTRAN and E-UTRAN systems. The PSHO is helpful for a CSFB strategy when the PS data session is active. In this case, the PS radio bearer is established earlier during the handover procedure, which helps to minimize the PS data interruption time during the transition to the other RAT, a clear benefit over the redirection case.

The CSFB call setup latency can be derived from the time when the ESR message is sent until the ALERTING message is received. This applies to any kind of CSFB, whether a redirection or PSHO. Additionally, the CSFB call setup delay is categorized into several KPIs to measure the factors that contribute to the delay. Table 3.2 illustrates these KPIs and the calculation methodologies.

3.4 VoLTE Call Flow and Relevant KPIs

One of the significant changes introduced in LTE is that when the mobile device connects to the network, it also implicitly gets an IP address, known as the EPS Default Radio Bearer (DRB). With the default EPS bearer activation, the packet call is established at the same time when

Table 3.2 KPIs impacting the CSFB call setup latency.

KPI	CSFB Stage	From Event	To Event
C1	LTE ESR to redirection of PSHO command + tune to UMTS	ESR	RRC Connection Release (redirection) or Mobility From E-UTRA (PSHO) + time to tune to 3G
C2	SIB read time including cell selection (redirection case only)	First SIB decoded	[Last SIB decoded (according to the total SIBs broadcast in a cell, typically MIB, SIB1, 2, 3, 5, 7, 11)] + [Cell selection completed]
C3	UMTS RRC latency (redirection case only)	RRC Connection Request	RRC Connection Setup Complete
C4	Non-access stratum end-to-end time	CM Service Request	ALERTING message
C5	Total call setup delay	ESR	ALERTING message

the UE attaches to the EPS (i.e., always on). Although, the default DRB is enough for the downlink and uplink data transfer in an EPS network, it comes without any guaranteed Quality of Service (QoS). For real-time applications such as voice, QoS is needed especially on the air interface. To exploit service differentiation, LTE has also introduced another EPS bearer known as the Dedicated EPS Data Bearer, which is initiated for any additional data radio bearer. To guarantee voice quality, the IP voice traffic needs to be carried over guaranteed bit rate (GBR) bearers with QoS class identifier (QCI = 1). The default bearer is a non-GBR bearer (i.e. with QCI = 9) and is used for the best-effort PS traffic [3]. The network resources associated with the VoLTE voice GBR bearer must be allocated when this bearer is established, and can be released once the voice call is ended. Additionally, another bearer is necessary to ensure that the IMS signaling part is transmitted with a different QoS. As a VoLTE call is initiated on a different Access Point Name (APN), the IMS signaling is mapped into a different QCI. The common configuration to transport the IMS signaling uses a default EPS bearer with QCI = 5.

For a VoLTE-capable device, the UE should be configured with the proper EPS bearers to transport LTE signaling and data, IMS signaling, and media traffic. The LTE signaling messages are mapped to the pre-allocated signaling radio bearers (SRBs). Therefore, a VoLTE-capable UE must support: SRB1 + SRB2 + 4x AM (Acknowledge Mode) DRB + 1x UM (Unacknowledged Mode) DRB [5, 13]. Figure 3.4 illustrates an example of the configurations and mapping for each EPS bearer during a VoLTE call across different LTE layers. The DRBs can be further mapped into layer-3 Radio Link Control (RLC) as AM or UM modes. The eNB links the choice of the RLC mode to certain QCIs in order to maintain the desired QoS. RLC AM mode ensures an in-sequence delivery of the packets subject to delay sensitivity (like

the PS data, LTE/IMS signaling), while RLC UM mode provides a reduction in processing time and overheads, suitable for VoLTE media packets.

The Session Initiation Protocol (SIP) is utilized in the IMS to manage all aspects of a session, including creation, modification, and termination. The UE and the network act as client and server using a standard set of requests, which are answered by a standard set of responses. The IMS in the client and server must support the SIP preconditions framework, as specified in [5]. Through an exchange of messages, both the originating and the terminating devices are aware of any preconditions associated with a specific session and their current status. The Session Description Protocol (SDP) is utilized within the SIP message to enable the characteristics of a session and the associated media to be specified. The SDP is referred to as an Offer/Answer model in which the client proposes a set of characteristics to which the server responds. There is no guarantee that the answer contains the same characteristics as the offer.

To initiate a VoLTE call, as illustrated in Figure 3.5, the UE sends a SIP INVITE with SDP information. The SDP information carries the media QoS requirement for the audio and its source transport address. The source of the transport address information for the IMS signaling bearer can be determined from the SIP INVITE. The terminating UE responds with a SIP 100 TRYING, acknowledging that the SIP INVITE has been received successfully. Now the MT UE sends a 183-Session progress message that includes the SDP answer (to indicate the selected codec – a set of wideband or narrowband adaptive multi-rate (AMR)). The IMS server then triggers the setup of a dedicated bearer to carry the voice call payload for both MO and MT UEs. The MO UE sends a provisional acknowledgment to the MT UE via the IMS core network to confirm that

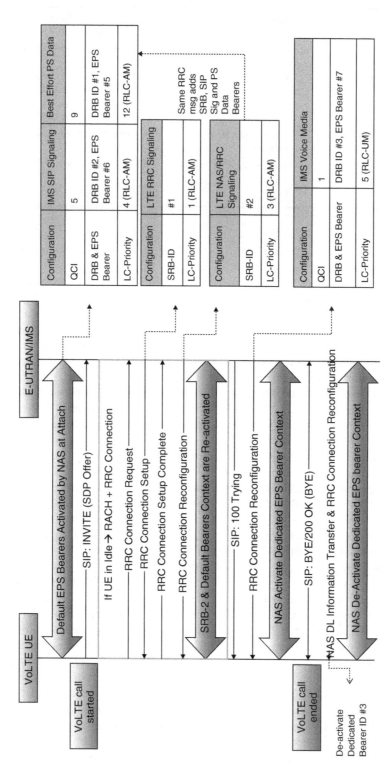

Figure 3.4 EPS bearers during a VoLTE call and at call end.

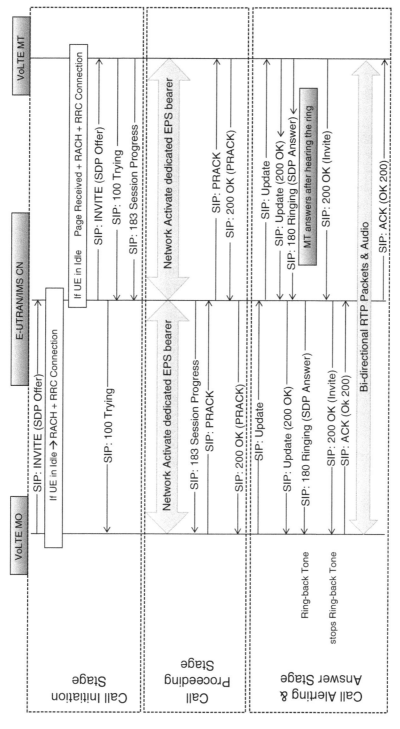

Figure 3.5 VoLTE to VoLTE call setup flow with QoS-aware devices, preconditions enabled, and network-initiated QoS.

codec selection is complete. Once the establishment of the media-dedicated bearer for the originating UE is complete, it sends a SIP UPDATE to the terminating UE to indicate that resource reservation has also been completed. Upon receipt of the SIP UPDATE, the terminating UE generates a user alert and responds with a 200 OK (for the SIP UPDATE message). The terminating UE then sends a 180-ringing SIP message to the originating UE, which triggers a ring-back tone to the originator. The terminating UE then sends a 200 OK for the original SIP INVITE and, after this point, the voice call path can be considered fully established.

The VoLTE call setup delay can be estimated from the SIP INVITE to the SIP 180-ringing message. It is important to measure the perceived delay after the user answers the call. This is because a VoLTE call can experience inactivity that triggers an RRC state transition to Idle while the call is being set up (VoLTE is a data session). Hence, we add the additional delay from when the SIP 200 OK is sent until the user receives the first downlink packet. Table 3.3 illustrates the VoLTE-related KPIs and their methods of calculation. The main reason to include the first downlink packet after call setup is to check a unique issue in VoLTE that is not there in CSFB or legacy CS calls. The VoLTE SIP messages are all IP and hence the eNB does not see the actual SIP message content. Therefore, if the MT party takes a long time to answer the call, the MO can go to Idle mode due to the expiry of the user-inactivity timer (i.e., there is no more data activity, which moves the UE from Connected to Idle mode).

3.5 VoLTE Handover and Data Interruption Time

In this section, we will revisit the VoLTE handover with and without PS data. The data interruption time will be defined in order to evaluate the impact on VoLTE services. Figure 3.6 illustrates the VoLTE handover

mechanism without a PS data session. The control-plane (C-plane) interruption times are represented in two KPIs, as follows:

Delay 1: From the RRC Connection Reconfiguration message until the RCC Connection Reconfiguration Complete. This delay impacts the interruption time.
Delay 2: From the RRC Connection Reconfiguration message until all mandatory SIBs have been decoded by the UE. This delay does not impact the interruption time but can affect the handover performance and stability, as SIBs on the target cell are needed for the UE to declare it is fully camped on the cell.

The user-plane (U-plane) interruption time is typically higher than the C-plane interruption time and starts from the last RTP packet on the source eNB to the first RTP packet on the target eNB.

3.5.1 Recommendations on Call Re-establishment

Call re-establishment is essential in VoLTE calls to avoid RTP timeouts. We should audit call re-establishment for both QCI = 1 and QCI = 9. In general, the call re-establishment procedure is the same for both QCI = 9 and QCI = 1 from air interface perspectives. From previous LTE trials, we have not seen frequent re-establishment reject on QCI = 9. Hence, investigation is recommended to see if QCI = 1 is impacted by any other issues.

It is essential for VoLTE to target quick RLF with a stable re-establishment procedure to ensure VoLTE continuity with less mute time. This can be done with RLF parameters and good RF planning, a similar concept to the legacy AMR situation in 3G, where carriers target quick RLF and stable re-establishment to avoid a high mute time and one of the callers ending the call. If call re-establishment is not stable, it is preferable to delay the RLF, as the RTP performance can still be acceptable in degrading RF conditions. Figure 3.7 summarizes the RLF recovery procedure.

Table 3.3 KPIs impacting VoLTE call setup latency.

KPI	VoLTE Stage	From Event	To Event
V1	SIP INVITE or session start to dedicated bearer setup	SIP: INVITE	Dedicated bearer established (QCI = 1)
V2	Dedicated bearer to ringing	Dedicated bearer established	SIP: 180 ringing
V3	Conversation delay after call answer	SIP: 200 OK (INVITE)	First DL packet after call setup
V4	Total call setup delay	*From* [SIP: INVITE] *to* [SIP: 180 ringing] + *From* [SIP: 200 OK] *to* [First DL packet after call setup]	

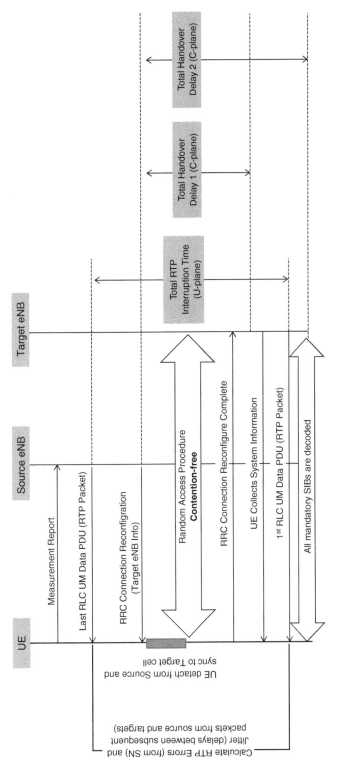

Figure 3.6 Handover procedure and VoLTE interruption time during handover.

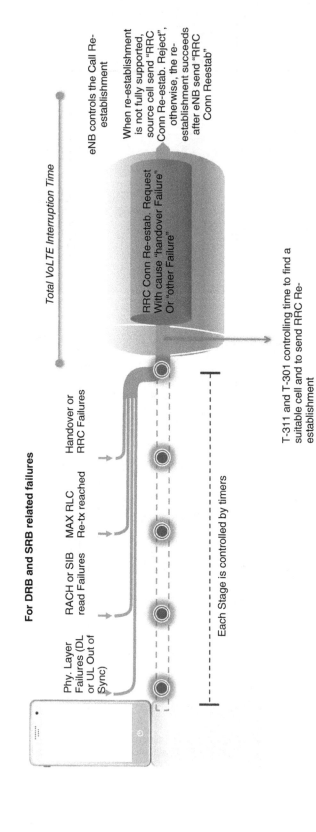

For DRB and SRB related failures

Phy. Layer Failures (DL or UL Out of Sync)

RACH or SIB read Failures

MAX RLC Re-tx reached

Handover or RRC Failures

Total VoLTE Interruption Time

eNB controls the Call Re-establishment

When re-establishment is not fully supported, source cell send "RRC Conn Re-estab. Reject", otherwise, the re-establishment succeeds after eNB send "RRC Conn Reestab"

RRC Conn Re-estab. Request With cause "handover Failure" Or "other Failure"

T-311 and T-301 controlling time to find a suitable cell and to send RRC Re-establishment

Each Stage is controlled by timers

Figure 3.7 RLF recovery procedure.

Table 3.4 Recommended settings for RLF timers for QCI = 1 and QCI = 9. Ue Timer Const (SIB2 applying to all services).

Parameter	Sample Guideline for QCI = 9	Sample Guideline for QCI = 1
T301	1000 ms	600 to 1000 ms
T310	1000 ms	1000 ms (600 ms if the call re-establishment is stable)
N310	10	5
T311	3000 ms	1000 ms
N311	1	2

Table 3.5 RLF parameters for **QCI = 1**. Enable RlfTimerConstGroup (Connected mode applying to certain QCI/service).

Parameter	Value	Comments
T311	1000 ms	Maximum time allowed for the UE to find a suitable E-UTRA cell or a cell using another RAT after radio link failure is declared.
T301	600 to 1000 ms	Typical duration between a re-establishment request and reject messages, varies between 25 ms and 70 ms.
T310	1000 ms (600 ms if the call re-establishment is stable)	Recovery timer before RLF is declared → target is to drop quickly and re-establish quickly. Can be set to lower value for VoLTE to do quicker re-establishment rather than higher mute time. It depends on how stable the re-establishment process is.
N310	5	Current setting (as in Table 3.4) prolongs RLF declaration. RTP losses and timeouts (typical RTP link aliveness timer is 10–20 seconds), can be reduced if RLF is declared sooner and re-establishment is stable.
N311	2	Current setting (as in Table 3.4) makes RLF recovery easy. If re-establishment is unstable, this value can be set as 1 to delay RLF, as RTP interruption may be higher due to the UE going to Idle if re-establishment does not succeed.

As per the 3GPP specifications, we can set different RLF parameters for both QCI = 1 and QCI = 9. RLF parameters (T301, T310, N310, T311, N311), according to the 3GPP (36.331), can be set in such a way that allows distinction to be made between the SIB2 RLF timers and the Connected mode RLF timer which can be set per QCI. New IE RLF-timers and constants, including RLF-related timers and counters, have been introduced, and it is possible for these to be set per UE. The IE is to be signaled (included) within a Radio-Resource-Config.-Dedicated message. Table 3.4 provides recommended parameters for QCI = 1 and QCI = 9. The QCI = 9 parameters are set well to delay RLF in the LTE for QCI = 9. However, for VoLTE-related services, it may cause a high RTP interruption and hence RTP timeout or longer mute period (one-way audio). More specifically, the QCI = 9 timers are designed to delay the RLF as long as possible to improve call drop rate KPIs but may not be suitable for VoLTE traffic. Therefore, we can keep QCI = 9 parameters unchanged and enable the Rlf Timer Const Group to differentiate RLF timers for QCI = 1, as shown in Table 3.5.

3.6 Single Radio Voice Call Continuity (SRVCC)

LTE is based completely on the PS domain; therefore, voice services are provided using a VoIP platform such as the IMS. Most of the LTE introduction strategy involves gradual expansion based on traffic load and capacity forecasts. Therefore, if LTE coverage is not available, a handover is necessary to 2G/3G CS networks to maintain the voice call. This entails a handover of a PS-based VoIP call to a CS-based voice call in a 2G/3G network. 3GPP introduced the SRVCC feature to support seamless handovers between PS VoIP calls and CS voice calls [14]. SRVCC combines optimized handovers defined between LTE and legacy 2G/3G networks, and VCC is

defined in the IMS core network [15]. SRVCC requires UE to support the ability to transmit and receive on two networks (PS-based and CS-based) simultaneously. SRVCC went through different stages in the standard in order to reduce the voice interruption time that impacts the user experience and improve the call setup success rate at different stages of the VoLTE call. As illustrated in Figure 3.8, 3GPP started to support SRVCC in Release 8/9 and then enhanced the mechanism to support enhanced SRVCC (eSRVCC) in Release 10 [16]. The main target of eSRVCC is to reduce the voice interruption during the inter-technology handover. eSRVCC targets an interruption of < 300 ms. Most LTE networks will introduce VoLTE over Release 10 devices only to offer better user experience using eSRVCC. This is controlled from the provisioning and handset perspectives. Therefore, in this analysis we will assess eSRVCC only.

An important SRVCC feature enabling PS to CS SRVCC access transfer of a call in the alerting phase, referred to as aSRVCC, has been introduced [17]. As illustrated in Figure 3.8, aSRVCC refers to a SIP session for which existing dialogs created by the SIP INVITE request initiating the session are early dialogs and the final SIP response has not been received yet, while a SIP 180 (ringing) response has already been received in existing early dialogs [16]. In addition, SRVCC procedures for both video calls (vSRVCC) and reverse SRVCC from UTRAN/GERAN systems to an E-UTRAN system (rSRVCC) are introduced in Release 11, as indicated in Figure 3.8.

Release 10 specifies support for PS to CS SRVCC access transfer of a call in the alerting phase [16]. However, this release does not provide support for PS to CS SRVCC access transfer of a call in the pre-alerting phase (i.e. after the SRVCC UE has received a SIP 183 (session progress) response containing the SDP answer and before the SIP 180 (ringing) response has been received). If the UE receives early media or announcements from the network in the pre-alerting phase, the network will not perform SRVCC of the IMS session in the pre-alerting phase, and this will impact the user experience. To address this case, PS to CS SRVCC of the originating call in the pre-alerting phase has been specified in Release 12 [17, 18]; this is referred to as bSRVCC, as shown in Figure 3.8.

Figure 3.9 illustrates the eSRVCC signaling procedure. Figure 3.10 demonstrates the handover mechanism in eSRVCC. The main idea of the eSRVCC scheme is to anchor a VoIP call (media session) on an IMS network element near the local end. This way, only the branch of the anchor point to the local end needs to be modified without the need for modifications on the remote end, and thus the handover delay can be shortened during the session transfer. In order to support eSRVCC, two new IMS entities have been introduced to anchor the media session: the Access Transfer Control Function (ATCF) and the Access Transfer Gateway (ATGW). Adopting the ATGW avoids path switching between the LTE and 2G/3G media gateways for the terminating UE when the IMS originating UE is transferring the IMS VoLTE call from PS to CS. From the LTE signaling point of view, eSRVCC works in a manner similar to the PSHO call flow described in Figure 3.3. The PSHO steps from 4 to 14 in Figure 3.3 also apply on the eSRVCC call flow during the handover from LTE to UMTS while the VoLTE call is active.

The call flow of aSRVCC in the alerting phase with QCI = 1 is demonstrated in Figure 3.11a. This phase refers to a SIP session for which all possible existing dialogs created by the SIP INVITE request initiating the session are early dialogs, for which no final SIP response has been received yet and for which a SIP 180 (ringing) response has already been received in existing early dialogs. The call flow of bSRVCC in the pre-alerting phase with QCI = 1 is illustrated in Figure 3.11b. This phase refers to a SIP session for which all possible existing dialogs created by the SIP INVITE request initiating the session are early dialogs, for which no final SIP response has been received yet and for which a SIP

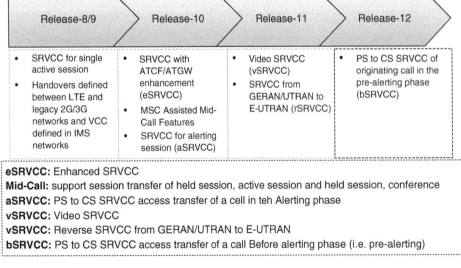

Figure 3.8 SRVCC standard evolution in 3GPP.

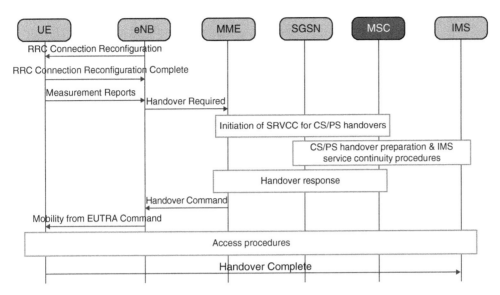

Figure 3.9 eSRVCC signaling procedure.

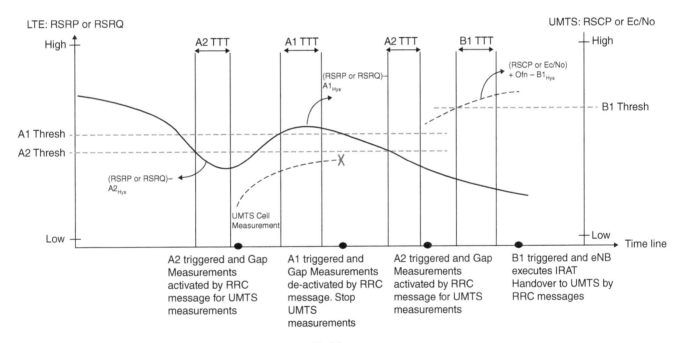

Figure 3.10 Handover mechanism and relevant parameters in eSRVCC.

180 (ringing) response has not been received yet in any existing early dialogs.

The SRVCC scenario in the pre-alerting phase *without* QCI = 1 establishment is demonstrated in Figure 3.11c. This scenario refers to a SIP session for which all possible existing dialogs created by the SIP INVITE request initiating the session are early dialogs for which no final SIP response has yet been received and neither a SIP 180 (ringing) response nor a SIP 183 (session progress) has been received in any existing early dialogs; QCI = 1 is not yet established either. For many vendors, in this case if B1/B2 is triggered, redirection is initiated, as this is

considered a PS data call from the eNB side. Without the establishment of QCI = 1 and with early SIP INVITE only, the eNB treats the call as a PS call and can trigger redirection as a response to B1 or B2.

3.7 Performance Analysis

In this section, the CSFB and VoLTE performance are assessed from two perspectives, as follows:

Call setup performance. The focus is on the MO side where the user can perceive the main delays in call

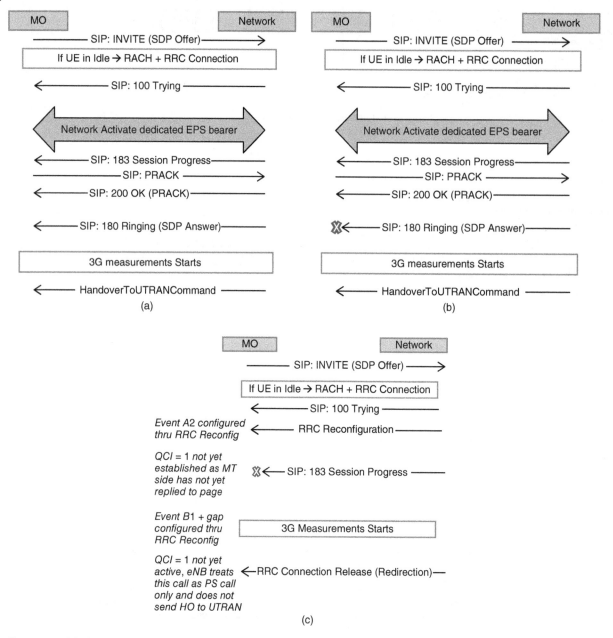

Figure 3.11 (a) aSRVCC in the alerting phase with QCI = 1; (b) bSRVCC in the pre-alerting phase with QCI = 1; (c) bSRVCC in the pre-alerting phase without QCI = 1.

establishment. Therefore, mobile-to-mobile (M2M) calls, either CSFB to CSFB or VoLTE to VoLTE, are assessed in this section. A mobile-to-mobile (M2M) call test case has been chosen to provide an end-to-end call setup delay that closely matches the user's experience with high-end smartphones.

VoLTE in-call mobility performance when LTE coverage degrades and a VoIP call is active. CSFB in-call performance assessment is not discussed in this section because the voice call falls back to UMTS and the performance afterwards becomes similar to that in legacy UMTS-only voice calls.

The data were processed from field measurements with a large sample size (for example, 100 CSFB/VoLTE calls for each scenario) and results were averaged over two different LTE access networks from two different suppliers and two smartphones with two different chipsets. The performance testing was conducted in collocated LTE/HSPA commercial networks in the UAE (i.e., the same physical sites were used). The LTE was deployed with 1800 MHz (20 MHz channel) and the UMTS with 2100 MHz (2 × 5MHz). The different CSFB scenarios listed in Table 3.1 were tested in the same exact cluster after modifying the scenario-relevant parameters (i.e.,

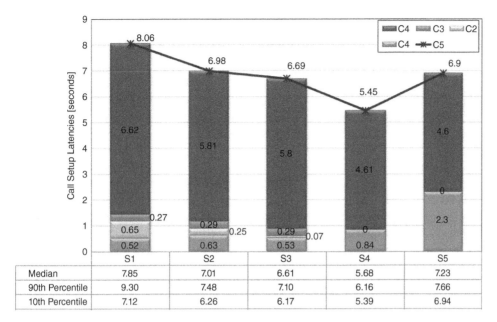

Figure 3.12 CSFB call setup delay performance in mobility – MO side.

changes were made in the RNC for 3G-related parameters and in the eNB for LTE-related parameters like SIB tunneling). The VoLTE testing was conducted in the same cluster with the same device supporting all CSFB strategies and VoLTE (Release 10 device). The KPIs were derived from the device side through post-processing scripts to the collected logs from the device modem.

3.7.1 CSFB Performance Analysis

Typically, several challenges are raised when CSFB is deployed in a network:

CSFB call setup delay
CSFB setup success rate
Data packet delay during fallback
Speed of camping back on LTE after the voice call is
  released.

The third factor is of slightly less importance because the smartphone user receiving or making a CS call is not expected to pay attention to the data session being partially interrupted in the background during the fallback process. Typically, the fourth factor is driven by the deployment strategy in the network: either camp back on LTE using the cell reselection procedure or use a network-based fast return to LTE. In this section, we benchmark the first and fourth factors and discuss the aspects related to improving the third factor.

As shown from the CSFB call flow in Figure 3.2, the process of CSFB can involve different stages, where each contributes exponentially to the call setup delay. Unlike the call setup performed in UMTS-only calls, the fallback mechanism requires steps that add extra delays to the

call setup. Therefore, it is quite challenging to ensure the same user experience between CSFB and legacy CS voice calls. Figure 3.12 provides the average along with the median, 90th and 10th percentiles for CSFB KPIs defined in Table 3.2 for the different scenarios defined in Table 3.1.

Figure 3.12 demonstrates that the fourth scenario (S4) (i.e., PSHO without measurements) experiences the lowest call setup delay. The redirection with SIB tunneling scenario (S3) is ranked second after S4. It can also be noted that the performance of the PSHO with measurement scenario (S5) is worse than the S4 scenario. The PSHO without measurements scenario performs the best in terms of call setup delay due to factors such as the UE not decoding the SIBs as well as notable reductions in NAS end-to-end signaling between the UE and the core network when establishing the CS call. The UE is not required to decode the SIBs because the UE directly enters the 3G dedicated channel in Connected mode (Cell_DCH state) and hence the cell parameters can be conveyed to the UE later through dedicated RRC messages. The NAS end-to-end delay in PSHO is generally lower than other scenarios because the UE is assigned the PS data bearer in the handover message itself and this reduces the delay in establishing the PS bearer afterwards. The time observed between ESR and handover command in S5 is higher than the blind redirection/handover cases. This is related to the need to configure the UE with 3G neighbors and IRAT measurements prior to the handover preparation by the eNB, which is not required in the blind redirection/handover scenarios. In all cases, the average time to tune to UTRAN after the redirection or PS handover command is about ~ 45 ms.

The blind PSHO mechanism can impact the call setup stability and, accordingly, the CSFB call setup success rate. Specifically, call setup failures can happen in a fast-changing RF environment or if the neighbor cell targeted for a blind handover is loaded (i.e. low EcNo ≤ −18 dB with good received signal code power (RSCP) ≥ −90 dBm). Therefore, if this strategy is adopted, careful optimization of the neighbor relationship is needed; typically, LTE and UMTS are deployed in different bands (such as LTE 1800 MHz and UMTS 2100 MHz). This may lead to different RF propagation characteristics at the cell edge. In this scenario, a neighbor relationship becomes harder to maintain when the reference signal received power (RSRP) of LTE 1800 at the cell edge is better than the collocated UMTS 2100 RSCP [3, 19].

On the other hand, redirection with SIB tunneling (S3) reduces the call setup delay because the SIBs are broadcast to the UE through the RRC connection release. As soon as the UE goes into 3G Idle mode, it acquires the SIBs directly from the RRC message and hence the UE only needs to search for the suitable cell/frequency indicated by the RRC release message, then it performs the cell selection procedure. The same concern about call stability needs to be considered here. Similar to PSHO without measurements, SIB tunneling CSFB works in such a way that multiple neighbors (cell ID and carrier frequency) must be defined in the RRC release message. Once the UE falls down to 3G Idle mode on the specified carrier, the UE selects the best cell from its own search and then looks to see if the best cell was part of the neighbor list in RRC redirection in order to skip those SIBs. If the best cell is not on this list, the UE is required to read all of the SIBs and proceed with the CSFB call setup. Hence, if the neighbor relationship is not optimized carefully, the SIB tunneling mechanism may potentially produce similar call setup delays to Release 8 CSFB strategies [3].

The redirection procedure with basic functions (S1) performs the worst, because the UE decodes all of the mandatory SIBs. However, the same strategy with enhanced SIB scheduling (S2) improves the SIB decoding stage. This is because the SIB periodicities of those mandatory SIBs (i.e. SIB1, SIB3, SIB5, and SIB7) are scheduled with lower intervals. In this scenario, the SIB periodicity of SIB3 is reduced from 1280 ms to 160 ms, SIB1/5 periods from 640 ms to 320 ms, while SIB7 remains at 160 ms. As a result, the UE quickly acquires those SIBs and then moves directly to the RRC connection setup stage. Scenarios S1 and S2 can be more stable in terms of CSFB success rate (compared to S3, S4, and S5) because the UE typically selects the best cell out of the explored 512 scrambling codes when the specific

cell is not defined for the UE within the RRC release message (only the frequency carrier is specified).

The PSHO with measurements (S5) in this study was exercised with three pre-configured UMTS frequency carriers, with a total of 32 UMTS neighbors on each carrier. This effectively means that the device needs to measure all neighbor cells before reporting the measurement values to the eNB as part of even B1 evaluation (i.e. the inter-RAT neighbor becomes better than a configurable threshold). This adds a significant delay to the stage between the ESR and the handover to 3G command. On average, each carrier with 32 neighbors would require ~ 600 ms to complete the measurement phase and then start the evaluation of event B1 to update the eNB with regard to the measured 3G cell signal level that assists in triggering the handover to the strongest cell. With a number of 3G neighbors ≤ 10, the UE can complete the measurements within ~ 300 ms while camped on LTE. Therefore, if the PSHO with measurement strategy (S5) is chosen for CSFB deployment, it is essential to optimize the neighbor list and the number of frequency carriers to be measured within the LTE gap. This strategy impacts not only on the CSFB setup delay but also the VoLTE eSRVCC delay, discussed later in this section. eSRVCC follows a similar mechanism to PSHO with measurements. Therefore, applying an optimization on the neighbor list and speeding up the measurements can improve the CSFB setup latency as well as the VoLTE eSRVCC mechanisms.

Once the CSFB call has been terminated, it is important to measure how quickly the UE camps back on the LTE system. Several methods can be utilized to camp back on LTE after CSFB call release:

1. Using the cell reselection mechanism
2. Using a fast return to LTE.

Cell reselection alone does not guarantee a fast return to LTE. This is because the UE may not move directly to Idle/paging channel (Cell_PCH) states where the reselection can be initiated. Therefore, a fast return to LTE (FRTL) allows the RNC to redirect the UE from UMTS to LTE after a CSFB call is released. In this mechanism, the RNC sends an RRC Connection Release message with the redirection information to the LTE network. The FRTL can be deployed with measurements or as a blind redirection. The most common mechanism deployed now is the blind redirection due to implementation simplicity.

Table 3.6 illustrates the time delay from when a CSFB call is released until the UE camps successfully back on LTE. The delays shown in the table are estimated with the common methods described above. These methods apply to any CSFB regardless of which strategy is used (redirection or PSHO). The delays in this table are calculated starting from when the UE releases the CSFB call

Table 3.6 Camping on LTE after CSFB call release.

Strategy	Average Delay
Cell reselection from UMTS to LTE	10.10 seconds
Network-based fast return to LTE (with measurements)	5.71 seconds
Network-based fast return to LTE (blind)	1.65 seconds

until the tracking area update is completed between the UE and the EPC.

The results in Table 3.6 indicate that the fast return method with blind redirection provides a faster way to camp back on LTE, which positively improves the end-user experience. The fast return method with measurements experiences higher delay in camping back on LTE because the UE needs sufficient time to measure the LTE cells during the 3G compressed mode before reporting the best cell to trigger a redirection to LTE. The drawback of using the blind redirection to LTE is that the UE may remain unreachable if it tries to camp on the LTE network while the LTE coverage is weak. However, this can be handled if a fast return is triggered by the RNC where there is a neighbor relationship with the current serving UMTS cell. Additionally, there are some protection mechanisms defined in [20] that allow the UE to camp back on UMTS within a certain timer if no suitable cell is found on LTE.

To conclude the observations in this section, the choice of a specific CSFB deployment strategy is vital. While each strategy has its own pros and cons, a mix of strategies is considered the recommended option. We should consider this option, since not all devices support PSHO. Therefore, CSFB with redirection should still be considered an option. When this strategy is deployed, it is preferred to follow the implementation of reducing SIB periodicity. Even in networks with a majority of PSHO-based devices, a mix of redirection and PSHO is still necessary. Typically, this is needed in the case when a CSFB call is triggered without PS activity. In this scenario, it is important to segregate the usage of PSHO from redirection, depending on whether the PS data are active or not (i.e. can be indicated simply if CSFB is initiated from Idle or Connected mode). This is to avoid establishing the PS RAB unnecessarily, once falling back to UMTS when PS data are not present. This implementation allows a more stable call setup and in-call performance by avoiding a 3G multi-RAB call (simultaneous CS + PS), which reduces the footprint of CS coverage. Additionally, redirection can be used for exceptional handling of anomalies to the PSHO fallback. In this scenario, the UE can be notified to do a redirection if the PSHO is not triggered. More specifically, if

the UE does not report a suitable UMTS cell during the gap measurements within a specified duration, the eNB can then autonomously redirect the UE to a pre-defined RAT and RF frequency. Once this timer has expired and there is no cell reported by the UE (due to the lack of 3G coverage, for example), the eNB can trigger a redirection procedure to another RAT (e.g. GSM). This, however, means an additional delay incurred by a user bounded by this timer threshold.

3.7.2 VoLTE Call Setup Performance Analysis

As discussed in the VoLTE call flow in Figure 3.4, VoLTE call setup involves different stages where each stage contributes to the call setup latency. Since a VoLTE call will remain on the LTE network, it is important to assess the call setup under different radio conditions. Three scenarios were considered to analyze the VoLTE call setup delay: near cell stationary, far cell stationary, and the mobility scenario. The near-cell condition was performed with Reference Signal Received Power (RSRP) of around −70 dBm (SINR = 22), and the far-cell condition was conducted with RSRP of around −110 dBm (SINR = 8). The mobility test was conducted in the same cluster of CSFB and with a typically loaded commercial LTE network where average RSRP = −81dBm and average SINR = 18.

Figure 3.13 summarizes the average call setup delay KPIs for VoLTE call setup in different RF conditions using the KPIs explained in Table 3.3, focusing on VoLTE-to-VoLTE call setup. The data were calculated from field measurements with a large sample size (100 calls for each RF scenario). Compared to the CSFB strategy, the call setup delay for VoLTE calls is lower than CSFB in all RF condition scenarios. This is to be expected, since the VoLTE call setup is conducted within the same radio access network and there is no need to fall back to UMTS at the call setup stage. Additionally, the signaling speed in LTE on the radio interface is faster than that in 3G and fewer signaling messages are needed to establish the call.

In the far cell stationary and mobility scenarios, the call setup delays are mainly observed when the MT side misses the paging message. The maximum observed VoLTE call setup delay when the MT misses the first paging attempt is 11.5 seconds. If the MT is in LTE Idle mode and the MO side initiates a call, the EPC pages the MT for an incoming call. If the MT misses the first page due to RF conditions, as in far cell or mobility scenarios, the MO perceives a higher setup delay at the stage V1, defined in Table 3.3. Typically, the MO will not be assigned a dedicated bearer with QCI = 1 unless the MT starts establishing the call and sends SIP: 183-Session Progress. Based on the paging repetition

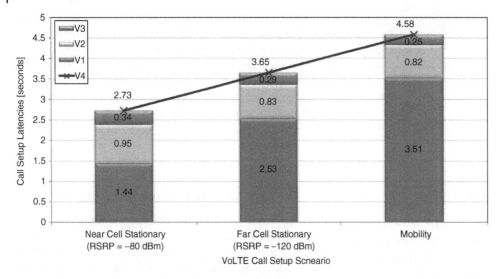

Figure 3.13 VoLTE call setup delay performance – MO side.

timer in the core network as well as the Idle discontinuous reception (DRX) cycle length [21], the re-paging may impose an additional time that contributes to the total call setup delay. The re-paging mechanism and the interval between each paging attempt are important for reducing this delay. Since a VoLTE call is treated as a PS call, the network parameters typically need to relax the re-paging mechanism to save the paging resources. This is beneficial for the paging dimensioning of a PS call, however, it can increase the VoLTE call setup delay, as explained. Therefore, the re-tuning of the paging mechanism is recommended to ensure a good VoLTE call setup delay in weak RF conditions. In an LTE network with VoLTE support, the re-paging mechanism can be increased to three re-paging attempts with an interval of 3–4 seconds between each attempt.

Another factor that impacts the call setup delay is the effect of the inactivity timer settings [21]. The inactivity timer controls how long the UE is allowed to remain in LTE Connected mode without any data activity. If data activity stops within this timer, the eNB requests the UE to move to Idle mode to save resources and reduce the device battery consumption. This timer is typically chosen to be 5–10 seconds in data-centric LTE network deployment, i.e., only data with voice via CSFB. Once VoLTE is deployed, the setting of this timer impacts the overall VoLTE call setup delay. The inactivity timer setting can lead to higher call setup delays that are perceived by both the MT and MO sides. During a VoLTE call initiation, the inactivity timer can kick in if there is no SIP activity for the duration of this timer. For example, if the MT takes a few extra seconds to establish the VoLTE call (i.e. the call is not immediately answered by the MT user), no SIP response is sent to the IMS to be forwarded to the MO. As a result, the MO will not detect any data activity and, accordingly, the eNB moves the MO into

Idle mode while the call is being established. Once the SIP response becomes available, the EPC pages the MO device again as it is in Idle mode. Therefore, moving the MO to Idle mode during a SIP inactivity period leads to extra paging to transmit SIP responses and establish the VoLTE call. Moreover, there is a risk that the MO may miss the paging message if it is experiencing weak RF conditions, which can also cause call setup failures; this is a drawback that needs to be addressed by the industry, as the MO should remain active while the VoLTE call is being established. A workaround solution may be implemented from the UE side but it would be preferable to have a consistent feature from 3GPP.

We observed an average of 1.3 seconds extra VoLTE call setup delay due to the impact of the inactivity timer. Therefore, we can enable the Connected-state DRX (C-DRX) feature and increase the inactivity timer. This feature allows the UE to stay in LTE Connected mode for a longer period and, at the same time, improves the device battery consumption without the need to push the UE directly to Idle mode for a short period of data inactivity [21]. Nonetheless, re-tuning the C-DRX parameters is required for VoLTE services to avoid additional inter-packet delay and increased jitter. Table 3.7 provides the recommended C-DRX parameters for VoLTE services (QCI = 1) and PS data services (QCI = 9) [21]. The C-DRX parameters for QCI = 9 remain the same as in [21] where a practical methodology is used to estimate the optimal C-DRX parameters that extend the battery life with slightly higher packet latencies. On the other hand, the VoLTE C-DRX parameters recommended in Table 3.5 are estimated to reduce the latencies and therefore improve VoIP audio performance. A detailed analysis of C-DRX parameter tuning for PS services can be found in [21]. A similar practical

Table 3.7 C-DRX parameters for VoLTE and PS data.

C-DRX Parameters	VoLTE (QCI = 1)	PS Data (QCI = 9)
Default paging cycle (s)	1.28	1.28
User inactivity timer (s)	30	20
On duration timer (ms)	≤ 5	10
DRX inactivity timer (ms)	≤ 10	200
DRX retransmission timer (ms)	4 to 8	2
Long DRX cycle (ms)	40	320
Short-cycle DRX-supported Indication	OFF	ON
Short DRX cycle (ms)	OFF	20
DRX short cycle timer	OFF	2

procedure is exploited to obtain the optimal C-DRX parameters for VoLTE services with QCI = 1.

As indicated in Table 3.7, the C-DRX mechanism for VoLTE is deployed with a long DRX cycle of 40 ms, which is appropriate for the VoLTE voice coder (vocoder) packets that are typically generated over an interval of 20 ms. During talk spurts, the incoming VoIP audio frames are typically placed in a buffer (a de-jitter buffer) and then processed by the IMS stack, ensuring a minimum perceived delay to the end user (i.e. a mouth-to-ear delay). Even though the vocoder packets flow every 20 ms, a long DRX cycle of 40 ms is still suitable for the end-to-end audio delay where two packets can be bundled over 40 ms and processed in the same sequence without impacting the end-user experience. Both PS data and VoLTE C-DRX parameter settings are configured to co-exist in the same eNB. More specifically, if the ongoing call is data-only (only the QCI = 9 service is activated), then the C-DRX parameters for this service can be sent by the eNB to the UE. Once the VoLTE call

is initiated with QCI = 1, the C-DRX parameters for this service are re-sent to the UE through RRC messages. The process of switching between different C-DRX parameters is dynamic and adopted by 3GPP [22]. Typically, there are multiple active bearers and the C-DRX parameters related to the QCI with highest priority (QCI = 1 in this case) are selected and sent to the UE.

The last experiment was performed to benchmark the VoLTE-to-3G and 3G-to-3G call setup latencies and compare them with VoLTE to VoLTE, as demonstrated in Figure 3.13. Testing devices were located in the same location during testing and call setup delays were calculated at different RSRP/RSCP (Receive Signal Code Power) values from the near cell (i.e. −80dB) to the edge of coverage (i.e. −120dB). As shown in Figure 3.14, the VoLTE-to-3G call setup latency is higher than that in VoLTE to VoLTE, even in the near-cell condition, but better than the CSFB call setup delay. A VoLTE-to-3G call can experience higher delays due to the CS part of the call. In general, VoLTE-to-VoLTE and VoLTE-to-3G calls, shown in Figure 3.13 and Figure 3.14, respectively, confirm the VoLTE introduction's added value in terms of better call setup delay, especially when compared with the 3G-to-3G voice calls under different RF conditions. It is interesting to see that the call setup delay starts to increase significantly when RSRP decreases beyond −110 dBm. Therefore, it is recommended to design the link budget for VoLTE services to be within an RSRP of −110 dB to ensure minimal call setup performance.

3.7.3 VoLTE Handover Performance Analysis

We initially demonstrate the performance of the VoLTE handover without PS data session. Figure 3.15 and Figure 3.16 demonstrate the CDF and PDF for Delay 1 and Delay 2, respectively. Delay 1 and Delay 2 are illustrated in Figure 3.6 for a typical LTE network on the

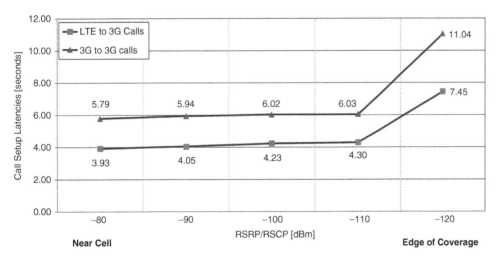

Figure 3.14 Call setup delay performance for VoLTE to 3G and 3G to 3G – MO side.

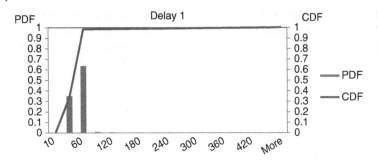

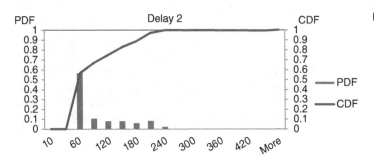

Figure 3.15 Handover execution time up to RRC Complete [ms].

Figure 3.16 Handover execution time up to SIB read [ms].

Table 3.8 Summary of Delay 1 and Delay 2 for C-plane without PS data.

KPI	Delay #1	Delay #2
Total # of handovers		410
Average/Min/ Max [ms]	36.4/19.0/923.0	92.9/31.0/944.0
Observations	Good values in general.	Good values, but higher than Delay #1 depending on SIB read time to complete the handover.

1800 MHz band. Table 3.8 summarizes the results in Figures 3.15 and 3.16. Delay 1 and Delay 2 represent the C-plane handover performance.

The U-plane is evaluated in terms of RTP interruption time and the jitter during the handover. The PDFs for U-plane RTP interruption time and the jitter during handover are illustrated in Figures 3.17 and 3.18. The results are obtained from a live LTE network. Table 3.9 provides a summary in terms of average, maximum, and minimum RTP interruption time and jitter during the handover.

The DL interruption time is calculated from RTP packets received, as RTP packets are decoded when a DL grant is assigned for a PDSCH transmission opportunity. Jitter during handover is the inter-arrival delay between consecutive packets from the last on the source and the first on the target for RTPs without losses. The handover execution time is within expectations and is similar to the PS best-effort performance. The VoLTE interruption time is acceptable in this exercise and is similar to the PS best-effort interruption, which is typically in the range of 60–70 ms. A typical U-plane interruption time with a contention-free RACH procedure is higher than that for C-plane delays by approximately 10–20 ms; contention-free RACH delays contribute to approximately 18 ms of delay. There are a few cases with higher than usual RTP delays on the DL. A typical cause of

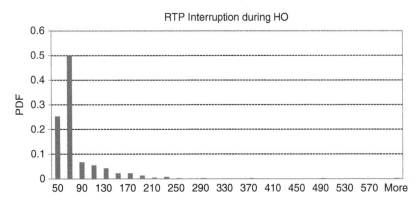

Figure 3.17 RTP interruption [ms].

Figure 3.18 Jitter during the handover [ms].

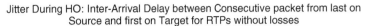

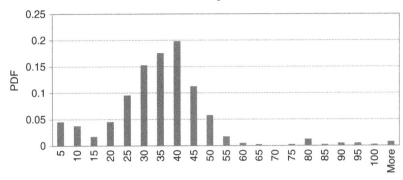

Jitter During HO: Inter-Arrival Delay between Consecutive packet from last on Source and first on Target for RTPs without losses

Table 3.9 Summary of U-plane performance without PS data.

KPI	# RTP Errors During HO	RTP Interruption Delays [ms]	Relative DL Jitter [ms]
Average/Min/Max	0.33/0/21	73.8/10.0/942.0	35.6/2.0/922.0
Observations	Up to 21 RTP packets lost during HO. Median = 0 as most cases did not observe RTP errors.	Values are similar to data interruption time of PS calls.	Time difference between jitter and RTP interruption on average = 38 ms. This means that there is ∼ 38 ms delay in DL scheduling during HO.

Figure 3.19 UL data interruption [ms].

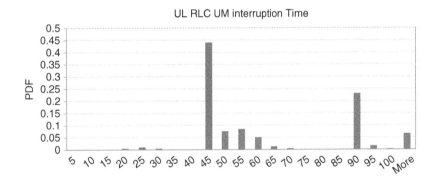

UL RLC UM interruption Time

higher than expected U-plane delay is areas of low SINR during handover.

The UL RTP performance during handovers is demonstrated in Figure 3.19 and the summary is provided in Table 3.10. The factors that may impact the handover delay are summarized in Table 3.11.

As handover execution time, on average, is 36.4 ms then, excluding this time, the new serving cell delays the UL grant needed to send the UL RTP packet by

Table 3.10 Summary of UL interruption time without PS data.

KPI	UL RTP Interruption
Average/Min/Max [ms]	64.1/16.0/225.0

approximately 28 ms on average, after the handover is completed. This delay also contributes to the relative jitter at the receiving end; however, this is an acceptable delay given that the UE would send UL RTP packets in intervals of 20 ms. After handover, it is observed that, in general, the new eNB makes the UL grant available in consecutive subframes, however, with low TBS size (typically 18 bytes), which requires the UE to send one complete RTP packet in multiple TBSs in consecutive subframes. Note that the UL interruption time is calculated from the RLC, as RLC PDUs for RTP packets are only sent when the UL grant is assigned for a PUSCH transmission opportunity.

The VoLTE handover performance with PS data is summarized in Tables 3.12 to 3.14. These results correspond to the results in Tables 3.8 to 3.10 that represent VoLTE

Table 3.11 Factors impacting handover delay.

Factors Increasing HO Delays	Delayed Handover	Core Delays	PDSCH BLER in Source Cell	RACH	PDSCH/PUSCH BLER in Target Cell	Delayed Scheduler
U-plane delay	Weak RF coverage on serving cell and neighbor cells. Handover parameters.	Delays from IMS or LTE core in delivering packets on DL.	Downlink RRC Handover message retransmissions or high BLER on source cell data packets.	RACH preamble retransmissions	Uplink RRC Handover Complete message retransmissions or high BLER on target cell DL data packets.	Delay in scheduling DL packets during talk spurts within handover region.

Table 3.12 Summary of Delay 1 and Delay 2 for C-plane with PS data.

Handover Execution Time (C-plane)		
Test	Delay #1	Delay #2
Without PS data (Average/Min/Max) [ms]	36.4/19.0/923.0	92.9/31.0/944.0
With PS data (Average/Min/Max) [ms]	32.7/26.0/137.0	118.4/47.0/390.0

Table 3.13 Summary of U-plane performance with PS data.

DL Data Interruption Time (U-plane)			
Test	# RTP Errors During HO	RTP Interruption Delays [ms]	Relative DL Jitter [ms]
Without PS data (Average/Min/Max) [ms]	0.33/0/21	73.8/10.0/942.0	35.6/2.0/922.0
With PS data (Average/Min/Max) [ms]	0.40/0/15	68.8/20.0/456.0	31.6/1.0/136.0

Table 3.14 Summary of UL interruption time with PS data.

UL Data Interruption Time (U-plane)	
Test	UL RTP Interruption
Without PS data (Average/Min/Max) [ms]	64.1/16.0/225.0
With PS data (Average/Min/Max) [ms]	71.1/16.0/225.0

handover performance without PS data. As is evident from these tables, there is no visible impact on VoLTE performance during the handover stage when PS data are active.

3.7.4 eSRVCC Performance Analysis

To benchmark the overall eSRVCC performance, we start by evaluating the voice interruption during LTE intra-frequency handover. LTE intra-frequency handover occurs when VoLTE UE is moving between different cells within the same network inside the LTE coverage area. This is calculated from the last downlink/uplink voice packet received on the source LTE cell to the first downlink/uplink voice packet received on the target LTE cell. This KPI is calculated for LTE cells deployed with the X2 interface [3].

The eSRVCC voice interruption time is calculated from the last downlink/uplink voice packet received on the source LTE cell to the first downlink/uplink voice packet received on the target UTRAN cell.

On the other side, we evaluate the impact of the signaling delay on the overall voice packet latency. The signaling delay in LTE intra-frequency handover is calculated from the handover command received by the UE (i.e., the RRC Reconfiguration message) until the UE sends the handover complete (i.e., the RRC Reconfiguration Complete message). The eSRVCC signaling delay is derived from the handover command (i.e., the Mobility From E-UTRA command) until the handover complete is sent by the UE.

Figure 3.20 illustrates the UL/DL voice interruption and signaling delay in mobility conditions for both eSRVCC to UTRAN and LTE intra-frequency handover for two different carriers. The results were obtained from field testing with voice activity of 100%. To measure the interruption time accurately, we generated 100% RTP packets where each RTP packet corresponded to a 20 ms voice frame. We connected the testing device to a source of voice loop-back to generate UL and DL RTP packets continuously. Using this setup, we ensured that the measured interruption was accurate. Otherwise, the measured values may have been inaccurate with a typical voice activity factor of approximately 50%. In this scenario and during the intra-frequency handover or eSRVCC, the voice interruption delays may include

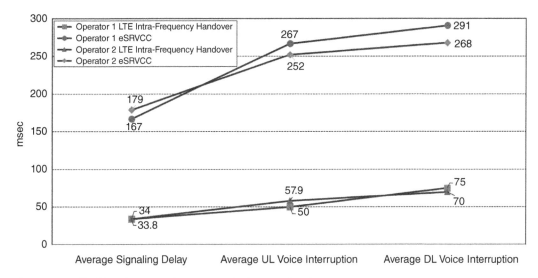

Figure 3.20 eSRVCC and LTE intra-frequency voice interruption time.

silent indicators (generated every 160 ms). The target was to see if the voice interruption during eSRVCC (IRAT handover) is within the 3GPP requirements (i.e., 300 ms) and also to compare it with the voice interruption during LTE intra-frequency handovers.

As shown in Figure 3.20, it is obvious that the eSRVCC experiences higher voice interruption and signaling delay than the LTE intra-frequency handover case. This is due to the change in radio access and the extra requirements of the routing of the voice packets within different core networks. However, it can be observed that the voice interruption fits within the expected 300 ms delay [16] for both carriers. The voice interruption of 300 ms is within the accepted range for mouth-to-ear delay, indicating an overall user satisfaction with the conversational quality.

eSRVCC requires the same level of optimization discussed for PSHO CSFB in order to improve its performance. Reducing the time needed to measure 3G cells in LTE can improve the conversational quality, as it will expedite the handover while the LTE coverage is degrading. However, the major concern becomes the reliability of the eSRVCC handover. A quick handover to 3G improves the time taken to leave degrading LTE coverage, but it must also ensure that the 3G cell quality is acceptable in order to proceed with the voice call until another handover occurs in 3G. This requires careful handover optimization at both ends: in LTE prior to the handover and in 3G after the handover.

We can observe a difference in the eSRVCC performance between the two carriers in Figure 3.20 while the LTE intra-frequency handover is very much similar. Carrier 2 experiences a higher signaling delay (still within an acceptable range) but better voice interruption on the UL and DL. This indicates that the network

can be optimized from the EPS network side (MME to eNB to UE) to handle signaling delays but the IMS is still a key factor in voice interruption on the media side (not signaling only impacts the interruption time). From the signaling side, Carrier 1 is optimized to trigger inter-RAT handover with less delay (better parameters and less delay between the MME and the eNB). The IMS in Carrier 2 seems to be better optimized in terms of the delays related to PS–CS path switching compared to Carrier 1, even though the EPS side of Carrier 1 is better optimized compared to Carrier 2. Therefore, it is important to handle and optimize voice interruption from the EPS and IMS perspective.

It has been observed from field testing that if the 3G network maps SRB over the HSPA channel during the handover, the measurement control message can be delivered to the UE as quickly as within 500 ms, allowing the addition of another strong 3G cell sooner. However, when the SRB is mapped into Release 99 DCH channels, the SRB rate is typically lower, such as 3.4 kbps or 13.6 kbps. The low SRB rate leads to a slower delivery of the 3G intra-frequency neighbor list to the UE and can cause radio link failure in 3G right after eSRVCC. We measured the time taken from the eSRVCC complete (i.e. handover complete sent by the UE) to the first measurement control message carrying the 3G neighbor to be within 2 seconds, with SRB mapped to DCH. This delay can impact the overall eSRVCC success rate and reduce the reliability of such inter-system handovers. Therefore, we need to re-tune fundamental 3G handover reporting and operations in order to secure better eSRVCC performance for VoLTE calls. The tuning and optimization of eSRVCC is one of the major challenges to deploying successful VoLTE services.

Similar challenges can also occur in the case of call setup. As discussed previously, SRVCC can occur at the alerting or pre-alerting phase of the VoLTE call. The main issue of initiating SRVCC during VoLTE call setup is that the eNB is not aware of the session phase status: alerting, pre-alerting, or post-alerting. The eNB determines that a UE is running a voice service if the eNB detects a service with QCI = 1 running on the UE. As QCI = 1 is typically established before the 180-ringing, the eNB can instruct the UE to do IRAT measurements and trigger SRVCC while the call setup has not been completed.

Since the IMS and Mobile Switching Center (MSC) may not support the aSRVCC or bSRVCC procedures, the call setup will fail in this scenario. The failure symptom is that the Mobile Switching Center (MSC) cannot proceed with SRVCC in the alerting or pre-alerting phase, and hence sends a "Disconnect" message to the UE as soon as it falls back to the 3G or 2G system after SRVCC is complete. There are two problematic scenarios in SRVCC during VoLTE call setup (in the alerting or pre-alerting phase):

1. The UE supports any of these two features (a/bSRVCC) but the IMS does not support either of them.
2. The IMS supports any of these features but the UE does not support either of them.

In both scenarios, SIP messages are transparent to the eNB and hence it cannot delay the SRVCC procedure if either the UE or the IMS does not support a/bSRVCC. Solutions are not yet widely available in the EPS for these scenarios, and hence it is important to validate these scenarios in the early stage of VoLTE deployment. The situation becomes worse with the mix of devices in the network, since these features are introduced in two different 3GPP releases. One workaround by delaying the SRVCC procedure after QCI = 1 is established for a fixed duration of time. However, this can lead to call drops instead of call setup failure if the LTE signal degrades quickly after the call setup and SRVCC is not yet initiated based on this fixed timer solution. Another workaround involving the Mobile Switching Center (MSC) detects that a/bSRVCC is not supported and the procedure cannot be completed, and hence the Mobile Switching Center (MSC) can request the cancellation of the SRVCC procedure itself and avoid further failures until the call has been established successfully. Finally, parameter optimization can be undertaken to relax the SRVCC trigger, but the tradeoff can hit the VoLTE in-call performance, where more call drops can occur with relaxed SRVCC parameters. Therefore, SRVCC thresholds can be adjusted in a smart way to accommodate both cases that greatly impact the VoLTE KPIs: call drops due to delayed SRVCC or call setup failure due to faster SRVCC.

3.8 Latency Reduction During Handover

Release 13 on latency reduction [23] has the following high-level objective:

To study enhancements to the E-UTRAN radio system in order to:

Significantly reduce the packet data latency over the LTE Uu air interface for an active UE.

Significantly reduce the packet data transport round trip latency for UEs that have been inactive for a longer period (in Connected state).

The study area covers resource efficiency, including air interface capacity, battery lifetime, control channel resources, specification impact, and technical feasibility. Both FDD and TDD duplex modes are considered. Further, the study item in [23] suggests that the following aspects are considered. First, the potential gains, like reduced response time and improved TCP throughput due to latency improvements on typical applications and use cases, should be identified and documented. Latency reductions may be achieved by protocol enhancements as well as shortened TTIs. In [24], the latency aspects related to the protocol enhancements, particularly to the handover procedure, for active UEs are considered. Packet data latency is one of the key performance metrics for vendors and operators. As part of HetNet mobility, several features were introduced to improve inter-eNB coordination (eICIC) to minimize interference, to ensure robust mobility with a low probability of radio link failure, handover failure, and ping-pongs, as well as efficient recovery from errors via successful RRC re-establishments, and signaling of UE mobility information from the UE to the network upon transition from RRC Idle to RRC Connected. Most of these features aimed to improve mobility and reduce interference.

One particular aspect of latency is the service interruption time during mobility. Improvements related to reducing user data interruption time during a handover procedure have not yet been well addressed in the recent 3GPP work. A recent study in [3] and [19] suggests that handover latency is a serious concern. Figure 3.21 summarizes the PS handover time and data interruption time in [19], and we provide recent results in Figure 3.22 and Figure 3.23 for the C-plane and U-plane, respectively. Figures 3.15 to 3.19 provide corresponding results for the VoLTE handover scenario. These X2-based handovers were performed in an asynchronous commercial network.

It can be seen from Figure 3.21 that the median handover interruption time is 50 ms, with values reaching up to 80 ms for 10% of the cases. Data interruption during handover can have a significant impact on the performance of higher-layer transmission protocols such

Figure 3.21 A study of interruption time during handover for an LTE system [19].

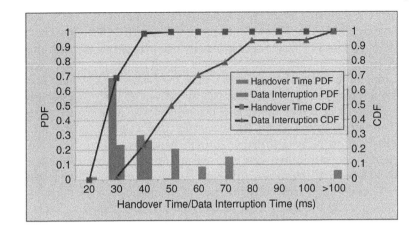

Figure 3.22 LTE C-plane handover delay distribution (Figure 3.37 in [3]).

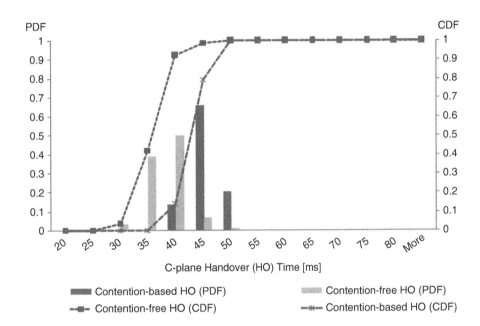

Figure 3.23 LTE U-plane interruption delay distribution during handovers (Figure 3.38 in [3]).

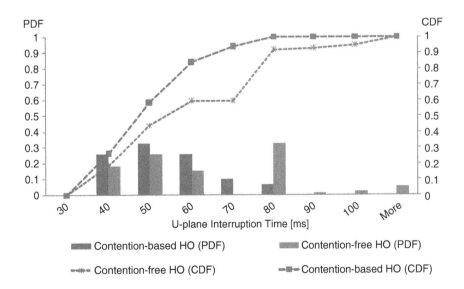

as TCP. Reducing handover interruption time improves latency and can therefore reduce the probability of TCP timeouts and TCP congestion-avoidance being triggered. A reduction in data interruption time during handover is one of the potential targets to reduce the packet data latency over the LTE Uu air interface for active UEs [24].

Figure 3.22 shows the LTE C-plane handover delay distribution and Figure 3.23 shows the LTE U-plane interruption delay distribution during handovers [3]. In the contention-free case, the mean C-plane delay is 36.4 ms and the mean U-plane delay is 62.3 ms. In the contention-based case, the mean C-plane delay is 43.4 ms and the mean U-plane delay is 50.0 ms. One strange thing about the results is that the mean C-plane delay in the contention-free case is shorter than that in the contention-based case, but the mean U-plane delay in the contention-free case is longer than that in the contention-based case. Since these are field results, different vendors and networks showed different results.

Table 3.15 shows a summary of causes for higher than expected C-plane and U-plane delays. Extra C-plane delay and U-plane delay are caused by RACH preamble retransmissions. Additionally, in the source cell, extra U-plane delay is caused by DL RRC Handover Command message retransmissions or high BLER of data packets, and in the target cell, extra U-plane delay is caused by UL RRC Handover Complete message retransmissions or high BLER of data packets [3]. The typical/average service interruption time in handover is longer than the typical/average handover delay from the field test results, and the extra service interruption time is caused by DL HO command retransmissions, UL HO complete retransmissions, or high BLER of data packets [25].

3.8.1 Components of Handover Latency

For LTE X2-based handover, when the UE sends a measurement report message over the RRC layer to the source eNB, the source eNB sends a handover request to the target eNB that includes a list of the bearers that will be transferred and possibly a DL data-forwarding proposal. The source eNB then sends an RRC Connection Reconfiguration message to the UE over the RRC layer,

after which the source eNB forwards any pending DL packets to the target eNB. The UE then synchronizes with the target eNB using a random access procedure, and, once successful, the UE sends an RRC Connection Reconfiguration Complete message over the newly established RRC. The UE then starts collecting the SIBs, which carry the required information for the UE about the cell-level configuration parameters, from the target eNB. The target eNB sends a UE Context Release message to the source eNB, confirming the successful handover and enabling source eNB resources to be released [3, 24].

Table 3.16 shows typical delays associated with the legacy X2-based handover procedure, including message and processing cost [24].

As can be seen, the delay in user data starts soon after a handover procedure (i.e. UE RRM) and continues until well after the handover procedure has been completed (i.e. RRC Reconfiguration Complete message). From Figure 3.23, the total user data interruption time could be twice as much as the time it takes for a handover procedure to complete. Considering the timing estimates in Table 3.16 and the call flow in Figure 3.24, the user data interruption time could be as high as 80 ms. A delay of 80 ms in the user data is significant and it is obviously desirable to minimize the interruption time as much as possible, especially in a VoLTE scenario [24].

Based on Table 3.16, we can observe that the overall delay for a typical X2-based handover is anywhere between ∼ 40 and ∼ 45 ms (excluding RRM measurement reporting delay). However, it must be noted that the user data interruption during the handover procedure can be much larger, as shown in Figure 3.24 and also in Figure 3.15 for the VoLTE handover scenario.

3.8.2 Synchronous Handovers

In the current 3GPP specifications, after receiving a handover command message, a UE immediately executes a handover. It stops communicating with the source eNB and performs synchronization with the target eNB, accessing the target cell via RACH. Because of the break-before-make handover procedure, data

Table 3.15 Summary of causes for higher than expected C-plane and U-plane delays (Table 3.20 in [3]).

Factors Increasing HO Delays	PDSCH BLER in Source Cell	RACH	PDSCH/PUSCH BLER in Target Cell
C-plane delay	—	RACH preamble retransmissions.	—
U-plane delay	Downlink RRC handover message retransmissions or high BLER on source cell data packets.	RACH preamble retransmissions.	Uplink RRC handover complete message retransmissions, high BLER on target cell data packets.

Table 3.16 Handover latency components – typical values.

Component	Description	Time (ms)
1	RRM measurement reporting from UE to eNB	~ 5 ms
2	Source cell RRM processing, build/send HO request to target cell	~ 7 ms
3	X2 backhaul delay for HO request message from source to target cell	~ 3 ms
4	Target cell HO request processing, build/send HO request reply	~ 7 ms
5	X2 backhaul delay for HO request reply message from target to source cell	~ 3 ms
6	Source cell processing of HO request reply, build/send RRC reconfiguration message to UE	~ 7 ms
7	Acquiring synchronization and communication w/target cell including RACH	~ 10 ~ 12 ms
8	UE sending RRC Reconfiguration Complete to target cell	~ 5 ms
	Total handover delay	**~ 50 ms**

Figure 3.24 PS handover delay components.

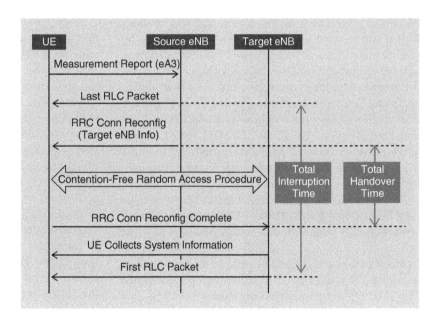

communication between the UE and the network could cease between breaking communication with the source cell and establishing communication with the target cell. However, if a UE is allowed to continue communicating with the source cell until the UE has accessed the target cell, service interruption time can be reduced [25].

One possible way to decrease handover latency is to reduce the delay between receiving the handover command from the source cell and being able to receive data from the target cell. This could be done by utilizing the properties of network synchronization. Several features recently accepted in the standards require synchronized networks (for example, feICIC, NAICS, eMBMS, and small cell enhancements), so, with the proliferation of these features, it would seem sensible to consider the opportunities to reduce latency by decreasing user data interruption time during the handover procedure (at least for synchronized networks). Therefore,

opportunities exist to take advantage of synchronized networks to reduce latency during handovers [24].

In a synchronized handover scenario, the time when the source eNB stops sending data to the UE and the time when the UE disconnects from the source cell can be synchronized. After receiving the handover command, the UE does not execute a handover immediately, unlike in current 3GPP specifications, but communicates with the source eNB before some pre-defined event. In addition, the source cell does not stop DL transmission upon receiving the UE Measurement Report message, as is the case in the legacy handover procedure. Instead, the DL transmission continues from the source cell until the exact time instance (e.g. SFN). If the pre-defined event is triggered, the UE sends a handover indication notifying the source eNB of an immediate handover execution. On receipt of the HARQ ACK to the handover indication, the UE can detach from the source cell and execute a handover to the target cell. After transmitting the

HARQ ACK to the handover indication, the source eNB can stop sending data to the UE and start forwarding data to the target eNB. Therefore, the time of handover execution in the UE and the source eNB can be synchronized [25]. As a second step, the random access to the target cell could also be skipped due to the network synchronization. Instead, when accessing the target cell, the UE would simply listen to the PDCCH and wait for a scheduling command. This would allow a reduction in the data interruption time, which, as shown in the call flow above, can be quite long.

This synchronization is achieved between the two eNB cells over X2 signaling and the UE via RRC signaling. The fact that all three nodes are in sync means that the source cell stops DL transmission to the UE, the target cell provides an uplink grant to the UE, and the UE acquires the target. Synchronous handovers without random access can be used to improve the user experience by minimizing the overall handover delay. A RACH attempt procedure during handovers typically takes ~10~12 ms. An average handover procedure takes ~40~45 ms to complete. Continuous transmission of user data from the source cell well after the handover procedure starts, and the elimination of ~10~12 ms of RACH delay during a handover procedure can significantly minimize the data interruption during handovers and improve the user experience.

The UE can derive the Timing Advance (TA) value to be used in the target cell. With the TA value known, the UE can acquire the target cell without performing a random access procedure, as is done today in legacy handovers. In order to derive the TA to be used in the target cell, it is assumed that the UE can measure the time difference (T_{DIFF}) in the signals received from the source ($T_{RX,SRC}$) and target cells ($T_{RX,TGT}$) while connected to the source cell (Figure 3.25).

$$T_{DIFF} = T_{RX,SRC} - T_{RX,TGT} \qquad (3.1)$$

The timing of the target cell can be derived from, for example, measurements of the reference signals transmitted by the corresponding cell. The TA to be used in the target cell (TA_{TGT}) is then calculated based on the TA

used in the source cell (TA_{SRC}). The timing advance compensates for the round trip time between the eNB and the UE; therefore, the required value is $TA_X = 2\ T_X$. This gives:

$$TA_{SRC} - TA_{TGT} = 2\ T_{RX,SRC} - 2\ T_{RX,TGT} \qquad (3.2)$$

$$TA_{TGT} = TA_{SRC} - 2\ T_{DIFF} \qquad (3.3)$$

Nokia proposed the consideration of synchronous handover as a candidate technique for reducing latency during handover [24]. One possible area of improvement in which several existing and new applications could benefit is minimizing latency during handovers for a moving mobile.

3.8.3 Synchronized Handover with Early Handover Command (EHC)

The synchronized handover solution can be combined with an EHC solution. With an EHC solution, after receiving the "early" handover command, the UE does not execute a handover immediately, unlike in current 3GPP specifications, but communicates with the source eNB and performs measurements continually before some pre-configured event. In addition, the source eNB keeps sending data to the UE until that event. If the pre-configured handover execution event is triggered, the UE sends a handover indication notifying the source eNB of an immediate handover execution. On receipt of the HARQ ACK to the handover indication, the UE can detach from the source cell and execute a handover to the target cell. In turn, the source eNB, after transmitting the HARQ ACK to the handover indication, can stop sending data to the UE and start data forwarding to the target eNB. With this solution, the service interruption time is the same as for a synchronized handover.

The strength of this solution is that it is helpful not only in terms of latency reduction but also in terms of mobility robustness. With regard to mobility robustness, this solution can theoretically achieve a zero handover-failure rate regardless of the UE velocity and the size of the handover region. This solution can solve the tradeoff between the handover failure rate and the ping-pong rate, achieving a zero handover-failure rate without increasing the ping-pong rate. Furthermore, if the solution is extended to keep fast-moving users out of small cells, the solution can accomplish a zero handover-failure rate and a zero ping-pong rate simultaneously. For more details on EHC, refer to [25] and the references therein.

Figure 3.26 [24] captures the typical interruption time for legacy handover compared to gains achieved by synchronous handover. As can be seen from the figure, the interruption time is much higher for UEs with more speed and a higher number of small cells under a macro

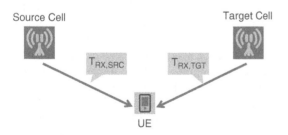

Figure 3.25 Propagation delays between the UE and the source and target cells.

Figure 3.26 Average percentage of interruption time.

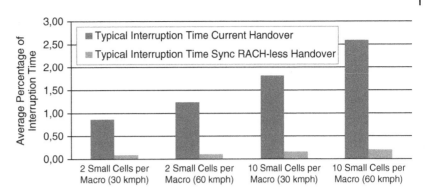

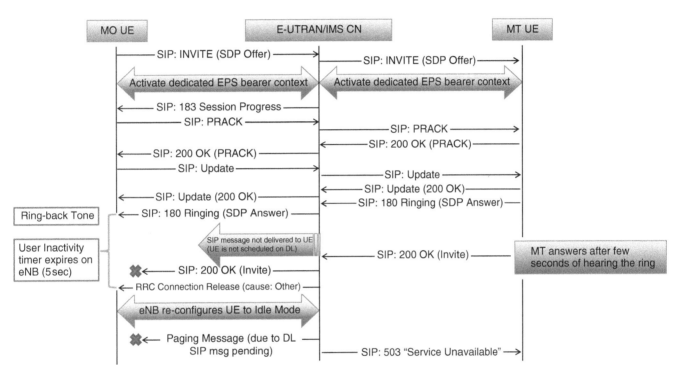

Figure 3.27 State transition at call setup.

cell. This is because if a UE is moving at higher speed, it will experience more frequent handovers than a UE with a lower speed. The increased number of handovers, each introducing some interruption time, adds up to a higher overall delay. However, the interruption time is significantly lower for the case of synchronized handover. The fact remains that the higher number of handovers a UE experiences, the higher will be the data interruption time. However, comparing the two scenarios, the benefits are obvious with synchronized handover.

3.9 Practical Use Cases and Recommendations

In this section, we will provide use cases and issues related to call setup and handover, and we will provide recommendations and resolutions for these cases.

3.9.1 State Transition at Call Setup

Figure 3.27 illustrates this use case. The possible solutions may be summarized as follows:

Possible impacting scenarios:
    VoLTE to VoLTE
    VoLTE to landline
    VoLTE to 3G/2G
    VoLTE to CSFB → will be more severe due to call setup delays on the CSFB side.
Network solution:
    User inactivity timer for PS data (e.g. QCI = 9) can be short in order to improve UE battery and network resources.
    Increasing user inactivity timer for VoLTE with QCI = 1 is recommended.
UE solution (feature, if implemented):

The UE can implement some features in order to send small background traffic in order to keep the user plane active without RRC transition from Connected to Idle mode.

3.9.2 SIP Timeout at Call Setup

Figure 3.28 illustrates this use case. The possible solutions may be summarized as follows:

Possible impacting scenarios:
VoLTE to VoLTE if the MT side experiences paging failures.
VoLTE to 3G/2G → more severe due to call setup delays on the paging side if the TAS does PS page hunting.
VoLTE to CSFB → more severe due to call setup delays on paging + CSFB side.
In some networks, issues related to SIP timeout = 6% were observed.
Network solution:
For a VoLTE user, the IMS can use SIP: Options to query if the UE is reachable on the IMS or not, and this way, the TAS can manage to page the UE quickly on the right domain and reduce PS page hunting.
Also, operators need to define the UE timer used for SIP-timeout measurement based on the paging re-tries in the network and other considerations (e.g. coverage).
UE solution (feature, if implemented):

The UE can implement CSFB if there is no "SIP response" for X seconds.
A timer can be pre-defined based on network implementation (10 seconds is the default used in the majority of networks), but can increase if the network experiences spotty LTE coverage or a TAS page-hunting implementation query is suboptimal.

3.9.3 UE Loses VoLTE due to an IMS Registration Issue

Figure 3.29 illustrates this use case. The possible solutions may be summarized as follows:

Possible impacting scenarios:
UE disables IMS
UE enables IMS
Call this UE quickly before it registers to the IMS, and hence the UE will do a CSFB call.
Check after UE comes back to LTE if it is IMS registered or not. Why?
After the network sends the register challenge (401 Unauthorized), the response to the challenge is not received by the IMS. Eventually, the UE tries to do a new registration but the IMS is still waiting for the response to the register challenge and responds with 403 Forbidden – this means no more IMS registration is allowed until reboot.
Network solution:
Disallow the IMS PDN (QCI = 5) bearer from being carried into 3G; this way the UE will always have to do IMS registration after camping back to LTE from

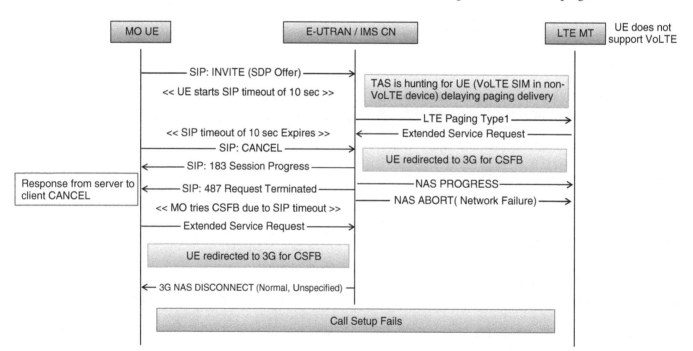

Figure 3.28 SIP timeout at call setup.

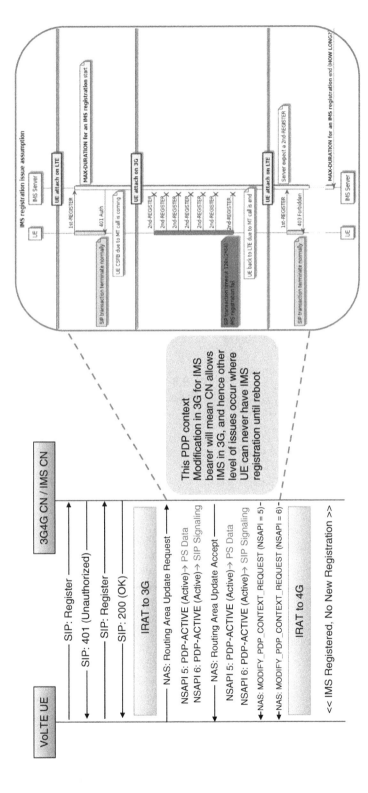

Figure 3.29 The UE loses VoLTE due to an IMS registration issue.

3G → there is VoIP in 3G, hence no need to keep PDP active for IMS PDN.

If a server response of 403_Forbidden because of attempted IMS registration is too frequent, it could follow 34.229 TC10_1, responding with 503 Service Unavailable and retry-after = 60s. The UE would delay for a period of time according to the value of "retry-after" and would then trigger the next IMS registration.

UE solution (feature, if implemented):

The UE can locally deactivate IMS PDN (with proper IE in 3G/2G Routing Area Update Request) once it moves to 3G/2G, which will not allow the core to modify the PDP corresponding to SIP signaling, and this way, the UE will always undertake IMS registration upon coming back to LTE.

3.9.4 bSRVCC Collision Leading to Subsequent Setup Failure

Figure 3.30 illustrates this use case. The possible solutions may be summarized as follows:

Possible impacting scenarios:

MO UE makes a call and triggers bSRVCC to 3G (or aSRVCC).

Then the MO hangs up before the call gets connected, and the RNC triggers a fast return to LTE.

The UE makes a subsequent VoLTE call that fails, and instead could trigger a CSFB call.

Network solution:

Graceful handling by the P-CSCF is required in order to inform the UE about resetting the TCP connection by sending a TCP RST to the UE prior to leaving 4G.

UE solution (feature, if implemented):

The UE can locally establish a new TCP connection and use it to send an INVITE SIP message for the subsequent IMS call.

3.9.5 Call Failure due to bSRVCC, Even if Supported

Figure 3.31 shows the call failure scenario due to bSRVCC, even when it is supported.

3.9.6 Network Planning Challenges

Figure 3.32 demonstrates network planning challenges for PS data and VoLTE.

As observed in Figures 3.33 and 3.34, the probability of triggering a measurement report before the IMS receives the 183-Session Progress is high.

Network solution:

Optimize thresholds for SRVCC.

Possible features such as bSRVCC-Flash SRVCC (network initiates CSFB instead of bSRVCC).

Another possible feature is HO Delay Timer, where the eNB avoids the bSRVCC phase until call setup is complete (timer-bounded).

3.9.7 IRAT Redirection During VoLTE

Figure 3.35 illustrates this issue and possible resolution. The case is addressed in 3GPP Release 14 – Signaling Enhancement to Support Redirection. The case may be summarized as follows:

Redirection is mainly useful in cases where there are no available measurement results in the eNB.

In this case, it makes sense for the eNB to perform redirection instead of handover.

During the redirection procedure, the eNB can only indicate the frequency layer, which can be independent from measurement results to the UE.

The UE can then select and access a suitable cell on the indicated frequency layer during redirection.

These principles of redirection should also be applied to voice services. However, redirection for VoLTE will cause call drops, which needs to be addressed.

3.9.8 No Response to SIP Message, Leading to Call Drop

This issue is illustrated in Figure 3.36. The issue typically arises during mobility and after several repetitive runs due to race conditions in the IMS server.

3.10 Conclusions

In this chapter we have analyzed the different aspects of the voice solutions offered by the LTE system – CSFB, VoLTE, and SRVCC. All of the results presented in the chapter are based on commercial, live LTE/HSPA+ networks. The KPIs defined in this chapter were adopted to evaluate the CSFB and VoLTE call setup delay; several techniques were then discussed in order to improve the overall call setup delay and voice interruption during eSRVCC.

The CSFB call flow and relevant call setup KPIs were established and evaluated. The call setup KPIs for different scenarios were analyzed. Based on detailed field testing and analysis, the best CSFB method was found to be blind PSHO, which minimizes the radio delays when moving from LTE to UMTS and also reduces core network delays when establishing the CS call, given that the PS bearer is activated during the handover itself. PSHO with measurements minimizes the core network delays but can potentially increase air interface delays due to

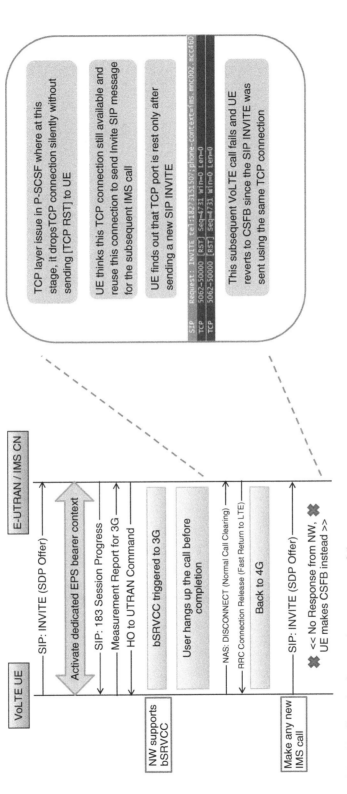

Figure 3.30 bSRVCC collision leading to subsequent setup failure.

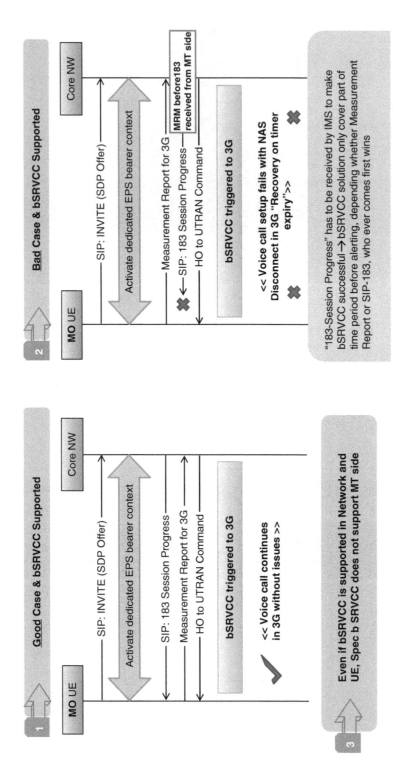

Figure 3.31 Call failure due to bSRVCC, even if supported.

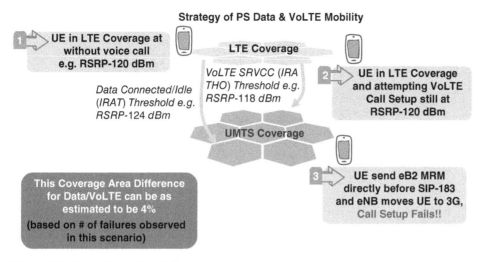

Figure 3.32 Network planning challenges.

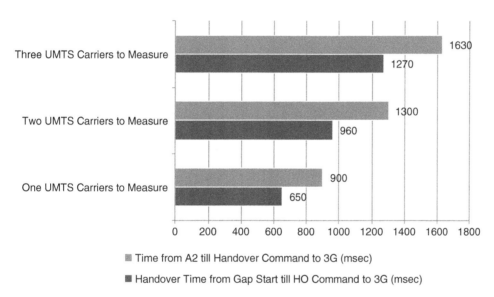

■ Time from A2 till Handover Command to 3G (msec)

■ Handover Time from Gap Start till HO Command to 3G (msec)

Figure 3.33 Impact of number of UMTS carriers on handover delay.

Figure 3.34 Average time from INVITE until 183-Session Progress in different scenarios.

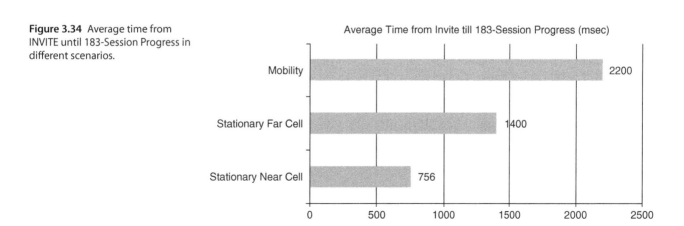

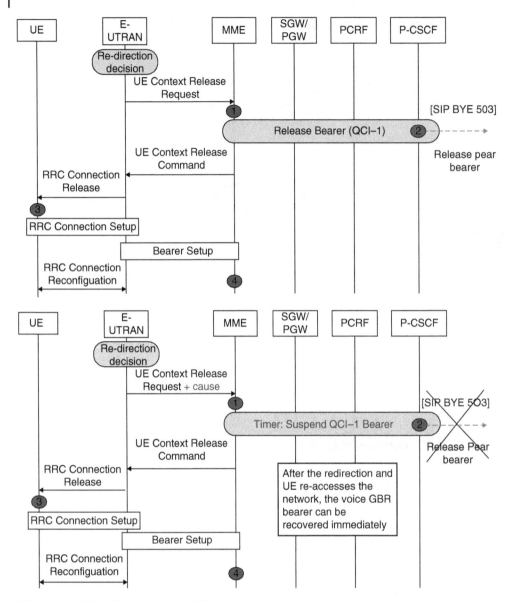

Figure 3.35 IRAT redirection during VoLTE.

the need to measure the 3G cells prior to the handover. The chapter highlighted some key techniques for reducing call setup delay and improving PSHO with measurements. Moreover, the chapter also provided techniques for improving CSFB call setup delay.

VoLTE call setup delay was analyzed in a similar manner. The relevant VoLTE KPIs were established and evaluated for different radio conditions, including near cell, far cell, and mobility. It was demonstrated that VoLTE provides a better end-user experience in terms of call setup delay. The VoLTE call setup delay and success rate can still be improved further by adopting the techniques discussed in this chapter, including paging repetition, re-tuning of inactivity timers, and enabling the C-DRX mechanism with a modified set of parameters targeting a lower packet latency.

eSRVCC performance in terms of voice interruption time during the IRAT handover to UTRAN was analyzed and compared to the interruption time for LTE intra-frequency handover. The eSRVCC voice interruption is worse than the LTE intra-frequency interruption by ∼ 200 ms, but still within the range for acceptable audio quality, i.e., 300 ms. The chapter provided several techniques for improving the voice interruption and the handover success rate for eSRVCC. Future work involves VoLTE voice quality and in-call performance, including a detailed evaluation of in-call jitter, delays, the packet loss error rate, and quality with concurrent PS and VoLTE calls.

We then discussed the user data interruption delay during the handover procedure for PS and VoLTE scenarios. The typical interruption time during handover

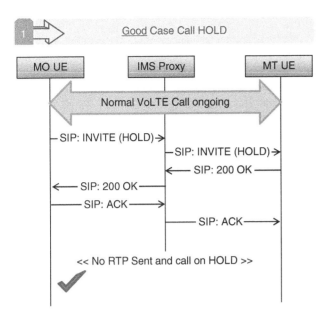

Figure 3.36 No response to SIP message, leading to call drop.

is 50 ms and can be as high as 80 ms. 3GPP considers the reduction of data interruption time during handover to be one of the potential targets for reducing packet data latency over the LTE Uu air interface for active UEs. Nokia proposed synchronous handover as a candidate technique for reducing the latency during handover. The synchronized handover solution can be combined with an EHC solution for mobility robustness.

Finally, practical use cases and issues were outlined along with recommendations and possible resolutions.

References

1 3GPP TS 22.228 V12.9.0: "Service requirements for the Internet Protocol (IP) Multimedia core network Subsystem (IMS)."

2 3GPP TS 22.173 V9.5.0 (2010-03): "IP Multimedia Core Network Subsystem (IMS) Multimedia Telephony Service and supplementary services; Stage 1," (Release 9)

3 Elnashar, A., Al-saidny, M. and Sherif, M. (2014) *Design, Deployment, and Performance of 4G-LTE Networks: A Practical Approach*, John Wiley & Sons, May.

4 3GPP TS 29.118 V10.13.0 (2014-06): "Mobility Management Entity (MME) – Visitor Location Register (VLR) SGs interface specification."

5 GSMA, IR.92 IMS Profile for Voice and SMS, Version 7.0, 3 March 2013.

6 GSMA, IR.94 IMS Profile for Conversational Video Service, Version 5.0, 4 March 2013.

7 3GPP TS 23.221 V10.0.0 (2011-03): "Architectural requirements."

8 Vargas Bautista, J. E. et al. (2013) "Performance of CS Fallback from LTE to UMTS," *IEEE Communications Magazine*, pp. 136–143, September.

9 Liou, R.-H. et al. (2014) "Performance of CS Fallback for Long Term Evolution Mobile Network,"

IEEE Transactions on Vehicular Technology, **63**(**8**), 3977–3984, October.

10 Lin, Y.-B. (2014) "Performance Evaluation of LTE eSRVCC with Limited Access Transfers," *IEEE Transactions on Wireless Communications*, **13**(**5**), May.

11 Jia, Y.-J. et al. (2015) "Performance Characterization and Call Reliability Diagnosis Support for Voice over LTE," *ACM Mobicom'15*, pp. 452–463.

12 3GPP TS 23.272 V10.3.1 (2011-04): "Circuit Switched (CS) fallback in Evolved Packet System (EPS)."

13 3GPP TS 24.299 V8.12.0 (2010-06): "IP multimedia call control protocol based on Session Initiation Protocol (SIP) and Session Description Protocol (SDP)."

14 3GPP TS 23.206: "Voice Call Continuity (VCC) between Circuit Switched (CS) and IP Multimedia Subsystem (IMS); Stage 2."

15 3GPP TS 23.216: "Single Radio Voice Call Continuity (SRVCC)."

16 3GPP TR 23.856 V10.0.0 (2010-09): "Single Radio Voice Call Continuity (SRVCC) enhancements."

17 3GPP TS 24.237 V12.7.1: "IP Multimedia (IM) Core Network (CN) subsystem IP Multimedia Subsystem (IMS) service continuity; Stage 3."

18 3GPP TS 24.008 V12.7.0: "Mobile radio interface Layer 3 specification; Core network protocols; Stage 3."

19 Elnashar, A. and El-Saidny, M. A. (2013) "Looking at LTE in Practice: A Performance Analysis of the LTE System based on Field Test Results," *IEEE Vehicular Technology Magazine*, **8**(**3**), 81–92, September.

20 3GPP TS 25.331 V10.18.0: "Radio Resource Control (RRC); Protocol specification."

21 Elnashar, A. and El-Saidny, M. A. (2014) "Extending the Battery Life of Smartphones and Tablets: A Practical Approach to Optimizing the LTE Network," *IEEE Vehicular Technology Magazine*, Issue 2, pp. 38–49, June.

22 3GPP TS 36.331: "Evolved Universal Terrestrial Radio Access (E-UTRA); Radio Resource Control (RRC); Protocol specification."

23 RP-150465: "SI proposal: Study on Latency reduction techniques for LTE."

24 R2-154259: "Latency reduction during handover," Nokia Networks.

25 R2-156412: "Synchronized Handover for Latency Reduction," ETRI.

4

Comprehensive Performance Evaluation of VoLTE

4.1 Overview

Voice over LTE (VoLTE) has been witnessing a rapid deployment by network carriers worldwide. During the phases of VoLTE deployment, carriers typically face challenges in understanding the factors affecting the VoLTE performance and then optimizing it to meet or exceed the performance of the legacy circuit-switched network (i.e., 2G/3G). The main challenges in terms of VoLTE service quality are the LTE network optimization and the performance aspects of the service in different LTE deployment scenarios. In this chapter we present a detailed, practical performance analysis of VoLTE based on commercially deployed 3GPP Release 10 LTE networks. The analysis evaluates VoLTE performance in terms of the Real-time Transport Protocol (RTP) error rate, RTP jitter and delays, the Block Error Rate (BLER) for different radio conditions, and VoLTE voice quality in terms of Mean Opinion Score (MOS). In addition, the chapter evaluates key VoLTE features such as Robust Overhead Compression (RoHC) and Transmission Time Interval (TTI) bundling. The chapter provides guidelines for best practice in VoLTE deployment as well as practical performance evaluation in commercial networks.

4.2 Introduction

VoLTE is an IP Multimedia System (IMS)-based voice service over the LTE network [1]. The IMS supports various access and multimedia services and has recently become the standard architecture of the Evolved Packet Core (EPC) [1, 2]. 3GPP has adopted the GSMA IR.92 IMS profile for voice and SMS [3] and the GSMA IR.94 IMS profile for conversational video [4] to provide high-quality, IMS-based telephony services over the LTE radio access network. The profiles define optimal sets of existing 3GPP-specified functionalities that network vendors, service providers, and handset manufacturers can use to offer compatible LTE voice/video solutions. Therefore, the commercial deployment of VoLTE mandates extensive testing between terminals and networks,

including the LTE radio access network (i.e., eNodeB), LTE EPC, and IMS. In addition to these challenges, the VoLTE optimization for different radio/loading conditions to find an acceptable tradeoff between the user's experience and network deployment complexities has led to a substantial delay in the widespread adoption of the VoLTE service. In this chapter, we address these challenges by providing the best practices for VoLTE-related features and practical performance evaluation based on field-testing results from commercial LTE 3GPP Release 10 networks.

The deployment of VoLTE brings a variety of benefits to telecom operators, as voice is still the main source of revenue. Hence, telecom carriers need voice evolution to compete effectively with over-the-top (OTT) Voice over IP (VoIP) applications that create a significant load on the mobile broadband networks and, accordingly, affect other services. VoLTE also improves the spectral efficiency and reduces network costs compared to legacy circuit-switched (CS) networks. Moreover, the spectral efficiency of LTE networks is higher than the traditional GSM/UMTS networks, which makes VoLTE a really suitable voice solution in 4G networks [5, 6]. For the same channel bandwidth, an LTE cell offers twice the cell-edge throughput of a UMTS cell [5, 6].

VoLTE enhances the end-user experience by providing a better Quality of Experience (QoE). VoLTE is equipped with numerous sets of integrated features that improve QoE aspects, such as better call setup time, higher efficiency in deep-coverage conditions, and lower battery consumption. Moreover, VoLTE supports a different range of speech codec rates, i.e., Adaptive Multi-Rate (AMR) with both wideband (AMR-WB) and narrowband (AMR-NB) and the Enhanced Voice Services (EVS) codec. While high-definition (HD) voice is being rolled out by telecom operators using AMR-WB with audio bandwidth of up to 7 kHz, 3GPP Release 12 introduced the EVS codec to offer up to 20 kHz audio bandwidth [7]. EVS at 13.2 kbps provides super-wideband (SWB) voice quality at a comparable bit rate to AMR and AMR-WB, and offers high robustness to jitter and packet losses.

Practical Guide to LTE-A, VoLTE and IoT: Paving the Way Towards 5G, First Edition. Ayman Elnashar and Mohamed El-saidny.
© 2018 John Wiley & Sons Ltd. Published 2018 by John Wiley & Sons Ltd.

Feasible solutions for providing voice call and service continuity over LTE-based networks, a comparison between various aspects of these solutions, and a possible roadmap that mobile operators can adopt to provide seamless voice over LTE are provided in [8]. A comprehensive evaluation and validation of VoLTE Quality of Service (QoS) is provided in [9]. The results in [9] give clear evidence that the VoLTE service fulfills the ITU-R and 3GPP standard requirements in terms of end-to-end delay, jitter, and packet loss rate. The VoLTE performance in terms of QoS is evaluated and validated using the OPNET modeler wireless suite in [9]. Railway voice communication based on VoLTE is proposed in [10]. The simulation results in [10] indicate that VoLTE is a viable option for providing railway voice communication, i.e., GSM-R. Therefore, LTE is proven to be a strong candidate for the future communication network on railways [11]. The feasibility of Semi-Persistent Scheduling (SPS) for VoIP is analyzed in [12], which evaluates its performance in terms of throughput of random access delays and traffic channels. A methodology to evaluate the VoIP capacity and performance of Orthogonal Frequency Division Multiple Access (OFDMA)-based systems is provided in [13]. This methodology can also be applied to VoLTE. A method for estimating cell capacity from network measurements in a multi-service, long-term evolution (LTE) system is presented in [14].

The performance of circuit-switched fallback (CSFB), VoLTE, and enhanced Single Radio Voice Call Continuity (eSRVCC) is key to guaranteeing a seamless voice experience within LTE/UMTS/2G networks [15–20], especially the concept that a mix of VoLTE and CSFB devices can co-exist in the same LTE network. The three features mentioned above were analyzed in [6, 20] to evaluate the call setup delays and the details behind the interruption time within the LTE network or over eSRVCC. It is demonstrated in [20] that VoLTE provides a better end-user experience in terms of call setup delay. However, the eSRVCC voice interruption is higher than the LTE intra-frequency interruption by ~ 200 ms, but still within the acceptable audio quality range of 300 ms. A comprehensive performance evaluation of Robust Overhead Compression (RoHC) for VoLTE by means of a testbed implementation is presented in [21]. In [22], it is envisaged that the quality of a call measured through the Mean Opinion Score (MOS) is always better in LTE compared with the UMTS network. The VoLTE MOS condition only becomes unstable upon reaching very poor RF conditions – where the Reference Signal Received Power (RSRP) < −117dBm and the Reference Signal Received Quality (RSRQ) < −12dB. The implementation of the Transmission Time Interval (TTI) bundling feature helps to improve the uplink coverage and minimizes the BLER. It is proposed in [22] that the serving cell SRVCC threshold RSRP = −108dBm for

networks without TTI bundling and RSRP = −117dBm for networks with TTI bundling deployment. However, this chapter demonstrates that RSRP = −110 dBm is the optimal threshold to maintain voice quality even with TTI bundling deployment, as Real-time Transport Protocol (RTP) performance can be impacted adversely beyond this level.

In this chapter, we present comprehensive practical analysis of VoLTE performance based on commercially deployed 3GPP Release 10 LTE networks. The analysis demonstrates VoLTE performance evaluation in terms of RTP error rate, RTP jitter and delays, BLER, and VoLTE voice quality in terms of MOS. In addition, this chapter evaluates key VoLTE features such as RoHC, TTI bundling, and SPS. The remainder of the chapter is organized as follows: VoLTE principles are summarized in Section 4.3. The main VoLTE features along with deployment best practices are outlined in Section 4.4. The testing environment is explained in Section 4.5. Practical performance analysis is provided in Section 4.6. EVS coding and voice evolution are covered in Section 4.7 and TTI bundling performance evaluation is provided in Section 4.8. BLER impact on voice quality is discussed in Section 4.9. Scheduler performance and VoLTE KPIs are analyzed in Sections 4.10 and 4.11, respectively. Finally, use cases and conclusions are provided in Sections 4.12 and 4.13, respectively.

4.3 VoLTE Principles

The IMS server carries all sources of VoLTE traffic and provides functions including subscribers' registration, authentication, control, routing, switching, media negotiation, and conversion. The voice codec type and rate of VoLTE service are negotiated at call setup by the UE and the IMS. The eNB provides only the media-plane bearer channel and the IMS signaling is transparent to the air interface. The VoLTE call includes an end-to-end voice/media flow transmitted on a dedicated Guaranteed Bit Rate (GBR) bearer with QoS class identifier (QCI) = 1 through the RTP protocol, User Datagram Protocol (UDP), or IP protocol. Another default bearer for Session Initiation Protocol (SIP) signaling is established beforehand using QCI = 5, through the UDP protocol, the Transmission Control Protocol (TCP), or the IP protocol. The IMS network and Evolved Packet System (EPS) transfer the IMS Access Point Name (APN) as well as the IMS PDN connection that hosts the two IMS bearers. Packet-switched (PS) services continue to use the default packet data network (PDN) connection with a non-GBR bearer (e.g. QCI = 9). Figure 4.1 illustrates the radio bearer concept for VoLTE and PS services within the EPC and IMS.

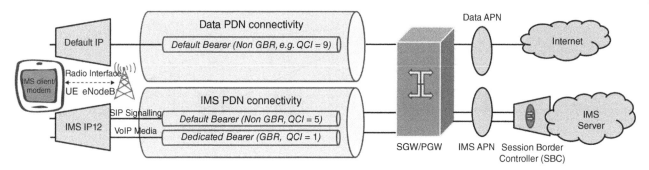

Figure 4.1 VoLTE bearers within the EPC/IMS.

The IMS clients on both sides of the VoLTE call determine the speech codec scheme (i.e., AMR type and codec rate). The IMS server is an optional entity in this process. During the attach procedure, the UE negotiates with the Mobile Management Entity (MME) and determines whether to select VoLTE as the voice policy or restrict the calls to CSFB calls. Radio bearers with QCIs of 1, 2, and 5 are established between two VoLTE UEs to carry conversational voice, signaling, and video, respectively. The eNB performs admission and congestion control for conversational voice (QCI = 1), signaling (QCI = 5), and video (QCI = 2). Moreover, the eNB performs dynamic or SPS scheduling and uses power control policies that are suitable for dynamic or SPS scheduling.

Several protocol interfaces are used between the UE's IMS client and the network's IMS server. They are the language between the IMS client/server to exchange either signaling information or actual media packets (i.e. voice packets). A summary of these protocols can be given as follows:

SIP: Delivers IMS signaling to negotiate a media session between two users.
RTP: Delivers IMS media packets.
RTP Control Protocol (RTCP): Used to synchronize streams by providing feedback on QoS information.
IP Security (IPSec): Used to carry the authentication and key agreement (AKA) as a secured tunnel for IMS clients.

The SIP protocol is used to create, modify, and terminate multimedia sessions between two users. It is a text-based client/server protocol and does not actually deliver IMS media. Every SIP request begins with a starting line that includes the name of the method (request type). Typically, a response is needed to complete the process. The actual SIP message describing the request/response is called the Session Description Protocol (SDP). The session requires certain information to be shared between the sender and the receiver, which goes through the SDP protocol. SDP information may also include contact information and bandwidth requirements for the session. Both SIP and SDP contents

are transferred between the IMS client and server to complete the signaling part of the IMS call. After the call has been established and the media packets (i.e. voice packets) start flowing, the RTP protocol is used in the IMS to carry the media. The RTP is a standardized packet format used for carrying the media streams (i.e., audio, video). The User Datagram Protocol (UDP) provides a transport layer mechanism for the RTP stream between two IP end points. In the case of VoLTE, this is between the IMS voice client in the UE and a media gateway in the IMS core network.

After the VoLTE call has been established and the media packets start flowing, the eNB performs dynamic scheduling and uses power control policies that are suitable for such scheduling. The eNB selects a modulation and coding scheme (MCS) index and physical resource blocks (PRBs) for voice users, similar to the mechanism in scheduling the PS data. The main purpose of the VoLTE scheduling technique is to maintain continuous transmission on the uplink and downlink in a way that minimizes packet delays. The data packets for voice services have a relatively small and fixed size, therefore, scheduling RTP packets on the radio and core network requires a stringent delay budget to control the inter-arrival time that minimizes jitter.

Due to delays in the VoLTE network from different network elements (i.e., eNB, EPC, IMS, and the transport network), the RTP packets' inter-arrival time can vary in time. The RTP packets during talk spurts are generated every 20 ms. However, the packets do not arrive precisely in that exact interval. This means that the VoLTE packets cannot be played out as they arrive due to variance in packet arrivals. Jitter is defined as a statistical variance of the RTP data packet inter-arrival time. In PS networks, the occurrence of variable delay is much higher than the values in CS networks. Figure 4.2 illustrates the concept of jitter and the delay requirements to keep the inter-arrival time within an acceptable range for better jitter buffer management. Jitter buffers are used to change asynchronous packet arrivals into a synchronous stream by tuning variable network delays into constant delays at the destination. The role of the jitter buffer

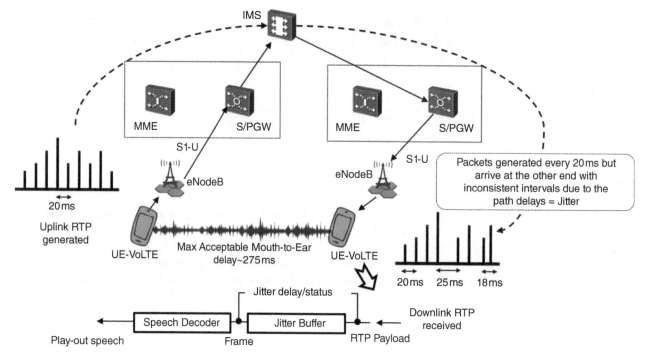

Figure 4.2 The concept of jitter and jitter buffer management.

is to set a tradeoff between delay and the probability of interrupted playout because of late packets. Late or out-of-order packets are discarded. IMS clients should implement adaptive jitter buffers to overcome these issues by dynamically tuning the jitter buffer size to the lowest acceptable value. Increasing the buffer size can increase latency. Even if the RTP packets remain in the correct sequence and there is zero packet loss, large variations in the end-to-end transmission time for the packets may cause degradation of audio quality that can only be fixed through the use of a jitter buffer. The typical RTP error rate is < 1% for audio and < 0.1% for video [23]. On the other hand, the radio network scheduler has a key role by securing sufficient scheduling resources in all radio conditions in order to keep the RTP error rate, jitter, and end-to-end delays within a range that can be handled efficiently by the jitter buffer.

4.4 Main VoLTE Features

Typically, the eNB scheduler is designed to schedule the UEs efficiently during the DL and UL transmissions for either small or large packet sizes. However, the VoLTE media stream has small voice packets with fixed inter-arrival intervals. Therefore, the eNB scheduler works with various special features specific to VoLTE in order to enhance the coverage, capacity, and quality of the voice calls. These eNB features are RoHC, TTI bundling, and SPS, which are all designed to assist

the VoLTE in-call performance. These features will be presented in the sections below.

4.4.1 Robust Header Compression

RoHC is a header compression protocol originally designed for real-time audio/video communication in a wireless environment. As VoLTE media packets are transported as IP traffic, the generated headers of the IP protocols can be massive. RoHC compresses the RTP, UDP, and IP headers to reduce the overall size of the voice packets. This decreases the radio resource utilization on the cell edge and therefore improves the overall cell coverage on both UL and DL. In addition, it reduces the number of voice packet segments, which improves the BLER associated with these smaller-size transmissions. This improves the VoLTE end-to-end delays and jitter.

The RoHC operation is illustrated in Figure 4.3. The RoHC function in LTE is part of Layer 2 at the user plane of the UE and eNB. Both UE and eNB behave as a compressor and decompressor for the user-plane packets on the UL and DL. The compression efficiency depends on the RoHC operating mode and the variations in the dynamic part of the packet headers at the application layer. A header can be compressed to one byte with RoHC, which efficiently reduces the voice packet size. RoHC in LTE operates in three modes: U-mode, O-mode, and R-mode (Unidirectional, Bidirectional Optimistic, and Bidirectional Reliable, respectively). The

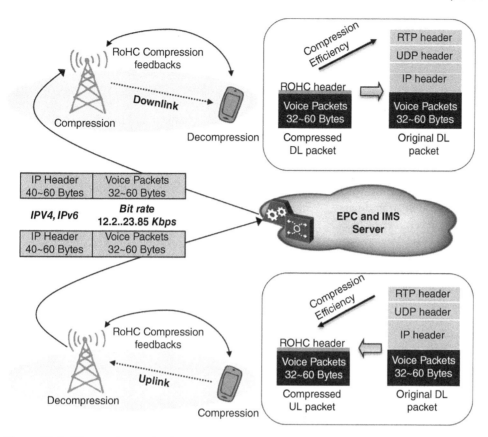

Figure 4.3 RoHC operation mechanism.

reliability of these modes and overheads used for transmitting feedback are different. In U-mode, packets can only be sent from the compressor to the decompressor, with no mandatory feedback channel. Compared with O-mode and R-mode, U-mode has the lowest reliability but requires the minimum overhead for feedback. In O-mode, the decompressor can send feedback to indicate failed decompression or successful context update. Therefore, it provides higher reliability than U-mode but it generates less feedback compared to R-mode. In R-mode, context synchronization between the compressor and decompressor is ensured only by the feedback. That is, the compressor sends the context updating packets repeatedly until acknowledgment is received from the decompressor. Therefore, R-mode provides the highest reliability but generates the maximum overhead due to the mandatory acknowledgment.

4.4.2 Transmission Time Interval Bundling

TTI bundling enables a data block to be transmitted in consecutive TTIs, which are packed together and treated as the same resource during the scheduling process. TTI bundling makes full use of Hybrid Automatic Repeat Request (HARQ) gains and therefore reduces the number of retransmissions and the Round Trip Time

(RTT). When the user's channel quality is degraded or the transmit power is limited, TTI bundling is triggered to improve the uplink coverage at the cell edge, reduce the number of different transmission segments at the Radio Link Control (RLC) layer and the retransmission overhead. The main advantage of TTI bundling is that it enhances VoLTE uplink coverage when the UE has limited uplink transmit power. Thus, it guarantees VoLTE QoS for cell-edge users. In a conventional scheduling mechanism, if the UE is not able to accumulate sufficient power to transmit a small amount of data, like a VoIP packet, the data packets can be segmented into smaller-size packets that fit within the UE's transmit power. Each segment will be transmitted with a separate HARQ process. However, this segmentation mechanism increases the amount of control information that needs to be transmitted. Therefore, the control channel load increases with the amount of segments, since every segment requires new transmission resources on these channels. Additionally, the probability of HARQ feedback errors increases with the number of segments, causing higher BLER at the cell edge. Therefore, the need to utilize a better segmentation method, like TTI bundling, is important.

For UEs at a cell edge, RLC segmentation is done first, then TTI bundling transmits the same packets four times in one scheduling period to extend the coverage by increasing uplink transmission reliability. The eNB decides when to activate or deactivate the TTI bundling for certain users based on the measured Signal-to-Interference Noise Ratio (SINR) and PRB usage on the uplink. The data block is in a bundle of TTIs, where the chunk of each bundle (up to four chunks) is modulated with different redundancy versions within the same HARQ identity. In the case of unsuccessful de-coding of the HARQ identity, the eNB sends a negative acknowledgment to the UE, which requires retransmission. The resource allocation during this operation is restricted to a certain number of PRBs and transport block size (TBS) in order to improve the probability of de-coding at lower data rates. The mechanism to transmit the same packets four times in one scheduling instance expands the coverage by increasing the uplink transmission reliability with a better success

rate gain. In addition, it guarantees that VoLTE packets are transmitted at a cell edge, when resources are limited, to improve the latencies in bad radio conditions. TTI bundling is estimated to provide 2–4 dB uplink coverage improvement for VoLTE services, which extends the cell radius for VoLTE services [24]. Figure 4.4 describes the TTI bundling mechanism and provides comparison with a scheduling mechanism that depends only on RLC segmentation procedures [25].

4.4.3 Semi-persistent Scheduling (SPS)

SPS reduces the control signaling overhead on the air interface while increasing the overall system capacity by means of scheduling the UE to receive the regular PRB resources in a fixed period with no scheduling grant on the Physical Downlink Control Channel (PDCCH). The feature is necessary in scenarios where voice service UEs and other data UEs co-exist in the same cell. Thus,

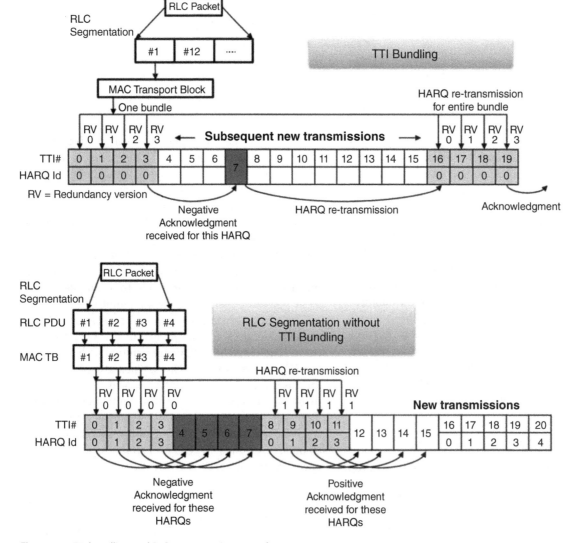

Figure 4.4 TTI bundling and RLC segmentation procedures.

the increase in the number of PRBs allocated to voice service UEs will cause a decrease in the number of PRBs available to other data UEs, and consequently, the cell throughput (capacity) will decrease. In SPS, the allocated traffic channel is released when a certain number of empty transmission slots are detected on the allocated data traffic channel. This is effectively achieved without changing the amount of resources or the packet size (i.e. MCS), at the beginning of the SPS allocation period (e.g. 10, 20, 32 … 640 subframes). With this reduction mechanism on the air interface, SPS scheduling offers up to 2.5x capacity improvement over the conventional dynamic scheduling in the limited PDCCH scenario, i.e., 5 PDCCH [26]. However, the dynamic scheduling mechanism is adopted in most deployed networks unless VoLTE load reaches the required SPS activation threshold (i.e. there is a significant increase in load with suboptimal control channel capacity). Therefore, this chapter does not present practical results for SPS since it is not widely deployed.

Table 4.1 Network and VoLTE parameters.

Configuration	DL/UL
LTE system bandwidth	20 MHz
UE category	6
MIMO configuration	MIMO 2 × 2, TM3
Mobility speed	80 km/h
RF Conditions in Mobility	**Average Values**
Serving cell RSRP [dBm]	−83.8
Serving cell RSRQ [dB]	−8.7
Serving cell RSSI [dBm]	−54.9
Serving cell SINR [dB]	20.2
VoLTE-relevant Parameter	**Value**
RoHC	ON
TTI bundling	ON
Dynamic scheduling	ON
SPS	OFF
C-DRX for VoLTE configuration	ON (LongDrxCycle = 40 ms)

4.5 Testing Environment and Main VoLTE KPIs

In this section, the VoLTE performance is assessed in terms of RoHC efficiency, RTP error rate, jitter, DL/UL BLER, handover delays, and voice quality in terms of MOS. The data were processed from field measurements with a large sample size (i.e., long VoLTE calls during mobility) and the results were averaged over two different LTE access networks (i.e., two different clusters) from two different suppliers and using two different smartphones to address the main trend and remove network and handset impact. The LTE networks in this trial are deployed with 1800 MHz (20 MHz channel) and are collocated with UMTS at 2100 MHz and GSM900/1800 MHz. The KPIs are estimated from the device side through post-processing scripts to the collected logs from the device modem. Table 4.1 summarizes the LTE network parameters, the average RF conditions during mobility, and VoLTE-related features. The testing was conducted in live, commercial LTE networks with normal loading, i.e., ~ 50%. The testing was conducted with and without PS data sessions to evaluate the impact of concurrent services (voice and data) on the VoLTE performance. The testing methodology is illustrated in Figure 4.5.

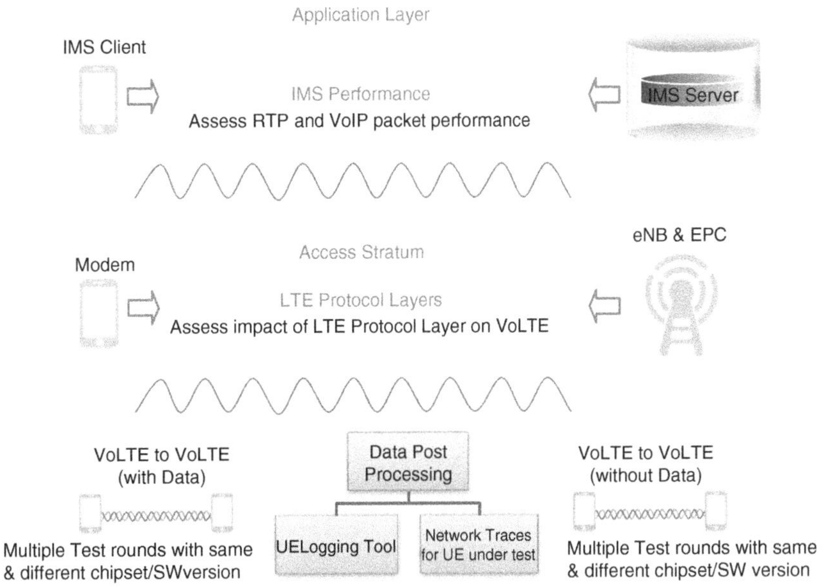

Figure 4.5 Testing methodology summary.

The main VoLTE KPIs demonstrated in this chapter may be summarized as follows:

Relative jitter: The inter-arrival time of subsequent RTP payloads calculated with reference to the previous RTP packet received in sequence during a talk spurt. The jitter is calculated per RTP stream during talk spurts as follows:

$$\text{Jitter} = (\text{IMS time of current RTP}$$
$$- \text{IMS time of previous RTP})$$
$$- \text{m}(\text{Time of current RTP}$$
$$- \text{Time of previous RTP}) \quad (4.1)$$

Then, the relative jitter $j(t)$ at time t can be calculated as follows:

$$j(t) = \text{abs}((s(t) - s(t-1)) - (r(t) - r(t-1))) \quad (4.2)$$

where $s(t)$ is the RTP timestamp embedded inside the recent received RTP packet, which is the actual timestamp of the RTP packet; $s(t-1)$ is the RTP timestamp embedded inside the previously received RTP packet, which is the actual timestamp of the RTP packet; $r(t)$ is the timestamp of the recently received RTP packet, i.e., the timestamp of arrival of the current RTP packet; and $r(t-1)$ is the timestamp of the previously received RTP packet, i.e., the timestamp of the previous RTP packet.

RTP DL Error Rate: The percentage of the RTP packets that are not received by the UE based on the RTP sequence number. The number of lost packets, "E", is calculated per RTP flow by adding the number of RTP packets lost (i.e. the gaps in the RTP sequence number). Similarly, the number of RTP packets successfully received, "N", is calculated per RTP flow by counting the number of RTP sequence numbers and payload received in order. Then, the RTP downlink error rate is calculated as follows:

$$\text{RTP DL error rate} = E/(E+N) \quad (4.3)$$

Both jitter and RTP error rate are calculated with reference to all RTP packets received over a certain interval (e.g. 1 second).

4.6 VoLTE Performance Evaluation

In this section, the VoLTE performance is evaluated in terms of RoHC, TTI, RTP error rate, jitter, BLER, handover delays (C-plane and U-plane), and voice quality in terms of MOS. All the results were obtained from commercial networks, as explained in the previous section.

4.6.1 RoHC Efficiency and Performance Evaluation

To evaluate the maximum capability of the RoHC, we tested two scenarios: the first scenario for concurrent VoLTE and data connections and the second scenario for a VoLTE standalone call. In both scenarios, the voice activity was continuous with minimum silent periods. In the first scenario, the packets at the radio side were typically multiplexing between both PS data and IMS within the same transmission time interval. However, because compression took place at the upper layers, the impact on the actual size of the radio packet (efficiency) was not much compared to the second scenario (IMS call only).

Table 4.2 provides practical results for RoHC header compression efficiency from real network deployment, as described in Section 4.4 based on IPv4. The table illustrates DL/UL header size and average compression efficiency in a mobility scenario for a long VoLTE-to-VoLTE call covering 100 eNBs. As observed, in both scenarios (i.e., with and without PS data), RoHC is capable of offering significant gain to the radio resources by reducing the packet size and compressing the headers, with an average efficiency of 81% to 92% and an overall average efficiency of 86.7%. Therefore, RoHC is very beneficial for VoLTE

Table 4.2 RoHC compression efficiency from field testing.

AMR-NB with 12.65 kbps and IPv4	Concurrent VoLTE and Data Sessions		VoLTE Session Only		Grand Average
	UL	DL	UL	DL	
AMR payload (bytes)	33	33	33	33	33
Average original header size (bytes)	40	40	40	40	40
Other protocol overhead (L1/MAC/RLC/PDCP) (bytes)[a]	8	8	8	8	8
Average compressed header size (bytes)	3.9	7.5	3.2	6.5	5.3
Average compression efficiency (%)	90.1	81.2	91.9	83.6	86.7
Required channel data rate after RoHC (kbps)	18.0	19.4	17.68	19.0	18.5

a) L1: Physical layer; MAC: Medium access control layer; PDCP: Packet data conversion protocol.

traffic transmitted alone or alongside other data traffic, which is a typical case in a smartphone (i.e. background data are ongoing while the user is on a voice call).

In terms of channel rate saving, and using the practical values in Table 4.2, it is obvious that RoHC can reduce the transmission data rate on the radio interface from:

Physical channel data rate

$$= (P + H + O) * 8/I = 32.4 \text{ kbps} \qquad (4.4)$$

where P = AMR payload, H = average original header size, O = other protocol headers, and I = RTP packet interval (i.e., 20 ms), to:

Physical channel data rate

$$= (P + RH + O) * 8/I = 18.5 \text{ kbps} \qquad (4.5)$$

where RH is the average of the compressed RoHC headers in all scenarios listed in Table 4.2, which is 5.3 bytes.

This indicates that RoHC boosts the air interface resources by almost twofold. This air interface saving is converted directly to enhanced capacity and coverage provided by RoHC. It is expected that RoHC will even offer higher gain with IPv6, since the header is 60 bytes [27].

As is evident from these results, the original voice packet sizes on the UL and DL are the same; however, the compression rate is different. It can be observed that the UL has slightly better compression efficiency than the DL in both scenarios, which is attributed to the different compression methods used by the UE and by the eNB. In addition, the volume of information carried by the compressed data packets varies with the state in which the data packets are compressed. The decision about compression state transitions (from sending uncompressed data to compressing data with maximum compression efficiency) is made by the compressor based on many factors, such as variations in the static part or dynamic part of packet headers and the acknowledgment feedback status from the decompressor.

Due to the robust performance of RoHC in any scenario, there are ongoing discussions now that may go into 3GPP later. More specifically, RoHC can be added into PDCP packets carrying data connections (see below). RoHC applies to VoLTE packets only at the moment, and the other way to compress data is being designed under proprietary solutions in the industry, now called "Uplink Data Compression (UDC)".

Table 4.3 provides the gain in DL and UL data compression for different data types. As indicated in the table, the gain in DL compression is minimal since the data are already compressed in the DL, while the gain in the UL is significant. There is ongoing discussion now in the 3GPP to apply the UL compression to PDCP packets carrying data connections. The discussion also concerns whether we can apply the compression to the header only, similar to VoLTE, or to the header and the payload.

4.6.2 TTI Bundling

In general, TTI bundling can achieve very good coverage and reliability [22]. However, careful optimization is required, especially in the choice of the codec rate for VoLTE calls. For example, a VoLTE call with AMR-WB codec rate of 23.85 kbps may face voice quality challenges even when TTI bundling is applied. In this scenario, every RTP packet sent on the UL will require an extra 8 ms to be transmitted. Assume that the maximum TBS of 63 bytes is granted to the UE during the TTI bundling operation (the typical size during this operation) and that when RoHC is not used (either it is not configured or it is not applied due to RoHC state transition), the PDCP protocol data unit (PDU) size (including AMR payload and IPv4 RTP/UDP/IP headers) will be 102 bytes. In this scenario, one AMR payload cannot fit within one TTI and hence, segmentation is needed. If RoHC is applied with the best compression efficiency (as demonstrated previously), the header size can be reduced to 3 bytes, meaning the PDU arrives at the MAC layer with a size of 65 bytes. In both cases, the TTI bundling grant will not be sufficient over a single 4 ms bundle to transmit a complete AMR payload. Therefore, it requires more than one bundle to transmit one AMR payload (i.e. one RTP packet); hence, it can take ≥ 8 ms, which increases delays and deteriorates the voice quality at a cell edge. In all cases, it is obvious that the usage of RoHC and TTI bundling together at a cell edge engenders the optimal performance in terms of coverage and voice quality [21, 26, 27].

The lower codec rates provide larger coverage and better radio robustness because fewer voice information bits need to be sent over the air. Since media packets are generated in 20 ms intervals, the total RTP packet size (TPS) (i.e., one voice frame) can be estimated as follows:

$$TPS = \text{Payload} + \text{AMR header}$$
$$\text{RTP header} + \text{protocol header} \qquad (4.6)$$

Then, the total RTP packet sizes for 12.65 kbps and 23.85 kbps codec rates can be estimated, respectively, as follows:

$$TPS = 12.65 \times 20 \text{ ms} + 11 \text{ bits}$$
$$+ 96 \text{ bits} + 64 \text{ bits} = 424 \text{ bits} \qquad (4.7)$$
$$TPS = 23.85 \times 20 \text{ ms} + 11 \text{ bits}$$
$$+ 96 \text{ bits} + 64 \text{ bits} = 648 \text{ bits} \qquad (4.8)$$

Table 4.3 DL and UL compression gain on the PS domain.

Scenario	Video	Social Network
DL compression gain	3%	4%
UL compression gain	392%	86%

When TTI bundling is enabled, the resource allocation size is restricted to a maximum of three PRBs and the modulation scheme must be QPSK [28]. Therefore, the selected MCS index cannot be greater than 10. After TTI bundling is enabled, the maximum available TBS is as large as 504 bits, which can be bundled into four TTIs (i.e., 504 LTE bits sent every 4 ms). Therefore, one RTP packet of 424 bits can fit within the bundled TTI every 4 ms for 12.65 kbps, while we need to send one RTP packet bundled every 8 ms in the case of 23.85 kbps (i.e. two RTP packets bundled every 4 ms with the size of 648 bits). Since the VoLTE service is delay-sensitive, if higher-layer data are not transmitted within the specified delay budget, the voice quality deteriorates. This implies that RTP throughput is cut in half, impacting the jitter and voice quality at a cell edge, in addition to the negative impact on the uplink coverage. On the other hand, TTI bundling does not apply to the DL; instead, methods like RLC segmentation are used to accommodate larger RTP packets within smaller protocol layer packets at cell edges. Therefore, for a codec rate like 23.85 kbps, packets will be segmented into more, smaller-size RLC layer PDUs than for 12.65 kbps, and hence, higher voice delays can be observed on the downlink as well.

Based on the above discussion, a reduction in the codec rate to 12.65 kbps is recommended to gain more uplink coverage with an AMR payload that can fit within a bundled packet. This is another reason why the high-definition codec rate of 23.85 kbps provides better voice quality in the center of a cell, while lower rates like 12.65 kbps provide better voice quality at cell edges, as demonstrated later in this section. Adaptive switching between different codec rates based on radio conditions or network load is still not widely applied. During an ongoing call, the mobile-originated UE and mobile-terminated UE have the ability to modify the codec mode and rate. The radio condition and user experience can be taken into account in this procedure. For the network load, the network congestion can be indicated by the explicit congestion notification (ECN) mechanism via the RTP protocol using the RTCP as a feedback mechanism. However, ECN usage is very limited as it can only indicate the occurrence of congestion at the E-UTRA side without further information on its level. As for an indication of radio conditions between the UE and the eNB that can be used for codec rate adaptation, this mechanism is not clearly standardized at this point in 3GPP. Therefore, developing proper optimization processes is recommended to provide a consistent user experience in all radio conditions. More details on the voice quality comparison between different codec rates are given later in this section. TTI bundling performance evaluation is also provided later in the chapter.

4.6.3 RTP and Jitter Evaluation in Different VoLTE Call Scenarios

We conducted a VoLTE to VoLTE long call under mobility conditions as summarized in Table 4.1 and evaluated the performance of the KPIs mentioned in Section 4.5. We tested three different scenarios: concurrent VoLTE and data connections with RoHC, a VoLTE standalone call with RoHC, and a VoLTE standalone call with RoHC turned off from the eNB side. In all scenarios, the voice activity was continuous with minimum silent periods, and we performed full download using an FTP server when a data session was present in the first scenario.

The probability density function (PDF) and cumulative distribution function (CDF) of the RTP error rate for all tested scenarios are illustrated in Figure 4.6. With RoHC activated, the average downlink RTP error is within the accepted range of $\leq 1\%$. However, due to the presence of data packets in parallel with RTP voice packets, the RTP errors observed are slightly higher. This can happen especially in cases when both data and voice packets are multiplexed within the same TTI, as explained earlier. In this case, the HARQ scheduled with a higher number of bits can jeopardize the VoLTE performance, causing more air interface BLER. However, as will be shown later, the impact of higher BLER is more obvious on the jitter than the RTP error rate. On the other hand, when RoHC is disabled, the RTP error rate degrades significantly. This is because the VoLTE packets are transmitted with uncompressed headers, leading to high packet sizes on the radio side. In turn, this leads to more RTP errors and holes in IMS transmission (i.e. out of sequence) that impacts the overall voice quality. It is evident that enabling RoHC has a positive impact on the performance alongside improvements to the capacity and coverage aspects, as discussed previously.

From an RTP jitter perspective, Figure 4.7 illustrates the PDF and CDF of the downlink-relative jitter for the same three scenarios. The presence of a concurrent PS data session significantly degraded the average relative jitter by 40%, even while RoHC was enabled. Additionally, the RTP error increased by 50%, as shown in Figure 4.6; however, it was still within the accepted range. In the case where RoHC was disabled during a VoLTE standalone call, the relative jitter was 20% higher compared to the same scenario with RoHC enabled. This stresses the importance of the RoHC feature to the overall VoLTE performance.

4.6.4 Scheduler Evaluation for VoLTE

The presence of PS data alongside a VoLTE call obviously degrades the overall RTP performance. It is therefore important to consider applying techniques to mitigate this negative impact. One of the options is to handle both data and VoLTE sessions in parallel at the eNB

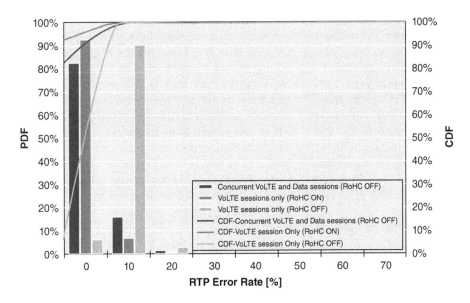

Test Scenario for RTP Error rate	Avrg	Median	Min	Max
Concurrent VoLTE and Data sessions (RoHC ON)	0.74%	0%	0%	51.2%
VoLTE session Only (RoHC ON)	0.34%	0%	0%	54.2%
VoLTE session Only (RoHC OFF)	7.77%	7.84%	0%	34.7%

Figure 4.6 Distribution of the RTP error rate under mobility conditions.

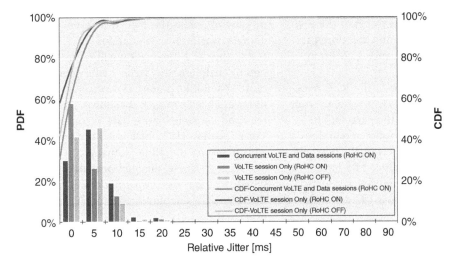

Test Scenario for Relative Jitter	Avrg	Median	Min	Max
Concurrent VoLTE and Data sessions (RoHC ON)	3.14 ms	1.0 ms	0 ms	34.7 ms
VoLTE session Only (RoHC ON)	1.87 ms	1.5 ms	0 ms	25.6 ms
VoLTE session Only (RoHC OFF)	2.32 ms	1 ms	0 ms	220 ms

Figure 4.7 RTP DL jitter with RoHC enabled, and without and with PS data.

scheduler level. In this context, the eNB scheduler can consider sending VoLTE packets in a given TTI without any data packets multiplexed, especially in bad radio conditions. When the downlink scheduler tends to utilize the same physical layer transmission to send both voice and PS data in the same TTI, the TBS can increase and this leads to higher BLER. Another technique is to use 2×2 MIMO codewords (or higher-order MIMO) for splitting the VoLTE and PS data into two different streams with Rank 2 spatial multiplexing, i.e. segmentation of packets at the MIMO codeword level. This can improve the spectral efficiency and minimize jitter and overall BLER.

So far, we have analyzed the relative jitter in different scenarios with concurrent PS data, during VoLTE standalone calls, and also with and without RoHC. In this section, we will analyze the relative jitter versus BLER at the radio interface. Figure 4.8 demonstrates the relative jitter versus DL BLER with and without a data session. As is evident from Figure 4.8, there is a clear relationship between relative downlink jitter and downlink BLER at the physical layer, as there is a trend for jitter to increase with an increase in retransmission at lower layers. It is also obvious that there are high fluctuations in jitter when PS data are present. More specifically, the jitter reached an average of 3.5 ms at a BLER of 10% with PS data, while the jitter reached an average of 2 ms at the same BLER level with a VoLTE standalone call. This implies that DL BLER and the BLER target at the eNB can affect the end-to-end RTP delays and therefore careful implementation of BLER convergence algorithms at the eNB scheduler is mandatory.

Due to delays at different interfaces, the RTP packets transfer with different inter-arrival intervals. From the eNB side, the convergence to the targeted BLER requires a stable flow of packets in order to maintain a suitable tracking time to achieve the target. Typically, the eNB scheduler selects the downlink TBS based on the reported CQI, which controls the MCS selection. The amount of data to be scheduled in a TTI determines the number of PRBs to be scheduled for UEs. Based on the feedback mechanism between the UE and the eNB, the scheduler keeps tracking the BLER measured versus the target BLER and hence starts the adjustment of the MCS and PRB to this UE. This mechanism enables the scheduler to allocate resources so as to maximize the resource utilization efficiency. However, in some cases where the packets flow irregularly, the tracking mechanism cannot converge, which can cause fluctuations in BLER, as observed in Figure 4.8. To maintain a good tradeoff between BLER and the scheduled resources, the scheduler should be designed to minimize the TBS while maintaining spectral efficiency in terms of good throughput.

To evaluate the scheduler behavior, we compared the downlink throughput of three cases: VoLTE with and without RoHC and over-the-top (OTT) VoIP using a commercial application. Typical OTT applications include Skype and FaceTime, where the IP packet transmission is handled without the network management of an operator. In this case, the OTT application utilized QCI = 9 for a normal PS data session carrying VoIP packets. Table 4.4 provides a comparison between VoLTE (with and without RoHC) and the OTT application in terms of resource utilization at different layers. As shown in Table 4.4, the OTT application consumed the most scheduling resources in the time domain, which means it requires more TTI utilization compared to VoLTE in general, while the PRB utilization in the frequency domain was almost similar for VoLTE and OTT. As a result, the OTT application will require higher throughput to maintain the voice quality and that consumes more scheduling resources from the network, affecting the overall cell capacity. As shown, the OTT application required six times more data throughput compared to VoLTE with RoHC, which means more resources mainly consumed in the time domain. On the other hand, VoLTE with RoHC enabled provided the most efficient scheduling and the lowest data rate. This indicates that when VoLTE is deployed with optimized features, it will

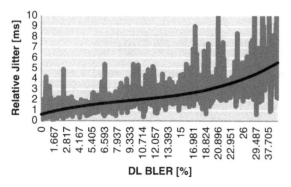

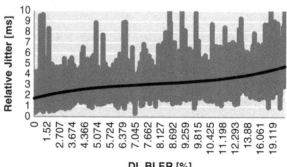

Figure 4.8 RTP relative jitter versus DL BLER.

Table 4.4 Downlink scheduling and throughput comparison between VoLTE and OTT.

Near Cell Conditions (VoLTE Scenarios Operating at 12.65 kbps and IPv4)	OTT	VoLTE (RoHC OFF)	VoLTE (RoHC ON)
Average time domain scheduling rate (%)	6	2	2
Average MAC layer TBS (bytes)	360.8	344.4	163.1
Average scheduled resource blocks	4.2	5.4	4.1
Average bit rate (kbps)	155.1	54.7	25.4

outperform OTT in cell capacity aspects. However, it should be noted that the TBS in all cases fairly exceeded the actual RTP payload (i.e. 72 bytes without RoHC and ~ 37 bytes with RoHC). This indicates that the scheduler still strives to meet the proper scheduling size when IMS packets transfer irregularly. This is an observation that may require further investigation in terms of the scheduler behavior and further work in terms of the evolution of VoLTE. We suggest reviewing the PRB to MCS mapping tables in 3GPP to add more granularity in order to absorb short and infrequent packet sizes so as to maximize the LTE spectrum efficiently.

4.6.5 RTP and Jitter Evaluation vs. Radio Conditions

In this section, we evaluate the RTP performance versus radio conditions. Figures 4.9 to 4.11 demonstrate the average RTP error rate and jitter as a function of RSRP, RSRQ, and SINR, respectively. The figures illustrate that the jitter and RTP error rate tend to increase when the RF conditions degrade. However, the most impacting factor is the RSRQ (loading and interference indicator), where the jitter and RTP error rate exhibit significant degradation (i.e., jitter > 10 ms and RTP error rate

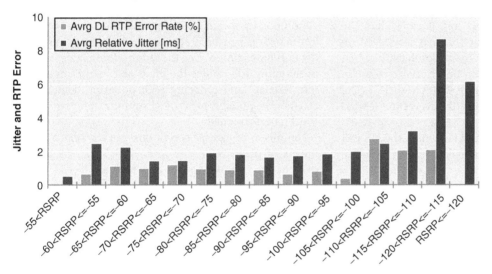

Figure 4.9 Average RTP and jitter versus RSRP.

Figure 4.10 Average RTP and jitter versus RSRQ.

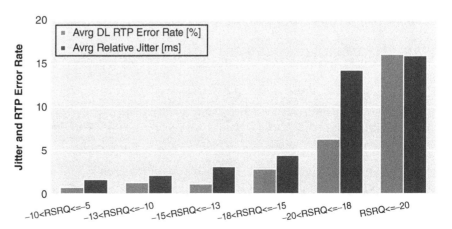

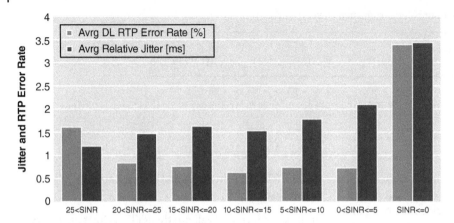

Figure 4.11 Average RTP and jitter versus SINR.

> 10%) when the RSRQ is degraded. This indicates that high interference or high loading at the eNB will directly affect the scheduling of the RTP packets and thus increase RTP delays. It is therefore important to optimize the handover parameters based on RSRQ as well as RSRP to improve the overall radio conditions in VoLTE performance. The eNBs can be configured with different handover parameters for VoLTE calls compared to other data sessions, since the eNB is aware of the VoLTE and data bearers and it can define the handover parameters differently for each service. More importantly, jitter starts to increase significantly when the RSRP < −110 dBm. This can be a decisive factor in setting the threshold for triggering eSRVCC to 3G or 2G to maintain good voice quality. With this recommended setting, the coverage of the VoLTE service will reach up to an RSRP of −110 dBm, beyond which the UE will be handed over to 3G/2G using eSRVCC in order to enhance voice quality.

RSRQ is a typical quality measure of the loading and interference in Connected mode. In a scenario where the operator has with multiple bands, RSRQ can be used as a trigger to distribute the load uniformly across the bands. In the case where one band/carrier is loaded, the RSRP will reflect only the channel condition, e.g. if the UE is near or far cell, however, the RSRQ can reflect better whether the cell on that carrier is loaded or not. Hence, the RSRQ can be used to trigger an inter-frequency handover from a loaded carrier to an unloaded one. This can be done through measurement events such as event A2 and A4, explained hereunder. As the voice service is more sensitive to the radio and loading conditions, improving the handover trigger based on RSRQ can be beneficial to VoLTE calls.

Based on the analysis of Figures 4.9 and 4.10, the RSRQ and RSRP (radio conditions) have a direct impact on the VoLTE performance, and therefore improving the radio conditions by means of RF optimization is needed. RF optimization can be categorized into: improving handover regions (improving the lack of dominance, as pilot overlapping will reduce the SINR and hence impact the interference) and reducing pilot pollution. Figure 4.12 clarifies the possible areas of RF optimization to improve VoLTE performance.

Handover triggering conditions can be improved by means of measurement reports, which can be configured by the eNB depending on whether a VoLTE call is active or not. Table 4.5 presents the events that need to beconfigured to the UE and the underlying parameter for each.

4.6.6 Handover Impact on VoLTE Call Performance

During VoLTE calls, it is expected that users will be in mobility conditions where handovers between different eNBs will occur frequently. Therefore, evaluating

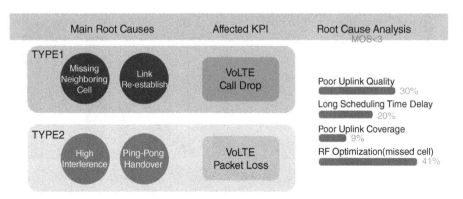

Figure 4.12 Possible RF optimization to improve VoLTE performance.

Table 4.5 3GPP events for improving handover.

A1	Serving cell becomes better than a threshold. UE measurements can be based on RSRP and/or RSRQ.
A2	Serving cell becomes worse than a threshold. UE measurements can be based on RSRP and/or RSRQ.
A3	Neighboring cell becomes offset better than the serving cell. UE measurements can be based on RSRP and/or RSRQ.
A4	Neighboring cell becomes better than a threshold. UE measurements can be based on RSRP and/or RSRQ.
A5	Serving cell becomes worse than threshold 1 and neighboring cell becomes better than threshold 2. UE measurements can be based on RSRP and/or RSRQ.
B1	Inter-RAT neighbor becomes better than a threshold. UE measurements can be based on other RAT measurement values (i.e. WCDMA CPICH RSCP and/or Ec/No).
B2	Serving cell becomes worse than threshold 1 and inter-RAT neighbor becomes better than threshold 2. UE measurements can be based on LTE and other RAT measurements.

the delays of RTP packets during the VoLTE call is important to ensure consistent voice quality in mobility conditions. The authors in [20] evaluated the performance of VoLTE calls in mobility conditions for both intra- and inter-system handovers. The results showed that inter-system handovers incurred higher delays than intra-system handovers. They evaluated both the user and control plane delays associated with these handover types. In this section, we will evaluate the impact of intra-frequency handover on the characteristics of a VoIP call represented by jitter and RTP errors that could occur during mobility conditions. Figure 4.13 demonstrates the intra-frequency handover procedures and the associated delays affecting the RTP interruption and jitter.

As explained in Figures 4.9 to 4.11, the radio conditions have a direct impact on the VoLTE performance. We will go one step further by evaluating the jitter and interruption during the handover procedure. Handover is essential during mobility to move the user from one eNB to another. If the parameters are set to delay the handover, then higher jitter and errors can be observed, as discussed previously in terms of RSRP and RSRQ. The

analysis in [20] presents the details of the voice interruption time where the actual RTP packets are suspended during the handover execution. Here, we evaluate the distribution of the RTP interruption and jitter during the handover execution procedure, as shown in Figure 4.14. As observed in the figure, the average jitter during handover is higher than the normal average observed in Figure 4.7. As estimated in [20], the average interruption time is ∼ 75 ms, which means that the jitter is directly impacted by the delay in transferring the VoLTE context from one eNB to another over the X2 interface. Given that the RTP packets are not lost, as is evident from the RTP error rate in Figure 4.14, the time difference between the jitter and RTP interruption is ∼ 40 ms. This means that there is ∼ 40 ms delay in downlink scheduling of two consecutive RTP packets during handover. The reduction of such interruption time is critical, especially for certain applications that utilize VoIP services. This topic requires further research for enhancements and it may be proposed as a contribution in LTE-Advanced Pro (3GPP Release 13 and beyond).

It is important to know that the IMS server and client maintain timers to detect any timeouts in RTP packet

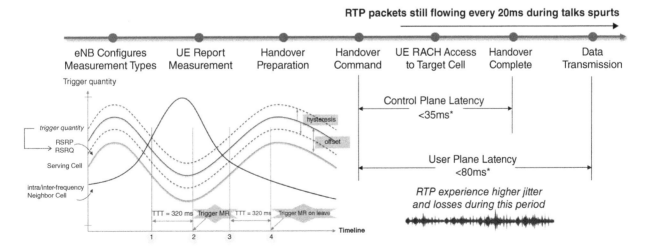

Figure 4.13 LTE intra-frequency handover procedures. *Refer to [20] for more details.

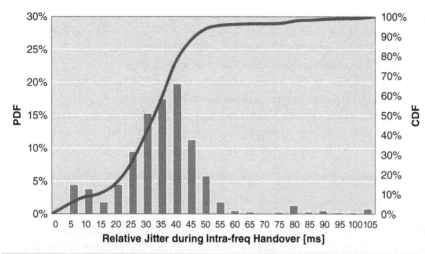

Figure 4.14 RTP performance during intra-frequency handover execution.

KPIS	Avrg	Min	Max
Number of RTP Packet lost during handover	0.33	0 ms	21
Relative DL Jitter during handover [ms]	35.6	2	922 ms

transfer, after which the VoLTE call can be released. In this context, the UE IMS client detects that DL RTP packets have not been detected within a certain time window, e.g. for 10 seconds, so the IMS client terminates the voice call to prevent the call from being active without audio. Therefore, optimizing the handover performance is essential to keeping the VoLTE call established with good retainability and consistent performance.

4.6.7 VoLTE Voice Quality Evaluation

In this section, we will evaluate the VoLTE voice quality in terms of the POLQA (Perceptual Objective Listening Quality Analysis) MOS score [28] and compare it with an OTT application (as described earlier in the chapter), 3G with AMR-WB and 2G with AMR-NB. POLQA was adopted in 2011 as ITU-T Recommendation P.863 [29]. For this series of tests we used the same clusters and testing methodology as described in Section 4.5. The testing

was conducted based on mobility with an average speed of 80 km/h. We used the Spirent Nomad voice quality tool with four channels for testing. In each test scenario, we used four devices: two for UL and two for DL. In the first session, we conducted 2G and 3G tests and in the second session we used two VoLTE calls with different coding rates, i.e., 23.85 kbps and 12.65 kbps. The test devices were locked on the tested network, i.e., 2G, 3G, or LTE with no handover or SRVCC. The OTT application was tested separately and the device was locked on LTE. The calls were long, i.e, around 16 minutes, which is around 50 cycles of the MOS testing device. We averaged the DL and UL and averaged the results over ten times for each round.

Figure 4.15 provides the average MOS values for VoLTE with a codec of 23.85, VoLTE with a codec of 12.65, OTT, 3G with AMR-WB of 12.65, and 2G with AMR-NB. The figure indicates that VoLTE engenders

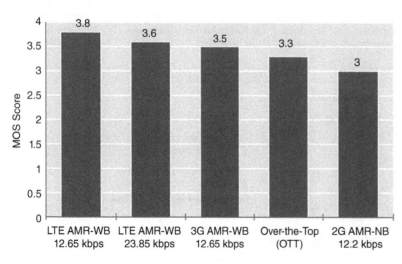

Figure 4.15 Voice quality measurements for different technologies (mobility)

the best voice quality compared to OTT and CS voice calls. In addition, the average MOS of VoLTE with the 12.65 kbps codec rate is better than VoLTE with the 23.85 kbps codec rate. This is because the 12.65 kbps rate is more robust at cell edges. Therefore, in mobility scenarios and since MOS values are averaged, the overall measured MOS of VoLTE with 12.65 kbps codec rate is better than VoLTE with 23.85 kbps codec rate. It should also be noted that LTE with 12.65 kbps rate falls within the range of "good quality", specified in ITU-T P.863 as $3.6 \leq MOS \leq 4.0$. On the other hand, 3G and OTT fall into the range of "acceptable quality", specified as $3.1 \leq MOS \leq 3.6$, while 2G falls into the "poor quality" range of $2.6 \leq MOS \leq 3.1$, as it is not high definition.

Figures 4.16 and 4.17 provide comparisons between VoLTE with the two codec rates in near-cell and far-cell scenarios, respectively. In addition, we considered VoWiFi in this exercise, as shown in Figure 4.18. The VoWiFi was tested based on an indoor environment. The near-cell criterion for VoWiFi (indoor) is RSSI −46 dBm and for LTE (outdoor) is RSRP $-60 \sim -65$ dBm. The cell-edge criterion for VoWiFi (indoor) is RSSI −73 dBm and for LTE (outdoor) is RSRP $-102 \sim -105$ dBm. The VoWiFi test was based on session border control (SBC)

Figure 4.16 Average VoLTE MOS values at near cell with different codec rates.

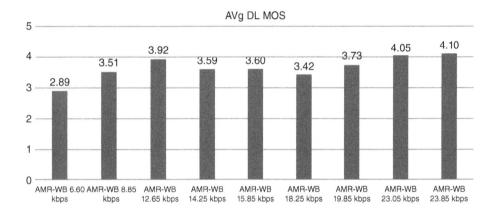

Figure 4.17 Average VoLTE MOS values at cell edges with different codec rates.

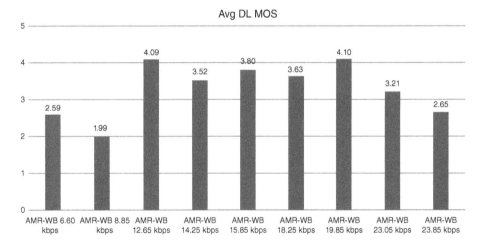

Figure 4.18 Voice quality for different VoLTEcodec rates in different RF conditions.

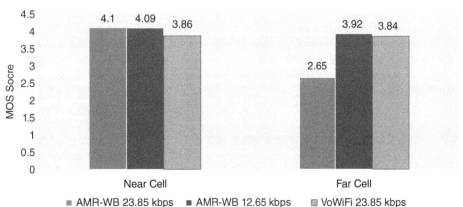

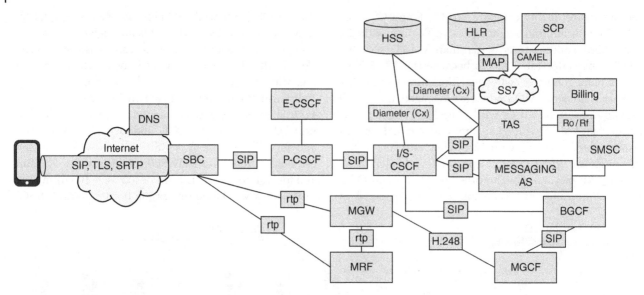

Figure 4.19 Non-roaming architecture for VoWiFi.

Wi-Fi calling. This service is a suite of communication services that emulate regular cell service using radio access technology of unlicensed Wi-Fi. The service is built using the IMS technology following the standard implementation of MMTel, as dictated by 3GPP and GSMA IR.92 guidelines. This Wi-Fi calling service (WFC) does not currently support seamless on-call mobility between (e)UTRAN and Wi-Fi (as a RAT). Figure 4.19 illustrates the topology for this VoWiFi service.

It is evident that VoLTE with the 23.85 kbps codec offers bad voice quality at cell edges, as explained earlier in the TTI bundling section. Therefore, using the speech rate of 12.65 kbps is recommended in order to guarantee a consistent user experience. In addition, the 23.85 kbps rate would consume more resources compared to the 12.65 kbps rate (due to the higher payload size), however with a minimal improvement in the voice quality in the near-cell scenario and with highly degraded quality at cell edges. VoWiFi offers a consistent voice quality experience almost comparable to VoLTE at 12.65 kbps. This proves that Wi-Fi calling can be offered along with VoLTE to complement the operator's VoIP service. In addition, VoWiFi does not need to use the operator's Wi-Fi network and may use any public Wi-Fi network. However, since VoWiFi is a VoIP service, it may be blocked by some service providers to avoid voice service cannibalization.

4.7 EVS Coding and Voice Evolution

The EVS codec has not been evaluated in this chapter in terms of practical performance, as it is not yet deployed in commercial networks. However, it is anticipated that EVS with SWB will outperform AMR-WB, even with mixed and music content, and will offer a 1.2 MOS improvement at comparable bit rates [30].

The new EVS codec continues the strong contribution from 3GPP in developing voice and audio codecs for mobile communication for GSM and 3GPP systems (3G WCDMA and LTE).

The following voice codecs have been developed:

The GSM EFR (enhanced full-rate) codec in 1996.
The AMR-NB (AMR narrowband) codec – also known simply as AMR – in 1999.
The AMR-WB (AMR wideband) codec in 2001.
The AMR-WB+ (extended AMR-WB) codec in 2004.
The new EVS codec in 2014.

The EFR codec is used in the 2G GSM system. It is also the highest quality rate in the multi-rate AMR-NB codec. The AMR-NB codec is the default codec in all 3GPP voice services in 3G and beyond (WCDMA and LTE). HD Voice is based on the AMR-WB codec, which is a wideband audio evolution of the AMR-NB codec. The extended AMR-WB (AMR-WB+) codec, with modes for coding stereo signals, is designed for generic audio for non-conversational applications such as music streaming. The EVS codec is the next step in 3GPP codec evolution. Table 4.6 summarizes the voice coding evolution [31].

The new EVS codec brings substantially enhanced voice quality, improved error resilience, and increased coding efficiency for narrowband (NB) and wideband (WB) audio bandwidths. Further quality enhancement is brought by the introduction of super wideband (SWB) and fullband (FB) audio. Figure 4.20 demonstrates the different bandwidths for voice codecs. Compared to HD Voice, the EVS codec can provide (a) substantially

Table 4.6 3GPP codecs.

	GSM/GERAN EFR (Enhanced Full-Rate)	AMR-NB (Adaptive Multi-Rate – Narrowband)	AMR-WB (Adaptive Multi-Rate – Wideband)	Extended AMR-WB (AMR-WB+)	EVS (Enhanced Voice Services)
Standardized	1996	1999	2001	2004	2014
Audio bandwidth	Narrowband	Narrowband	Wideband	Fullband	Fullband Super wideband Wideband Narrowband
Use in 3GPP	Used in the GSM system.	Default codec for voice in 3G and beyond (WCDMA, LTE).	The HD Voice codec. Default codec for wideband voice in 3G and beyond (WCDMA, LTE).	Recommended codec for generic audio in 3G and beyond (PS services over WCDMA, LTE).	The evolved HD Voice codec for LTE.
Bit rate(s) [kbps]	12.2	4.75–12.2	6.6–23.85	6–48	5.9–128

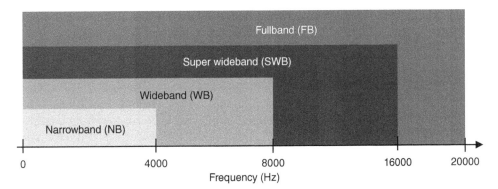

Figure 4.20 Theoretical audio bandwidths: narrowband, wideband, super wideband, and fullband. (The actual bandwidths in use may be somewhat narrower than these; for example, for narrowband, the bandwidth is typically about 300–3400 Hz.)

improved quality at similar bit rates, (b) quality raised to an unprecedented level at increased bit rates, and (c) significantly improved network capacity while maintaining the same quality as in HD Voice.

An important goal in EVS codec development was to enable the further-evolved HD Voice (brought by the EVS codec) to interact with the HD Voice service. The resulting EVS codec is fully interoperable with the AMR-WB codec of HD Voice. The EVS codec therefore represents a continuous evolution of HD Voice to even higher quality and more natural communication; it is a complementary, and not a competing, solution to reach even higher levels of voice and audio quality [31].

The features of the EVS codec are as follows [32]:

Support for input–output sampling at 8, 16, 32, 48 kHz independent of coded audio bandwidth.
Support for narrowband (NB), wideband (WB), super wideband (SWB), and fullband (FB) coded bandwidths.
Support for bit rates of 5.9 (VBR), 7.2, 8, 9.6, 13.2, 16.4, 24.4, 32, 48, 64, 96, and 128 kbps for the *EVS primary modes.*

Support for 9 AMR-WB bit rates for the *EVS AMR-WB interoperable modes.*
A jitter buffer management (JBM) solution conforming to TS 26.114 [33].
Rate switching at arbitrary frame boundaries.
Packet loss concealment.
Discontinuous transmission (DTX) operation for rates up to 24.4 kbps.

Table 4.7 shows the EVS primary modes and the signal bandwidth supported for each codec bit rate. Discontinuous transmission (DTX) operation is supported for each bit rate of the standardized codec (primary modes and interop modes). Source codec bit rates for the EVS AMR-WB interop (IO) modes are shown in Table 4.8.

Extensive testing has been performed within 3GPP to verify the performance of the EVS codec over a wide range of operating points and content types [32], including multi-bandwidth tests conducted with the P.800 DCR method [34]. Three experiments, M1, M2, and M3, have been conducted to evaluate the mixed bandwidth performance of the EVS codec. Across a wide range of bit rates, the mixed-bandwidth tests may effectively

Table 4.7 Source codec bitrates for the EVS primary modes.

Source Codec Bit Rate (kbps)	Signal Bandwidths Supported
5.9 (VBR)	NB, WB
7.2	NB, WB
8.0	NB, WB
9.6	NB, WB, SWB
13.2	NB, WB, SWB
13.2 (channel aware)	WB, SWB
16.4	NB, WB, SWB, FB
24.4	NB, WB, SWB, FB
32	WB, SWB, FB
48	WB, SWB, FB
64	WB, SWB, FB
96	WB, SWB, FB
128	WB, SWB, FB

Table 4.8 Source codec bitrates for the EVS AMR-WB IO modes.

Source Codec Bit Rate (kbps)
6.6
8.85
12.65
14.25
15.85
18.25
19.85
23.05
23.85

characterize the codec, for example, encoding up to SWB bandwidth relative to encoding up to only NB or WB. While experiments M1 and M2 used clean/noisy speech, experiment M3 used mixed/music content for evaluating extended bandwidth coding performance of the EVS codec beyond NB/WB. All three mixed-bandwidth experiments used the ITU-T P.800 DCR subjective test methodology.

Experiment M1 (DCR): NB/WB/SWB clean speech in North American English language under clean channel conditions.

Experiment M2 (DCR): NB/WB/SWB speech in Finnish language under background noise (car noise at 20 dB SNR) under clean channel conditions.

Experiment M3 (DCR): NB/WB/SWB music and mixed content in North American English language under clean channel conditions.

Experiment M1 was conducted to evaluate the EVS codec multi-bandwidth NB/WB/SWB clean speech performance under clean channel conditions. This experiment was conducted in North American English.

Figure 4.21 shows the EVS codec NB/WB/SWB performance at bit rates ranging from 5.9 to 64 kbps. For benchmarking purposes, the previously standardized codecs AMR-NB (7.4 to 12.2 kbps) and AMR-WB (8.85 to 23.85 kbps) were also tested in the same mixed-bandwidth test.

Figure 4.21 provides the results for Experiment M1. Observations from this characterization experiment with clean speech include:

The EVS-NB codec performance is significantly better than the AMR-NB across a wide range of bit rates. In

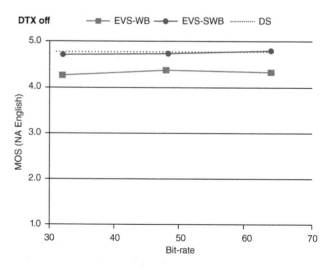

Figure 4.21 EVS-NB, WB, SWB mixed-bandwidth test, DTX on, with clean speech in North American English.

particular, the EVS-NB coding at its lowest bit rate VBR 5.9 kbps already achieves the subjective quality comparable to that of AMR-NB at its highest bit rate of 12.2 kbps coding in clean speech. The EVS-NB coding at its highest bit rate of 24.4 kbps starts converging to the subjective quality region of AMR-WB at its lowest bit rate of 8.85 kbps.

The EVS-WB codec performance is significantly better than the previously standardized AMR-WB codec. In particular, the EVS-WB coding at 5.9 kbps achieves subjective quality better than the AMR-WB at 8.85 kbps and comparable to that of AMR-WB at 12.65 kbps in clean speech. The subjective quality of EVS-WB coding starting at 9.6 kbps is significantly better than the AMR-WB coding at its highest bit rate of 23.85 kbps. There is a steady progression of EVS-WB subjective quality from 5.9 kbps to 16.4 kbps and tending towards the region of DMOS saturation for WB at higher bit rates.

The EVS-SWB codec performance is significantly better than the previously standardized AMR-WB codec as well as the corresponding bit rates of EVS-WB. The subjective quality of EVS-SWB coding at 9.6 kbps is better than the AMR-WB at 23.85 kbps as well as EVS-WB at 24.4 kbps. There is a further significant quality increase starting from 13.2 kbps as compared to 9.6 kbps.

From the mixed-bandwidth experiment, M1, the EVS-SWB clean speech quality at 13.2 kbps already approaches that of the "Direct Source" quality based on the ITU-T P.800 DCR test methodology.

Experiment M2 was conducted to evaluate the EVS codec mixed-bandwidth NB/WB/SWB noisy speech performance (in a noisy car at 20 dB SNR) under clean channel conditions in the Finnish language. Figure 4.22 show the EVS codec NB/WB/SWB performance at bit rates ranging from 5.9 to 64 kbps. For benchmarking purposes, the previously standardized codecs AMR-NB (7.4 to 12.2 kbps) and AMR-WB (8.85 to 23.85 kbps) were also tested in the same mixed-bandwidth test.

Figure 4.22 provides the results for Experiment M2. Observations from this characterization experiment with noisy speech include:

The EVS-NB codec performance is significantly better than the AMR-NB across a wide range of bit rates. In particular, the subjective quality of EVS-NB coding at 7.2 kbps is comparable to that of AMR-NB at its highest bit rate of 12.2 kbps coding. There is a steady increase in EVS-NB subjective quality performance from 5.9 kbps to 24.4 kbps.

The EVS-WB codec performance is significantly better than the previously standardized AMR-WB codec. In particular, the EVS-WB coding at its lowest bit rate at 5.9 kbps achieves subjective quality comparable to that of AMR-WB at 8.85 kbps. The subjective quality of EVS-WB coding at 13.2 kbps is comparable to that of AMR-WB coding at its highest bit rate of 23.85 kbps in noisy speech. There is a consistent progression of EVS-WB subjective quality in DTX from 5.9 kbps to 24.4 kbps (Figure 4.22) and tending towards the region of DMOS saturation for WB at higher bit rates tested in the DTX-off configuration [32].

The EVS-SWB codec performance is significantly better than the previously standardized AMR-WB codec as well as the corresponding bit rates of EVS-WB. The subjective quality of EVS-SWB coding at 13.2 kbps is significantly better than the AMR-WB at 23.85 kbps. There is a significant progression of EVS-SWB subjective quality in DTX from 9.6 kbps to 24.4 kbps and tending towards the region of DMOS saturation for SWB at higher bit rates tested in the DTX-off configuration [32].

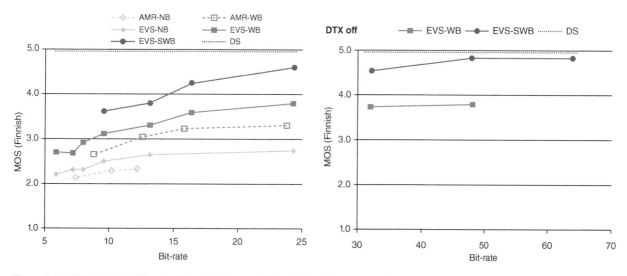

Figure 4.22 EVS-NB, WB, SWB mixed-bandwidth test, DTX on/off, with noisy speech (car noise at 20 dB SNR) in the Finnish language.

Experiment M3 was conducted to evaluate the EVS codec mixed-bandwidth NB/WB/SWB mixed/music coding performance under clean channel conditions. This experiment was conducted in North American English with mixed/music content. Figure 4.23 shows the EVS codec NB/WB/SWB performance at bit rates ranging from 5.9 to 64 kbps. The VBR mode is designed to achieve the average data rate (ADR) of 5.9 kbps for active speech. In order to further evaluate and confirm the performance of the VBR mode in music/mixed content, this experiment included the VBR condition in NB and WB. While achieving the ADR of 5.9 kbps for active speech, the VBR mode may result in a different ADR between 5.9 and 8 kbps for music/mixed content; the ADR values in this experiment were 7.01 kbps for NB and 7.53 kbps for WB for North American English music/mixed content. For benchmarking purposes, the previously standardized codecs AMR (7.4 to 12.2 kbps) and AMR-WB (8.85 to 23.85 kbps) were also tested in the same mixed-bandwidth test.

Figure 4.23 provides the results for Experiment M3. Observations from this characterization experiment with mixed/music content include:

The EVS-NB mixed/music codec performance is significantly better than the AMR across a wide range of bit rates. In particular, the subjective quality of EVS-NB coding at its lowest constant bit rate coding of 7.2 kbps is significantly better than that of AMR 7.4 kbps coding. Starting from 9.6 kbps, EVS-NB is significantly better than AMR at any bit rate. EVS-NB up to 13.2 kbps performs equal to or better than AMR-WB at a similar bit rate.

The EVS-WB codec performance is significantly better than the previously standardized AMR-WB codec at any comparable bit rate. In particular, the EVS-WB coding starting at 13.2 kbps achieves subjective quality that is significantly better than that of AMR-WB at any bit rate; for North American English music and mixed content, EVS-WB coding starting at 9.6 kbps achieves better quality than AMR-WB at any bit rate.

The EVS-SWB codec performance is significantly better than the previously standardized AMR-WB codec at any comparable bit rate. The subjective quality of EVS-SWB coding starting at 13.2 kbps is significantly better than the AMR-WB at its highest bit rate of 23.85 kbps. For North American English music and mixed content, EVS-SWB coding starting at 9.6 kbps achieves better quality than AMR-WB at its highest bit rate. There is a steady progression of EVS-SWB subjective quality in DTX from 9.6 kbps to 24.4 kbps and tending towards the region of DMOS saturation for SWB at higher bit rates tested in the DTX-off configuration [32].

For North American English music and mixed content, EVS-SWB coding performs significantly better than EVS-WB at the same bit rate.

EVS is the next-generation codec in 3GPP and provides advantages over existing 3GPP codecs in terms of:

- Extended audio bandwidth (super wideband, full-band).
- Improved performance for narrowband and wideband speech.
- Improved robustness against transmission errors.
- Lower average bit rate through discontinuous transmission and through source-controlled variable bit rate operation for active speech.

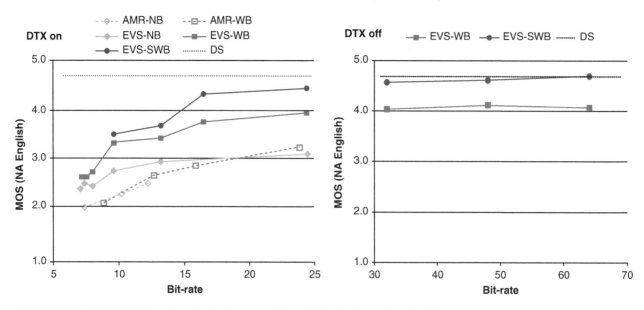

Figure 4.23 EVS-NB, WB, SWB music and mixed content, DTX on/off, in North American English.

- Better performance for music and mixed content in all bandwidths.
- Backward interoperability to AMR-WB by inclusion of EVS AMR-WB IO modes.

The fixed-point EVS codec was tested rigorously using the ITU-T P.800 [34] methodology with native listeners. The standardization successfully delivered the EVS codec standard with greatly enhanced performance as compared to current codec standards from 3GPP, ITU-T, and the IETF [33]. EVS is currently the best available codec for all mobile and VoIP applications. The performance of the EVS codec excels, especially at low bit rates of up to 24.4 kbps, a feature of utmost importance for the deployment of cost-effective mobile services, a cornerstone of mobile operator businesses.

4.8 TTI Bundling Performance Evaluation

In order to evaluate the TTI bundling gain at a cell edge, we conducted a field test to measure the average MOS at a cell edge with and without TTI bundling. Figure 4.24 provides the average MOS versus RSRP for the two scenarios. As depicted in the figure, the TTI bundling improved MOS when RSRP > 117dBm. This is the threshold that may be set to trigger TTI bundling. Forcing TTI at RSRP < 117dB will unnecessarily repeat the packet four times, which increases jitter and accordingly reduces the MOS. TTI bundling extended the cell edge to RSRP = −129dBm, while without TTI bundling, the cell radius was limited to RSRP =∼ −123 dBm (i.e., the call was dropped). Therefore, TTI bundling extended the coverage by 6 dB. This gain can be beneficial for new technologies such as the 3GPP narrowband Internet

of Things (NB-IoT), which relies on extended coverage concepts and requires a 20 dB increase in coverage to reach underground sensors and deep indoor coverage [35]. The NB-IoT will be discussed in Chapter 7.

In the uplink transmission performance, PUSCH BLER and PHICH ACK rate condition against DL path loss were evaluated. The proposed threshold was set in high DL PL condition (RSRP < −110dBm) in order to leverage LTE's better call quality. A 10% UL BLER evaluation was used to obtain the optimal threshold for SRVCC. In the actual network test scenario, note that the reference signal power in LTE was set to 18 dBm.

From the field tests, TTI bundling starts to be utilized in high PL conditions where there is not enough power headroom for the UE − DL path loss = 120dB. It is noticeable that TTI bundling improves the UL BLER, which increases the UL coverage as well as the PHICH ACK rate by approximately 5% (see Figure 4.25) [22].

Considering that the network implemented the TTI bundling feature in line with VoLTE, at DL PL = 138dB, the UL BLER condition starts to degrade abruptly above 10%, which is the limit of UL BLER for data transmission. It is concluded that beyond this PL condition, the quality of the voice call is degraded, as indicated by the UL BLER and UL data ACK rate. At PL = 135dB ,equivalent to RSRP = −117dBm − LTE, the MOS instability point obtained from Figure 4.24, the UL BLER condition is still below the 10% safe line. In this book, it is proposed that an RSRP of −117dBm is used as the serving cell SRVCC threshold. With the improvement in transmission performance due to TTI bundling, we also recommend deploying the TTI bundling feature to aid VoLTE performance improvement [22].

Figure 4.24 TTI bundling impact on MOS.

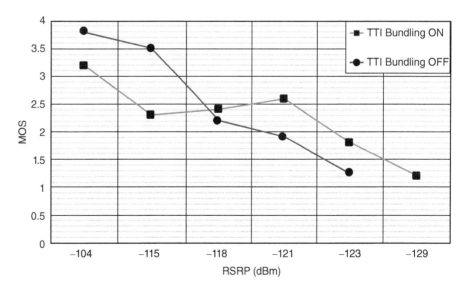

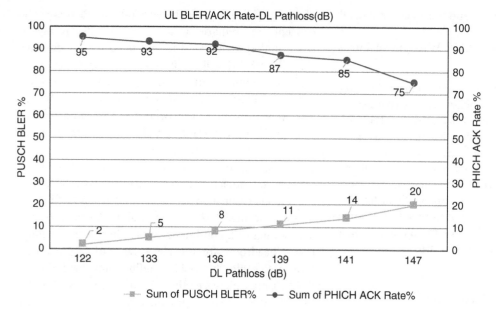

Figure 4.25 PUSCH BLER, PHICH ACK rate versus DL PL condition with TTI bundling feature deployed.

4.9 BLER Impact on Voice Quality

We conducted another experiment to demonstrate the BLER impact on voice quality. In this experiment, we enabled a feature that allows the scheduler to trigger eSRVCC (i.e. the UE goes into a 3G network) if the BLER is higher than 20%, a feature called "Quality Triggered Handover" (we did not use 10% BLER because this is a typical threshold used in all networks). We tested the average voice quality in four different routes with and without this feature. As indicated in Figure 4.26, the voice quality deteriorates before eSRVCC triggering because of high BLER (i.e. there are more dropped voice packets) and the long radio transmission delay caused by the PDCP packet discard. This justifies the case for maintaining good BLER in VoLTE.

4.10 Scheduler Performance

Table 4.9 provides a scheduler comparison between the two infra-vendors used in our testing for the case of concurrent VoLTE and data sessions. The first scheduler does not tend to utilize the same PDSCH transmission to send both IMS media and PS data in the same TTI. This means that more VoLTE media packets need to be transmitted alone, which can increase waiting time in the buffer and hence the overall jitter. Since multiplexing is not used, the TBS scheduled is low and many TBSs have many padded bits, which wastes the PRB/MCS utilization and overall capacity. In addition, this scheduler does not seem to utilize MIMO for IMS media packets and there is no RLC segmentation over two codewords. This conservative deployment can impact jitter and overall BLER. The second scheduler tends to utilize the

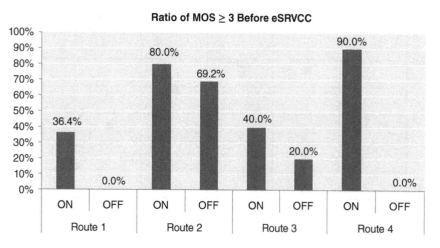

Figure 4.26 Impact of BLER on voice quality.

Table 4.9 Scheduler implementation comparison.

Scheduler Behavior	Scheduler 1	Scheduler 2
IMS media packets sent without data (i.e., packets are not multiplexed) (%)	88	53
IMS media + PS data multiplexed in same TTI (%)	12	47
IMS packets transmitted using MIMO two codewords (%)	0	5.42
Inter-TTI scheduling delay (i.e., delay between two MAC layer packets scheduled by eNB carrying IMS media) (ms)	34.3	20.9
Average TBS scheduled (bytes)	585	1291
Average padding per TBS at MAC (bytes)	302	41

same PDSCH transmission to send both IMS media and PS data in the same TTI. Since multiplexing is used, the TBS scheduled is higher in general and the scheduler tends to minimize the padding and hence improve the PRB/MCS utilization and the overall capacity. In addition, this scheduler utilizes MIMO Rank 2 for IMS media packets and, in this use case, RLC segmentation over two codewords. This improves the spectrum efficiency as well.

4.11   VoLTE KPI Evaluation

In this section, we will provide VoLTE performance assessment based on the network KPIs. The data were collected from a live LTE 3GPP Release 10 network with VoLTE enabled. The KPIs were collected from a loaded network with four LTE bands: band 3 (10 MHz), band 7 (15 MHz), band 8 (5 MHz), and band 20 (10 MHz). VoLTE services were enabled on all bands. The KPIs that will be analyzed in this section include: VoLTE call attempts, VoLTE call setup performance, VoLTE call drop performance, and VoLTE throughput requirement.

Figure 4.27 illustrates the overall VoLTE to CSFB call attempts ratio versus the cell ID in a busy hour. For every 40 voice-initiated calls, there is one VoLTE call. Therefore, the average number of CSFB call attempts is 120 and the average number of VoLTE call attempts is three during busy hours on all cells. From the attempts, VoLTE penetration can be estimated to be 2.5% of the overall number of LTE devices. There is still room to increase VoLTE penetration to retain more LTE users, since LTE capacity is higher than 3G in general.

Figures 4.28 and 4.29 provide the daily RAB establishment success rate per LTE band in a busy hour for QCI=1 and QCI=5, respectively. The figures demonstrate a stable VoLTE RAB setup success rate for both QCI=1 (media) and QCI=5 (IMS registration), as follows:

The median PS data E-RAB establishment success rate is 99.86%, while it is 99.6% for VoLTE.

On average, there are three VoLTE call setup attempts (QCI=1) and 86 IMS registration attempts (QCI=5).

There is a lower QCI=1 RAB establishment success rate on band 20, which may be related to more attempts on this loaded band.

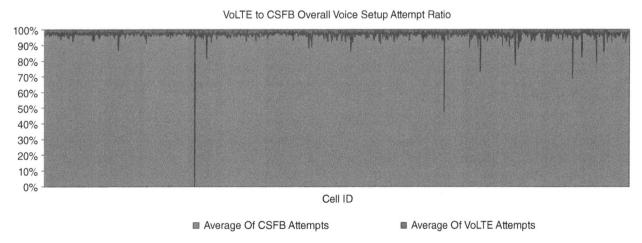

Figure 4.27 VoLTE to CSFB overall call attempts ratio versus cell ID.

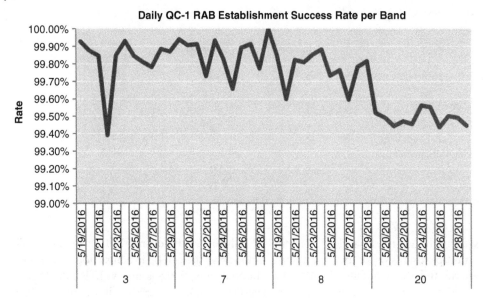

Figure 4.28 Daily RAB establishment success rate per LTE band for QCI = 1.

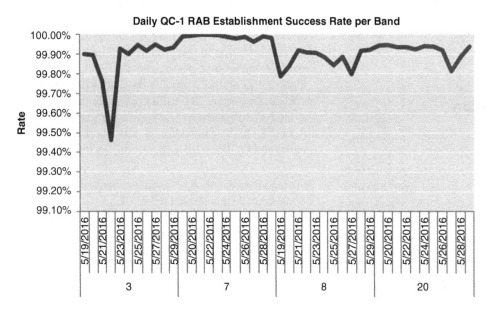

Figure 4.29 Daily RAB establishement success rate per LTE band for QCI = 5.

Figures 4.30 and 4.31 provide the daily drop rate per LTE band in a busy hour for QCI=1 and QCI=5, respectively. Figure 4.32 provides the PS daily call drop rate per LTE band. The figures illustrate a higher VoLTE call drop rate and IMS registration drop rate than PS data call drop rate, as follows:

The PS data average call drop rate is 0.07% while the VoLTE average call drop rate is 0.45%.

The QCI=5 drop rate (i.e. call drops during the IMS registration procedure) is still higher than PS data, with an average of 0.11%.

A review of VoLTE deployment is important. As there is low penetration at the moment, it may not be a current

cause for alarm, but it is something we recommend monitoring.

There is a higher VoLTE drop rate on band 8 due to the load on this band.

Figure 4.33 shows VoLTE DL and UL throughput demand in busy hours for different bands. As expected, VoLTE requires higher throughput on the UL than the DL. It is therefore essential for a good VoLTE deployment to take good care of UL optimization. On average, DL throughput = 189 kbps and UL throughput = 333 kbps. In general, this shows that VoLTE does not consume much resource from the network and will be a good deployment choice to improve

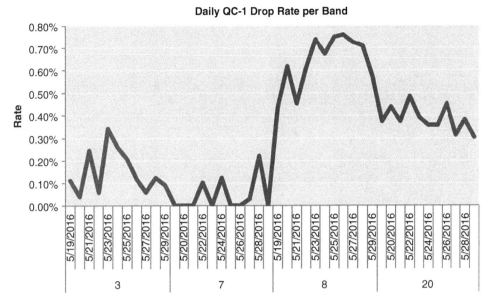

Figure 4.30 Daily drop rate per LTE band for QCI = 1.

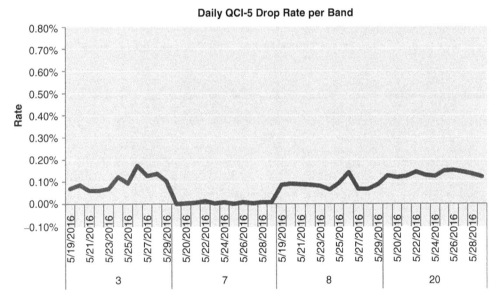

Figure 4.31 Daily drop rate per LTE band for QCI = 5.

network capacity if the strategy is to retain more users on LTE. Theoretically, if the AMR-WB 12.65 kbps codec rate is used, the expected DL throughput with RoHC for IPv4 will be ~ 15 kbps and for the AMR-WB 23 kbps codec rate it will be ~ 26 kbps.

Figure 4.34 and Table 4.10 illustrate VoLTE DL and UL resource utilization in busy hours in terms of PRB utilization at different LTE bands for the subject network. As expected, VoLTE requires very low resource block utilization from the overall resources available. In general, Figure 4.34 shows that VoLTE does not consume much UL/DL resource and will be a good deployment choice to improve network capacity if the strategy is to retain more

users on LTE. From field testing, it was observed that if the AMR-WB 12.65 kbps codec rate is used, the expected PRB utilization with RoHC for IPv4 will be ~ 4% (near cell stationary with 100% voice activity). PRB utilization will depend on voice activity and RTP packet segmentation in different RF conditions.

4.12 Use Cases and Recommendations

In this section, we will provide use cases and issues related to VoLTE performance and we will provide recommendations or resolutions for these cases.

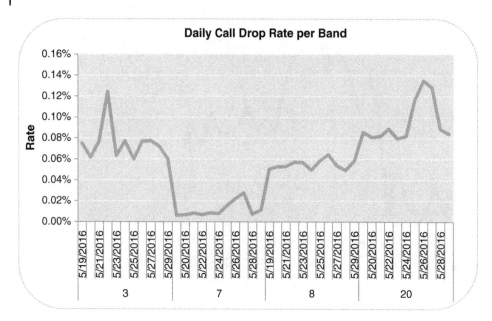

Figure 4.32 PS daily call drop rate per LTE band.

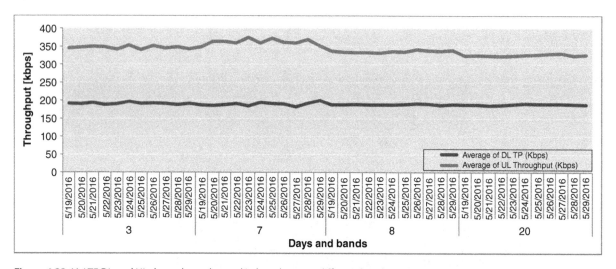

Figure 4.33 VoLTE DL and UL throughput demand in busy hours at different bands.

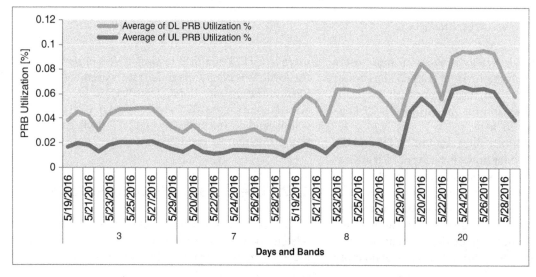

Figure 4.34 VoLTE DL and UL resource utilization at different LTE bands and in busy hours.

Table 4.10 Average VoLTE DL and UL PRB utilization at different LTE bands for different QCIs.

Band	QCI=9 Avrg DL PRB Utilization %	QCI=9 Avrg UL PRB Utilization %	QCI=1 Avrg DL PRB Utilization %	QCI=1 Avrg UL PRB Utilization %
3	5.78	10.58	0.042	0.018
7	3.75	8.35	0.027	0.012
8	7.15	12.21	0.055	0.017
20	20.44	12.65	0.080	0.054
Grand total	12.88	12.14	0.07	0.04

4.12.1 Traffic vs. User vs. Throughput Distribution in Busy Hours

We have used the same network summarized in Section 4.11 for the analysis in this section. Figure 4.35 shows the percentage of RRC connected users and DL traffic, and the number of cells and average number of connected users versus DL user throughput.

As is evident from Figure 4.35, 54% of the cells generate 89% of the traffic for 89% of the users to reach an average DL user throughput of < 2 Mbps. The overall DL user throughput during a busy hour is 2.7 Mbps and DL cell throughput = 10.4 Mbps. In addition, 59% of connected users are camping on only 23% of cells and generating 59% of DL traffic with DL user throughput < 0.5 Mbps. As a result, DL user throughput is suffocated due to band/bandwidth planning where the user has limited access to the LTE eNB, or the user has access but to a very loaded band/cell.

4.12.2 Daily Per-band Share of Average RRC Connected Users – All Hours

Figure 4.36 illustrates the average number of connected users at different LTE bands on a daily basis. Even though band 8 is deployed in ∼ 41% of the cells, the average number of daily users is much lower than band 3 and band 7,

which, combined, cover 14% of the cells. This creates a high load on the other blanket layer (band 20) that covers 45% of the cells, and ∼ 50% of the users are connected to cells in band 20, where 25% are connected to band 3 (which is not deployed widely). This impacts the end-user performance in general, and shows that less penetration of band 3 is deployed in highly used areas.

4.12.3 Main KPIs for Different LTE Bands in Busy Hours

Table 4.11 summarizes the main KPIs for different LTE bands. The following is a summary of the main observations:

MIMO observations (as a function of SINR):

B-20 is showing lower than expected for a typical low-band performance, due to higher interference with a higher load.

B-8 performs the best due to lower load and interference.

B-7 is impacted the most due to being a high band where SINR may not be very good, especially for indoor users.

B-3 has good performance but still lower than typical → many sites observed 0% R2 reporting.

CQI (as a function of RF and load):

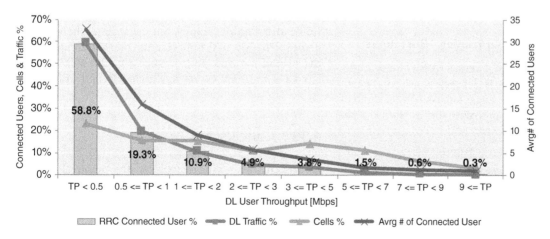

Figure 4.35 RRC connected users, DL traffic, number of cells, and average number of connected users versus DL user throughput.

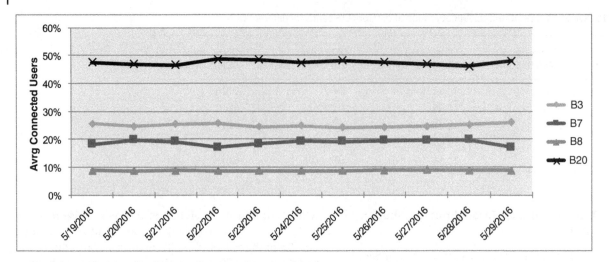

Figure 4.36 Average number of connected users at different LTE bands on a daily basis.

Table 4.11 Main KPIs for different LTE bands in busy hours.

Band	Bandwidth [MHz]	Count of Cells	Avrg # of Connected Users	Avrg MIMO R2 [%]	Avrg CQI	Avrg DL Cell Throughput [Mbps]	Avrg DL Connected User Throughput [Mbps]	Avrg UL Connected User Throughput [Mbps]	Avrg DL BLER [%]
3	10	260	11.49	43.81	12.38	14.05	3.67	0.47	6.85
7	15	35	9.81	21.76	12.91	15.68	5.50	0.68	5.58
8	5	841	4.01	62.73	11.07	9.75	4.33	0.77	5.97
20	10	924	21.36	37.20	8.67	9.79	0.94	0.16	8.27

B-20 is impacted the most due to higher load, while all other bands report good CQI.

Improving CQI in general is recommended through RF optimization.

Cell and user throughput:

B-20 is impacted the most due to the higher load → PRB utilization will be higher, which will cause lower scheduling for all users.

4.12.4 CQI vs. MIMO KPI Correlation – Busy Hours

Figure 4.37 demonstrates CQI versus MIMO Rank 2 utilization in busy hours. There is clear relationship between MIMO Rank 2 and CQI, as expected. As MIMO Rank 2 is dependent on DL SINR, improvements to SINR can be made in many ways, as follows:

Poor SINR correlates with poor RSRP.

Improving handover regions (improving lack of dominance, as pilot overlapping will reduce SINR).

Reducing pilot pollution and modifying OTA parameters for better performance.

Better load-balancing techniques when multiple carriers are deployed.

4.12.5 Per-band DL Throughput vs. CQI and MIMO – Busy Hours

Figures 4.38 and 4.39 demonstrate DL throughput versus CQI and MIMO Rank 2 respectively for different LTE bands. As is evident from the figures, the two main scheduling aspects that control the throughput (CQI and MIMO) are impacted on band 20 more than any other bands due to the high load and disproportionate traffic distribution among the cells. Since band 20 is deployed heavily, it is impacted the most and impacts on the overall network performance. On the MIMO side, band 3 is impacted the most, with less Rank 2 utilization.

4.13 Conclusions

In this chapter, we have analyzed the practical performance of VoLTE based on commercially deployed 3GPP Release 10 LTE networks. The evaluation demonstrates VoLTE performance in terms of RoHC, RTP error rate, RTP jitter and delays, BLER, and VoLTE voice quality in terms of MOS. This chapter provided best practices

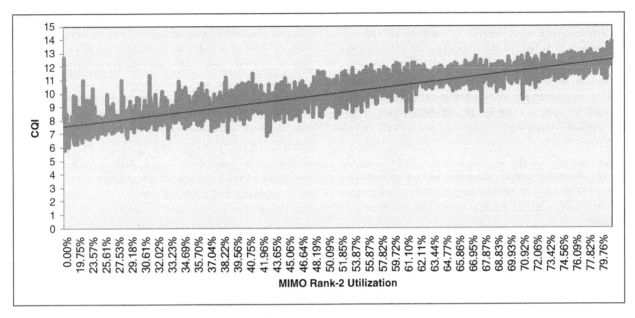

Figure 4.37 CQI versus MIMO Rank 2 utilization in busy hours.

Figure 4.38 CQI versus DL throughput.

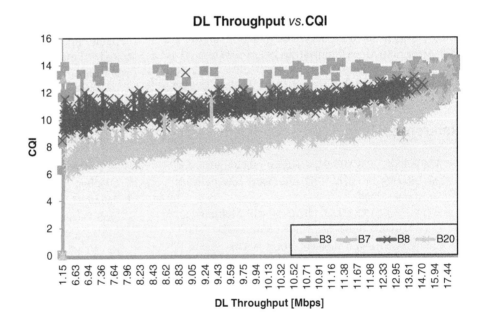

Figure 4.39 MIMO Rank 2 utilization versus DL throughput.

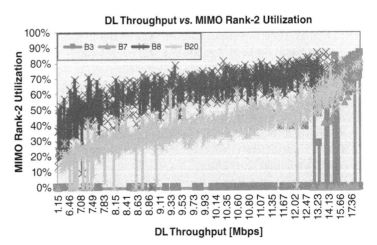

for VoLTE deployment and VoLTE-related features. We demonstrated that RoHC is capable of offering significant gain to the radio resources by reducing the packet size and compressing the headers, with an average efficiency of 81% to 92% and overall average efficiency of 86.7%. Accordingly, RoHC boosts the air interface resources by almost twofold. In addition, RoHC offers $\sim$ 7% improvement in RTP error rate and $\sim$ 24% reduction in jitter. The use of RoHC and TTI bundling together at cell edges will be the best in terms of improving the coverage and voice quality. However, reducing the codec rate to 12.65 kbps is recommended in order to gain more uplink coverage with an AMR payload that can fit within a bundled packet.

Concurrent data and VoLTE sessions cause 40% degradation in jitter and a 50% increase in the RTP error. However, they are still within the accepted ranges. In addition, jitter reached an average of 3.5 ms at a BLER of 10% with PS data while jitter reached an average of 2 ms at the same BLER level with a VoLTE standalone call. The presence of PS data alongside a VoLTE call has an obvious impact on the overall RTP performance. We provided techniques to mitigate these drawbacks. The degradation in the RSRQ leads to significant degradation in jitter and RTP error rate. More importantly, jitter starts to increase significantly when the RSRP < −110 dBm. Therefore, it is recommended that the coverage of VoLTE service should be limited to RSRP = −110. There is $\sim$ 40 ms delay in downlink scheduling of two consecutive RTP packets during handover. Reducing such interruption time is critical, especially for certain applications that utilize VoIP services.

We evaluated VoLTE in terms of voice quality using POLQA MOS. It was demonstrated that VoLTE engenders the best voice quality compared to CS voice calls (2G/3G). In addition, the average MOS of VoLTE with 12.65 kbps is better than VoLTE with 23.85 kbps. It was highlighted that VoLTE with 23.85 kbps offers very bad voice quality at cell edges. Therefore, it is recommended that the codec rate should be fixed at 12.65 kbps to guarantee a consistent user experience. Future work may include proposing techniques to reduce the RTP data interruption of 40 ms during handover and evaluating the adaptive codec selection to guarantee best voice quality and minimize radio resource utilization.

VoLTE performance was also evaluated based on network KPIs. VoLTE performance was evaluated in terms of VoLTE call attempts, VoLTE call setup performance, VoLTE call drop performance, and VoLTE throughput requirements.

Finally, practical use cases and issues were demonstrated along with recommendations and possible resolutions.

References

1 3GPP TS 22.228 V12.9.0: "Service requirements for the Internet Protocol (IP) Multimedia core network Subsystem (IMS)."

2 3GPP TS 22.173 V9.5.0 (2010-03): "IP Multimedia Core Network Subsystem (IMS) Multimedia Telephony Service and supplementary services; Stage 1," (Release 9).

3 GSMA, IR.92 "IMS Profile for Voice and SMS," Version 7.0, 3 March 2013.

4 GSMA, IR.94 "IMS Profile for Conversational Video Service," Version 5.0, 4 March 2013.

5 Elnashar, A. and El-Saidny, M. A. (2013) "Looking at LTE in Practice: A Performance Analysis of the LTE System based on Field Test Results," *IEEE Vehicular Technology Magazine*, **8**(3), 81–92, September.

6 Elnashar, A., Al-saidny, M. and Sherif, M. (2014) *Design, Deployment, and Performance of 4G-LTE Networks: A Practical Approach*, John Wiley & Sons.

7 3GPP TS 26.441: "Codec for Enhanced Voice Services (EVS); General overview". Online: http://www.3gpp.org/DynaReport/26441.html

8 Tabany, M. R. and Guy, C. G. (2013) "Performance Analysis and Deployment of VoLTE Mechanisms over 3GPP LTE-based Networks," *International Journal of Computer Science and Telecommunications*, **4**(**10**), October.

9 Tabany, M. R. and Guy, C. G. (2014) An End-to-End QoS Performance Evaluation of VoLTE in 4G E-UTRAN-based Wireless Networks, *ICWMC* **2014**, pp. 90–97.

10 Calle-Sánchez, J., Molina-García, M., Alonso, J. I. and Fernández Durán, A. (2013) "Long term evolution in high speed railway environments: feasibility and challenges," *Bell Labs Technical Journal*, **18**(2), 237–253.

11 Sniady, A., Sonderskov, M. and Soler, J. (2015) "VoLTE Performance in Railway Scenarios: Investigating VoLTE as a Viable Replacement for GSM-R," *IEEE Vehicular Technology Magazine*, **10**(3), 60–70.

12 Seo, J. and Leung, V. (2012) "Performance modeling and stability of semi-persistent scheduling with initial random access in LTE," *IEEE Transactions on Wireless Communications*, **11**(12), 4446–4456.

13 Bi, Q., Vitebsky, S., Yang, Y. Yuan, Y. and Zhang, Q. (2008) "Performance and capacity of cellular OFDMA systems with voice-over-IP traffic," *IEEE Transactions on Vehicular Technology*, **57**(6), 3641–3652.

14 Fernández-Segovia, J. A., Luna-Ramírez, S., Toril, M. and Sánchez-Sánchez, J. J. (2015) "Estimating Cell Capacity From Network Measurements in a Multi-Service LTE System," *IEEE Communications Letters*, **19**(3), 431, March.

15 Sanchez-Esguevillas, A., Carro, B., Camarillo, G., Lin, Y.-B., Garcia-Martin, M. A. and Hanzo, L. (2013) "IMS: the new generation of Internet-protocol-based multimedia services," *Proceedings of the IEEE*, **101**(8), 1860–1881.

16 Vargas Bautista, J. E. et al. (2013) "Performance of CS Fallback from LTE to UMTS," *IEEE Communications Magazine*, pp. 136–143, September.

17 Liou, R.-H. et al. (2014) "Performance of CS Fallback for Long Term Evolution Mobile Network," *IEEE Transactions on Vehicular Technology*, **63**(8), 3977–3984, October.

18 Lin, Y.-B. (2014) "Performance Evaluation of LTE eSRVCC with Limited Access Transfers," *IEEE Transactions on Wireless Communications*, **13**(5), May.

19 3GPP TS 23.272 V10.3.1 (2011-04): "Circuit Switched (CS) fallback in Evolved Packet System (EPS)."

20 Elnashar, A., El-Saidny, M. A. and Mahmoud, M. (in press) "Practical Performance Analyses of Circuit Switched Fallback (CSFB) and Voice over LTE (VoLTE)," accepted for publication in *IEEE Transactions on Vehicular Technology*.

21 Maeder, A. and Felber, A. (2013) "Performance Evaluation of ROHC Reliable and Optimistic Mode for Voice over LTE," *Proceedings of the 77th IEEE Vehicular Technology Conference (VTC Spring)*.

22 Villaluz, M. et al. (2015) "VoLTE SRVCC Optimization as Interim Solution for LTE Networks with Coverage Discontinuity," *International Conference on Information and Communication Technology Convergence (ICTC)*, pp. 212–216.

23 3GPP TS 26.114: "IP Multimedia Subsystem (IMS); Multimedia telephony; Media handling and interaction."

24 NSN (2013) *From voice over IP to voice over LTE,* White Paper, November.

25 3GPP TS 36.321 V12.5.0: "Medium Access Control (MAC) protocol specification," (Release 12).

26 Ozturk, O. and Vajapeyam, M. (2013) *Performance of VoLTE and data traffic in LTE heterogeneous networks,* IEEE GLOBECOM.

27 Venmani, D. P. et al. (2012) "Impacts of IPv6 on Robust Header Compression in LTE Mobile Networks," *ICNS 2012: The Eighth International Conference on Networking and Services*, pp. 175–180.

28 ETSI CTI Plugtests Report, VoLTE QoS Assessment, 0.1.0 (2013-12).

29 ITU-T POLQA Recommendation P.863. Online: http://www.itu.int/rec/T-REC-P.863/en

30 Fraunhofer Institute for Integrated Circuits IIS, Enhanced Voice Services (EVS) Codec, Technical Paper. Online: http://www.iis.fraunhofer.de/content/dam/iis/de/doc/ame/wp/FraunhoferIIS_Technical-Paper_EVS.pdf

31 Nokia Networks (2015) *The 3GPP Enhanced Voice Services (EVS) codec,* White Paper.

32 3GPP TR 26.952 V13.1.0 (2016-03): "Codec for Enhanced Voice Services (EVS); Performance Characterization."

33 3GPP TS 26.114: "IP Multimedia Subsystem (IMS); Multimedia telephony; Media handling and interaction."

34 ITU-T Recommendation P.800, "Methods for Subjective Determination of Transmission Quality," August 1996.

35 TR 45.820 V13.1.0 (2015): "Cellular system support for ultra low complexity and low throughput internet of things," November.

5

Evaluation of LTE-Advanced Features

5.1 Introduction to LTE-Advanced Features

The standardization in 3GPP Release 8 defines the first specifications of LTE. The Evolved Packet System (EPS) is defined, mandating the key features and components of both the radio access network (E-UTRAN) and the core network (Evolved Packet Core, EPC). Orthogonal Frequency Division Multiplexing is defined as the air interface with the ability to support multi-layer data streams using Multiple-Input, Multiple-Output (MIMO) antenna systems to increase spectral efficiency. LTE is defined as an all-IP network topology differentiated over the legacy circuit switch (CS) domain. However, Release 8 specification makes use of the CS domain to maintain compatibility with 2G and 3G systems by utilizing the voice call Circuit-Switch Fallback (CSFB) technique for any of those systems. Other significant aspects defined in this initial 3GPP release are Self-Organizing Networks (SONs) and Home Base Stations (home eNodeBs), aiming to revolutionize heterogeneous networks. Moreover, Release 8 provides techniques for smartphone battery saving, known as Connected-mode Discontinuous Reception (C-DRX).

LTE Release 9 provides improvements to Release 8 standards, most notably enabling improved network throughput by refining SONs and improving eNodeB (eNB) mobility [1]. Additional MIMO flexibility is introduced with multi-layer beamforming. Furthermore, CSFB improvements are introduced to reduce voice call setup time delays.

The International Telecommunication Union (ITU) has created the term IMT-Advanced (International Mobile Telecommunications-Advanced) to identify mobile systems whose capabilities go beyond those of IMT-2000. In order to meet this new challenge, 3GPP's partners have agreed to expand specification scope to include the development of systems beyond 3G's capabilities. Some of the key features of IMT-Advanced are: worldwide functionality and roaming, compatibility of services, interworking with other radio access systems,

and enhanced peak data rates to support advanced services and applications with a nominal speed of 100 Mbps for high mobility and 1 Gbps for low-mobility users.

Release 10 defines LTE-Advanced (LTE-A) as the first standard release that meets the ITU's requirements for Fourth Generation, 4G [2]. The increased data rates up to 1 Gbps in the downlink and 500 Mbps in the uplink are enabled through the use of scalable and flexible bandwidth allocations up to 100 MHz, known as Carrier Aggregation (CA). Additionally, improved MIMO operations have been introduced to provide higher spectral efficiency. The support for heterogeneous networks and relays added to this 3GPP release also improves capacity and coverage. Lastly, a seamless interoperation of LTE and WLAN networks is defined to support traffic offload concepts.

Release 11 continues the evolution towards the LTE-A requirements. Enhanced interference cancellation and CoMP (Coordinated Multi-Point) transmission are means for further improving the capacity in 4G networks. Furthermore, 3GPP Release 12 provides network operators with new options for increasing capacity, extending battery life, maximizing cost efficiency, supporting diverse applications and traffic types, enhancing backhaul, and providing customers with a richer, faster, and more reliable experience. This is done by introducing main categories of enhancements, such as LTE small cell and heterogeneous networks, enhancements to MIMO and beamforming, and, most importantly, the addition of more carrier aggregation combinations on the uplink and downlink including the FDD–TDD joint operation concept.

3GPP Release 13 is the first release to be officially called LTE-Advanced Pro (LTE-A Pro). In this release, LTE in unlicensed spectrum and carrier aggregation with Wi-Fi are introduced. The main topics introduced include aggregation of a primary cell, operating in licensed spectrum to deliver critical information and guaranteed Quality of Service, with a secondary cell operating in unlicensed spectrum to boost the data rate opportunistically, defined as Licensed-Assisted Access

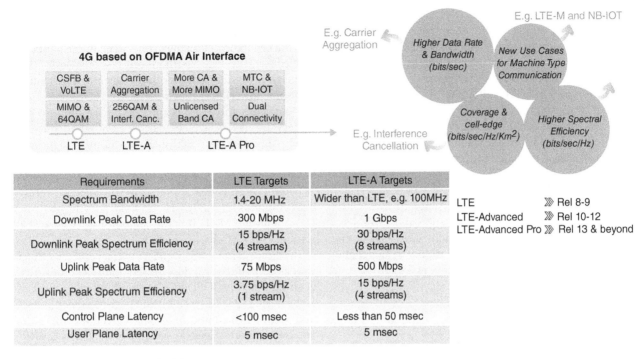

Requirements	LTE Targets	LTE-A Targets
Spectrum Bandwidth	1.4-20 MHz	Wider than LTE, e.g. 100MHz
Downlink Peak Data Rate	300 Mbps	1 Gbps
Downlink Peak Spectrum Efficiency	15 bps/Hz (4 streams)	30 bps/Hz (8 streams)
Uplink Peak Data Rate	75 Mbps	500 Mbps
Uplink Peak Spectrum Efficiency	3.75 bps/Hz (1 stream)	15 bps/Hz (4 streams)
Control Plane Latency	<100 msec	Less than 50 msec
User Plane Latency	5 msec	5 msec

Figure 5.1 Main LTE and LTE-A (Pro) features.

(LAA). In addition, LTE+Wi-Fi aggregation is added as a feature referred to as LWA (LTE/Wi-Fi Aggregation). LTE-A Pro also enhances the carrier aggregation to be up to 32 Component Carriers (CCs), and hence provides a major leap in the achievable data rates for LTE as well as in the flexibility to aggregate large numbers of carriers in different bands. The other major achievement of LTE-A Pro is the introduction of new use cases and traffic promised for Internet-of-Things (IOT) markets. The IOT is defined in Release 13 with both LTE-MTC (LTE for Machine-Type Communications) and Narrowband Internet-of-Things for LTE (NB-IoT). The target of these technologies is to provide Low-Power, Wide-Area (LPWA) communication for IoT use cases, targeting wider coverage networks and low-power consumption devices that can last for years and are designed to be cost-effective solutions for sensor-like networks.

Figure 5.1 illustrates IMT-Advanced requirements and how they are achieved from Release 8 (LTE) to Release 10 (LTE-A). An operator's decision to evolve into the advanced LTE releases depends largely on factors like device capabilities, network vendor roadmaps, and the modeled business costs.

This chapter describes the main features added in LTE-A and LTE-A Pro, covering carrier aggregation types, Higher-Order Modulation (HOM), and interference cancellation techniques as part of Releases 12 and 13. The Internet of Things is discussed in detail in Chapter 7.

5.2 Carrier Aggregation in LTE-A and LTE-A Pro

Carrier aggregation in 3GPP is not a new topic, and was introduced for HSPA+ in Release 8. In addition to improvements in the data rate and latencies by combining multiple frequencies (carriers) during downlink (or later uplink) scheduling, there have been other motivations. In particular, greater trunking efficiency with joint scheduling of radio resources over multiple frequency carriers can greatly improve the cell capacity. If the number of users stays the same, aggregating of carriers can provide load balancing across the carriers by reducing the unused resources on each. This is mainly the case for users with higher-burst data rates for bursty applications such as HTTP Web browsing, gaming, etc.

In practice, bursty traffic is more common than full-buffer traffic. In this case, all the UEs may not be actively downloading data continuously. Instead, the UEs download data in "bursts," and the data burst size is more like a truncated log-normal distribution.

In HSPA+, multiple carrier aggregation options are available in 3GPP, where Release 8 has introduced Dual-Cell HSDPA (DC-HSDPA). DC-HSDPA combines two 5 MHz adjacent carriers, hence the name "dual-cell." The majority of HSPA+ networks worldwide have deployed the feature. From there, 3GPP has been continuously evolving carrier aggregation with different combinations, in terms of bandwidth and bands. Aggregating more carriers (up to 8x5 MHz) increases the data rate, whilst expanding the carriers being aggregated across different bands would encourage operators of different spectrum to deploy. Hence, DC-HSDPA across different bands has been referred to as Dual-Band, Dual-Cell HSDPA (DB-DC-HSDPA) and multi-carrier HSDPA referred to as MC-HSDPA. The combination of different bands supported has been limited to those most common in market deployments. For example, Release 9 supports a combination of UMTS 2100 and 900 MHz.

LTE has utilized the same concept to meet the requirements of IMT-Advanced. Carrier aggregation in LTE-A (shortened to "CA" for LTE) allows the radio interface to combine any number of carriers, up to five, with several combinations of bandwidth and bands. LTE has already provided the flexibility for bandwidth usage up to 20 MHz. LTE-A, hence, adds the flexibility to aggregate different LTE bandwidths within different bands.

5.2.1 LTE TDD+TDD or FDD+FDD Carrier Aggregation

The downlink and uplink carrier aggregation can be configured independently. However, 3GPP allows the number of uplink carriers being aggregated to be equal to or less than the downlink.

Additionally, CA facilitates efficient use of fragmented spectrum for contiguous and non-contiguous spectrum aggregation. The bandwidth of carriers being aggregated can vary from 1.4, 3, 5, 10, 15, to 20 MHz [3]. This is particularly helpful for operators looking for higher data rates but who do not have contiguous 20 MHz spectrum. Therefore, 3GPP defines the term Component Carriers (CCs), which can have bandwidths of 0.4, 3, 5, 10, 15, or 20 MHz, sticking to the Release 8 numbering.

Given that LTE-A supports up to 100 MHz (i.e. 5x20 MHz), the maximum number of CCs allowed is five. The UE may simultaneously receive or transmit one or multiple CCs depending on its capabilities. Each CC supports backward-compatible operation, ensuring interoperability between CA and non-CA deployments. Only CCs that belong to the same eNodeB may be aggregated, and they are assumed to be synchronized (i.e. a single timing advance command to control the UE's uplink transmission timing for all UL CCs).

Similar to DC-HSDPA, LTE CA defines Primary Cell (PCell) and Secondary Cell (SCell) concepts. At least one of the CCs must have a control channel (PDCCH) where the RRC indicates which CC the UE will camp on for control messages. Additionally, carrier aggregation does not apply for a UE in RRC Idle mode. Hence, Idle mode mobility procedures of LTE Releases 8 and 9 apply in a network deploying CA.

A UE that is configured for CA connects to one PCell and up to four SCells. The PCell is the one operating on the primary frequency, on which the UE either performs the initial connection establishment procedure or initiates the connection re-establishment procedure, or it is the cell indicated as the primary cell in the handover procedure. The PCell is therefore responsible for EPS security procedures, upper layer system information, and some lower-layer functions.

Meanwhile, the SCell is the one operating on a secondary frequency, which may be configured once an RRC connection is established and is used to provide additional radio resources. At any point during the call, SCells can be activated or deactivated by MAC signaling. When an SCell is deactivated, the UE does not receive data, monitor the PDCCH, or send CQI feedback. A UE's identity (C-RNTI) is the same in the PCell and all its configured SCells.

The linkage between primary and secondary cells as well as component carriers on the uplink and downlink is also defined in 3GPP. PCC is the PCell's CC and SCC is the SCell's CC. The relationship between downlink CC and uplink CC, composing a serving cell (i.e. one PCell and one or more SCells), is signaled to the UE by the RRC (i.e. SIB2). The number of downlink SCCs configured is always larger than or equal to the number of UL SCCs. Figure 5.2 illustrates CA concepts.

If the UE is configured with one or more SCells, there can be multiple DL-SCH and there may be multiple UL-SCH per UE. Hence, there may be one DL-SCH and UL-SCH on the PCell, one DL-SCH and zero or one UL-SCH for each SCell.

With CA's flexibility of deployment, several types are defined in 3GPP. Figure 5.3 illustrates the types of CA bands.

The intra-band contiguous type of CA is expectedly deployed where a contiguous bandwidth wider than 20 MHz is available in the same frequency band. This type is rare in today's spectrum allocation.

When operators acquire fragmented LTE spectrum, inter-band CA becomes an important choice of deployment. Depending on the spectrum allocated, the operator can choose from the intra-band non-contiguous type or inter-band. In the former, the operator may have some bandwidth but fragmented in the same frequency band. In the latter, the bandwidth is allocated within different frequency bands.

Inter-band carrier aggregation may get a full set of combinations not only among bands, but also among different bandwidths. The carrier aggregation bandwidth class alone may not be able to give forward compatibility whenever a new set of band and bandwidth combinations is added in later releases. As more band combinations and also different bandwidth combinations are being added, there is a need for additional signaling to indicate which of the set of bandwidths the UE supports and the eNB can operate on. 3GPP-supported CA band combinations and related sets of supported uplink and downlink

Figure 5.2 Carrier aggregation concept.

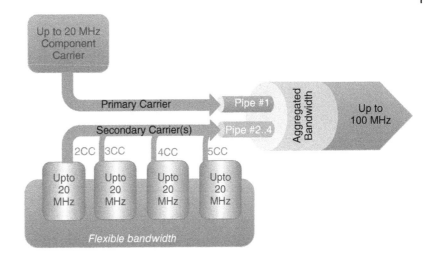

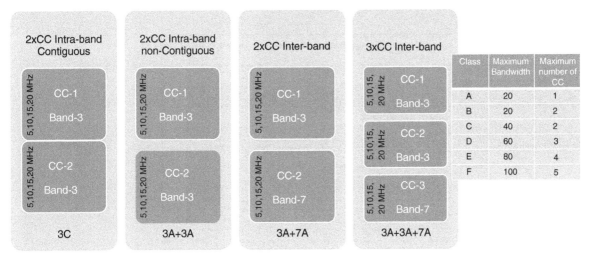

Figure 5.3 Types of carrier aggregation.

bandwidth combinations are specified in TS36.101. UE capability signaling is specified in TS36.331. With the existing signaling, it is not possible to indicate support for additional bandwidth combinations for already-existing band combinations, assuming that such additions in future releases are to be expected. Therefore, a new field, BandwidthCombinationSet (bitmap), is introduced per band combination as a non-critical extension for Release 10. This allows a Release 10 UE to signal support for new bandwidth combinations introduced in later releases. For example, for 3A–20A, there are two bandwidth combination sets (BCSs): BCS-0 and BCS-1. BCS-0 was added to Release 11, supporting up to 30 MHz aggregation (up to 20 MHz on B-3 and 10 MHz on B-20). BCS-1 was added into Release 12, supporting up to 40 MHz aggregation (up to 20 MHz on B-3 and 20 MHz on B-20). This way, the same class A (up to 40 MHz) can support both BCSs added in different 3GPP releases.

Carrier aggregation is expected to enhance both the capacity and user experience, as shown in Figures 5.4 and 5.5. In terms of capacity, a field trial conducted on a network supporting 2xCC CA with bands 20+3 showed that 49% was the average gain to the overall capacity in sites with carrier aggregation. In addition, three times better end-user DL throughput for carrier aggregation users over a single carrier was observed in the same trial.

5.2.2 FDD–TDD Joint Operation

Efficient TDD and FDD spectrum usage and utilization of different technologies jointly are becoming more and more important for future LTE deployments in order to cope with increased throughput and capacity needs. This increases the need to support LTE TDD–FDD joint operation such that both spectrum resources can be fully utilized to improve system performance and user experience. The overall structure of TDD+FDD joint carrier operation is shown in Figure 5.6. As shown, a choice of deployment depends on the network topology, but it is expected that both deployments will be available in different markets because network operators may

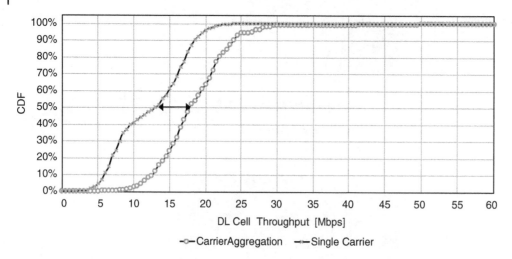

Figure 5.4 Capacity gains of carrier aggregation.

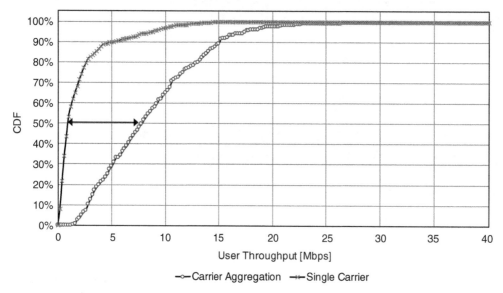

Figure 5.5 User experience gains of carrier aggregation.

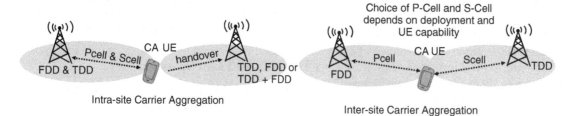

Figure 5.6 Types of TDD–FDD joint carrier aggregation.

have different phases of FDD and TDD deployments (i.e. TDD could have been deployed in a network before FDD due to spectrum availability). Inter-site carrier aggregation is discussed in later sections.

The following benefits can be considered for TDD–FDD joint operation [4]:

User throughput
  The UE can achieve higher throughput by simultaneously receiving and/or transmitting from both TDD and FDD carriers.
Load balancing
  The UE can experience higher throughput by using a less loaded carrier.

The network can utilize the resource more efficiently between TDD and FDD carriers.

Coverage and mobility

It is of interest from the operators' perspective that a carrier with lower carrier frequency provides sufficient coverage/mobility while another carrier with higher carrier frequency is used to improve throughput offloading, which may also provide mobility robustness.

Network capacity

It would be desirable from the operators' perspective to increase the frequency utilization.

The capacity gains are now realized effectively in all networks where carrier aggregation is deployed because of the joint scheduling of radio resources over multiple frequency carriers that can greatly improve the cell capacity. In practice, bursty traffic is more common than full-buffer traffic due to the nature of smartphone traffic. In this case, all the UEs may not be actively downloading data continuously. Instead, the UEs download data in "bursts," and the data burst size is more like a truncated log-normal distribution. Carrier aggregation can therefore provide the opportunity for the scheduler to utilize the resource more efficiently between different carriers deployed on different frequencies, hence balancing the load better, increasing the resource block utilization across different carriers, and improving the capacity.

The introduction of TDD–FDD joint operation will not involve significant change on the protocol layers, where much of the architecture stays the same as for FDD+FDD carrier aggregation; however, it requires some additional features, as summarized in Table 5.1 [5].

From RRC perspectives in [5], the RRC layer signaling informs the UE about the status of CA enablement through "dl-CarrierFreq-r10" IE sent in any radio bearer reconfiguration, based on the UE's band capability that it advertises through the UE Capability Information message (i.e. the presence of support for TDD+FDD

band combinations). From the UE side, in addition to advertising the proper TDD+FDD band combinations, the UE signals whether it is capable of supporting TDD PCell (primary cell, anchor) and/or FDD PCell in any supported band combination including at least one FDD band and at least one TDD band as part of the IE "tdd-FDD-CA-PCellDuplex-r12". The first bit is set to "1" if the UE supports the TDD PCell. The second bit is set to "1" if the UE supports the FDD PCell. This field is included only if the UE supports band combinations including at least one FDD band and at least one TDD band. This will give the flexibility for the UE to advertise its support for a PCell from the TDD carrier. Therefore, it is expected that the network will assign the PCell accordingly (either on FDD or TDD, depending on the UE capability). Additional IE "simultaneousRx-Tx" is used to define whether the UE supports simultaneous Rx and Tx for TDD+FDD CA operation, for which the UE shall support simultaneous reception and transmission on different bands for each band combination including at least one FDD band and at least one TDD band.

The 3GPP specification allows either TDD or FDD to be configured as the PCell, with the flexibility given to the UE depending on its capability, as explained above. The deployment of PCell as FDD or TDD depends on the band allocation of the TDD and FDD carriers and the network deployment status at the time of enabling the feature. For example, if the FDD band is deployed in the lower bands, then the FDD carrier is more suitable to serve as PCell to improve the coverage, especially on the UL (if UL CA is not enabled, the PCell will be the anchor carrier for the UL), and the TDD carrier can be configured as SCell (secondary cell) in order to enhance the downlink throughput. On the other hand, in deployments where TDD has been maturely deployed before FDD (FDD carriers are being rolled out with coverage holes), then a TDD carrier is expected to be configured by the scheduler as PCell.

The mobility concepts in TDD+FDD CA have the same procedure with standard inter-frequency handover in the same duplex LTE mode (i.e. FDD to FDD handover or TDD to TDD handover). Depending on the UE's radio capability, some events (e.g., inter-frequency) may require the eNB to enable measurement gaps, Feature Group Indicators (FGIs) indicate whether the UE supports inter-frequency measurements and handover. This support is very common among commercial devices now.

There are two modes of inter-frequency measurement: gap-based measurement and gapless measurement. For the UE to report the inter-frequency cells based on gap measurements, the eNB needs to configure the UE with a measurement gap. If measurement gaps are required, a likely approach is to configure inter-frequency measurements when the coverage of the serving LTE cell

Table 5.1 General 3GPP requirements for TDD–FDD joint operation.

Area	3GPP Requirements
Simultaneous Rx/Tx capability	Must support if TDD–FDD CA is supported in Release 12.
Duplex mode of PCell	Either TDD or FDD could be PCell.
Cell synchronization	TDD and FDD cells are synchronized.
UE capability assumed	UE supports DL CA and allows not to support UL CA.
PUCCH on PCell or SCell	PCell only.

reaches some pre-defined minimal value. In this way, the overhead associated with measurement gaps is minimized. The gap duration is always 6 ms with a repetition period of either 40 or 80 ms (corresponding to 15 or 7.5% of available subframes, respectively, and referred to as gp0 or gp1) [6]. During the time the gap is configured to the UE by higher layers, the eNB does not schedule data on the downlink for the UE and the UE does not transmit data on the uplink either, thus, when the gap is enabled for an extended duration, LTE performance can be impacted. TDD, in particular, can observe suboptimal performance when gaps are enabled due to the subframe sharing among UL and DL. Therefore, gapless measurement (also known as automatic measurement gaps) has been introduced in 3GPP, a specifically helpful technique for UEs supporting carrier aggregation with dual receivers [7]. The UE can advertise the capability of gapless measurement per band or band combination, for which, the eNB does not need to deliver gap configurations to UEs. This can improve the inter-frequency handover and shorten the data interruption time.

5.2.2.1 FDD–TDD Joint Operation Throughput Analysis

As in all carrier aggregation techniques, TDD+FDD CA provides significant improvements to the user throughput. Figure 5.7 shows the throughput expected in different RF conditions. As expected, TDD+FDD can benefit from the presence of multiple carriers with higher bandwidth. The user throughput gains are mostly observed when CQI is higher than 5; this is the expected CQI to start enabling carrier aggregation by the eNB (most eNBs configure carrier aggregation by higher layers based on CQI and traffic volume, and CQI = 5 is typically used in deployed networks).

The network on which the field testing was performed has bands 3 and 40 with 15 MHz and 20 MHz bandwidth, respectively. These bandwidth configurations and TDD UL/DL configuration = SA2 were used.

5.2.2.2 FDD–TDD Joint Operation Coverage Analysis

TDD+FDD joint carrier operation provides multiple advantages in addition to an increase in throughput. On the uplink, the FDD uplink improves TDD because the PCell sends UL acknowledgments (HARQ processes ACK), then the downlink throughput on the TDD SCell is improved compared to a TDD-only user (at a cell edge). On the other hand, TDD downlink improves FDD, because a TDD SCell is being aggregated with an FDD PCell on the DL, thus the combined throughput of TDD+FDD is better than an FDD-only user (in a cell center). It is also expected that TDD coverage will increase significantly, because at cell edges, the TDD-only user can lose its connectivity, but the TDD+FDD user will benefit from the presence of an FDD carrier.

Figure 5.8 shows the data analysis from the trial. In this figure, analysis of the throughput comparison is made between TDD+FDD CA, FDD-only, and TDD-only in terms of distance from the eNB. The aim is to assess the coverage when TDD+FDD CA is used, and the impact of the uplink and downlink on the overall performance compared with legacy carrier aggregation mechanisms.

Based on Figure 5.8, we can make the following observations:

The coverage comparison is made with the UE operating in FDD-only mode and TDD-only mode with a single carrier each. The third UE implements TDD+FDD CA. Therefore, it is a comparison of a single carrier

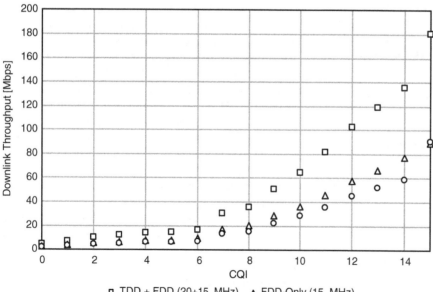

Figure 5.7 Downlink throughput as a function of channel quality indicator.

Figure 5.8 TDD + FDD carrier aggregation coverage analysis.

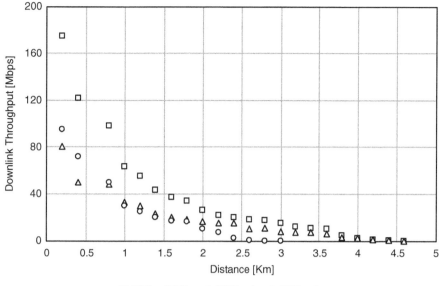

deployed on either FDD or TDD only with TDD+FDD CA (the PCell is always from FDD and the SCell from a TDD carrier).

- As FDD is deployed with 15 MHz and TDD with 20 MHz, the TDD throughput is slightly better in near-cell conditions, and the TDD+FDD throughput is almost doubled.

- At 1.5 km distance, the TDD-only UE with a higher band loses uplink coverage quicker and hence the DL throughput of this UE is impacted (DL TP reaching 5 Mbps). In the exact same location, the TDD+FDD CA device is operating with the UL from FDD and hence, the DL throughput is better because the UL is still active. At this location, TDD+FDD CA performs better than the FDD-only device as well, because the DL is aggregated from both PCell and SCell.

- At 2.5 km, the TDD-only UE loses UL coverage due to being deployed on a higher band and hence, DL throughput for the TDD-only device reaches 0 Mbps. At the same location, the TDD+FDD CA device is operating with the UL from FDD (better coverage in band 3), therefore, the DL throughput is maintained. In this location, TDD+FDD is better than FDD-only because the DL of the TDD layer is still active (i.e. CA is still active by the higher layers, and TDD CQI is still suitable to transmit transport blocks but with low MCS/PRB).

- At 4 km, the TDD-only device has 0 Mbps throughput, since the coverage on both UL and DL is totally lost. The TDD+FDD is operating with the UL from FDD and hence, the DL throughput can be maintained, similar to that in the FDD-only device. Beyond this point, the eNB disables the carrier aggregation SCell at higher layers because the TDD CQI is very low.

Therefore, it can be concluded that the coverage, in terms of user throughput, is better for the TDD+FDD device compared to the TDD-only device. At cell edges, the TDD+FDD coverage will be better than TDD-only because the FDD layer is active.

5.2.2.3 Additional Considerations

Inter-site Carrier Aggregation and Dual Connectivity The coverage footprint difference and the deployment of FDD and TDD on different eNBs may require additional considerations. The typical deployment of carrier aggregation is that with intra-eNB coverage areas, where the eNB is FDD-only, TDD-only, or is deployed with both FDD and TDD bands.

However, intra-eNB carrier aggregation limits the deployment of technologies like TDD+FDD where all sites require the collocation of bands allocated to FDD and TDD. Due to coverage difference (FDD in low bands and TDD in high bands), the network operator may need to deploy TDD in separate sites than FDD and also to reduce the number of sites, hence saving on investment costs. Therefore, it is expected that TDD+FDD joint carrier operation will be deployed as inter-eNB (inter-site) carrier aggregation. This can further introduce issues such as:

Scheduler delays between the two different eNBs.

Additional X2 interface latencies.

Re-ordering of RLC layer packets efficiently, including the segmented packets between the two eNBs.

Synchronization of time and phase between the multiple eNBs.

Several solutions can handle the inter-site carrier aggregation deployment, and these may be summarized as follows:

Inter-site TDD and FDD carriers with ideal backhaul by fiber connection.
Non-co-site TDD and FDD carriers with non-ideal backhaul.

In 3GPP 36.932, the backhauling is defined as ideal or non-ideal in terms of latency and capacity bandwidth. The ideal backhaul provides delay less than 2.5 µs with a bandwidth up to 10 Gbps, while all other types of backhaul are non-ideal. Examples of the latency and capacity requirements of backhaul are given in Table 5.2.

One of the solutions to inter-site carrier aggregation is Dual Connectivity (DuCo), which allows UEs to receive data simultaneously from different eNBs in order to boost the performance. DuCo may be designed to be applicable regardless of the combination of duplex modes (i.e., FDD+FDD, TDD+TDD, and TDD+FDD). DuCo can be deployed in inter-eNB scenarios including those covering TDD and FDD joint operation with both ideal and non-ideal backhaul. The main target of DuCo (also added in 3GPP Release 12) is to extend the LTE-A carrier aggregation functionality to allow a UE to receive data simultaneously from both a macro and a small cell eNB in a heterogeneous network. It can be utilized as a technique for inter-site TDD–FDD joint operation with non-ideal backhaul, e.g. Ethernet or microwave. It can also extend the feature to support TDD–FDD joint carrier operation provided by different vendors (in cases where TDD and FDD are deployed from different infra-vendors) with non-ideal backhaul. In general, DuCo requires UE implementation changes

in different protocol layers; in addition, eNBs involved in inter-eNB CA must be synchronized in time and phase. As a result, it is expected to be some time before we see it at a commercial level.

The other solution is to enable a relaxed backhauling technique to serve the same purpose, utilizing the existing non-ideal backhaul used in currently deployed networks. This technique, as shown in Figure 5.9, uses backhaul links characterized by lower latency, and utilizes the independent scheduler functionalities residing in each eNB. The common implementation of this technique is to utilize relaxed backhauling in which the eNBs are interconnected with a one-way delay ≤ 10 ms round trip time (best performance of 4–5 ms). From the scheduler perspective, inter-site CA generates RLC PDUs and transmits them to the SCell in advance, then transmits RLC PDU segments to fit within the scheduled MAC PDU size in order to minimize backhaul delay performance degradation.

Compared to DuCo, inter-site CA with relaxed backhaul performs the aggregation at the MAC layer level within the eNB, while DuCo performs the aggregation at higher layers (e.g. the PDCP layer) within the S-GW (or the master eNB in small-cell deployment), where traffic splitting can be implemented with a different mechanism.

TDD–FDD Joint Operation Bands in 3GPP Band combinations for carrier aggregation are increasing rapidly based on the requirements of each region and operator. 3GPP up to Release 14 has introduced ~356 band combinations for inter-band CA with two carriers, ~160 combinations for inter-band CA with three carriers, ~25 combinations for inter-band CA with four carriers, and ~65 combinations for intra-band CA.

Table 5.2 Ideal and non-ideal backhaul latency and capacity categories.

Type	Backhaul Technology	Latency (One way)	Throughput
Non-ideal backhaul	Fiber access 2	5–10 ms	100–1000 Mbps
	Wireless backhaul	5–35 ms	10–100 Mbps typical
Ideal backhaul	Fiber access 4	less than 2.5 µs	Up to 10 Gbps

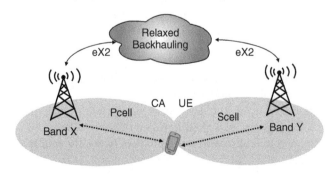

Figure 5.9 Inter-site TDD + FDD carrier aggregation with non-ideal backhaul.

Table 5.3 Inter-band combinations for TDD+FDD carrier aggregation.

FDD Bands → TDD Bands ↓	1	3	5	7	8	11	19	20	21	25	26	28
38	40	40	40					40				
39					55							
40	40	60	50	60	30			40				80
41	60	100			80	50				80	55	50
42	60	60		60	50	50	55	60	55			60

For TDD+FDD CA, the downlink combinations for inter-band CA with up to two bands are shown in Table 5.3. The band combinations vary in terms of bandwidth and the maximum aggregated component carriers. The table shows the maximum allocated bandwidth for each band combination, in MHz, from 3GPP 36.101. Some entries show up to 100 MHz of aggregation, which implies contiguous and non-contiguous aggregation. For example, bands 1+41 can work with an aggregation up to 60 MHz – 20 MHz on band 1 and 40 MHz contiguous on band 41.

As bands and band combinations are increasing significantly in different regions, some design requirements at the RF front-end (RFFE) side need to be taken into consideration:

The frequency duplex gap between uplink and downlink represents the RFFE filtering sharpness. Some bands require a minimum gap between DL and UL (e.g. band 13 with a gap of 12 MHz). This gap makes duplex filter design difficult and may need additional transmit power reduction with a relaxed receive sensitivity requirement.

Frequency band allocation is also a potential case for consideration. The digital divided band 28 has relatively wide range, low-band frequency: uplink 703–748 MHz and downlink 758–803 MHz. The duplex filter has limited relative bandwidth (frequency gap/center frequency = relative bandwidth). Therefore, the band 28 duplexer is usually separated into two: band 28A and 28B. This can result in a co-existence issue that can occur not only between cellular bands, but also between cellular bands and other air interfaces, such as GPS, Wi-Fi, Bluetooth, or FM. Generally, the solutions require more complex filtering, additional maximum power reduction, or relaxation of expected performance. Because bands 7, 38, 40, and 41 are the bands closest to the ISM band, the RFFE design needs to contain a filter to avoid interference, which impacts the transmit path link budget and also increases the switch usage for band selection.

From TDD+FDD CA perspectives, some bands may not be suitable for aggregation even if supported by the same operator. For example, band 7 and band 38 are adjacent FDD and TDD bands (within 2600 MHz), and supporting TDD+FDD carrier aggregation for them can lead to challenges in the UE's RFFE design. In this respect, the receiver of one band can be blocked by the transmitter of the other band, impacting the size and cost of the device. Therefore, if the deployment is designed in such a way that there is no uplink on band 7 and band 38 simultaneously, the in-device co-existence issue can be avoided. For example, a third band can be aggregated as the PCell, while band 7 and band 38 are only used for SCells to supplement the downlink transmissions. As a result, 3GPP has specified a combination of band 7+38 but with respect to three-carrier aggregation, where 3A-7A-38A has been added for UEs of Category 9 and above. In this three-band CA, band 3 is always used for the PCell with both uplink and downlink transmission with the UE, while both bands 7 and 38 provide a link to the UE as SCell downlink aggregation without uplink transmissions.

5.2.3 LTE Licensed Assisted Access (LAA)

As mobile broadband services expand, users expect radio networks to deliver larger bandwidth and higher throughput. This can be achieved by making good use of low-cost, unlicensed spectrum. Licensed assisted access (LAA) taps into unlicensed spectrum as an adjunct to licensed spectrum, aiming for substantial uplifts in mobile networks' data rate and user experience [8]. With LAA, cellular networks can greatly offload data traffic to unlicensed spectrum, which delivers the same robustness in mobility and QoS as licensed spectrum.

In a global context, unlicensed spectrum is mostly available on the 2.4 GHz and 5 GHz frequency bands. LAA is deployed on the 5 GHz band, which is further divided into multiple sub-bands. Table 5.4 lists these sub-bands and application scenarios in certain countries and regions.

Table 5.4 Unlicensed sub-bands on the 5 GHz band.

Sub-band (MHz)		5150–5250	5250–5350	5470–5725	5725–5825
Application scenario	US/Canada	Indoor/Outdoor	Indoor/Outdoor	Indoor/Outdoor	Indoor/Outdoor
	EU	Indoor	Indoor/Outdoor	Indoor/Outdoor	N/A
	Korea	Indoor	Indoor/Outdoor	Indoor/Outdoor	Indoor/Outdoor
	Japan	Indoor	Indoor	Indoor/Outdoor	N/A
	China	Indoor	Indoor	N/A	Indoor/Outdoor
	Australia	Indoor	Indoor/Outdoor	Indoor/Outdoor	Indoor/Outdoor
	India	Indoor	Indoor	N/A	Indoor/Outdoor

Carrier aggregation with at least one SCell operating in the unlicensed spectrum (e.g. 5 GHz) is referred to as licensed assisted access (LAA), as shown in Figure 5.10. In LAA, the configured set of serving cells for a UE always includes at least one SCell operating in the unlicensed spectrum, also called the LAA SCell. LAA SCells act as regular SCells and are limited to downlink transmissions in 3GPP Release 13.

LAA provides the following advantages in terms of network deployment:

Reduces operators' costs in extending LTE spectrum.
Builds on existing licensed LTE networks, making them easier to deploy and manage. On one hand, licensed LTE networks still manage user authentication, UE access, transmission, and security. On the other hand, extra unlicensed spectrum allows licensed LTE networks to accommodate more carriers and cells.
Improves the downlink throughput and mobile broadband experience of UEs.
Enables larger coverage areas and cell capacity under the same conditions thanks to the use of LTE physical layer techniques and multi-UE centralized scheduling as well as the PCell in CA.

In Release 13, if the maximum number of unlicensed channels on which the E-UTRAN can simultaneously transmit is equal to or less than four, the maximum frequency separation between any two carrier center frequencies on which LAA SCell transmissions are performed should be less than or equal to 62 MHz. Therefore, LAA downlink CA can involve up to three LTE and LAA component carriers (CCs) to deliver a maximum CA capability of 60 MHz bandwidth. LAA CA in the downlink supports three types of CC combination:

2CCs: One LTE CC plus one LAA CC.
3CCs: Two LTE CCs plus one LAA CC.
3CCs: One LTE CC plus two LAA CCs.

The LAA SCell configuration procedure is based on the UE measurement of the LTE PCell. An LAA cell operates in unlicensed spectrum, which may also be used by Wi-Fi cells. This calls for fair co-existence between LAA and other radio access technologies (RATs). The LAA eNB applies Listen-Before-Talk (LBT) before performing a transmission on an LAA SCell. When LBT is applied, the transmitter listens to/senses the channel to determine whether the channel is free or busy. If the channel is determined to be free, the transmitter may perform the transmission; otherwise, it does not perform the transmission. If an LAA eNB uses channel access signals of other technologies for the purpose of LAA channel access, it should continue to meet the LAA maximum energy detection threshold requirement.

Figure 5.10 Overall LTE LAA concept.

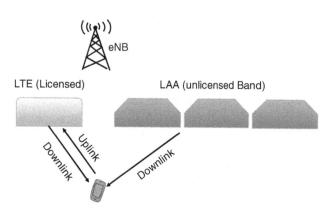

5.2.4 LTE and Wi-Fi Aggregation (LWA)

LWA is an approach introduced in 3GPP Release 13 which is capable of leveraging legacy devices and network architecture, allowing the aggregation of LTE and Wi-Fi traffic on the downlink in licensed and unlicensed bands, respectively.

For LTE traffic transmission, LWA uses unlicensed bands just like LAA, but transmission is made through Wi-Fi unlike LAA. This means LWA does not need new LTE-enabled 5 GHz hardware, and it can transmit LTE traffic through Wi-Fi Access Points (APs) connected to the LWA network. This provides an advantage in that the Wi-Fi APs can use LTE core network functions (e.g. authentication, security, etc.) without a dedicated gateway, depending on the type of deployment. The other advantages of LWA are:

Better control and utilization of resources on both LTE and WLAN links – hence good load balance across traffic in the network.

Increased aggregated throughput for a single user and multi-users, which improves the system capacity.

Fair co-existence between different technologies operating in different bands, where interference in the 5 GHz unlicensed band does not become an issue.

Cost savings, as new hardware may not be required, and existing eNBs and Wi-Fi APs can become LWA-enabled with a software upgrade. This depends on the deployment scenario, as will be described next.

E-UTRAN supports LTE-WLAN aggregation operation whereby a UE is configured by the eNB to utilize radio resources of LTE and WLAN [8]. Two scenarios are supported depending on the backhaul connection between LTE and WLAN, as illustrated in the overall architecture in Figure 5.11:

Non-collocated LWA scenario for a non-ideal backhaul.
Collocated LWA scenario for an ideal/internal backhaul.

In the collocated scenario, the LTE eNB and the Wi-Fi AP are integrated. The LWA eNB performs scheduling for PDCP packets on the PDCP layer, and transmits some over LTE and others over Wi-Fi via the APs after encapsulating them in Wi-Fi frames. This is referred to as the L2-forwarding mechanism. All the packets received, over either LTE or Wi-Fi, are then aggregated at the PDCP layer of the LWA UE. In the collocated LWA scenario, the interface between LTE and WLAN depends on the implementation. For LWA, the only required interfaces to the core network are S1-U and S1-MME, which are terminated at the eNB. No core network interface is required for the WLAN. The protocol layer architecture is shown in Figure 5.12.

In the non-collocated LWA scenario, as shown in Figure 5.12, the eNB is connected to one or more WTs via an Xw interface. The Xw user plane interface (Xw-U) is defined between the eNB and WT. The Xw-U interface supports flow control based on feedback from WT. The Xw-U interface is used to deliver LWA PDUs between the eNB and WT. For LWA, the S1-U terminates in the eNB and, if Xw-U user data bearers are associated with E-RABs for which the LWA bearer option is configured, the user plane data are transferred from the eNB to WT using the Xw-U interface. In the non-collocated LWA scenario, the Xw control plane interface (Xw-C) is defined between the eNB and WT. The application layer signaling protocol is referred to as Xw-AP (Xw Application Protocol). There is only one S1-MME connection per LWA UE between the eNB and the MME. Respective coordination between the eNB and WT is performed by means of Xw interface signaling. The S1-MME is terminated in the eNB; the eNB and the WT are interconnected via the Xw-C interface. The transport network layer of Xw-C is built on SCTP on top of IP. The application layer signaling protocol is referred to as Xw-AP [8].

A WLAN mobility set is a set of one or more WLAN Access Points (APs) identified by one or more SSIDs, within which WLAN mobility mechanisms apply while the UE is configured with LWA bearer(s), i.e., the UE may perform mobility between WLAN APs belonging to the mobility set without informing the eNB. The

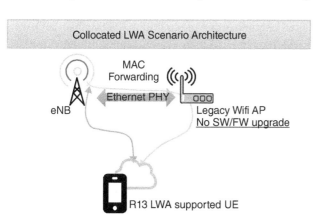

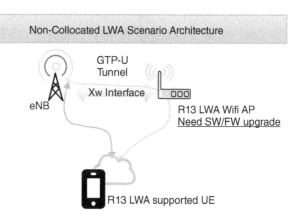

Figure 5.11 Overall architecture for each LWA deployment scenario.

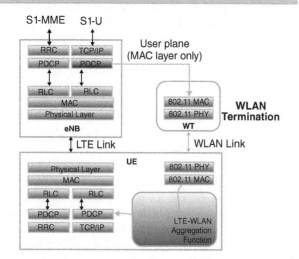
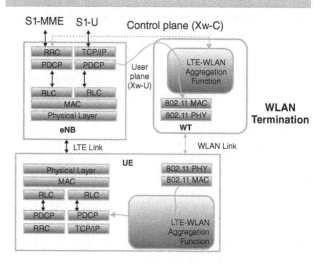

Figure 5.12 Overall protocol for each LWA deployment scenario.

eNB provides the UE with a WLAN mobility set. When the UE is configured with a WLAN mobility set, it will attempt to connect to a WLAN whose identifiers match those of the configured mobility set. UE mobility to WLAN APs not belonging to the UE mobility set is controlled by the eNB, e.g. updating the WLAN mobility set based on measurement reports provided by the UE. A UE is connected to at most one mobility set at a time.

All APs belonging to a mobility set share a common WT which terminates Xw-C and Xw-U. The termination end points for Xw-C and Xw-U may differ. The WLAN identifiers belonging to a mobility set may be a subset of all WLAN identifiers associated with the WT. A UE supporting LWA may be configured by the E-UTRAN to perform WLAN measurements. WLAN measurement objects can be configured using WLAN identifiers (e.g. SSIDs), WLAN channel numbers, and WLAN bands. WLAN measurement reporting is triggered using RSSI. A WLAN measurement report may contain RSSI, channel utilization, station count, admission capacity, backhaul rate, and a WLAN identifier. WLAN measurements may be configured to support the following functions that can be interchanged as over-the-air messages from eNB to UE through the Xw-C interface:

LWA activation
Inter-WLAN mobility set mobility
LWA deactivation.

5.3 Higher-order Modulation (HOM) for Uplink and Downlink

Cellular technology evolution depends greatly on the modulation and coding schemes in order to leverage more bits per symbol transmission depending on channel conditions. Common early digital modulation types include ASK (amplitude shift keying) and FSK (frequency shift keying). These schemes cause the carrier to assume one of two possible states depending on whether the system must transmit a binary 1 or a binary 0; each discrete carrier state is referred to as a symbol. Quadrature phase shift keying (QPSK) is another modulation technique, and this transmits two bits per symbol. In other words, a QPSK symbol does not represent 0 or 1, but 00, 01, 10, or 11. This two-bits-per-symbol transmission is possible because the carrier variations are not limited to two states. In ASK, for example, the carrier amplitude is either amplitude option A (representing a 1) or amplitude option B (representing a 0). In QPSK, the carrier varies in terms of phase, not frequency, and there are four possible phase shifts. Phase shift keying techniques provide advantages in terms of higher bandwidth and spectral efficiency that can be a factor in terms of the many bits transmitted per symbol. Beyond QPSK, higher-order modulations (HOM) like 16QAM, 64QAM, and 256QAM (QAM stands for Quadrature Amplitude Modulation) enable a larger access channel rate in high-SNR (Signal-to-Noise Ratio) areas, which better approximates the Shannon channel capacity in very good radio environments. These QAM techniques can transmit more bits per symbol but are bounded by the SNR of the radio channel. Hence, adaptive modulation and coding schemes are typically used in today's cellular networks like 3G and 4G. Adaptive modulation means that the network scheduler can adaptively schedule the device with a different modulation scheme depending on the channel condition feedback from the device.

For the data channels in LTE, QPSK, 16QAM, and 64QAM were supported up to 3GPP Release 11. In Release 12 and beyond, 3GPP introduced 256QAM modulation for both downlink and uplink data channels. Historically, GPRS technology evolved into Enhanced Data Rates for Global Evolution (EDGE), which introduced a higher-rate modulation scheme (8-PSK, phase shift keying) and further enhanced the peak data rate to 384 kbps. When 3G HSPA was introduced in Release 5, the peak data rates of 14.4 Mbps offered by this standard were made possible by introducing faster scheduling with shorter subframes and the use of a 16QAM modulation scheme. High-Speed Uplink Packet Access (HSUPA) was standardized in Release 6, with a maximum rate of 5.76 Mbps with QPSK modulation. Both of these standards, together known as HSPA (High-Speed Packet Access), were then upgraded in Release 7 of the 3GPP standard to HSPA+, which used many techniques to increase data rates on the uplink and downlink. Among these was adding an even higher modulation scheme (64QAM) on the downlink and 16QAM on the uplink.

In LTE, Release 8 came with 64QAM support for both uplink and downlink. However, the majority of devices available in the market up to 2016 did not have 64QAM on the uplink with UE category 4. As LTE started to evolve more and more towards higher speed, better spectral efficiency, and capacity requirements, a rather straightforward way of providing higher data rates within a given transmission bandwidth was the use of higher-order modulation schemes. Using higher-order modulation allows us to represent more bits with a single modulated symbol and directly increases bandwidth utilization. Thus, it is expected that more devices will start supporting 64QAM on the uplink and even higher modulation schemes like 256QAM on the downlink beyond 2016. Figure 5.13 shows the constellation of 16QAM, 64QAM, and 256QAM.

However, the higher bandwidth utilization through using HOM leads to a reduced minimum distance between modulated symbols with an increased sensitivity to noise and interference, inter-symbol interference, as shown clearly in Figure 5.13. Consequently, adaptive modulation and coding and other link adaptation strategies can be used in the eNB scheduler to decide when to use lower- or higher-order modulation based on the UE's feedback through CQI reports for the downlink and the SINR estimation on the uplink. By applying these adaptive methods, the scheduler can substantially improve the throughput and achievable data rates in a communication link for both downlink and uplink.

5.3.1 Downlink 256QAM

In 3GPP Release 12, 256QAM utilization is allowed in conjunction with carrier aggregation as part of UE Category 12, for example. Compared to carrier aggregation with UE Category 6, providing 64QAM only, the throughput can reach 391.6 Mbps at the physical layer for Category 12, as shown in Figure 5.14. Therefore, applying 256QAM provides an extra 90 Mbps increase in a single-user throughput in very good radio conditions. However, this gain depends greatly on the channel conditions.

Let's take a closer look at measurements from a live Release 11 network with 64QAM in the downlink as a maximum modulation scheme user. As shown in Figure 5.15, there is a clear relationship between the number of users in a cell and the average CQI reported by users. Additionally, 64QAM utilization can be very low depending on RF conditions and loading in cells. As shown in the figure, on average, the 64QAM utilization is 20% for 50% of the cells. In real-life deployment, if the average CQI is at 10, then 256QAM utilization will be very low in general. Modulation has a dependency on

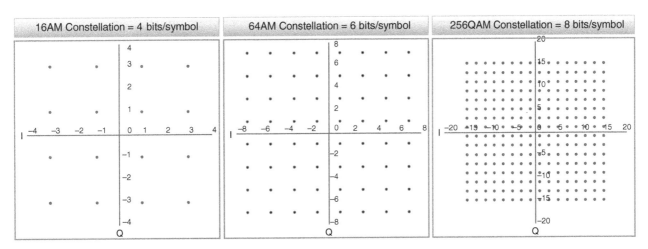

Figure 5.13 Higher-order modulation, QAM concepts.

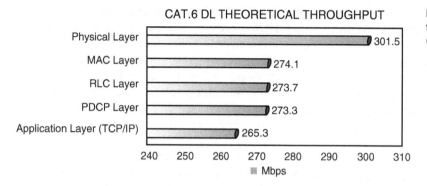

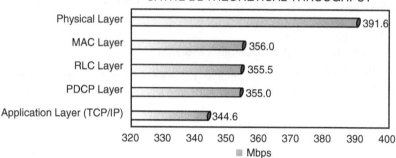

Figure 5.14 Theoretical downlink throughput for Category 6 (64QAM) and Category 12 (256QAM).

the MCS chosen based on the UE-reported CQI. Therefore, network optimization is crucial before deploying higher-order modulation like 256QAM.

In 3GPP Release 12, 256QAM utilization starts at CQI 12, as shown in Table 5.5.

This means that to realize higher utilization of 256QAM, the channel conditions need to be above CQI = 12. To evaluate this in terms of SINR levels, let's look at the live network data shown in Figure 5.16. As observed, there is a clear relationship between SINR and CQI, as expected. Improving channel conditions will improve the SINR and hence increase the CQI in order to realize higher utilization of 256QAM.

SINR improvements can be made in many ways:

Poor SINR correlates with poor RSRP.
Improving handover regions (improving lack of dominance, as pilot overlapping will reduce SINR).
Reducing pilot pollution and modifying OTA parameters for better performance.
Better load-balancing techniques in cases where multiple carriers are deployed.

Therefore, it is crucial to clean up the radio conditions when deploying such HOM techniques, or deploying 256QAM for indoor cells where interference and load are maintained.

Additional consideration should be given to applying interference cancellation techniques. These techniques are discussed later. One experiment is shown in Figure 5.17. In this deployment, without applying interference cancellation, the performance of 256QAM

is very similar to 64QAM in most radio conditions. However, after applying interference cancellation techniques and improving radio conditions in a way that improves the HOM utilization, higher throughput gains are observed. With 256QAM optimized to work in macro cells with proper RF planning and interference cancellation, a gain of 40% to throughput compared to 256QAM and 64QAM without RF planning can be seen. This signifies the importance of deploying HOM with improved RF planning, as discussed above.

5.3.2 Uplink 64QAM

Uplink enhancements have two strands: LTE-FDD and LTE-TDD. In LTE-TDD it becomes important to improve the uplink because of the frame structure where the uplink can be the bottleneck in general, as discussed earlier. 64QAM is part of the enhancements needed to improve the user and cell throughput.

In a smartphone-centric network, uplink traffic utilization is lower than the downlink due to the application usage in normal hours. Also, in LTE-FDD networks, in particular with high bandwidth like 20 MHz, the uplink SINR is not very high during normal hours. As shown in Figure 5.18 for a network deployed with up to 16QAM, the utilization of HOM is very high, reaching 54%. This is because high 16QAM shows high UL SINR in general, due to low uplink load. Therefore, this encourages the deployment of even more HOM like 64QAM.

When 64QAM is deployed, one expects to see a 40% mean throughput improvement observed for 30% system

Figure 5.15 CQI versus number of LTE connected users and downlink modulation distribution in the same network.

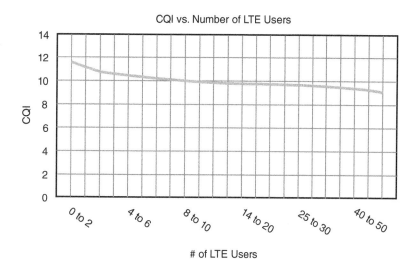

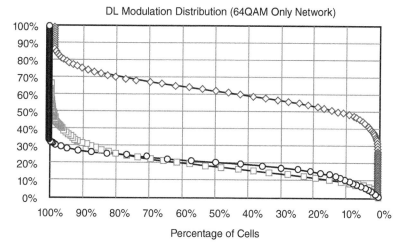

Table 5.5 CQI to modulation mapping in 3GPP.

Category 6 CQI Index	Category 6 Modulation	Category 12 Modulation
1	QPSK	QPSK
2	QPSK	QPSK
3	QPSK	QPSK
4	QPSK	16QAM
5	QPSK	16QAM
6	QPSK	16QAM
7	16QAM	64QAM
8	16QAM	64QAM
9	16QAM	64QAM
10	64QAM	64QAM
11	64QAM	64QAM
12	64QAM	256QAM
13	64QAM	256QAM
14	64QAM	256QAM
15	64QAM	256QAM

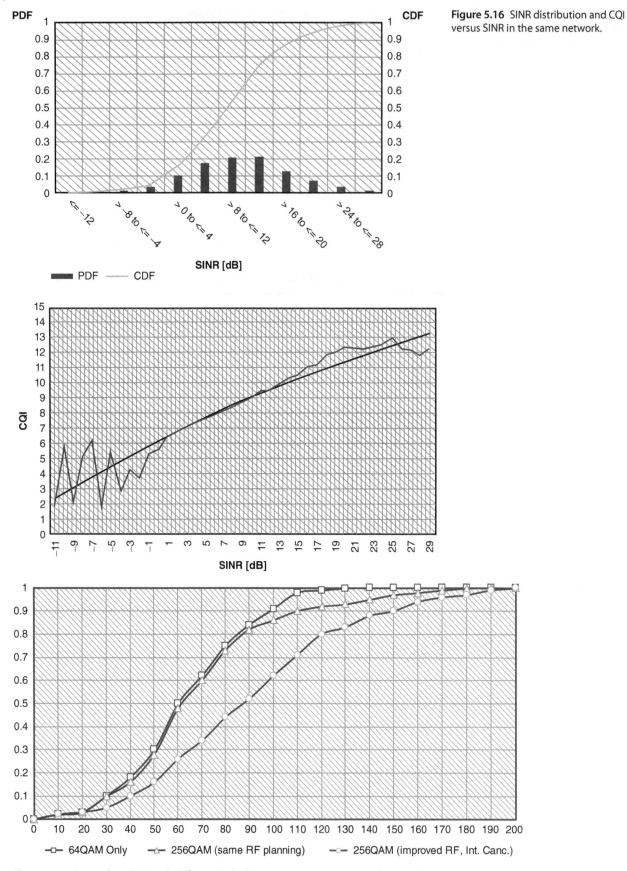

Figure 5.16 SINR distribution and CQI versus SINR in the same network.

Figure 5.17 Gains of 256QAM with different RF deployment strategies, compared to 64QAM only in the same network.

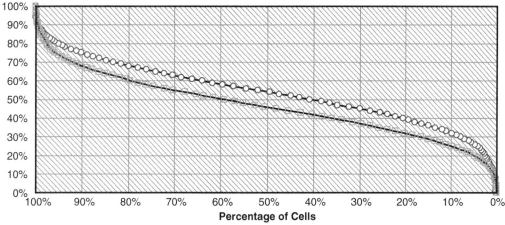

Figure 5.18 Uplink modulation distribution from network data for all sites.

loading. As the load increases, the gain decreases, as expected, because of lower SINR due to higher uplink interference. Thus, the advantages of 64QAM lessen as system loading increases. Figure 5.19 illustrates the gain under different loading conditions. It can therefore be concluded that to maintain the gain of better 64QAM utilization, it is better to have lower system load in general or apply very good uplink interference and cancellation when the load increases. This can be done by applying received diversity on the eNB side with more receivers, such as four-way Rx diversity. Figure 5.19 also shows that there is a significant probability of uplink 64QAM observed in macro eNB deployment:

37% in the 2Rx eNB case in the presence of mid-loading. More eNB Rx deployment can improve UL 64QAM performance.

43% for 4Rx eNB when mid-loading is observed.

Comparing techniques such as 2xCA+16QAM on the uplink with 64QAM on a single carrier, one would expect to see higher throughput on the UL with 2xCA. This is normal because with 2xCA+16QAM, the theoretical throughput can reach 100 Mbps, with each carrier reaching up to 50 Mbps. On the other hand, 64QAM on a single carrier can reach up to 75 Mbps. Figure 5.20 shows that 2xCC UL CA provides better mean UE throughput. The maximum improvement is 1.5 (4 bits → 6 bits per symbol) for 64QAM/1CC and 2 (20 MHz → 40 MHz) for 16QAM/2CC. However, one of the drawbacks for 2xCA compared to a single carrier with HOM is the battery consumption. Table 5.6 shows that carrier aggregation on the uplink requires an additional 8% current consumption for the intra-band case, and 17% for the inter-band aggregation case. This is one of the main tradeoffs to consider while deploying different uplink enhancement techniques.

5.4 LTE-A Feature Dependencies

5.4.1 UE Category and Feature Combinations Towards Giga LTE

With the introduction of many LTE-A features, such as more carrier aggregation, higher-order modulation, and advanced MIMO techniques, a combination of these three features may apply to devices based on their UE categories. Table 5.7 summarizes the UE categories.

It is important to understand that while some categories support a combination of carrier aggregation, 256QAM, and 4x4 MIMO, the given category can be limited to one of these factors. The maximum data rates in each category are presented as the MAX_Throughput {MIMO configuration, higher-order modulation, carrier aggregation combination} that provides whatever maximum throughput possible. Table 5.8 shows the breakdown of UE categories based on these combinations. For example, Category 11 supports up to 4xCC, 256QAM, and 4x4 MIMO. However, this category can achieve the targeted 600 Mbps throughput by:

4xCC with 2x2 MIMO and 64QAM; or
3xCC with 2x2 MIMO and 256QAM; or
2xCC with 4x4 MIMO and 64QAM.

Therefore, 4x4 MIMO can only be used with 64QAM and 2xCC in this UE category. If 256QAM is to be used, it can be done using the eNB to support it with up to 3xCC (60 MHz aggregation) and up to 2x2 MIMO.

UE Categories 15 and 16 provide throughput near 1 Gbps. Combining the LTE-A features, it can be seen in Figure 5.21 that the throughput achieved can be close to 1 Gbps with different combinations of bandwidth aggregation, modulation, and MIMO. This gives good flexibility

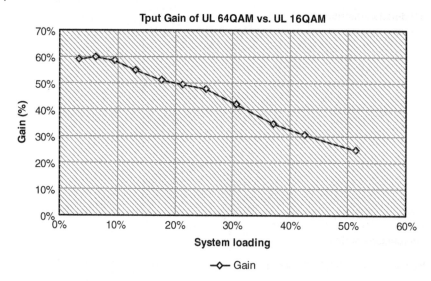

Figure 5.19 Gain of uplink 64QAM over 16QAM and overall 64QAM utilization under different loading conditions.

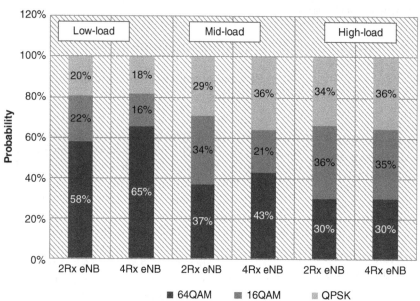

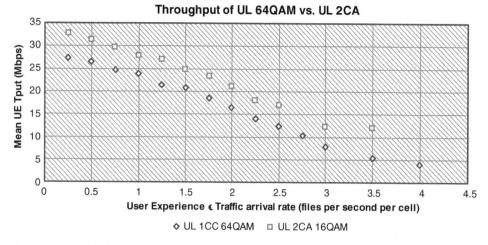

Figure 5.20 Uplink throughput comparison between 64QAM and uplink carrier aggregation.

Table 5.6 Power consumption comparison between different uplink schemes – UE side.

	Baseline: 16QAM/1CC (150 mA)		
	64QAM/1CC	16QAM/2CC intra-band	16QAM/2CC inter-band
Additional current (total)	1.5%	9.4%	18.6%

Table 5.7 LTE UE categories and corresponding technology for each category.

		Downlink			Uplink		
UE Category	3GPP Release	Max. Data Rate [Mbps]	Max. Number of Layers	Support for 256QAM	Max. Data Rate [Mbps]	Support for 64QAM	General Description
Category 0	Release 12	1	1	No	1	No	Internet of Things
Category 1	Release 8/9	10	1	No	5	No	
Category 2	Release 8/9	51	2	No	25	No	Non CA
Category 3	Release 8/9	102	2	No	51	No	Non CA
Category 4	Release 8/9	150	2	No	51	No	Non CA
Category 5	Release 8/9	300	4	No	75	Yes	Non CA
Category 6	Release 10	301	2 or 4	No	51	No	DL 2xCC
Category 7	Release 10	301	2 or 4	No	102	No	UL 2xCC
Category 8	Release 10	3000	8	No	1500	Yes	Non CA
Category 9	Release 11/12	452	2 or 4	No	51	No	DL 3xCC
Category 10	Release 11/12	452	2 or 4	No	102	No	DL 3xCC + UL 2xCC
Category 11	Release 11/12	603	2 or 4	Optional	51	No	DL 4xCC
Category 12	Release 11/12	603	2 or 4	Optional	102	No	UL 2xCC
Category 13	Release 12	391	2 or 4	Yes	150	Yes	DL 2xCC

to the operators in terms of mixing and matching these features to achieve the targeted throughput and capacity.

On the other hand, 1 Gbps throughput can be achieved by aggregating LTE with Wi-Fi through the LWA, as discussed earlier. With this licensed+unlicensed band aggregation, high throughput can be achieved depending on the capability of the Wi-Fi AP, as shown in Figure 5.22.

5.4.2 Interference Cancellation and Coordination Techniques

In LTE systems, the users occupy physical resource blocks in both the downlink (OFDMA) and the uplink (SC-FDMA) in a cell. The eNB ensures orthogonality in the frequency domain, and therefore, the intra-cell interference is typically low. However, the inter-cell interference can be relatively high, because all neighboring cells can provide services over the entire system band (i.e. the frequency re-use factor is 1). As a result, the inter-cell interference is severe for cell-edge users, in particular. Table 5.9 shows the impact of inter-cell interference. In this trial, it is illustrated that a single user camped on a single cell while all other neighboring cells are turned off can reach the maximum throughput on a

single carrier of 147 Mbps. When adding a neighboring cell within 10 dB of the serving cell, and when that neighbor does not schedule any user, the serving cell incurs higher interference, impacting the SINR. When the SINR is impacted, the single user camped on the serving cell starts to see lower scheduling, cutting the throughput down to 75 Mbps. This shows how critical inter-cell interference is in LTE networks. Adding a third cell further reduced the single user throughput to 60 Mbps. Since these three cells cannot be in soft handover (unlike a 3G system) and power controlled, there must be other ways to control this interference. This section describes some of the techniques to reduce the interference in either the uplink or the downlink.

We categorize the interference in LTE into three areas, as shown in Figure 5.23:

High interference mitigation efficiency through advanced receivers: Using IRC (Interference Rejection Combining), for example, offers higher interference mitigation gains than MRC (Maximum Ratio Combining). IRC is more efficient if there are more antennas and fewer interference sources in the directions opposite to the desired beam.

Table 5.8 Breakdown of UE category support for a combination of MIMO, HOM, and CA.

UE Category	Downlink		Maximum DL Throughput Calculations
	Max. Data Rate [Mbps]	Maximum Number of Bits of a DL-SCH Transport Block Received Within a TTI	
Category 6	301	75376 (2 layers, 64QAM) 149776 (4 layers, 64QAM)	• 2xCC with 2x2 MIMO and 64QAM = (75376 * 2 Codewords * 2 Carriers)/10^3 = 301 Mbps • 1xCC with 4x4 MIMO and 64QAM = (149776 * 2 Codewords)/10^3 = 300 Mbps Then, 4x4 can't be used with > 20 MHz because it would exceed the category rate.
Category 9	452	75376 (2 layers, 64QAM) 149776 (4 layers, 64QAM)	• 3xCC with 2x2 MIMO and 64QAM = (75376 * 2 Codewords * 3 Carriers)/10^3 = 452 Mbps • 1xCC with 4x4 MIMO and 64QAM = (149776 * 2-Codewords)/10^3 = 300 Mbps Then, 4x4 can't be used with > 20 MHz because it would exceed the category rate. So, if 4x4 MIMO is used, this category acts like Category 6.
Category 11	603	75376 (2 layers, 64QAM) 97896 (2 layers, 256QAM) 149776 (4 layers, 64QAM) 195816 (4 layers,256QAM)	• 4xCC with 2x2 MIMO and 64QAM = (75376 * 2 Codewords * 4 Carriers)/10^3 = 603 Mbps • 3xCC with 2x2 MIMO and 256QAM = (97896 * 2 Codewords * 3 Carriers)/10^3 = 587 Mbps • 2xCC with 4x4 MIMO and 64QAM = (149776 * 2 Codewords * 2 Carriers)/10^3 = 599 Mbps Then, 256QAM can't be used with > 60 MHz, and 4x4 can't be used with > 40 MHz because this would exceed the category rate, and maximum CA is 80 MHz but with 2x2 MIMO and 64QAM. A combination is also possible (e.g. 2xCC with 2x2MIMO + Third carrier with 4x4 MIMO)
Category 13	391	97896 (2 layers, 256QAM) 195816 (4 layers,256QAM)	• 2xCC with 2x2 MIMO and 256QAM = (97896 * 2 Codewords * 2 Carriers)/10^3 = 391 Mbps • 1xCC with 4x4 MIMO and 256QAM = (195816 * 2 Codewords)/10^3 = 391 Mbps Then, 256QAM can't be used with > 40 MHz, and 4x4 can't be used with > 20 MHz because this would exceed the category rate.

Interference cancellation techniques in 3GPP: Using different techniques offered in 3GPP Releases 11, 12, and 13. In contiguous coverage areas of intra-frequency E-UTRAN cells, cell-edge users experience decreased throughput because they receive strong cell-specific reference signal (CRS) and physical downlink shared channel (PDSCH) interference from neighboring cells. Techniques in this category enable the eNodeB to collect neighboring cell data and deliver assistance information about the neighboring cells to a UE. The UE then performs IC on the neighboring cells' reference signals and data signals transmitted in each subframe.

Interference avoidance and cell coordination: Using coordination techniques between eNBs to cooperate in scheduling uplink and downlink transmission to multiple users in a way that mitigates inter-cell interference by sharing channel status information and user data between cells.

In contiguous coverage areas of intra-frequency E-UTRAN cells, cell-edge users experience decreased throughput because they receive strong cell-specific reference signal (CRS) and physical downlink shared channel (PDSCH) interference from neighboring cells. This was illustrated in Table 5.9. Network-Assisted Interference Cancellation (NAIC) enables the eNodeB to collect neighboring cell data and deliver assistance information about the neighboring cells to a UE. The UE then performs interference cancellation on the neighboring cells' reference signals and data signals transmitted in each subframe. This feature helps to reduce the interference of intra-frequency neighboring

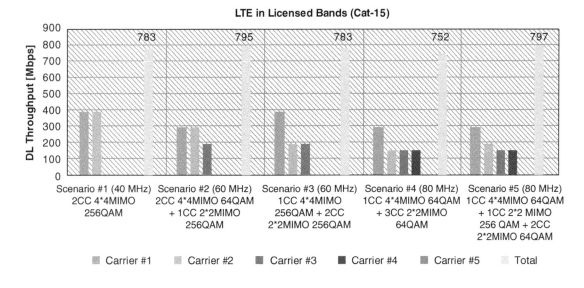

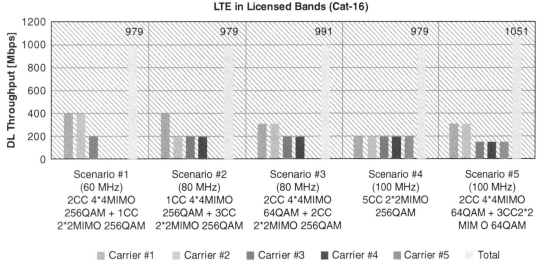

Figure 5.21 Different scenarios for achieving close to 1 Gbps downlink throughput in LTE-A Pro.

cells at the cell edge. Network-Assisted Interference Cancellation and Suppression (NAICS) is defined in 3GPP Release 12. It aims to mitigate CRS and PDSCH interference from neighboring cells. The basic principles and procedure of NAICS are the same as those of CRS-IC. Table 5.10 summarizes the differences between these two methods.

Uplink CoMP uses the antennas of two cells jointly to receive signals from the uplink PUSCH of a UE, and then combines the received signals. This macro diversity technique improves the uplink demodulation and thus improves cell-edge users' throughput. Uplink CoMP can also coordinate signals received from multiple users at the edges of different cells. By coordinating the uplink signal combining of multiple UEs served by different neighboring cells, meaningful uplink inter-cell interference is achieved and thus cell throughput and uplink

coverage are increased. The gain of uplink CoMP is shown in Figure 5.24. As shown in the figure, the PS data enhancements can be up to 15% improvement in cell-edge user throughput. On the other hand, UL CoMP is expected to provide gains to VoLTE. As discussed in previous chapters, the uplink is the main bottleneck to VoLTE coverage. Therefore, coordination of the scheduling between two cells at cell edges can provide good gains in VoLTE voice quality, which can improve the VoLTE uplink coverage by up to 2 dB.

In 3GPP releases earlier than 3GPP Release 11, PDSCH data of LTE cell-edge users cannot be sent with the help of neighboring cells to improve the performance of users at cell edges. This gain can be achieved by DL CoMP based on transmission mode 10 (TM10) introduced in 3GPP Release 11. DL CoMP enables multiple cells to cooperate to perform downlink transmission. The

LTE and Wi-Fi (LWA) Aggregation
DL Throughput [Mbps]
1400
1200
1000
800
600
400
200
0
450
920
866
1168
Scenario #1 (60 MHz)
3CC 2*2MIMO 64QAM
Scenario #2 (40 MHz)
2CC 2*2MIMO 64QAM
LTE Carrier #1
LTE Carrier #2
LTE Carrier #3
Wi-Fi
Giga Wi-Fi
Total

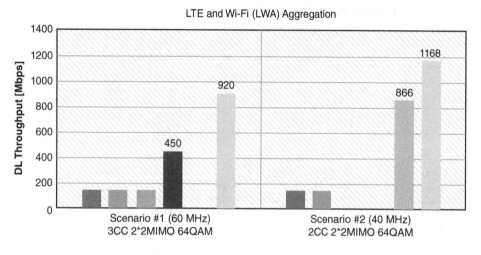

Figure 5.22 Expected aggregated throughput in LWA.

Table 5.9 Impact of inter-cell interference in LTE.

Case No	Scenario	User (DL FTP)	PCI	RSRP	RSRQ	MCS	DL Throughput (Mbps)
1	One cell active	1 active	39	−74	−10	28	147
2	Two cells active	1 active	39	−74	−10	21–22	75
		0 active	35	−86	−17		
3	Three cells active	1 active	39	−74	−10	19	60
		0 active	35	−84	−16		
		0 active	31	−85	−20		

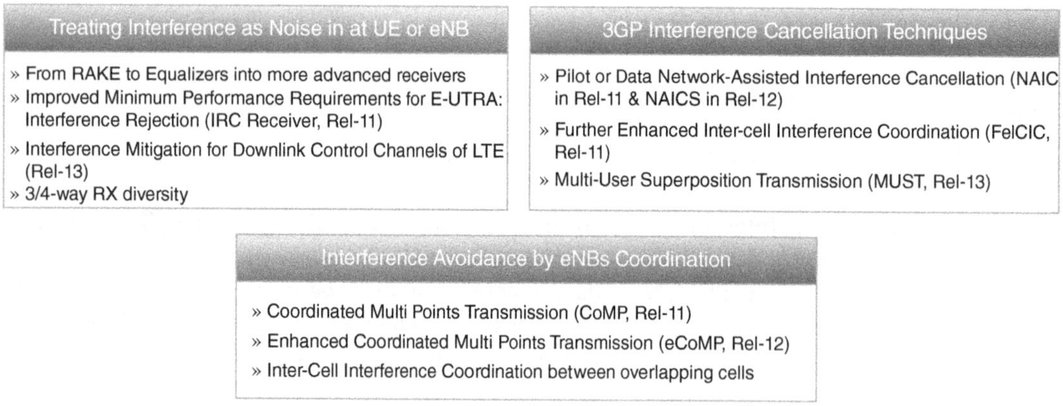

Treating Interference as Noise in at UE or eNB
» From RAKE to Equalizers into more advanced receivers
» Improved Minimum Performance Requirements for E-UTRA: Interference Rejection (IRC Receiver, Rel-11)
» Interference Mitigation for Downlink Control Channels of LTE (Rel-13)
» 3/4-way RX diversity
3GP Interference Cancellation Techniques
» Pilot or Data Network-Assisted Interference Cancellation (NAIC in Rel-11 & NAICS in Rel-12)
» Further Enhanced Inter-cell Interference Coordination (FeICIC, Rel-11)
» Multi-User Superposition Transmission (MUST, Rel-13)
Interference Avoidance by eNBs Coordination
» Coordinated Multi Points Transmission (CoMP, Rel-11)
» Enhanced Coordinated Multi Points Transmission (eCoMP, Rel-12)
» Inter-Cell Interference Coordination between overlapping cells

Figure 5.23 Categories of interference cancellation.

feature is yet to realize commercial launch and hence is not assessed in this book.

5.5 Other Enhancements Towards Advanced LTE Deployments

With the success of LTE deployment worldwide, the industry is seeing LTE networks as the main stream of investment. Operators, networks, and chipset vendors are driving new features in the LTE ecosystem to cover topics of performance, capacity, coverage, and enhanced spectral efficiency. As operators have different LTE radio deployments, TDD or FDD, they view the need for features from different perspectives. LTE-TDD operators look more into features that increase uplink efficiency and coverage, as these are the main bottlenecks. In LTE-TDD, the uplink can be a bottleneck due

Table 5.10 Comparison between NAIC and NAICS interference cancellation.

Network-Assisted Interference Cancellation (NAIC)	Network-Assisted Interference Cancellation and Suppression (NAICS)
Enables the identification of most interfering cells for a given UE, using X2 information from neighboring cells, and provides relevant cell information for the UE to cancel the Cell-specific Reference Signal (C-RS) from these neighboring cells.	Enables the cancellation of the interference of data channels (PDSCH) from other cells or from other users in the same cell.
Release 11 feature requiring support in UE and network.	Release 12 feature requiring support n UE and network.
Feedback given to UE from eNB: PCI and # of antenna port.	Higher-layer signaling of parameters related to interference PDSCH:
UE indicates its support for feature in UECapabilityInformation RRC message.	Resource allocation (PRB), system bandwidth, PCI, power control parameters.
	Blind detection for some other PDSC parameters by UE, to reduce scheduling restriction and signaling overhead.
Improvement of LTE DL performance for NAIC-capable terminals in macro–macro scenarios (FeICIC is the corresponding feature for HetNet scenarios.	Boosts the capacity in heavily loaded networks and improves cell spectral efficiency.
Reduces handover failures and call drops caused by downlink interference in bad coverage areas.	Co-channel interference, either from inter-cell or co-scheduled intra-cell users, is expected to improve with NAICS.

Figure 5.24 Uplink CoMP gains for PS data and for VoLTE.

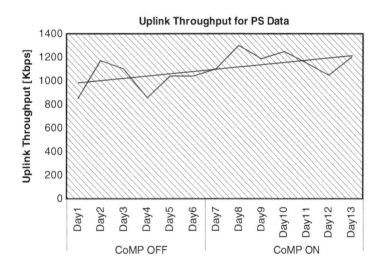

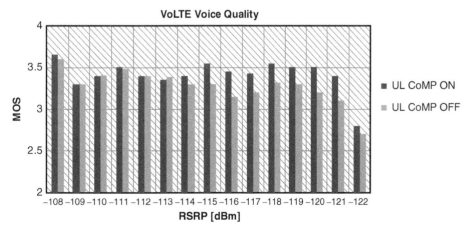

to the frame structure, and coverage can be suboptimal to FDD because of the high band deployment. For LTE-FDD operators, capacity and performance are key, and enhanced features like MIMO, modulation, and VoLTE are of more interest.

In this section, we will evaluate some of the features that are of specific interest to the industry. Some of these features are still under 3GPP review for standardization, and others have been standardized in previous 3GPP releases but are now of more interest due to the current deployment status.

5.5.1 High Power/Performance User Equipment (HPUE)

In 3GPP up to Release 11, UEs were specified to operate with the maximum uplink transmit power to 23 dBm ±2, called Power Class 3. The UL is the limiting link in LTE and the gap between DL and UL is ~5 dB, mainly due to differences in Tx power between the DL and UL, the number of antennas deployed in the eNB versus UEs, and many other factors. TDD systems have been deployed in many parts of the world to support existing low-frequency FDD network deployments. In such scenarios, a low-frequency FDD band is used for coverage and a TDD band for increased capacity needs, since high-frequency TDD bands have shorter signal propagation properties compared to low-frequency FDD bands. Increasing the UL Tx power on high-frequency bands could help utilize the capabilities of TDD bands to improve overall user experience.

With this objective, 3GPP introduced a new power class that allows the device to operate with a maximum transmit power of 26 dBm ±2, referred to as Power Class 2. This was introduced in Release 11 to operate with FDD band 14 (700 PS) for public safety applications. Later, this power class was extended to operate for LTE-TDD band 41 (TD 2500).

The main objectives in adding Power Class 2 to band 41 were as follows:

Increasing the device uplink transmit power would result in significant UL coverage extension with huge cost savings for wireless operators.

Improving the LTE-TDD competitiveness with LTE-FDD deployment, as HPUE will provide a 3 dB power increase, leading to similar performance as with the use of LTE-FDD deployed in a mid-frequency band.

Making minimal changes to the UE modem design. Power Class 2 UEs would be implemented using the same architecture as Power Class 3 UEs and with the same assumptions in terms of insertion losses. Additionally, power consumption is not expected to increase dramatically with the 3 dB power increase.

5.5.2 Uplink Data Compression (UDC)

Efficient spectrum usage and utilization for user applications are becoming more and more important for future LTE deployments in order to cope with increased throughput and capacity needs. This increases the need to support new features and methods such that spectrum, resources, and latencies can be improved in order to enhance the system performance and user experience. The explosion of data connectivity can put disproportionate pressure on either the uplink or downlink, depending on the usage of smartphones and applications. Figure 5.25 shows a trial involving live network traffic and smartphone profiling. As shown, users enjoy data-intensive streaming and social media services, which are the most common uses for mobile Internet. Applications can have different types of requirements for downlink throughput, uplink throughput, packet latencies, Quality of Service, and resource assignment from the network scheduler. Therefore, network operators are shifting the focus onto how to improve the end-user experience and the network capacity with as many different types of data usage as possible.

There are many different techniques and features available now in 3GPP that can be used to improve the areas of throughput (both uplink and downlink) and latencies, in addition to improving capacity, especially for users at cell edges. The categories of these features are summarized in Table 5.11; some have been deployed and others have been specified in 3GPP already. The categories listed in the table provide enhancements to throughput, spectral efficiency, interference, coverage, mobility, and battery consumption.

However, operators are looking into more enhancements coming from techniques deployed at upper layers. One of these is the data compression feature. Compressing the uplink data between a UE and the EPS (Evolved Packet System) can provide gains in terms of user throughput, latencies, and capacity by performing such compression for HTTP traffic.

5.5.2.1 Compression Methods in 3GPP Now

The radio spectrum resource is precious and has to be utilized economically. However, in the RTP (Real-time Transport Protocol)/UDP (User Datagram Protocol)/IP protocol structure of the services that are used widely, such as streaming services and VoIP, the header fields occupy 40 to 60 bytes for each payload that can be as small as 32 bytes. Transmitting the header fields of the RTP/UDP/IP protocol consumes a large amount of radio link resource. Therefore, 3GPP specified ways to compress the headers of RTP/UDP/IP packets. Typically, the PDCP (Packet Data Convergence Protocol) header compression undertakes this function.

Compression techniques are not new to 3GPP specifications. There have been several compression methods

Figure 5.25 Typical traffic type distribution by smartphone users.

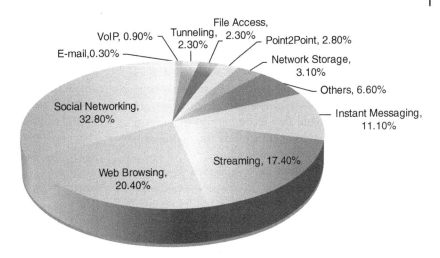

Table 5.11 General 3GPP LTE and LTE-A Pro feature categories.

LTE and LTE-A Pro Features	Brief Summary
Carrier aggregation	Aggregation up to five carriers including licensed and unlicensed bands for UL and DL to improve throughput, capacity, and latencies.
MIMO techniques	Improve spectral efficiency with many types of MIMO transmission mode, FD-MIMO, MU-MIMO, SU-MIMO.
Higher-order modulation	Improves spectral efficiency within both UL and DL.
Interference cancellation and suppression	Improves cell-edge throughput, capacity, and latencies including interference from pilot and data/control channels.
Short transmission time interval (TTI)	Improves latencies in ranges below 20 ms.
SON and mobility enhancements	Features aiming to improve user throughput in the handover region, and increase network efficiencies in mobility.
Connected mode discontinuous reception (C-DRX)	Improves battery consumption and resource utilization for bursty traffic.

introduced in 3G UMTS time, with a focus on headers, given the large size with IPv4 and IPv6 packets. Generally speaking, PDCP header compression has two types of algorithm:

IPHC (IP Header Compression): Can apply to PS data services.

RoHC (Robust Header Compression): Can apply to VoIP services.

IPHC uses the concept of the packet stream context. A context is a set of data which contains field values and value change patterns in the packet header. For each packet stream, the context is formed at the compressor and the decompressor. After the context is established on both sides, the compressor can compress the packets. The compressor and decompressor store most fields of this full header as context. The context consists of the fields of the header whose values are constant and thus need not be sent over the link at all, or whose values change little between consecutive headers, so that PDCP header compression uses fewer bits to send

the difference from the previous value compared with sending the absolute value.

The IPHC schemes are suitable for radio links with strong link-level checksums but are not robust enough to handle high Bit Error Rate (BER) and long Round Trip Time (RTT). Because high BER and long RTT can commonly impact delay-sensitive applications, an efficient and robust compression scheme is needed. In this regard, the RoHC technique was developed.

RoHC is a header compression protocol originally designed for real-time audio/video communication in a wireless environment. As VoIP media packets are transported as IP traffic, the generated headers of the IP protocols can be massive. RoHC compresses the RTP, UDP, and IP headers to reduce the overall size of the voice packets. This decreases the radio resource utilization on the cell edge and therefore improves the overall cell coverage on both UL and DL. In addition, it reduces the number of voice packet segments, which improves the BLER associated with these smaller-size transmissions. This improves the VoLTE end-to-end delays and jitter. On the LTE side, the biggest beneficiary of compression at the moment is VoLTE.

The RoHC operation is illustrated in Figure 5.26. The RoHC function in LTE is part of Layer 2 at the user plane of the UE and eNB. Both UE and eNB behave as a compressor and decompressor for the user-plane packets on the UL and DL. The compression efficiency depends on the RoHC operating mode and the variations in the dynamic part of the packet headers at the application layer. A header can be compressed to one byte with RoHC, which efficiently reduces the voice packet size. RoHC in LTE operates in three modes: U-mode, O-mode, and R-mode (Unidirectional, Bidirectional Optimistic, and Bidirectional Reliable, respectively). The reliability of these modes and overheads used for transmitting feedback is different. In U-mode, packets can only be sent from the compressor to the decompressor, with no mandatory feedback channel. Compared with O-mode and R-mode, U-mode has the lowest reliability but requires the minimum overhead for feedback. In O-mode, the decompressor can send feedback to indicate failed decompression or successful context update. Therefore, it provides higher reliability than U-mode but it generates less feedback compared to R-mode. In R-mode, context synchronization between the compressor and decompressor is ensured only by the feedback.

That is, the compressor sends the context updating packets repeatedly until acknowledgment is received from the decompressor. Therefore, R-mode provides the highest reliability but generates the maximum overhead due to the mandatory acknowledgment.

RoHC is an extensible framework consisting of the following profiles. Note that in terms of support, Profiles 3 and 4 are not very common at present:

RoHC Profile 0: Provides a way to send packets without any compression.
RoHC Profile 1: Compresses packets with IP/UDP/RTP protocol headers.
RoHC Profile 2: Compresses packets with IP/UDP protocol headers.
RoHC Profile 3: Compresses packets with Encapsulating Security Payload (ESP) and IP protocol headers.
RoHC Profile 4: Compresses packets with IP.

Given that the gains from RoHC are significant – we showed earlier in the book trials in which they reached around 92% for UL VoLTE traffic and 84% for the DL – there has been a push in the industry to extend the concept of compression into the user-plane data payload as well as the header part. Along the way, several priority solutions have been introduced by mobile networks related to 3G (e.g. Data Acceleration, DACC) or LTE (Uplink Data Compression, UDC). These methods

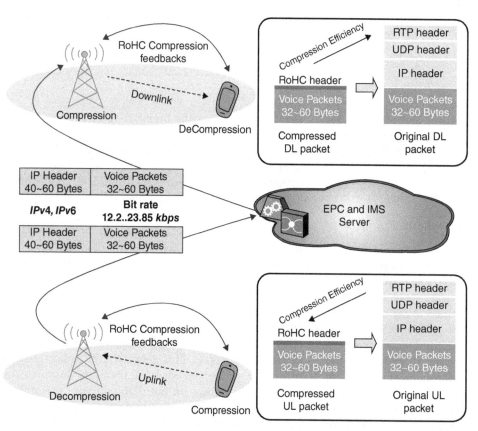

Figure 5.26 RoHC operation for VoLTE traffic.

focus on performing UL data compression between the UE and the radio access network. However, these algorithms have not yet made it into the 3GPP standard, which makes it difficult to scale them up for a wider deployment, given their proprietary nature. In brief, these algorithms target compression only for the payload of the TCP, UDP, and IPv4/v6 packets. Hence, the compression and decompression layers need to parse the TCP, UDP, and IPv4/v6 headers to identify the start and end of the payload. The type of application used also manifests difficulties.

These features can compress original IP-level packets into compressed packets. The compressed packets are sent on the air interface in the uplink and decompressed to reconstruct the original IP-level packets. IP-level traffic can consist of compressible data. The uplink can have repetitive (TCP ACKs) or textual data, which are good targets for compression. Other kinds of traffic, for example pictures or secure data connections, have little or no benefit from compression. Some data files, such as text files, picture files in the BMP format, and certain text-style database files, can be compressed to a large extent. Other types of files are not compressed well; for example, most multimedia files, as they exist in a highly compressed state. These file types use efficient techniques to compress the data they contain. Files that are already compressed cannot be further compressed to any significant extent.

Therefore, 3GPP started efforts to unify the type of data compression and what it can achieve for different types of files and applications. The main targets of the proposed algorithm are to:

Increase uplink capacity when the amount of data sent in the uplink is reduced.
Increase the uplink and downlink throughput by sending less uplink data with the same amount of information

(on the downlink the benefits come from shorter TCP ACKs, needed for the same amount of downlink data).
Improve uplink interference and coverage by reducing the uplink transmit power (Tx power) for the data channels when the amount of uplink data is reduced.
Improve latencies by sending the same amount of information with less round trip time.

5.5.2.2 Possible Applicable LTE Mode of Data Compression

In many applications deployed in today's networks, the compression is already performed at the application layer for most downlink traffic. For example, websites are designed in such a way to send HTML data in a compressed format and transmit all multimedia data (for example, images, video, and audio) in an encoded form. However, on the uplink most of the data are typically pieces of protocol information such as IP/TCP/HTTP headers, and these are quite compressible. Therefore, the benefits of data compression will come mainly for the uplink packets, while little or no benefit is expected from compression on the downlink.

From live FDD network trials in a 20 MHz channel (band 3), it has been observed that the traffic generated on the downlink can be ten times the traffic generated on the uplink. This information was gathered during busy hours. Additionally, due to the small-size packets (bursty traffic) generated on the uplink, the modulation and the overall block error rate are maintained very well. Figure 5.27 shows the overall traffic distribution in a typical FDD network deployed with smartphones. Of course, the uplink-to-downlink ratio can be totally different during mass events, where users tend to send heavy traffic on the uplink (for example, videos).

In terms of uplink interference in the same network trial (FDD), Figure 5.28 shows that UL interference during busy hours increases with higher UL PRB utilization

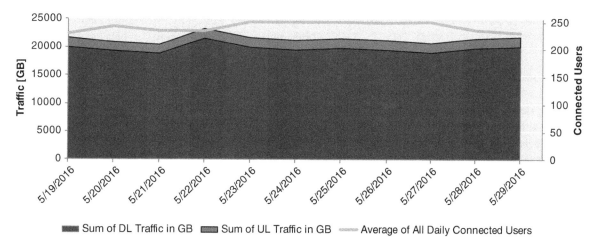

Figure 5.27 Uplink and downlink traffic distribution in a live network – from network counters.

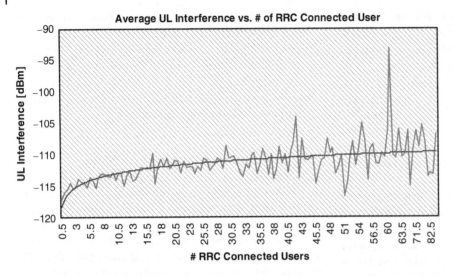

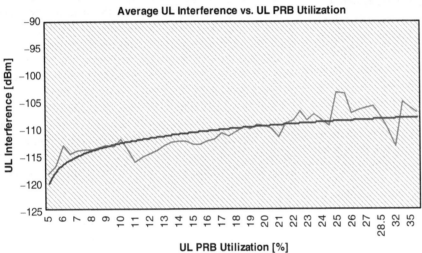

Figure 5.28 UL interference level during busy hours – from network counters.

and when the number of connected users increases. Even at higher utilization, the interference is still manageable; however, the variance increases with a higher number of connected users. In general, uplink interference management depends on the type of receivers implemented by the eNB. There are two types of uplink interference receiver:

Maximum ratio combining (MRC): If interference is spatially white, its performance is high. However, if interference with power spectral density is not evenly distributed over the entire frequency domain, MRC performance is relatively low. It can be applied to uplink physical layer channels: PUSCH, PUCCH, and PRACH.

Interference rejection combining (IRC): IRC uses interference mitigation technology to improve its performance further. It can also be applied to uplink physical layer channels: PUSCH, PUCCH, and PRACH.

The type of receiver shown in Figure 5.28 is MRC, and uplink interference is expected to improve when applying more advanced receivers in the eNB, like the IRC type.

However, in a network deployed with TDD, the complexity of capacity management increases. In TDD, the frame is shared between both uplink and downlink subframes, and therefore, there can be a clear impact on the end-user throughput. The data throughput for TDD from field test results is shown in Figure 5.29 for both UL and DL in band 40, with 10 MHz and Category 4 devices (compared to the theoretical LTE-FDD throughput in a 10 MHz channel). In addition to the TDD frame structure, other channels require extra resources in-band, with the TDD resources assigned for data channels. The scheduling is based on SIB (System Information Block) transmission. For example, SIB1 requires 9 PRBs (Physical Resource Blocks) and is repeated every 20 ms; SIB2/3 requires 15 PRBs repeated every 160 ms. If these resources are needed for multiple channels, this will impact the data throughput even further.

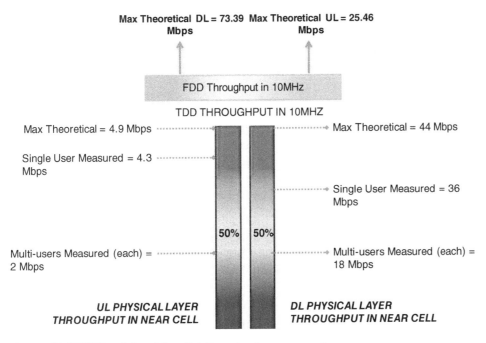

Figure 5.29 LTE-TDD uplink and downlink throughput measurements.

The radio and cell-loading conditions also have an impact on the scheduling for TDD. As per 3GPP, the UE may skip decoding a transport block in an initial transmission if the effective channel code rate is higher than 0.931. Therefore, the eNB scheduler may tend to avoid using a combination of MCS/PRBs that would lead to a scheduled TBS (Transport Block Size) of > 0.931. As a result, in different loading conditions where the control channel is assigned a higher CFI (Control Format Indicator; i.e. more resource is needed for the control channel PDCCH), the scheduler will limit the MCS even further, impacting the overall end-user throughput.

Therefore, applying data compression techniques in a TDD network can be more beneficial in terms of the targets of the feature discussed previously. For this reason, the industry has been considering applying data compression algorithms in TDD-only mode, especially in the first phase of standardization.

5.5.2.3 Uplink Data Compression Performance Analysis Using RoHC

We have studied applying RoHC into the headers of data services and not restricting it to VoLTE traffic. The method is expected to be applicable to most types of application and can be used as a means of achieving a higher compression ratio among different file transfers. As discussed previously, the effort to introduce a pure data compression method to the payload may be limited to a few applications, but standardizing header compression for PS data (not just VoLTE, as is currently the case in 3GPP), will bring significant gains and will be scalable

to a wide range of deployments, and not restricted to proprietary applications.

RFC 4995 has configuration parameters that are mandatory and that must be configured by upper layers between compressor and decompressor peers. These parameters define the RoHC channel. The RoHC channel is a unidirectional channel, i.e. there is one channel for the downlink and one for the uplink. There is thus one set of parameters for each channel, and the same values are used for both channels belonging to the same PDCP entity. This would follow the RoHC compression mechanism in the PDCP layer, similar to what is used for VoLTE traffic. Therefore, the network-side overhead from adding RoHC into data traffic and packets is not expected to be impacted, as most networks have already exercised it for VoLTE deployment.

5.5.2.4 Performance Aspects of Uplink Data RoHC

As RoHC has matured enough on the network side, it is expected that the network scheduler impact will be minimal. The compression and decompression operations are entirely implemented in the PDCP layer. Therefore, the only entities that require additional software changes as a result of this feature are the UE and the eNB. All other entities such as the mobile operating system (e.g. Android) and other network entities such as proxy servers, IP NATs, and firewalls are unaffected by the compression/decompression in this algorithm.

In this regard, the trials included different types of file transfer: video streaming, HTTP Web browsing, and text over instant messaging. The testing was extensive to cover hours of active loading of these applications. The

Table 5.12 Results for uplink data RoHC trial.

	Applications	Average Packet Size (Bytes)	TCP/IP Header Ratio (%)	Compression Gain (%)
DL	Video streaming	1422.77	3.66	3.18
	Web browsing	1156.54	4.49	3.96
	Instant messaging	1136.69	4.58	4.02
UL	Video streaming	54.77	95.09	392.25
	Web browsing	121.59	43.07	53.322
	Instant messaging	92.90	56.19	86.392

overall gains are given by the KPI related to the overall compression gain, as follows:

$$compression\ gain$$
$$= \left[\left(\frac{bytes\ before\ compression}{bytes\ after\ compression} \right) - 1 \right] \times 100\% \qquad (5.1)$$

Live network data show the gains in Table 5.12 across these applications. The trials were performed with LTE-TDD, 20 MHz (theoretical DL throughput = 110 Mbps, UL throughput = 10 Mbps).

The following points summarize the observations that can be made from the testing results shown in Table 5.12:

Observations on packet size distribution:
DL: More than 70% of packets are of a size larger than 1400 bytes.
UL: More than 90% of packets are smaller than 100 bytes.

Observations on RoHC performance evaluation:
UL RoHC (53%~392%) is more effective than DL RoHC (3%~4%).
Compression gain increases as the TCP/IP header ratio increases.

As expected, the gains will come mainly for the uplink, with little or no benefit from compression on the downlink. The reason is, as discussed previously, related to the fact that these services use efficient techniques to compress the data they contain. Files that are already compressed cannot be further compressed to any significant extent on the downlink.

Figure 5.30 summarizes the distribution of the contents of the uplink traffic carrying packets related to each service. The figure shows that the three services – video streaming, HTTP Web browsing, and instant messaging – have headers that contribute to 94.4%, 36.1%, and 53.9% of the overall content of the packets, respectively. As a result, RoHC for these IP data services showed that video streaming benefitted the most from uplink compression. The gains were substantial and would be expected to improve the end-user experience, especially in an LTE-TDD network deployment.

5.5.2.5 Highlights of UDC Development and Challenges

In RAN plenary #73 in September 2016, the Study on UL data compression in LTE (UDC) was proposed, highlighting the following:

As the number of LTE subscribers increases, the uplink interference level reaches 5~10 dB in practical networks, making uplink transportation in poor radio conditions difficult.

The typical UL/DL configuration in a TD-LTE network is configuration 2, i.e. 3DL: 1UL. Quite often, the uplink becomes a bottleneck in cases of, for example, file uploading.

Due to power limitations, RLC segmentation is a common way to extend uplink coverage.

For this 3GPP work item, the investigation of UDC solutions and the way forward was set, and it included, at a minimum: identifying compression algorithms or compressed data formats for the purpose of performance evaluation, evaluating the performance of the identified compression solutions with compressed data formats based on traffic characteristics in a practical network, and evaluating the probability of decompression failure and the performance impact on the application layer traffic based on agreed-upon traffic types and traffic modeling.

Until this point, there was no agreement on adopting UDC in 3GPP. The evaluation of all of the above points needs to be agreed by the standard body and this may take time; it could be introduced in Release 14. Therefore, at the moment, only proprietary solutions are applied with gains that vary depending on several factors: most importantly, whether the traffic generated between the client and server is encrypted at the application layer or not. Taking HTTP as a baseline for comparison, one of the major aspects is that the ratio of end-to-end encrypted traffic has risen sharply. The introduction of HTTP 2.0 further accelerates the development of more encrypted applications where operators can lose insight into real-time customer experience per application. Table 5.13 shows this trend between 2011 and 2016.

How does this trend impact data compression? Certain types of encrypted data files cannot be compressed

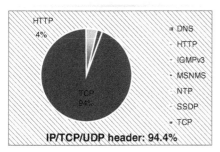

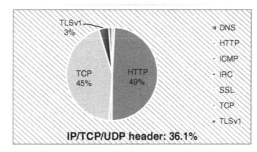

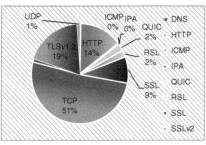

Figure 5.30 Uplink traffic volume distribution for different services.

Table 5.13 Encrypted traffic ratio.

Encrypted Traffic Ratio	2011	2016
Clear-text HTTP and TCP	85%	49%
HTTPS	8%	43%
UDP	7%	7%

very much. In practice, there are several methods of encryption used in HTTP-type traffic – application layer encryption or transport layer encryption – and the application of each method depends on the type of Web application used. Encryption turns data into high-entropy data, usually indistinguishable from a random stream. Compression relies on patterns in order to gain any size reduction. Since encryption destroys such patterns, the compression algorithm is unable to give much (if any) reduction in size if you apply it to encrypted data. Therefore, it is expected to have very minimal gains in terms of compression for such traffic, even if the compression takes place at the air interface (probably limited to compression of the GET Request and Response of the HTTP protocol only).

On the other hand, if any UDC type of algorithm is introduced in 3GPP, it is expected to have a significant change on the protocol layers in both the UE and the eNB. In practice, the RRC and PDCP layer must be impacted; new signaling IEs must be introduced to negotiate the data compression in addition to having new PDCP PDU types. RoHC, on the other hand, will utilize the same changes that have already been made

in the UE and eNB (inherited from VoLTE), and the RRC is the only side that could be impacted. Hence, the complexity of deployment lessens with RoHC applied to IP data bearers.

5.5.3 Enhanced VoLTE (eVoLTE)

Additional work is being added into Release 14 in order to improve VoLTE and ViLTE performance. The study in the specification considers enhancing performance in the following areas:

Enabling VoLTE/video codec mode and codec rate selection and change over E-UTRA.
Improving the VoLTE/video quality perceived by the user by reducing packet loss or allowing the use of higher codec rates.
Prioritizing VoLTE/video access and/or VoLTE/video-related signaling and reducing call drop probability.
Investigating mechanisms that are applicable to different codec types including AMR, EVS, and video in both downlink and uplink to enable:
    Codec mode and rate selection at call setup.
    Codec rate adaptation during an ongoing call.
    Coder adaptation can be triggered cell-wide or on a per-UE or per-DRB basis.
    Up/down-side tuning of the codec rate.

Enhancing the perceived VoLTE/video quality, focusing on:
    Identifying whether it is possible to re-use the coverage enhancement techniques that have been discussed in previous releases.

Identifying possible further enhanced HARQ for TTI bundling beyond Release 12 enhancements. Prioritizing enhancements to VoLTE/video access and/or VoLTE/video-related signaling and reducing call drop probability (e.g. potential call drop during mobility) by signaling enhancements for VoLTE/video.

After VoLTE was deployed, many issues were observed on live networks, impacting the end user and network performance. These main issues related to VoLTE are categorized in Figure 5.31, and a summary of the areas for improvement is given in Figure 5.32.

Voice services have high requirements in terms of end-to-end delay, or mouth-to-ear latency. The end-to-end delay threshold to achieve a "very satisfied" user experience is 200 ms, and that to achieve a "satisfied" user experience is 275 ms. VoLTE users become dissatisfied with voice quality when the end-to-end delay exceeds 275 ms. The recommended packet delay budget for the Uu interface is 50–80 ms. The delay budget between the EPC and eNodeB is 20 ms. If the transmission delay between the EPC and eNodeB is greater than 20 ms, voice quality may not be guaranteed after VoLTE is deployed on the eNodeB.

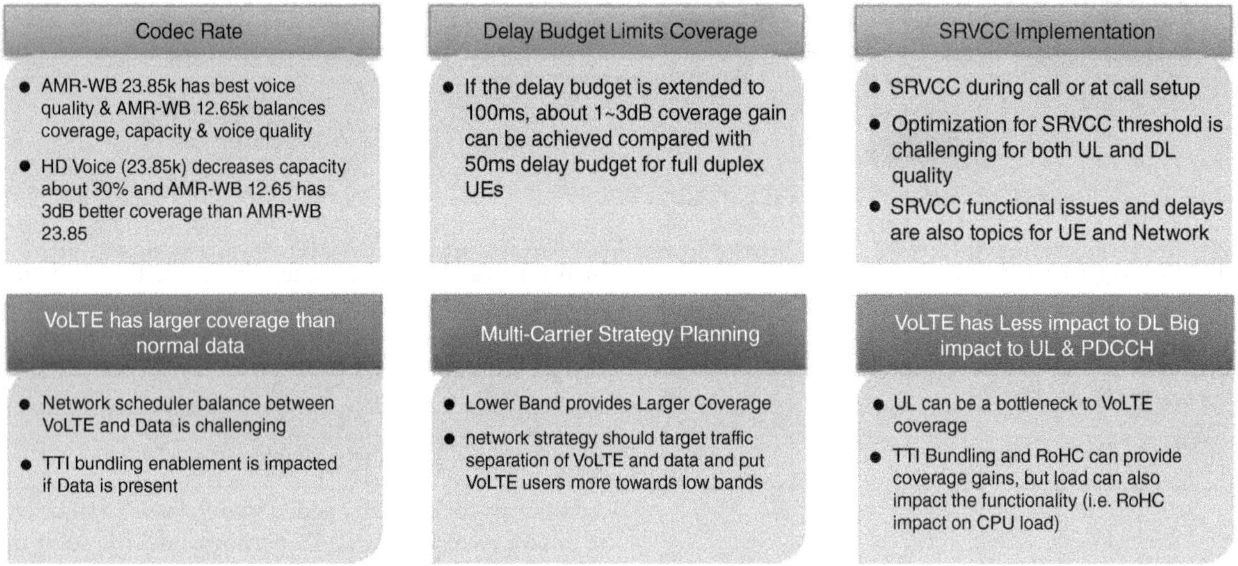

Codec Rate	Delay Budget Limits Coverage	SRVCC Implementation
• AMR-WB 23.85k has best voice quality & AMR-WB 12.65k balances coverage, capacity & voice quality • HD Voice (23.85k) decreases capacity about 30% and AMR-WB 12.65 has 3dB better coverage than AMR-WB 23.85	• If the delay budget is extended to 100ms, about 1~3dB coverage gain can be achieved compared with 50ms delay budget for full duplex UEs	• SRVCC during call or at call setup • Optimization for SRVCC threshold is challenging for both UL and DL quality • SRVCC functional issues and delays are also topics for UE and Network

VoLTE has larger coverage than normal data	Multi-Carrier Strategy Planning	VoLTE has Less impact to DL Big impact to UL & PDCCH
• Network scheduler balance between VoLTE and Data is challenging • TTI bundling enablement is impacted if Data is present	• Lower Band provides Larger Coverage • network strategy should target traffic separation of VoLTE and data and put VoLTE users more towards low bands	• UL can be a bottleneck to VoLTE coverage • TTI Bundling and RoHC can provide coverage gains, but load can also impact the functionality (i.e. RoHC impact on CPU load)

Figure 5.31 Concept challenges in VoLTE strategy and planning.

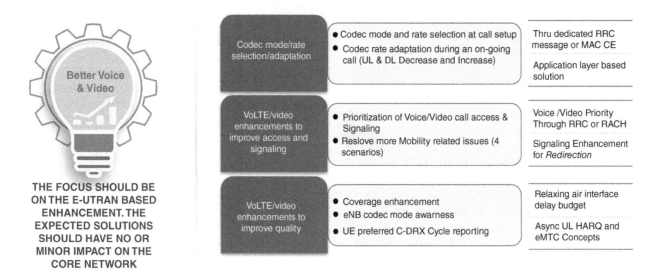

THE FOCUS SHOULD BE ON THE E-UTRAN BASED ENHANCEMENT. THE EXPECTED SOLUTIONS SHOULD HAVE NO OR MINOR IMPACT ON THE CORE NETWORK

Figure 5.32 Areas for improvement for eVoLTE in 3GPP Release 14.

eVoLTE is targeting an increase in the delay budget in order to extend the overall coverage of VoLTE services. In eVoLTE, an increase in the delay budget to 200 ms has been agreed, while changes are needed at the core and the radio side to keep the end-to-end delay within the acceptable range, as described above. The other key aspect of eVOLTE is that the codec rate changes dynamically during the call. Codec rates have been discussed in detail in previous chapters.

References

1 3GPP TS 36.913 V9.0.0 (2009) "3rd Generation Partnership Project; Technical Specification Group Radio Access Network; Requirements for further advancements for Evolved Universal Terrestrial Radio Access (E-UTRA)," (Release 9).

2 3GPP TR 36.913 V10.0.0 (2011) "3rd Generation Partnership Project; Technical Specification Group Radio Access Network; Requirements for further advancements for Evolved Universal Terrestrial Radio Access (E-UTRA) (LTE-Advanced)," (Release 10).

3 3GPP TR 36.101 V11.3.0 (2012) "3rd Generation Partnership Project; Technical Specification Group Radio Access Network; Evolved Universal Terrestrial Radio Access (E-UTRA); User Equipment (UE) radio transmission and reception," (Release 11).

4 3GPP TR 36.847 V12.0.0 (2013) "3rd Generation Partnership Project; Technical Specification Group Radio Access Network; Evolved Universal Terrestrial Radio Access (E-UTRA); LTE Time Division Duplex (TDD) – Frequency Division Duplex (FDD) joint operation including Carrier Aggregation (CA)," (Release 12).

5 3GPP TR 36.331 V12.3.0 (2014) "3rd Generation Partnership Project; Technical Specification Group Radio Access Network; Evolved Universal Terrestrial Radio Access (E-UTRA); Radio Resource Control (RRC); Protocol specification," (Release 12).

6 3GPP TS 36.133 V8.16.0 (2012) "3rd Generation Partnership Project; Technical Specification Group Radio Access Network; Evolved Universal Terrestrial Radio Access (E-UTRA); Requirements for support of radio resource management."

7 Elnashar, A., El-saidny, M. A. *et al.* (2014) *Design, Deployment, and Performance of 4G-LTE Networks: A Practical Approach*, John Wiley & Sons.

8 3GPP TS 36.300 V13.2.0 (2016) "3rd Generation Partnership Project; Technical Specification Group Radio Access Network; Evolved Universal Terrestrial Radio Access (E-UTRA) and Evolved Universal Terrestrial Radio Access Network (E-UTRAN); Overall description; Stage 2," (Release 13).

6

LTE Network Capacity Analysis

6.1 Overview

Network capacity analysis based on KPIs is essential for cellular network performance optimization. It aims to reflect network status and to eliminate network issues. Network capacity analysis based on KPIs can be done on a network-wide basis or a cell basis (via monitoring top cells). It not only helps to figure out and understand network data easily but also helps to align the performance with the pre-defined SLA (Service Level Agreements). By analyzing network KPIs, we can find out how the network is performing and plan/forecast network capacity.

Understanding user behavior and forecast analysis for different services are key factors during the deployment phase of different technologies. Network KPIs measure and monitor targeted performance while ensuring and maintaining user experience enhancement through a continuous optimization process. This chapter focuses on LTE network KPI analysis to evaluate and detect LTE performance. The analysis in this chapter can be used to benchmark LTE network performance.

The analyzed LTE counters in this chapter were collected for a commercially deployed 3GPP Release 10 LTE network. The analysis in this chapter includes LTE users (connected and active), LTE scheduling, LTE traffic downlink/uplink analysis, TTI utilizations, throughput analysis, physical resource block utilization analysis, modulation and codec scheme, channel quality indicator (CQI), "MIMO" analysis, data connection performance, CSFB performance, and packet data convergence protocol (PDCP) performance.

6.2 Introduction

The importance of network capacity analysis based on KPIs appears in the measurement and monitoring phase in order to keep the network performance within the targeted service level agreement. Before the network deployment phase, it is important to understand the user behavior and the traffic forecast per service and to estimate signaling to traffic utilization based on available resources. In addition, it is important to understand smartphone behavior and penetration of smartphones in the network, as this plays a key role in estimating signaling and traffic overhead in the network. For any network during the deployment phase there are optimization targets per technology, i.e. GSM, UMTS, LTE, and IRAT. These optimization targets should include different services per technology.

The necessary optimization actions will be highly dependent on user traffic; there could be significant differences between forecast and actual traffic patterns. New features may be needed to cope with the actual traffic trend compared to the original estimations, hence, continuous optimization attempts should be made to maximize the utilization of the deployed network while adequately planning for the necessary CAPEX investments.

In the measurement and monitoring phase, it is always important to find a balance between target performance KPIs and enhanced user experience. Target performance objectives also include other important factors such as reducing signaling overhead, increasing DL/UL traffic, reducing congestion, and improving resource utilization. From a user experience perspective, it is important to consider battery power consumption and to achieve a good mean opinion score (MOS) for voice quality per different technology, i.e., 2G/3G/VoLTE/VoWiFi. Call drop rate, call setup time, throughput, and connectivity counters are also core factors for monitoring and improving user experience. Figure 6.1 demonstrates the relationship between user behavior, network deployment, and targeted performance. The figure illustrates three phases: understand and forecast, design and optimize, and then measure and monitor KPIs. The key remaining factor is the analysis of the KPIs for the network capacity and dimensioning.

There are three main network deployment phases: the network planning phase, the initial optimization phase, and the ongoing optimization phase. Figure 6.2 illustrates these three phases in detail.

Practical Guide to LTE-A, VoLTE and IoT: Paving the Way Towards 5G, First Edition. Ayman Elnashar and Mohamed El-saidny.
© 2018 John Wiley & Sons Ltd. Published 2018 by John Wiley & Sons Ltd.

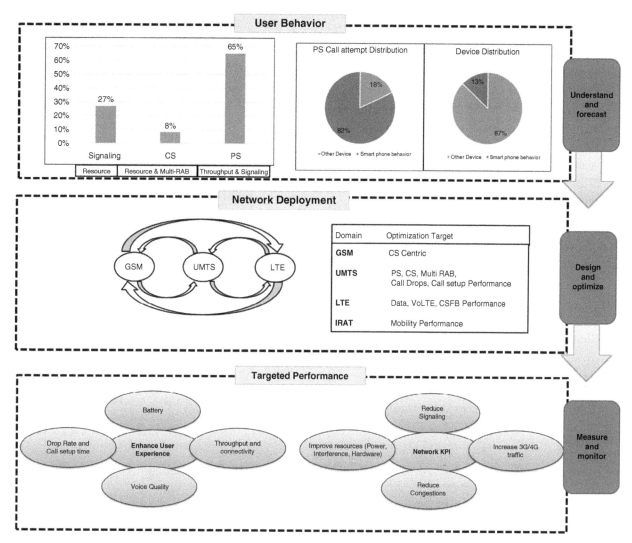

Figure 6.1 Relationship between user behavior, network deployment, and targeted performance.

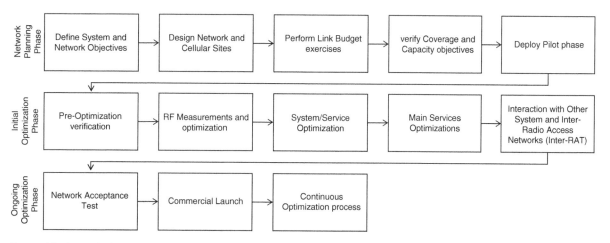

Figure 6.2 Network phases – network deployment, initial and ongoing optimization.

The network planning phase consists of multiple stages:

Stage 1: Definition of the system and network objectives.
Stage 2: Design the network and cellular sites.
Stage 3: Perform link budget exercises.
Stage 4: Verification of coverage and capacity objectives.
Stage 5: Deployment of pilot sites/cluster phase.

The initial optimization phase consists of the following stages:

Stage 1: Verification of the pre-optimization process.
Stage 2: RF measurements and optimization.
Stage 3: Service/system optimization.
Stage 4: Main services optimization.
Stage 5: Interaction with other systems and inter radio access networks (inter-RAT).

Finally, the ongoing optimization phase consists of the following stages:

Stage 1: Network acceptance test.
Stage 2: Commercial launch.
Stage 3: Continuous optimization process.

In this chapter, we focus on the LTE counters shown in Table 6.1.

Table 6.1 The analyzed LTE counters.

User traffic KPIs (connected and active users)	MIMO statistics
DL and UL traffic ratio	DL/UL resource block distribution
TTI utilization time domain	PDCCH performance
DL and UL cell throughput	RACH statistics
DL and UL scheduling	UL interference performance
CQI-related distribution	Connection and reliability KPIs
Modulation-related distribution MCS distribution	RRC connection success
DL/UL BLER statistics	E-RAB establishment and drop rate
PDCP performance	LTE CSFB performance

6.3 Users and Traffic Utilization

The statistics related to traffic behavior focus on user distribution, DL traffic ratio, UL traffic ratio, and TTI utilization, while the performance analysis consists of optimization concerns and suggested solutions, as shown in Figure 6.3.

6.3.1 RRC Connected User Distribution

The distribution of RRC Connected users can be calculated from the counter of Average Users Traffic, which is measured based on user state at the sampling time. The number of UEs in RRC Connected mode in a cell is sampled per second. At the end of a measurement period, the average sampling results are used as the values of the corresponding counter. Figures 6.4 and 6.5 show the overall and daily numbers of RRC Connected users, respectively, for the analyzed network. The network used for analysis is summarized in Table 6.2.

From the results, the number of RRC Connected users has a median value of 8.5 users and a peak average value of 81 users on the overall measurement period; in addition, 20% of cells have three users or fewer. Table 6.3 shows median and peak values for the number of RRC Connected users on weekdays, weekends, and overall. Peak numbers of RRC Connected users are higher at weekends, but the median is lower than on weekdays.

6.3.2 Active User Distribution

The distribution of active users can be calculated from Equation (6.1) as follows:

$$\text{Active user distribution} = \frac{[DL.THRP.Time\ (ms)]}{[Cell.DL.THRP.Time\ (ms)]} \quad (6.1)$$

DL.THRP.Time in Equation (6.1) is the DL throughput time in ms, which is the total transmit duration of downlink PDCP SDUs (Service Data Units) in a cell, the transmit duration of downlink PDCP SDUs transmitted to all UEs for each service with a specific QCI is sampled per ms in a cell. Cell.DL.THRP.Time in Equation (6.1) is

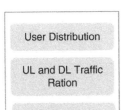
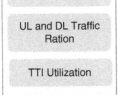

Figure 6.3 Traffic behavior and performance.

Figure 6.4 Overall number of RRC Connected users.

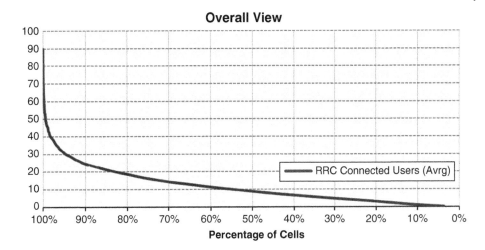

Figure 6.5 Daily numbers of RRC Connected users.

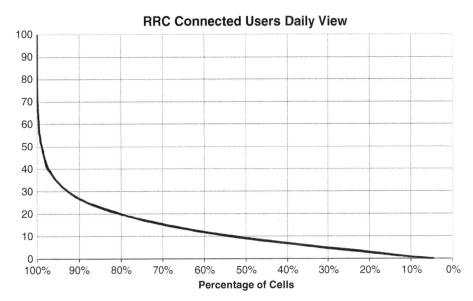

the DL cell throughput time in ms, which is the total duration of downlink data transmission with a precision of 1 ms in a cell.

Table 6.2 LTE network parameters.

Configuration	DL/UL
LTE system bandwidth	20 MHz
UE category	6
MIMO configuration	MIMO 2x2, TM3
Mobility speed	80 km/h
RF Conditions in Mobility	**Average Values**
Serving cell RSRP [dBm]	−83.8
Serving cell RSRQ [dB]	−8.7
Serving cell RSSI [dBm]	−54.9
Serving cell SINR [dB]	20.2
C-DRX for VoLTE configuration	ON (LongDrxCycle = 40 ms)

Table 6.3 RRC connected users – average and peak values for weekdays, weekends, and overall.

Median	Weekday	Weekend	Overall
	9.1 users	8.9 users	8.5 users
Peak	88.5 users	102 users	81.6 users

Figures 6.6 and 6.7 illustrate the overall and daily numbers of RRC Active users, respectively.

Figure 6.6 shows that the average number of active users is low (about one user), while from Figure 6.4, the average number of connected users is 8.5, which indicates that the users are connected with bursty traffic and, on average, there is one active user per 8.5 connected users. The same trend was observed during weekdays and at weekends. Table 6.4 shows the median and peak values for weekdays, weekends, and overall.

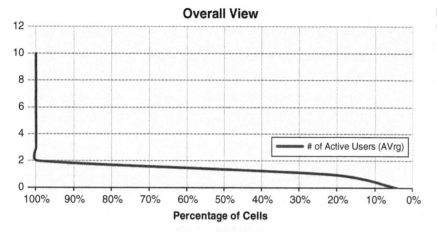

Figure 6.6 Overall number of RRC Active users.

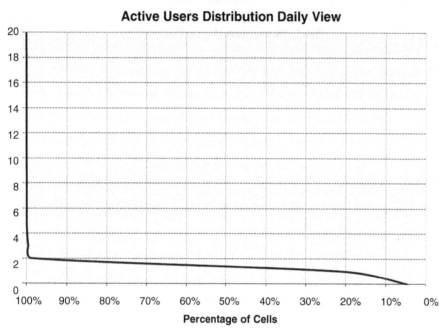

Figure 6.7 Daily number of RRC Active users.

Table 6.4 RRC Active users – average and peak values for weekdays, weekends, and overall.

	Weekday	Weekend	Overall
Median	1.1 users	1.1 users	1.1 users
Peak	4.3 users	18.6 users	8.7 users

6.3.3 DL to UL Traffic Ratio

In order to calculate the downlink traffic we can consider the number of DL throughput bits in a cell, which is the total downlink traffic volume for PDCP SDUs in this cell for services with a specific QCI ranging from 1 to 9 in a cell. DL traffic (DL.Traff) can be calculated as shown in Equation (6.2):

$$DL.Traff = \frac{[No.DL.THRP.Bits]}{[No.DL.THRP.Bits + No.UL.THRP.Bits]}$$
(6.2)

While UL traffic (UL.Traff) can be calculated as shown in Equation (6.3):

$$UL.Traff = \frac{[No.UL.THRP.Bits]}{[No.DL.THRP.Bits + No.UL.THRP.Bits]}$$
(6.3)

In the above equations, No.DL.THRP.Bits is the total downlink traffic volume for PDCP SDUs in a cell. This counter measures the total traffic volume of transmitted PDCP SDUs for services with a specific QCI ranging from 1 to 9 in a cell. The size of the payloads in the transmitted PDCP SDUs of each service with a specific QCI is sampled in a cell. The variable No.UL.THRP.Bits is the total uplink traffic volume for PDCP SDUs in a cell. This counter measures the total traffic volume of received PDCP SDUs for services with a specific QCI ranging from 1 to 9 in a cell. The size of the payloads in the received PDCP SDUs of each service with a specific

Figure 6.8 DL to UL traffic ratio.

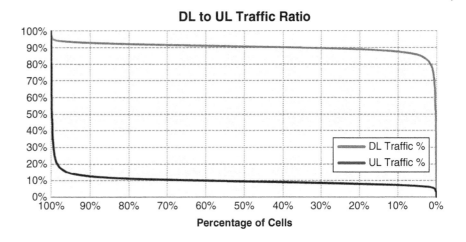

DL to UL Traffic Ratio

Figure 6.9 Average traffic volume per user.

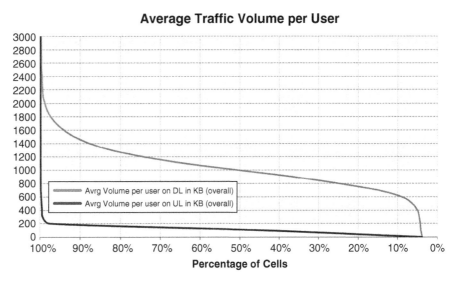

Average Traffic Volume per User

QCI is sampled in a cell. Figure 6.8 demonstrates the DL to UL traffic ratio.

From Figure 6.8, the average downlink traffic to uplink traffic ratio is 9 to 1. This means that LTE traffic is likely to be more asymmetric, which indicates a lot of streaming video usage and burst traffic applications (social networking), in general. From Figure 6.9, the median average traffic volume per user is 996 kB on the DL and 106 kB on the UL, which means that users are using data connections to generate ~900% more downlink traffic than uplink traffic.

6.3.4 TTI Utilization (Time Domain Utilization)

The DL Transmission Time Interval (TTI) utilization is calculated as shown in Equation (6.4):

$$DL.TTI.UTL = \left[\frac{Cell.DL.THRP.Time \text{ (ms)}}{3600000} \right] * 600$$

$$(6.4)$$

The variable Cell.DL.THRP.Time is the DL cell throughput time in ms, which is the total duration of downlink data transmission with a precision of 1 ms in a cell.

UL TTI utilization is calculated as shown in Equation (6.5):

$$UL.TTI.UTL = \left[\frac{Cell.UL.THRP.Time \text{ (ms)}}{3600000} \right] * 600$$

$$(6.5)$$

Here, Cell.UL.THRP.Time is the UL cell throughput time in ms, which is the total duration of uplink data transmission with a precision of 1 ms in a cell.

Figures 6.10 and 6.11 show the overall TTI occupancy for DL and UL, respectively.

It can be observed from Figures 6.10 and 6.11 that there is low average TTI utilization on the DL and UL overall. More specifically, the average value of TTI utilization is 6.4% on the DL and 6.1% on the UL for all sites in a three-week timeframe; moreover, 20% of cells have 2%

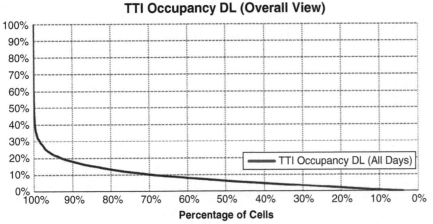

Figure 6.10 TTI occupancy – DL.

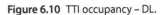

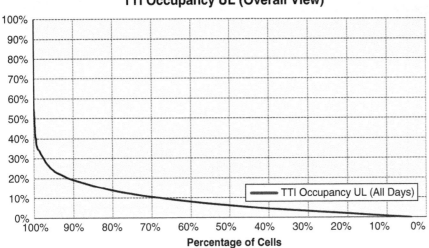

Figure 6.11 TTI occupancy – UL.

utilization or less, and the peak average TTI utilization on a daily basis per cell is 62% on the DL and 68% on the UL.

6.4 Downlink Analysis

The statistics related to the downlink focus on throughput, scheduling, and resource utilization, while the performance analysis consists of optimization concerns and suggested solutions, as shown in Figure 6.12.

6.4.1 DL Cell Throughput Performance

DL cell throughput is calculated from Equation (6.6):

$$DL.Cell.THRP = \frac{[No.DL.THRP.Bits]}{[Cell.DL.THRP.Time(\text{ms}) * 1000]}$$

(6.6)

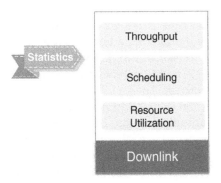

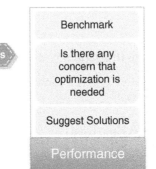

Figure 6.12 Downlink analysis and performance.

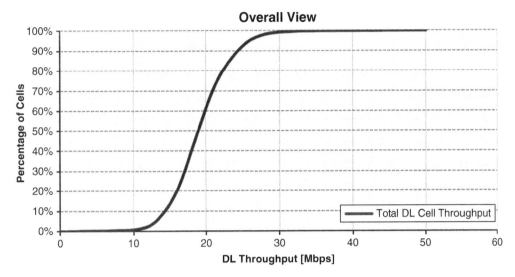

Figure 6.13 Downlink throughput (Mbps).

The variable No.DL.THRP.Bits is the number of DL throughput bits, which is the total downlink traffic volume for PDCP SDUs in a cell; it measures the total traffic volume of transmitted PDCP SDUs for services with a specific QCI ranging from 1 to 9 in a cell. The size of the payloads in the transmitted PDCP SDUs of each service with a specific QCI is sampled in a cell. Cell.DL.THRP.Time in Equation (6.6) is the total duration of downlink data transmission with a precision of 1 ms in a cell.

Figure 6.13 shows that the average DL cell throughput is 18.9 Mbps.

6.4.2 Downlink Bit Rate per Resource Block

The DL bit rate per resource block is calculated from Equation (6.7):

$$DL\,bit\,rate\,per\,resource\,block$$
$$= \frac{[No.DL.THRP.Bits]}{[1000000 * 3600 * Avg.Used.DL.PHYRB]}$$
$$(6.7)$$

The variable No.DL.THRP.Bits is the number of downlink throughput bits, which is the total downlink traffic volume for PDCP SDUs in a cell; it measures the total traffic volume of transmitted PDCP SDUs for services with a specific QCI ranging from 1 to 9 in a cell. The size of the payloads in the transmitted PDCP SDUs of each service with a specific QCI is sampled in a cell. In Equation (6.7), Avg.Used.DL.PHYRB is the average number of DL physical resource blocks used, which is the average number of PRBs used by the PDSCH and by the PDSCH DRBs in a cell. This is used to analyze the usage of downlink PRBs. The number of downlink PDSCH PRBs is determined by the downlink traffic and the quality of signals. During a measurement period,

the average number of PRBs used by the PDSCH and by PDSCH DRBs is measured per ms in a cell. Figure 6.14 shows the average DL bit rate per resource block.

The figure shows that the DL bit rate per RB is 0.22 Mbps with 20 MHz; the maximum bit rate per RB is 1.42 Mbps, so only 15.5% of a single RB throughput is utilized on the DL. This indicates that lower MCS, in general, is scheduled for each RB (resource blocks are available, but TBS scheduling is low, or time scheduling is low).

6.4.3 DL User Throughput

The downlink user throughput includes the downlink user throughput per connected user and the downlink user throughput per active user. The downlink user throughput per connected user is calculated as follows:

Firstly, in order to calculate the downlink user throughput per connected user, we should calculate the downlink time schedule per second using the formula:

$$DL.Time.SchpSec = \frac{[DL.THRP.Time\,(ms)]}{(3600 * 1000)} \quad (6.8)$$

Where DL.THRP.Time is the cell DL throughput time in ms, which is the total transmit duration of downlink PDCP SDUs in a cell. The transmit duration of downlink PDCP SDUs transmitted to all UEs for each service with a specific QCI is sampled per ms in a cell.

Secondly, the downlink volume (DL.VoL) is calculated:

$$DL.Vol = \frac{[No.DL.THRP.Bits]}{[1000000 * 3600]} \quad (6.9)$$

Where No.DL.THRP.Bits is the number of DL throughput bits, which is the total downlink traffic volume for PDCP SDUs in a cell. This counter measures the total traffic volume of transmitted PDCP SDUs for services with a specific QCI ranging from 1 to 9 in a cell.

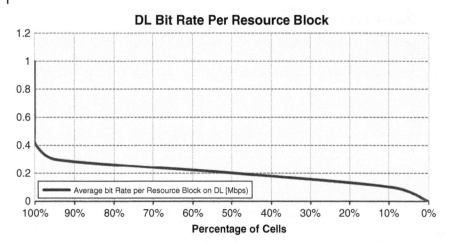

Figure 6.14 DL bit rate per resource block.

Thirdly, from Equations (6.8) and (6.9), the downlink throughput for all users is calculated from DL.Time. SchpSec and DL.Vol:

$$DL.THRP.for.all.users = \frac{[DL.Time.SchpSec]}{[DL.Vol]}$$
$$(6.10)$$

Where DL.Time.SchpSec is calculated from Equation (6.8).

Lastly, DL throughput per connected user is equal to DL throughput for all users divided by the average number of users:

$$DL.THRP.perConnUser = \frac{[DL.THRP.for.all.users]}{[Avg.Traff.Users]}$$
$$(6.11)$$

Where Avg.Traff.Users is the average number of users in a cell; the measurement is based on the user state at the sampling time. The number of UEs in RRC Connected mode in a cell is sampled per second. At the end of a measurement period, the average and maximum of these sampling results are used as the values of this counter.

The DL user throughput per active user is calculated as follows:

The number of active users is calculated from Equation (6.12):

$$Act.Users = DL.Time.SchpSec \Big/$$
$$\left[\frac{Cell.DL.THRP.Time(\text{ms})}{[60*60*1000]} \right] \quad (6.12)$$

Then, the DL throughput per active user is calculated from Equation (6.13):

$$DL.THRP.perActUser = \frac{[DL.THRP.for.all.users]}{[Act.Users]}$$
$$(6.13)$$

Figure 6.15 shows that the median DL active user throughput is 16.3 Mbps while the median DL RRC connected user throughput is 1.8 Mbps; the difference in this throughput indicates that there are many connected users with low data activity. As indicated previously, there is an average of one active user per 8.5 connected users, therefore, one user is getting high throughput while other users have lower throughput due to bursty traffic behavior while staying in RRC Connected mode.

Figure 6.15 DL throughput per active user and per connected user.

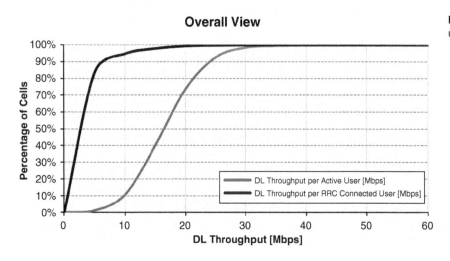

6.4.4 Scheduling Rate vs. DL Throughput

Figure 6.16 shows the user scheduling rate compared with the DL user throughput. It can be seen that bursty traffic in unloaded cells provides a better end-user experience. When the scheduling rate reduces, the user throughput increases, as users tend to send small amounts of data and finish the session quickly. For a higher scheduling rate, the user throughput decreases, which is an indication of more users sharing resources at the same time. From the cell side, Figure 6.17 shows TTI utilization compared to the DL cell throughput. As TTI utilization increases, the cell throughput increases, combining these two sides. The deployment of carrier aggregation is recommended when traffic increases, so that each carrier sees better TTI utilization to increase cell throughput. The user throughput on each carrier would improve, as the user's scheduling rate for each carrier would reduce.

Carrier aggregation performance will be discussed in the next section.

6.4.5 DL Carrier Aggregation Gains

As shown in Figure 6.18, aggregating higher bands results in an increase in the number of simultaneous scheduled users per TTI. From the user side, the bursty traffic nature in unloaded cells provides a better end-user experience. When the scheduling rate reduces, the user throughput increases, as users tend to send small amounts of data and finish the session quickly. For a higher scheduling rate, the user throughput decreases, an indication of more users sharing resources at the same time. From the cell side, as TTI utilization increases, the cell throughput increases. When traffic increases, each carrier sees better TTI utilization to increase cell throughput, while user throughput on each carrier would

Figure 6.16 User scheduling rate versus DL user throughput.

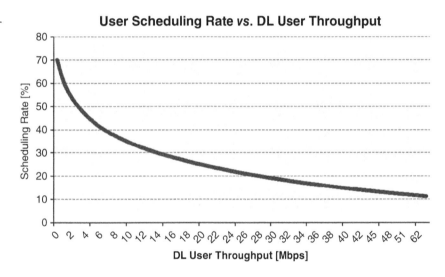

Figure 6.17 TTI utilization versus DL cell throughput.

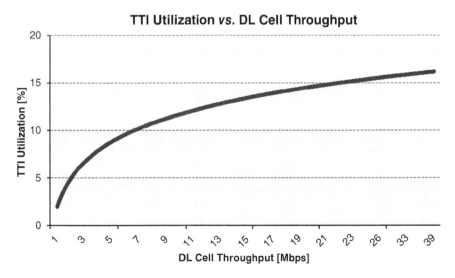

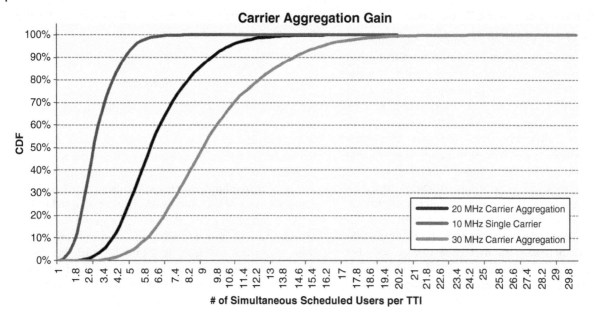

Figure 6.18 Carrier aggregation gain.

improve, as the user's scheduling rate for each carrier would reduce.

6.5 DL KPI Analysis

6.5.1 DL CQI Analysis and Modulation Performance

Before calculating the overall CQI distribution, let's first understand what is meant by a Channel Quality Indicator (CQI). A CQI indicates the downlink channel quality calculated by the UE. The UE sends the evaluation result to the eNB through the uplink channel. The CQI report is used as the input for MAC scheduling of the eNB. Counters measure the number of wideband CQI reports with a specific value ranging from 0 to 15 in a cell (if RI > 1, the CQI for each of the two codewords is calculated separately) and indicate the overall quality of the downlink channel.

The overall CQI distribution is calculated from Equation (6.14):

$$Mean\,CQI = \frac{Sum\,[(0 * N\_WCQI\_R\_0) \\ + (1 * N\_WCQI\_R\_1) + \ldots\ldots \\ + (15 * N\_WCQI\_R\_15)]}{Sum\,[(N\_WCQI\_R\_0) \\ + (N\_WCQI\_R\_1) + \ldots.. \\ + (N\_WCQI\_R\_15)]}$$

(6.14)

The variable N_WCQI_R_x stands for the number of wideband CQI reports with value x and the range of x is from 0 to 15. Figure 6.19 illustrates the average CQI per cell, which is about 10, while the overall average CQI for all cells is also about 10; this is evident from Figure 6.20.

Figure 6.19 Mean CQI per cell.

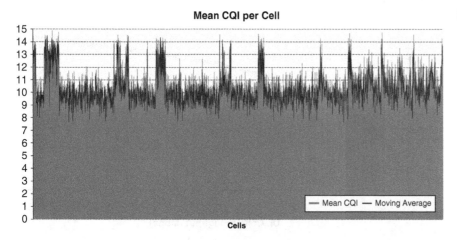

Figure 6.20 Overall CQI distribution.

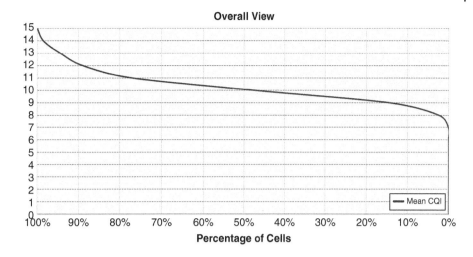

Figure 6.21 CDF daily CQI distribution.

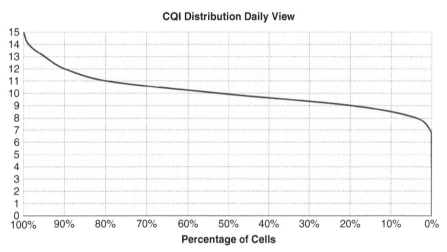

Figure 6.22 CQI versus number of LTE users.

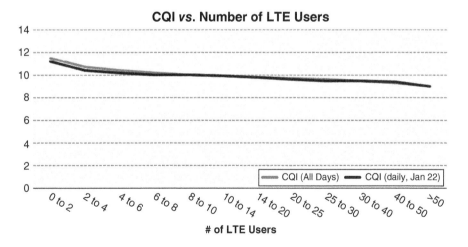

In general, the CQIs is in the vicinity of 10; a few sites show a median CQI ranging to the maximum (most likely indoor cells) while the majority of sites show a median CQI in the range [8,10]. A good CQI yields good modulation performance. The CQI reported on a daily basis (weekdays and weekends) is stable, as RF conditions do not change, but the load would impact the CQI distribution. The daily CQI distribution is shown in Figure 6.21.

CQI versus load (the number of LTE active users) is illustrated in Figure 6.22. As shown in the figure, the CQI is degraded (gracefully) with an increase in the number of connected users. Figure 6.23 illustrates the relationship between CQI and traffic on all cells.

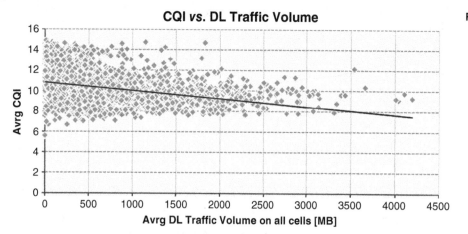

Figure 6.23 CQI versus traffic.

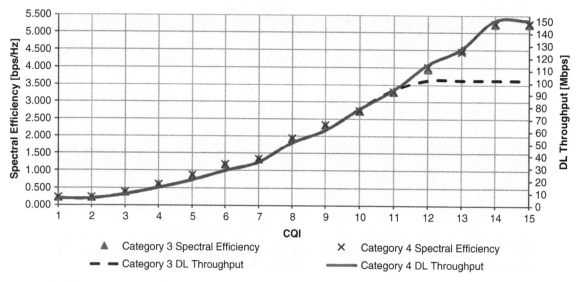

Figure 6.24 CQI versus spectral efficiency.

There is a clear relationship between the number of users in a cell and the average CQI reported by users in that cell. In addition, the CQI is degraded with an increase in the average downlink volume. Therefore, the load in terms of the number of connected users and/or average download volume traffic impacts the CQI.

CQI versus spectral efficiency and DL throughput for different UE categories are illustrated in Figure 6.24. The current median CQI = 10 yields a spectral efficiency of ~2.7 bps/Hz/cell. Improving the CQI would directly increase spectral efficiency and give even better throughput utilization for higher UE categories (like Category 4 versus Category 3). In addition, DL carrier aggregation can partially balance the load between two carriers, which can improve CQI per user and therefore the spectral efficiency.

6.5.2 DL Modulation Distribution

The DL modulation distribution contains different types of modulation: QPSK, 16QAM, 64QAM, and 256QAM.

The percentage of QPSK modulation used on the DL is calculated as shown in Equation (6.15):

$$\begin{aligned}
&Percentage\ of\ QPSK\ on\ DL \\
&= \frac{N\_TBS\_iTx\_DL\_QPSK}{\begin{aligned}&Sum[(N\_TBS\_iTx\_DL\_QPSK\\&+N\_TBS\_iTx\_DL\_16QAM\\&+N\_TBS\_iTx\_DL\_64QAM\\&+N\_TBS\_iTx\_DL\_256QAM)]\end{aligned}}
\end{aligned} \quad (6.15)$$

The variable N_TBS_iTx_DL_QPSK is the number of transmit blocks initially transmitted on the downlink SCH in QPSK modulation mode; N_TBS_iTx_DL_16QAM is the number of transmit blocks initially transmitted on the downlink SCH in

16QAM mode; and N_TBS_iTx_DL_64QAM is the number of transmit blocks initially transmitted on the downlink SCH in 64QAM mode.

Similarly, the percentage of 16QAM and 64QAM modulations used on the DL can be calculated as shown in Equations (6.16) and (6.17), respectively:

$$
\begin{aligned}
&\textit{Percentage of } 16 \textit{ QAM on DL} \\
&= \frac{N\_TBS\_iTx\_DL\_16QAM}{Sum[(N\_TBS\_iTx\_DL\_QPSK} \\
&\quad + N\_TBS\_iTx\_DL\_16QAM \\
&\quad + N\_TBS\_iTx\_DL\_64QAM \\
&\quad + N\_TBS\_iTx\_DL\_256QAM)]
\end{aligned}
\tag{6.16}
$$

$$
\begin{aligned}
&\textit{Percentage of } 64 \textit{ QAM on DL} \\
&= \frac{N\_TBS\_iTx\_DL\_64QAM}{Sum[(N\_TBS\_iTx\_DL\_QPSK} \\
&\quad + N\_TBS\_iTx\_DL\_16QAM \\
&\quad + N\_TBS\_iTx\_DL\_64QAM \\
&\quad + N\_TBS\_iTx\_DL\_256QAM)]
\end{aligned}
\tag{6.17}
$$

Figures 6.25 and 6.26 illustrate the overall and daily views of modulation types (QPSK, 16QAM, and 64QAM), respectively. 256QAM was not enabled on the network analyzed.

It can be observed from Figures 6.25 and 6.26 that QPSK has the highest mean values compared to the other modulation types (i.e. 16QAM and 64QAM). In

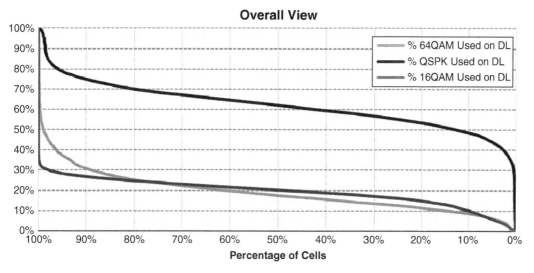

Figure 6.25 Overall view of modulation.

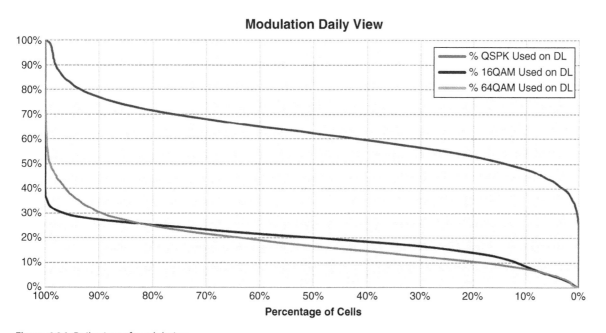

Figure 6.26 Daily view of modulation.

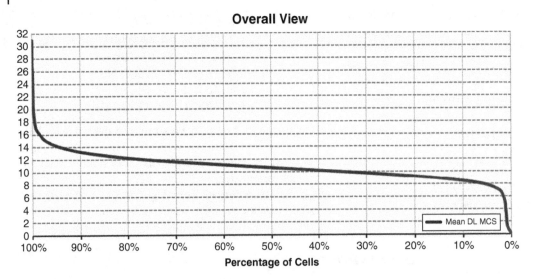

Figure 6.27 Mean DL MCS.

addition, it can be observed that 64QAM utilization is low, i.e. less than 20%. More specifically, the average usage of QPSK modulation is about 62%, while the average usage of 16QAM is about 21% and the average usage of 64QAM is about 17%. Typically, this is an indication of RF conditions, as 64QAM utilization is a function of CQI and SINR. It is worth noting that 50% of the sites indicate low 64QAM utilization in both overall and daily views.

6.5.3 DL MCS Distribution

The DL MCS distribution consists of the summation of DL modulation and coding scheme (MCS) index from MCS index 0 to MCS index 31, as defined in 3GPP 36.213 [1]. The MCS index value summarizes the modulation type and the coding rate that is used in a given PRB. Typically, a higher MCS index offers a higher spectral efficiency (which translates to a higher potential data rate) but requires a higher SINR to support it.

The overall MCS distribution is calculated as follows:

$$MCS\ distribution$$

$$= \frac{\begin{array}{l} Sum[(0 * N.MCS.0.Sch.PDSCH) \\ +(1 * N.MCS.1.Sch.PDSCH) + ... \\ +(31 * N.MCS.31.Sch.PDSCH)] \end{array}}{\begin{array}{l} Sum[(N.MCS.0.Sch.PDSCH) \\ +(N.MCS.1.Sch.PDSCH) + ... \\ +(N.MCS.31.Sch.PDSCH)] \end{array}} \quad (6.18)$$

The variable N.MCS.x.Sch.PDSCH is the number of times MCS with index x is scheduled on the PDSCH, where the range of x is from 0 to 31. The downlink MCS index is an output of downlink scheduling. The counters measure the number of times that the eNodeB schedules an MCS index for UEs in the downlink. Each codeword is configured with MCS; therefore, the MCS for each codeword is counted separately. The corresponding counter

is incremented by 1 each time the eNodeB schedules an MCS index for a UE in the downlink within a TTI. Figure 6.27 shows the mean DL MCS compared to the percentage of cells. The PDF for different MCS index based on the number of cells is provided in Figure 6.28.

It can be observed from Figures 6.27 and 6.28 that the median DL MCS index is 10.6. High usage of low MCS (i.e., MCS 0 and MCS 1) is due to bursty traffic (low TBS needed during a data session). This average MCS confirms why the throughput per RB is low (i.e., the throughput/RB is 0.22 Mbps). In conclusion, high usage of low MCS impacts the overall mean MCS per cell, and is most likely because of bursty traffic behavior.

6.5.4 DL BLER per Modulation Distribution

To calculate the overall DL BLER, first the downlink BLER per modulation distribution should be calculated, as outlined below.

The percentage BLER for QPSK is calculated from Equation (6.19):

$$BLER.QPSK.\% = \frac{[N.DL.Err.TBs.iTX.QPSK]}{[N.DL.TBs.iTX.QPSK]} \quad (6.19)$$

Where N.DL.Err.TBs.iTX.QPSK is the number of downlink error TBs after initial transmission in QPSK modulation mode, while N.DL.TBs.iTX.QPSK is the number of TBs initially transmitted on the downlink SCH in QPSK modulation mode.

The percentage BLER for 16QAM is calculated from Equation (6.20):

$$BLER.16QAM.\% = \frac{[N.DL.Err.TBs.iTX.16QAM]}{[N.DL.TBs.iTX.16QAM]} \quad (6.20)$$

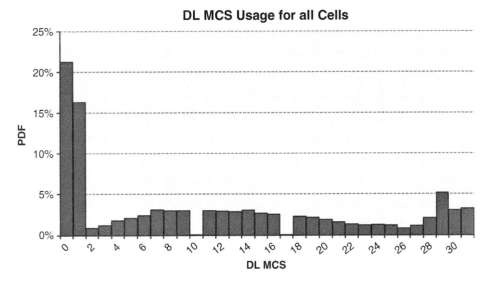

Figure 6.28 DL MCS usage for all cells.

Where N.DL.Err.TBs.iTX.16QAM is the number of downlink error TBs after initial transmission in 16QAM mode, while N.DL.TBs.iTX.16QAM is the number of TBs initially transmitted on the downlink SCH in 16QAM mode.

The percentage BLER for 64QAM is calculated from Equation (6.21):

$$BLER.64QAM.\% = \frac{[N.DL.Err.TBs.iTX.64QAM]}{[N.DL.TBs.iTX.64QAM]} \quad (6.21)$$

Where N.DL.Err.TBs.iTX.64QAM is the number of downlink error TBs after initial transmission in 64QAM mode, while N.DL.TBs.iTX.64QAM is the number of TBs initially transmitted on the downlink SCH in 64QAM mode.

The overall downlink BLER calculation is:

$$Overall\,BLER = \frac{\begin{array}{l}[N.DL.Err.TBs.iTX.QPSK]\\+[N.DL.Err.TBs.iTX.16QAM]\\+[N.DL.Err.TBs.iTX.64QAM]\end{array}}{\begin{array}{l}[N.DL.TBs.iTX.QPSK]\\+[N.DL.TBs.iTX.16QAM]\\+[N.DL.TBs.iTX.64QAM]\end{array}}$$

$$(6.22)$$

The above calculations do not consider 256QAM, which can be considered in a similar manner if it is enabled in the network.

Figure 6.29 illustrates the average value of BLER for each modulation scheme. The BLER for 64QAM has an

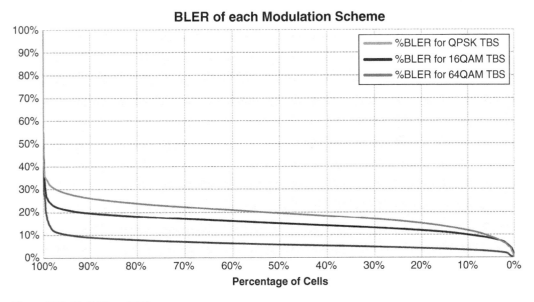

Figure 6.29 DL BLER per MCS.

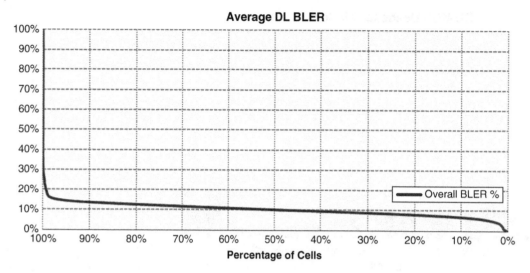

Figure 6.30 Average DL BLER per cell.

average value of 19.6%, while the BLER for 16QAM has an average value of 15.1% and the BLER for QPSK has an average value of 5.6%; the overall BLER for all modulation schemes is 10.1%, which is inline with the threshold for the BLER.

Figure 6.30 illustrates the average DL BLER versus the percentage of cells. As is evident, the average DL BLER for all cells is 10.1%. It can also be observed that the BLER for 64QAM is higher than for 16QAM and both are higher than for QPSK. This justifies why 64QAM is being used conservatively in the scheduler, due to the higher BLER. To converge to the overall BLER of 10%, the scheduler may tend to reduce MCS and hence push the modulation to a lower level.

6.5.5 DL Resource Block Utilization

The downlink physical resource block (PRB) utilization is calculated as shown in Equation (6.23) below:

$$DL.RB.UTL = \frac{Avg.Used.DL.PHYRB}{100} \quad (6.23)$$

*100 RBs for 20 MHz bandwidth.

The variable DL.RB.Ut is the downlink resource block utilization and Avg.Used.DL.PHYRB is the average number of PDSCH physical resource blocks used. The average number of PDSCH PRBs used is the counter that measures the average number of PRBs used by the PDSCH and by PDSCH DRBs in a cell. They are used to analyze the usage of downlink PRBs. The number of downlink PDSCH PRBs is determined by the downlink traffic and the quality of signals. During the measurement period, the average number of PRBs used by the PDSCH and by PDSCH DRBs is measured per millisecond in a cell. Figure 6.31 illustrates daily RB utilization for all cells. Only a few cells are utilizing 100 RBs on a

daily average basis, which means this LTE network is not heavily loaded.

Overall RB utilization is illustrated in Figure 6.32, where the median DL RB utilization is 6%, which indicates that the amount of data to be scheduled in a TTI is low, on average. A heavily loaded LTE network is considered to be one where RB utilization exceeds 25%.

6.5.6 Introduction to MIMO

MIMO techniques can be divided into two types: diversity and multiplexing. For beamforming or transmit diversity (Rank = 1), a single stream with high gain is sent for low SINR. Transmit diversity increases the SINR through diversity gain while beamforming increases the SINR through array gain by dynamically steering the beam towards a certain user.

In spatial multiplexing (Rank > 1), multiple streams are multiplexed with high gain for high SINR. Capacity enhancement is achieved through multiple increments in peak throughput of a single user. Thanks to rich, multi-path environments, two streams or more can be multiplexed using the same frequency to the same user (SU-MIMO) or multiple users with multi-user MIMO (MU-MIMO). The difference between transmit diversity and spatial multiplexing is illustrated in Figure 6.33.

6.5.6.1 MIMO Techniques: Open-loop MIMO and Closed-loop MIMO

In open-loop MIMO (OL-MIMO), the uplink feedback overhead is low, since only the Rank Indicator (RI)/CQI report is available from UEs. The pre-coding matrix indication (PMI) report is not required from UEs (channel phase information). Therefore, the channel information is not utilized to improve MIMO multiplexing performance. OL-MIMO is insensitive to the

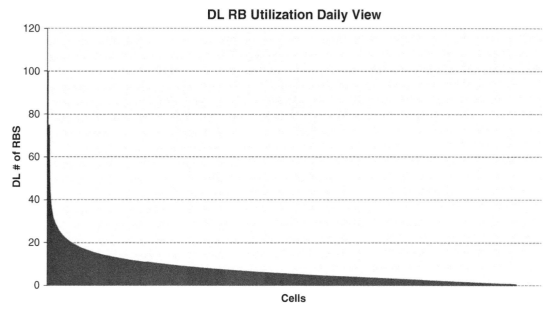

Figure 6.31 Daily RB utilization.

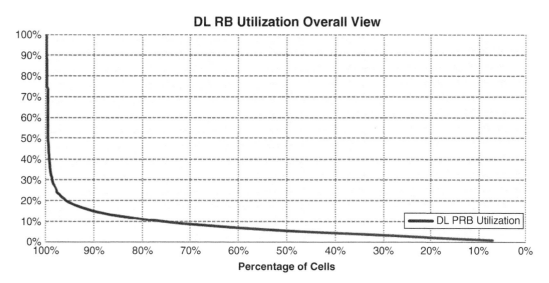

Figure 6.32 Overall RB utilization.

speed of movement and is applicable to low-speed and high-speed scenarios.

In closed-loop MIMO (CL-MIMO), the uplink feedback overhead is high, since UEs provide RI/CQI/PMI reports. UE channel information is fully utilized to improve MIMO multiplexing performance. CL-MIMO is sensitive to the speed of movement and is applicable to low-speed or static scenarios in order to be able to measure and report the channel information accurately. CL-MIMO depends on UEs and requires accurate PMIs from UEs. For fast-fading channels, it will be difficult to measure the channel information accurately and therefore, OL-MIMO will be a better choice.

It has been shown in [2] that ideal CL-MIMO provides a 2 dB theoretical performance gain over OL-MIMO.

Nevertheless, MIMO has proven to be an appropriate method of boosting user throughput, especially for low to medium system loads; moreover, the dynamic MIMO switch has proven to be very robust against variations in parameter settings. Figure 6.34 illustrates OL-MIMO versus CL-MIMO.

6.5.6.2 MIMO Techniques: Single-user MIMO and Multi-user MIMO

For single-user MIMO (SU-MIMO), one user occupies resources exclusively, a single user can transmit a single data stream or multiple streams, and it also supports only a single data stream in the LTE Release 8/9 uplink due to the single transmission antenna configuration of UEs.

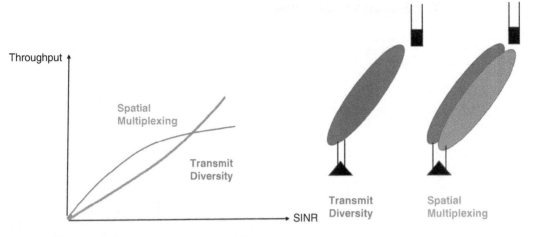

Figure 6.33 Transmit diversity versus spatial multiplexing.

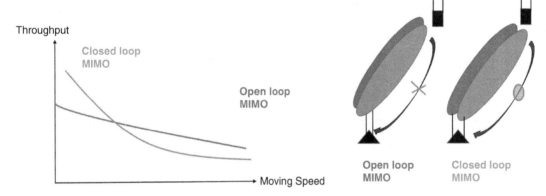

Figure 6.34 OL-MIMO versus CL-MIMO.

For multi-user MIMO (MU-MIMO), multiple users share resources in the downlink, and space diversity is exploited via Space Division Multiple Access (SDMA). The total cell throughput increases when there is a large number of users. Figure 6.35 illustrates the difference between single-user MIMO and multi-user MIMO.

6.5.7 MIMO Utilization

MIMO Rank 2 or higher utilization is the main KPI that indicates enhancement of the capacity due to MIMO multiplexing and is calculated as follows:

$$MIMO.Ut = \frac{\begin{bmatrix} Tot.N.DL.PRB.CL.R2 + \\ Tot.N.DL.PRB.BF.R2 + \\ Tot.N.DL.PRB.OL.R2 \end{bmatrix}}{\begin{bmatrix} Tot.N.DL.PRB.CL.R2 + \\ Tot.N.DL.PRB.BF.R2 + \\ Tot.N.DL.PRB.OL.R2 + \\ Tot.N.DL.PRB.CL.R1 + \\ Tot.N.DL.PRB.BF.R1 + \\ Tot.N.DL.PRB.OL.R1 \end{bmatrix}} \quad (6.24)$$

where Tot.N.DL.PRB.CL.R2 is the total number of used downlink PRBs in closed-loop Rank 2; Tot.N.DL.PRB.BF.R2 is the total number of used downlink PRBs in BF Rank 2 mode; Tot.N.DL.PRB.OL.R2 is the total number of used downlink PRBs in open-loop Rank 2; Tot.N.DL.PRB.CL.R1 is the total number of used downlink PRBs in closed-loop Rank 1; Tot.N.DL.PRB.BF.R1 is the total number of used downlink PRBs in BF Rank 1 mode; and Tot.N.DL.PRB.OL.R1 is the total number of used downlink PRBs in open-loop Rank 1.

Figure 6.36 shows overall MIMO utilization while Figure 6.37 shows daily MIMO utilization. It can be noted that the median MIMO Rank 2 utilization is 46%, which indicates good SINR levels in general. Lower cell overlap produces good SINR and hence, high MIMO Rank 2 reporting by UEs. More than 10% of cells have zero Rank 2 utilizations, which is usually the indoor sites, as only legacy IBS sites were deployed with single antennas and therefore it is difficult to deploy MIMO over these systems.

A comparison between OL-MIMO (transmit mode 3 (TM3)) and CL-MIMO (TM4) is shown in Table 6.5.

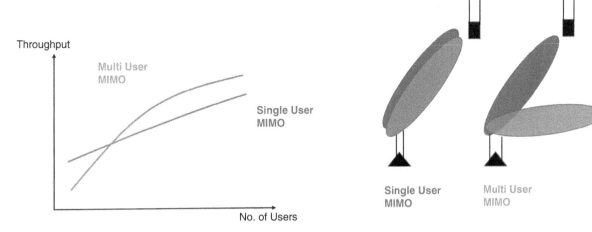

Figure 6.35 SU-MIMO versus MU-MIMO.

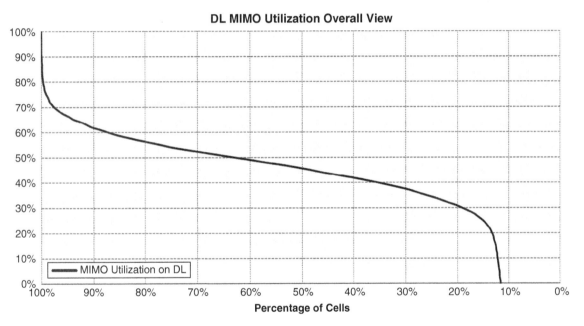

Figure 6.36 Overall DL MIMO Rank 2 utilization.

6.5.8 CL MIMO (TM4) Trial

In this section, a study has been undertaken to compare LTE KPI performance statistics for OL-MIMO (TM3) and CL-MIMO (TM4).

As illustrated in Figure 6.38, average CQI improved after the activation of CL-MIMO (TM4). In addition, as indicated in Figure 6.39, DL traffic increased by 25% after the activation of CL-MIMO (TM4). As illustrated in Figure 6.40, 64QAM utilization increased by an average of 5% after the activation of CL-MIMO (TM4). Finally, as illustrated in Figure 6.41, DL throughput increased by an average of 7% after the activation of CL-MIMO (TM4). As indicated in Figure 6.42, CL-MIMO Rank

2 decreased due to accurate reporting of the channel information.

These results are applicable to a specific scenario and for the network under test. Other scenarios could produce better or degraded performance.

In conclusion, there is a visible improvement in DL quality KPIs (i.e. CQI), which is being translated into a better user experience. There is a significant improvement in DL throughput, and better spectral efficiency is achieved without any additional resources or cost. The fact that double-codeword share is reduced compared to the special multiplexing indicates the effect of a delayed decision on bursty traffic which is not perceived by the user.

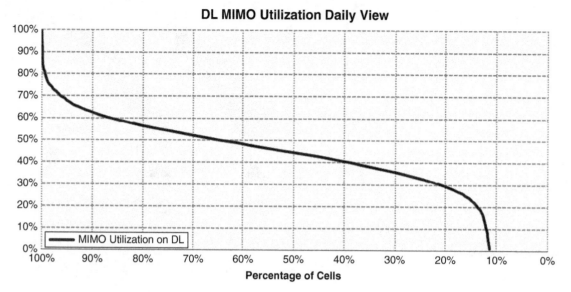

Figure 6.37 Daily MIMO utilization.

Table 6.5 OL-MIMO vs. CL- MIMO.

Transmission Mode 3			Transmission Mode 4		
Number of antenna ports	2		Number of antenna ports	2	
PDCCH DCI format	2A		PDCCH DCI format	2	
	Pre-coding information not sent			Pre-coding information decides on the number of layers	
Number of codewords	Number of layers	MIMO technique	Number of codewords	Number of layers	MIMO technique
One codeword when UE reports RI=1 and no PMI (fixed pre-coding)	2 layers	Transmit diversity	One codeword when UE reports RI=1 with unreliable PMI	2 layers	Transmit diversity
Two codewords when UE reports RI=2 and no PMI (fixed pre-coding)	2 layers	OL spatial multiplexing	Two codewords when UE reports RI=2 with reliable PMI (3 pre-coding choices)	2 layers	Rank 2 CL spatial multiplexing
			One codeword when UE reports RI=1 with reliable PMI (4 pre-coding choices)	1 layer	Rank 1 CL spatial multiplexing

6.5.9 PDCCH Performance

The Physical Control Format Indicator Channel (PCFICH) is used to inform the UE about the number of OFDM symbols used for the Physical Downlink Control Channel (PDCCH) in a subframe. Since PDCCH time domain allocation (depending on capacity) is not fixed, this channel indicates how the UE is scheduled. It indicates to the UE the CFI (Control Format Indicator) that determines the number of OFDM symbols assigned to the PDCCH in a subframe. Specifically, it defines the number of OFDM symbols that the DCI occupies

in a subframe. Table 6.6 shows the number of OFDM symbols assigned for the PDCCH in a subframe.

The PDCCH carries the scheduling assignments and other control information. It is the main channel used to schedule the UE for user data, signaling, paging, system information blocks, and random access (RACH) responses. In the frequency domain, the PDCCH is transmitted on an aggregation of one or several consecutive CCEs (Control Channel Elements) and one CCE corresponds to nine REGs (36 REs). In the time domain, the PDCCH occupies one to four symbols per subframe

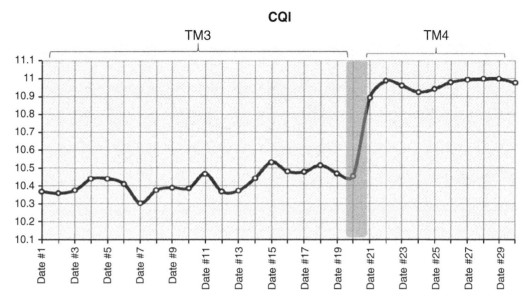

Figure 6.38 Mean CQI for OL/CL-MIMO.

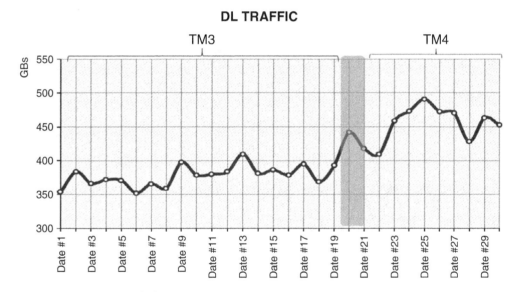

Figure 6.39 Downlink traffic for OL/CL-MIMO.

and the PCFICH is used to inform the UE of the PDCCH control region. The PDCCH supports multiple formats (called aggregation levels) which vary depending on the information sent in the channel:

PDCCH Format 0 → This consists of one CCE (9 REGs).

PDCCH Format 1 → This consists of two CCEs (18 REGs).

PDCCH Format 2 → This consists of four CCEs (36 REGs).

PDCCH Format 3 → This consists of eight CCEs (72 REGs).

Let us consider LTE PDCCH channel dimensioning as an example of LTE dimensioning. Once we know the number of resource elements (REs) available for the PDCCH, we can calculate how many available CCEs, and hence how many UEs, we can fit in a subframe, as shown in Figure 6.43.

In the analyzed network, the PDCCH symbol adaptation is done dynamically, which means the number of occupied symbols by the PDCCH is adjusted based on the required CCEs.

It can be observed from Figure 6.44 that the CFI usage is dynamic, and its mean is varying between 1~3, which is good in terms of allowing dimensioning of the PDCCH depending on the user's RF conditions and the total cell load.

It can be observed from Figure 6.45 that the mean PDCCH CCE usage is about 4. As the cell load is low, it

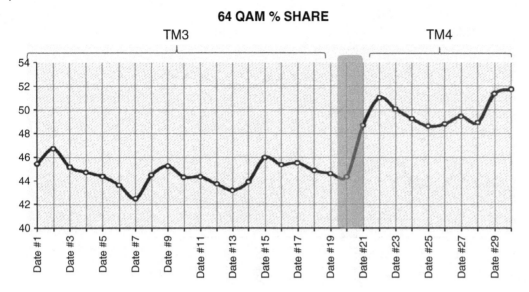

Figure 6.40 64QAM percentage share for OL/CL-MIMO.

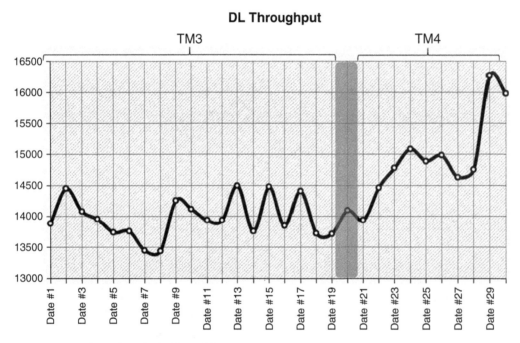

Figure 6.41 DL throughput for OL/CL-MIMO.

seems that CFI = 1 is mostly utilized. This is also evident by using more CCEs for the PDCCH (more resource elements); both high CFI and CCE would result in better user throughput but lower cell capacity.

A simulation of the maximum number of scheduled users on the PDCCH based on the mean CFI and CCE observed from counters is illustrated in Figure 6.46. Based on CFI and CCE usage, the average number of UEs scheduled simultaneously is five users, while the current configuration for the PHICH resource parameter is set to "ONE".

6.5.10   DL Summary of Observations

The points below summarize the observations for the DL analysis:

There is a low number of active and connected LTE users on average; on average, one active user per 8.5 connected users for 50% of the cells.

The number of RRC Connected users is a relevant factor in radio network dimensioning.

The share of LTE subscribers that are RRC Connected can provide useful information to estimate the radio

OL vs Closed Loop Share

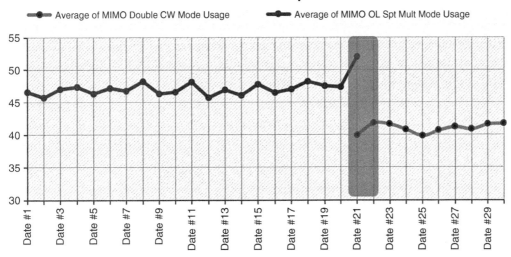

Figure 6.42 OL/CL-MIMO loop share.

Table 6.6 Number of OFDM symbols for the PDCCH in a subframe.

CFI Value	Number of OFDM Symbols Assigned to PDCCH in a Subframe	
	System bandwidth > 10 RBs	System bandwidth ≤ 10 RBs
1	1	2
2	2	3
3	3	4

Packet transmission is asymmetric on the DL and UL. Users are using data connections to generate ~10 times more downlink traffic than uplink traffic. Therefore, solutions for improving downlink efficiency are recommended. For example, smart schedulers for interference management, multi-band load balancing, carrier aggregation, and Quality of Service differentiation. In mass events, the traffic distribution can be completely different – the uplink traffic can be relatively higher from a smartphone network. Mass events can be assessed in future projects.

network requirements based on the number of LTE subscribers.

There is a low average TTI utilization on the DL and UL in the overall view; the median value is 6.4% on the DL and 6.1% on the UL for all sites in a three-week frame. The LTE network is obviously smartphone dominant with bursty traffic behavior.

6.5.11 Downlink Analysis Conclusion and Recommendations

A CQI adjustment feature is recommended to improve 64QAM utilization. The network currently uses a BLER target fixed at 10%, so pushing the BLER target to 30% adaptively would help to increase 64QAM utilization and

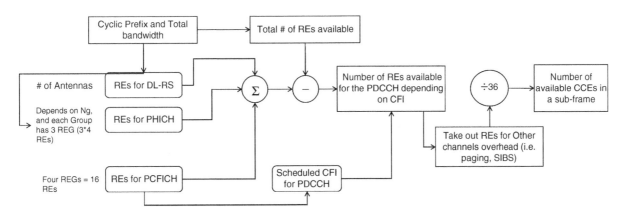

Figure 6.43 LTE dimensioning example.

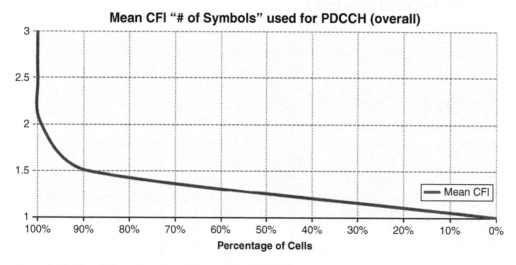

Figure 6.44 Mean CFI number of symbols used for all PDCCH.

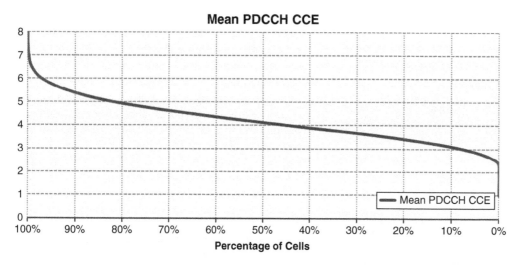

Figure 6.45 PDCCH performance.

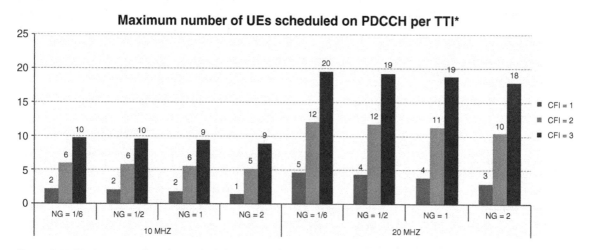

Figure 6.46 Maximum number of UEs scheduled on the PDCCH per TTI.

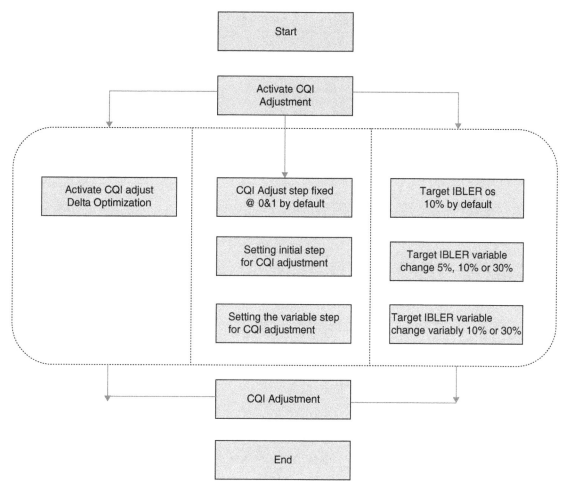

Figure 6.47 CQI adjustment.

target a less conservative scheduler. The process of CQI adjustment is illustrated in Figure 6.47.

DL Frequency-Selective Scheduling (DL FSS) is also recommended, as the basic concept of DL FSS is that the UE needs to estimate the channel quality and report the CQI of downlink sub-bands to the eNB, after which the eNodeB calculates the scheduling priorities based on sub-band CQIs reported by UEs and schedules UEs based on the scheduling priorities of UEs in each sub-band. In this mode, the UEs are scheduled to the sub-bands with the optimal channel quality. Each UE is allocated its individual best part of the spectrum. Figure 6.48 shows RBG assignment in DL frequency-selective scheduling.

Spectral efficiency usage can be improved through DL resource block allocation using interference randomization, which ensures that the RBs allocated for neighboring cells do not overlap, thereby reducing inter-cell interference. This feature is particularly beneficial for lightly loaded cells where it increases spectral efficiency, improves SINR, and increases MIMO

Rank 2 utilization. Figure 6.49 shows how interference randomization works.

6.6 UL KPI Analysis

The KPI analysis for the uplink focuses on throughput, scheduling, and resource utilization; the performance analysis consists of optimization concerns and suggested solutions. These elements, described in this section, are illustrated in Figure 6.50.

6.6.1 UL Cell Throughput Performance

The uplink cell throughput is calculated using Equation (6.25):

$$UL.Cell.TP = \frac{[No.UL.THRP.Bits]}{[Cell.UL.THRP.Time(\text{ms}) * 1000]}$$

$$(6.25)$$

The variable No.UL.THRP.Bits is the number of UL throughput bits, which is the total uplink traffic volume

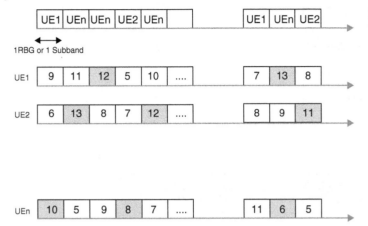

Figure 6.48 RBG assignment for DL FSS.

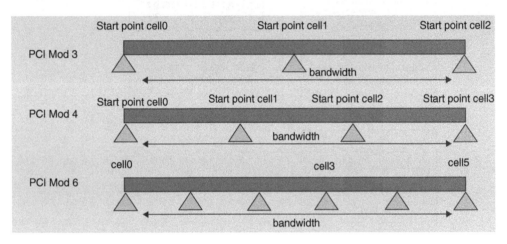

Figure 6.49 Interference randomization.

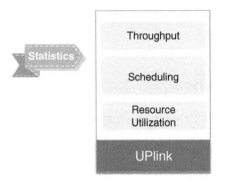

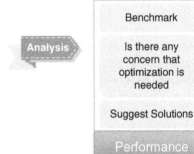

Figure 6.50 Uplink analysis and performance.

for PDCP SDUs in a cell. It measures the total traffic volume of transmitted PDCP SDUs for services with a specific QCI ranging from 1 to 9 in a cell. The size of the payloads in the transmitted PDCP SDUs of each service with a specific QCI is sampled in a cell. The variable Cell.UL.THRP.Time (ms) is the total duration of uplink data transmission with a precision of 1 ms in a cell.

The CDF of the UL throughput distribution is illustrated in Figure 6.51. As shown in Figure 6.51, the median UL cell throughput is ~2.0 Mbps.

6.6.2 UL Bit Rate per Resource Block

The uplink bit rate per resource block is calculated from Equation (6.26):

$$UL.BRPRB = \frac{[No.UL.THRP.Bits]}{[1000000 * 3600 * Avg.Used.UL.PHYRB]}$$
(6.26)

The variable UL.BRPRB is the uplink bit rate per resource block, while the Avg.Used.UL.PHYRB is the

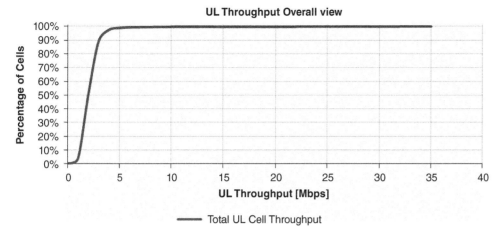

Figure 6.51 Uplink cell throughput.

average number of UL physical resource blocks used, which is the average number of PRBs used by the PUSCH and by PUSCH DRBs in a cell. They are used to analyze the usage of uplink PRBs. The number of uplink PUSCH PRBs is determined by the uplink traffic and the quality of signals. During a measurement period, the average number of PRBs used by the PUSCH and by PUSCH DRBs is measured per ms in a cell. Figure 6.52 shows the average UL bit rate per resource block.

From Figure 6.52, the UL bit rate per RB is 0.017 Mbps. Therefore, with a 20 MHz carrier, the maximum bit rate per RB is 0.52 Mbps. Only 3.3% of a single RB throughput is utilized on the UL, which means very low UL utilization. This indicates that lower MCS, in general, is scheduled for each RB (resource blocks are available, but TBS scheduled is low, or time scheduling is low).

6.6.3 UL MCS Distribution

We can calculate the overall UL MCS distribution as follows:

$$MCS\,distribution = \frac{Sum[(0 * N.MCS.0.Sch.PUSCH) + (1 * N.MCS.1.Sch.PUSCH) + \dots + (31 * N.MCS.31.Sch.PUSCH)]}{Sum[(N.MCS.0.Sch.PUSCH) + (N.MCS.1.Sch.PUSCH) + \dots + (N.MCS.31.Sch.PUSCH)]}$$

(6.27)

The variable N.MCS.x.Sch.PUSCH is the number of times MCS with index x is scheduled on the PUSCH, and the x value ranges from 0 to 31. The uplink MCS index is an output of uplink scheduling. The counters measure the number of times that the eNodeB schedules

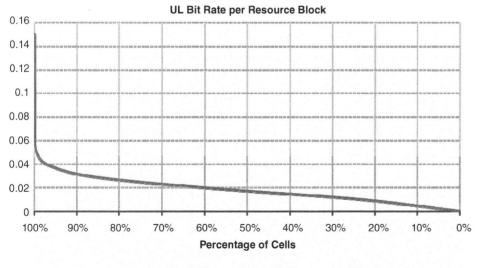

Figure 6.52 Uplink bit rate per resource block.

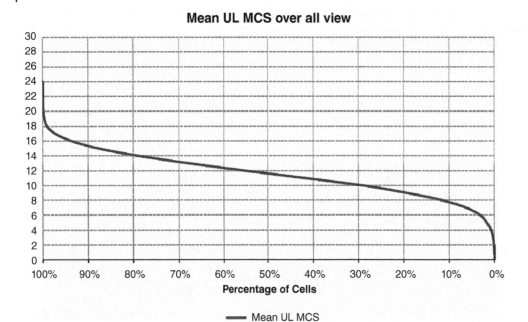

Figure 6.53 Mean UL MCS.

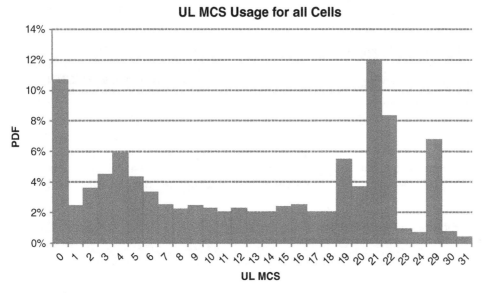

Figure 6.54 UL MCS usage for all cells.

an MCS index for UEs in the uplink, and the MCS index ranges from 0 to 31. Each codeword is configured with an MCS. Therefore, the MCS for each codeword is counted separately. The counters are used to analyze the uplink scheduling condition.

The corresponding counter is incremented by one each time the eNodeB schedules an MCS index for a UE in the uplink within a TTI. Figures 6.53 and 6.54 demonstrate, respectively, the mean UL MCS and the PDF of MCS index.

It can be observed from Figures 6.53 and 6.54 that the median UL MCS is 13.1. The scheduler seems to maximize UL MCS due to good UL SINR within the

session. This average MCS confirms throughput per RB is low due to a low scheduling rate combined with low MCS scheduled. With MCS 11.7, throughput/RB with 100% scheduling = 0.18 Mbps; in other words, with MCS 11.7, throughput/RB with 10% scheduling = 0.018 Mbps, which means UL bursty traffic in general.

6.6.4 UL User Throughput

The uplink user throughput includes the UL user throughput per connected user and the UL user throughput per active user. The uplink user throughput per connected user is calculated as follows:

Firstly, the uplink time schedule per second is calculated:

$$UL.Time.SchpSec = \frac{[UL.THRP.Time \text{ (ms)}]}{(3600 * 1000)} \quad (6.28)$$

Where UL.THRP.Time (ms) is the cell UL throughput time in ms, which is the total transmit duration of uplink PDCP SDUs in a cell; the transmit duration of uplink PDCP SDUs transmitted to all UEs for each service with a specific QCI is sampled per ms in a cell. Secondly, the uplink volume (UL.VoL) is calculated:

$$UL.Vol = \frac{[No.UL.THRP.Bits]}{[1000000 * 3600]} \quad (6.29)$$

Thirdly, from Equations (6.28) and (6.29), the uplink throughput for all users is calculated:

$$UL.THRP.for.all.users = \frac{[UL.Time.SchpSec]}{[UL.Vol]} \quad (6.30)$$

Where UL.Time.SchpSec is calculated from Equation (6.28).

Lastly, the UL throughput per connected user is equal to the UL throughput for all users divided by the average number of traffic users, as follows:

$$UL.THRP.perConnUser = \frac{[UL.THRP.for.all.users]}{[Avg.Traff.Users]} \quad (6.31)$$

Where Avg.Traff.Users is the average number of users in a cell. The measurement is based on the user state at the sampling time.

The number of UEs in RRC Connected mode in a cell is sampled per second. At the end of a measurement period, the average and maximum of these sampling results are used as the values of this counter.

The UL user throughput per active user can be calculated as follows:

Firstly, the number of active users is calculated:

$$Act.Users$$
$$= UL.Time.SchpSec \Big/ \left[\frac{Cell.UL.THRP.Time(\text{ms})}{[60 * 60 * 1000]} \right] \quad (6.32)$$

Then, the UL throughput per active user is calculated as follows:

$$UL.THRP.perActUser = \frac{[UL.THRP.for.all.users]}{[Act.Users]} \quad (6.33)$$

Figure 6.55 explains that the median UL active user throughput is 1.8 Mbps while the median UL RRC Connected user throughput is 0.19 Mbps. The difference in this throughput indicates that many connected users have low data activity (most likely DL only activity). As is evident, there is, on average, one active user per 8.5 connected users. Therefore, one user is getting high throughput while others have lower throughput due to bursty traffic behavior while staying in RRC Connected mode.

6.6.5 UL Modulation Distribution

The UL modulation distribution contains different types of modulation: QPSK, 16QAM, and 64QAM if it is deployed. The percentage of QPSK modulation used on the uplink is calculated as follows:

$$Percentage\,of\,QPSK\,on\,UL$$
$$= \frac{(N\_TBS\_iTx\_UL\_QPSK)}{\begin{array}{c}Sum[(N\_TBS\_iTx\_UL\_QPSK) \\ + (N\_TBS\_iTx\_UL\_16QAM)]\end{array}} \quad (6.34)$$

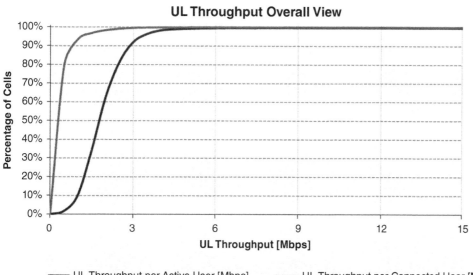

Figure 6.55 UL throughput.

The variable N_TBS_iTx_UL_QPSK is the number of transmit blocks initially transmitted on the UL SCH in QPSK modulation mode and N_TBS_iTx_UL_16QAM is the number of transmit blocks initially transmitted on the UL SCH in 16QAM mode. In the analyzed network, 64QAM was not deployed on the UL.

Similarly, the percentage of 16QAM used on the uplink is calculated as follows:

$$
\begin{aligned}
&Percentage\ of\ 16\ QAM\ on\ UL \\
&= \frac{(N\_TBS\_iTx\_UL\_16QAM)}{Sum[(N\_TBS\_iTx\_UL\_QPSK)} \\
&\quad + (N\_TBS\_iTx\_UL\_16QAM)]
\end{aligned}
\quad (6.35)
$$

Figure 6.56 demonstrates the utilization of QPSK and 16QAM. As shown, UL 16QAM is used more than UL QPSK – the overall amount of 16QAM used in the UL is about 54% while the overall amount of QPSK used in the UL is 46%. This indicates less UL interference and also indicates the potential of 64QAM in the UL. The expected utilization of 64QAM in the UL is about 10% and it will mainly improve UL TP in indoor scenarios.

Generally, 16QAM utilization is considered high, on average (MCS order is higher than 10, on average). From Figures 6.56 and 6.57, it can be observed that MCS is higher than 10 counts for 56% of the overall distribution while MCS is lower than 10 counts for 44%. MCS is a function of UL SINR, which means high 16QAM shows high UL SINR in general, due to low UL load. So, this network is not heavily loaded in terms of UL.

6.6.6 UL BLER per Modulation Distribution

The uplink BLER for each modulation distribution is calculated as follows:

The percentage of BLER for QPSK is calculated:

$$
BLER.QPSK.\% = \frac{[N.UL.Err.TBs.iTX.QPSK]}{[N.UL.TBs.iTX.QPSK]}
\quad (6.36)
$$

Where N.UL.Err.TBs.iTX.QPSK is the number of uplink error TBs after initial transmission in QPSK modulation mode, while N.UL.TBs.iTX.QPSK is the number of TBs initially transmitted on the uplink SCH in QPSK modulation mode.

The percentage of BLER for 16QAM is calculated:

$$
BLER.16QAM.\% = \frac{[N.UL.Err.TBs.iTX.16QAM]}{[N.UL.TBs.iTX.16QAM]}
\quad (6.37)
$$

Where N.UL.Err.TBs.iTX.16QAM is the number of uplink error TBs after initial transmission in 16QAM mode, while N.UL.TBs.iTX.16QAM is the number of TBs initially transmitted on the uplink SCH in 16QAM mode.

The overall uplink BLER is calculated:

$$
\begin{aligned}
&Overall\ BLER \\
&= \frac{\begin{array}{l}[N.UL.Err.TBs.iTX.QPSK] \\ + [N.UL.Err.TBs.iTX.16QAM] \\ + [N.UL.Err.TBs.iTX.64QAM]\end{array}}{\begin{array}{l}[N.UL.TBs.iTX.QPSK] \\ + [N.UL.TBs.iTX.16QAM] \\ + [N.UL.TBs.iTX.64QAM]\end{array}}
\end{aligned}
\quad (6.38)
$$

As explained in Figure 6.58, the average BLER for QPSK modulation is about 9.1% while the average BLER for 16QAM is 4.8%.

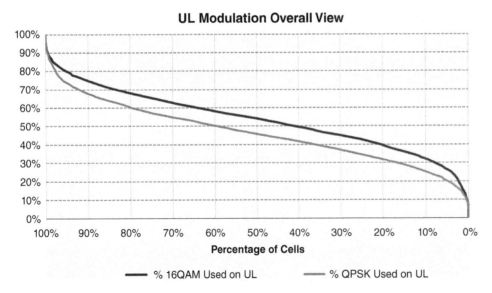

Figure 6.56 Overall view of modulation.

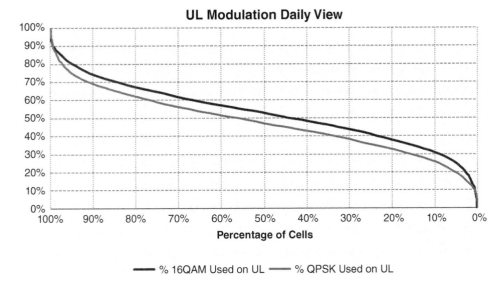

Figure 6.57 Daily view of modulation.

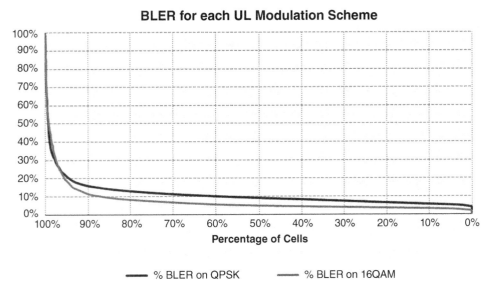

Figure 6.58 UL BLER per modulation scheme.

It can be observed from Figure 6.59 that the overall median BLER is 6.7%, BLER observed on 16QAM is lower than QPSK, and the overall BLER is not converging directly to 10%. This may show that the scheduler is being conservative in scheduling higher TBs as QPSK is achieving close to 10% BLER, and with a small uplink session duration, convergence does not kick off quickly. This depends on scheduler behavior and it varies from vendor to vendor. We may tune the scheduler parameters to push the BLER to 10%, however this may impact the overall performance.

6.6.7 UL Resource Block (RB) Utilization

The uplink resource block utilization is calculated as follows:

$$UL\ Resource\ Block\ (RB)\ Utilization$$
$$= \frac{Average\ number\ of\ used\ PUSCH\ PRBs}{100}$$

$$(6.39)$$

*100 RBs for 20 MHz bandwidth.

The Average number of used PUSCH PRBs is the counter that measures the average number of PRBs used by the PUSCH and by PUSCH DRBs in a cell. They are

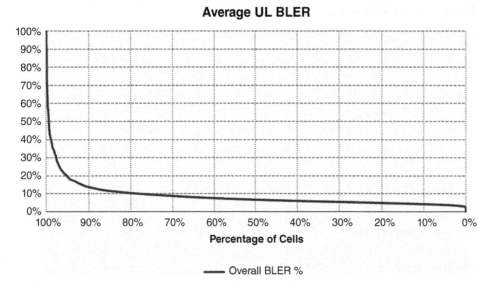

Figure 6.59 Average UL BLER for all cells.

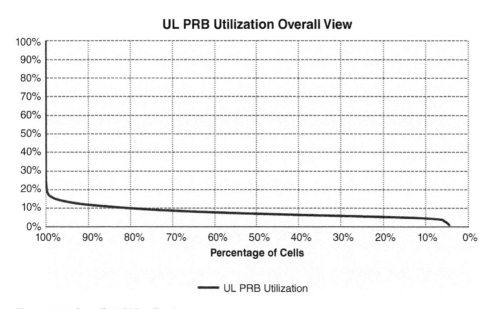

Figure 6.60 Overall UL PRB utilization.

used to analyze the usage of uplink PRBs. The number of uplink PUSCH PRBs is determined by the uplink traffic and the quality of signals. During a measurement period, the average number of PRBs used by the PUSCH and by PUSCH DRBs is measured per ms in a cell. Overall UL PRB utilization and daily UL PRB utilization are illustrated in Figures 6.60 and 6.61, respectively.

It can be observed from Figures 6.60 and 6.61 that the median UL RB utilization is 7.2%, which indicates a low load in general on the UL. Note that a heavily loaded network is considered to be one where RB utilization exceeds 25%. Moreover, there are no cells with 100 RBs utilized on a daily average basis.

6.6.8 RACH Performance

The random access procedure is used in various scenarios including initial access, handover, and call re-establishment. It is mainly used when a UE needs to initiate a connection with a cell to perform open-loop power control. The PRACH channel is used for the uplink to transmit preambles with certain power that is adjusted based on the path loss and number of retransmissions needed for each preamble to minimize the collision of preambles between users. LTE defines RACH procedures as contention-based or contention-free. Figure 6.62 illustrates contention-based random access and contention-free random access.

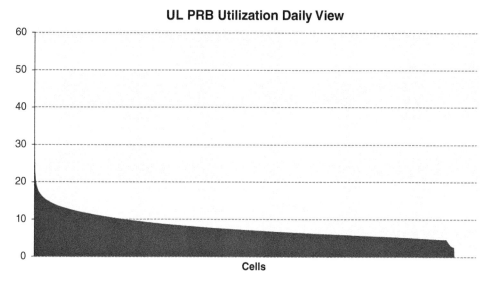

Figure 6.61 Daily UL PRB utilization.

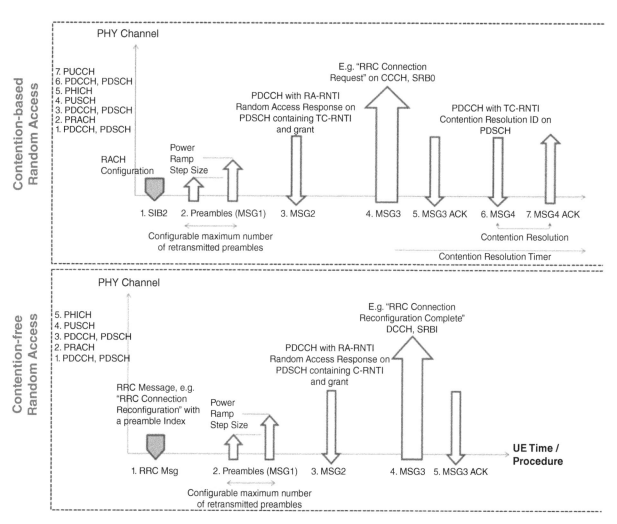

Figure 6.62 Contention-based random access versus contention-free random access.

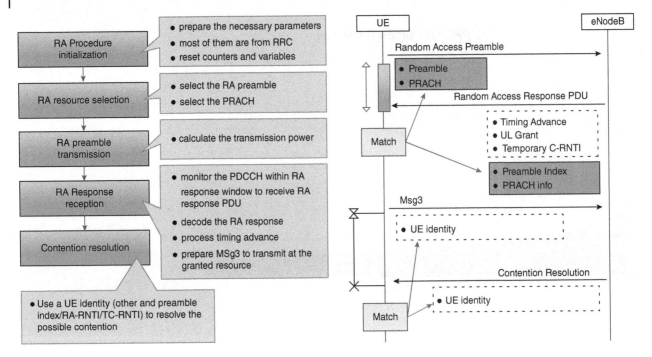

Figure 6.63 Random access procedure and signaling flow.

The general steps and call flow for the random access (RA) procedure are shown in Figure 6.63. It is evident that the random access procedure consists of RA procedure initialization, RA resource selection, RA preamble transmission, RA response reception, and contention resolution. The signaling flow between the UE and eNodeB is also explained in Figure 6.63.

It can be observed from Figure 6.64 that RACH attempts for HO are almost equal to contention-free

RACH and amount to about 5.6, while contention-based RACH is about 31.9, which means that RACH for a HO attempt is close to contention-free. It indicates that contention-free RACH is mainly used for handover only; on average, for 50% of a cell, there are six handover attempts per minute, which indicates high-mobility users. The median RACH contention-based usage in all cells is 83.7% while the median RACH contention-free usage in all cells is 16.3%, which indicates that RACH

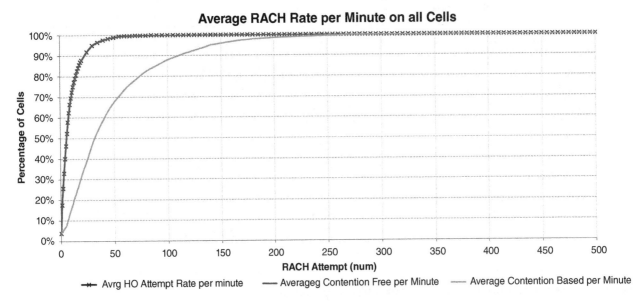

Figure 6.64 RACH type distribution.

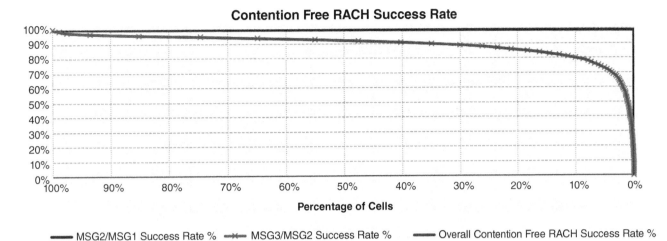

Figure 6.65 RACH type distribution.

is mainly used for reasons other than handover (i.e. UL data, RLF, etc.).

6.6.8.1 RACH Contention-free Performance

The non-contention-based RA procedure is used in a handover or in the recovery of uplink synchronization. The preamble used in the RA procedure is the non-contention-based preamble, otherwise known as the dedicated preamble. It is allocated by the eNodeB. During handover, the eNodeB sends the preamble to the UE with the RRC signaling message. In the recovery of uplink synchronization, the eNodeB sends the preamble to the UE with the DCI on the PDCCH. The counters measure the number of times the non-contention-based preamble is received in a handover or in the recovery of uplink synchronization in a cell. In order to evaluate the success rate of RA procedures during handover, Equation (6.40) shows the overall RACH contention-free performance.

$$Overall\ RACH\ Contention-free\ Performance$$
$$= \frac{[N.Msg3.Rsp.HO.Trig.CF.RA]}{[N.CF.Preamble.Rx]} \quad (6.40)$$

The variable N.Msg3.Rsp.HO.Trig.CF.RA is the number of times the UE Msg3 Response message is received in the handover-triggered, non-contention-based RA procedure, and N.CF.Preamble.Rx is the number of times the non-contention-based preamble is received in a handover or in the recovery of uplink synchronization in a cell.

Figure 6.65 shows the distribution of RACH types. It can be noted that the median RACH contention-free success rate is about 92%, while 20% of sites see a success rate of 86% or higher. As observed, the success rate of the Msg3 Response message being received in a cell after a Msg2 response is sent is 92% (100% success rate for

Msg2 sent for every Msg1 received[1]). This indicates that issues surround Msg2 mainly, and it is not easy to conclude from counters only whether it is a DL or an UL issue with Msg2; field testing is required.

6.6.8.2 RACH Contention-based Performance

There are two groups of RA preambles: Group A and Group B. Group A always exists and Group B exists only with a specific configuration in the SIB2 parameter. The determination of Group A and Group B is described in 3GPP 36.321 5.1.1 Random Access Procedure Initialization [3]. The overall RACH contention-based performance is calculated as shown in Equation (6.41):

$$Overall\ RACH\ Contention-based\ Performance$$
$$= \frac{\begin{bmatrix}(N\_CR\_Preamble\_GA)\\ + (N\_CR\_Preamble\_GB)\end{bmatrix}}{\begin{bmatrix}(N\_Times\_C\_Preamble\_GA\_Rec)]\\ + (N\_Times\_C\_Preamble\_GB\_Rec)\end{bmatrix}} \quad (6.41)$$

The variable N_CR_Preamble_GA is the number of times a cell sends a contention resolution message after receiving a preamble in Group A; N_CR_Preamble_GB is the number of times a cell sends a contention resolution message after receiving a preamble in Group B; N_Times_C_ Preamble_GB_Rec is the number of times the contention preamble in Group B is received; and N_Times_C_ Preamble_GA_Rec is the number of times the contention preamble in Group A is received.

From Figure 6.66, the median RACH contention-based success rate is 85.4% while the RACH success rate for preambles coming from [Group A, Group B] = [88%, 56%] and 20% of sites see a success rate of 66% or higher. The Msg2 to Msg3 success ratio is 99.5%. Therefore,

---

1 The counter would not show how many preambles were needed, as it is pegged on the cell side, which leads to inconclusive reasons as to why Msg2 is having issues.

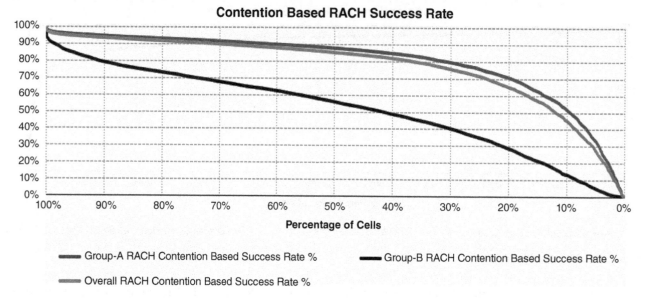

Figure 6.66 RACH type distribution.

the issue seems to be mainly related to Msg4 contention resolution. Total usage of Group B from overall contention-based RACH = 7.1%; even though Group B attempts are low, checking why the success rate is low is important. Field testing is required to capture these RACH-related issues.

6.6.9   UL Interference Performance

Figure 6.67 demonstrates the average UL interference, while Figure 6.68 provides the average UL interference versus average cell TTI utilization. It can be

observed from Figures 6.67 and 6.68 that there is no high UL interference, as the median value is −117 dBm. Also, UL interference increases with higher utilization of UL TTIs. Even at higher utilization, the interference is still manageable, with variance starting to become clear after TTI utilization in a cell becomes higher than 25%.

The same trend is shown in Figure 6.69, where UL interference increases with higher utilization of UL resource blocks. With current traffic behavior and UL interference levels, it does not seem that the network requires the deployment of UL carrier

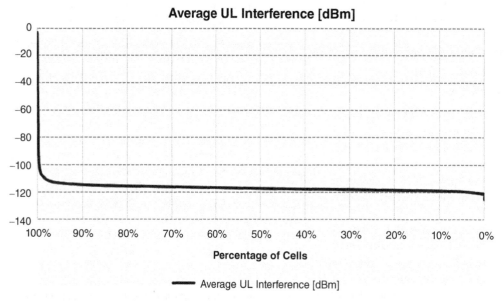

Figure 6.67 Average UL interference.

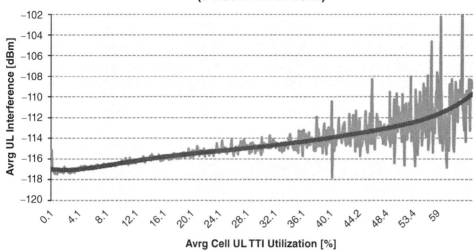

Figure 6.68 UL interference versus cell UL TTI utilization.

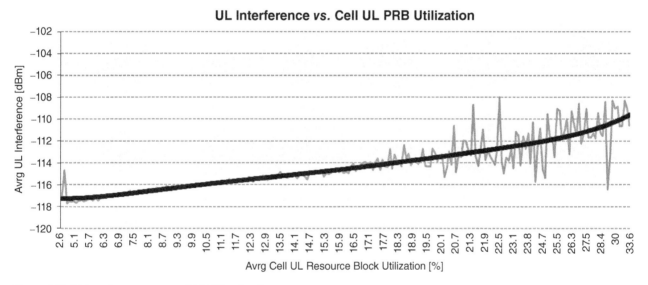

Figure 6.69 UL interference versus cell UL PRB utilization.

aggregation. UL interference, on average, is low, however the trend can change with higher loads (i.e. mass events or in single sites served by many users, like indoor sites).

6.6.10 Uplink Analysis Conclusion and Recommendations

16QAM usage can be improved through an adaptive target IBLER feature. The current network uses a BLER target fixed at 10%, but convergence does not happen because of low periods in each session and also because QPSK BLER is converging, which means the scheduler is satisfied with the BLER target initially. The decision as to whether to use the adaptive target IBLER function, in

which the target IBLER value for the user is adaptively adjusted to 1%, 5%, 10%, or 30%, is based on user distance from the eNodeB and channel quality fluctuations. If the SINR fluctuates significantly or the user is far away from the eNodeB, a large target IBLER value is selected; if the user is close to the eNodeB and the SINR is stable, a small target IBLER value is selected.

The uplink scheduler can be improved through enhanced UL frequency-selective scheduling. When this feature is enabled, the eNodeB also takes the number of UEs to be scheduled and the number of active UEs in a cell into consideration during PUSCH resource allocation. This enables more UEs to use frequency-selective scheduling with a small cost increase on the eNodeB. When frequency-selective scheduling is not performed, interference-randomization-based scheduling can reduce inter-UE interference and therefore increase the user throughput. As observed, the uplink interference increases with an increased number of users. Therefore, enabling this feature can be a good step prior to uplink carrier aggregation.

RACH performance can be improved through a RACH resource adjustment feature. As observed, the RACH success rate in the network is low. Smartphones may tend to generate more RACH attempts due to data, handovers, or connection establishment.

For a system bandwidth of 15 MHz or 20 MHz, the default PRACH occurrence interval is 5 ms; that is, there are two PRACHs in each radio frame. In this case, RACH resource adjustment is recommended only when contention-based random access is performed fewer than 50 times per second and non-contention-based random access is performed fewer than 30 times per second. The purpose of RACH resource adjustment is

to match PRACHs with loads. Though loads vary with time, RACH resource adjustment can be used as long as the loads within a period meet the conditions. In the current network, there are 29 contention-based RACH attempts every minute and there are 6 contention-free RACH attempts every minute.

6.7 Data Connection Performance

The statistics related to the data connection performance analysis focus on call setup, call drop, and CS fallback, while the performance analysis consists of optimization concerns and suggested solutions, as summarized in Figure 6.70.

Let us take a close look at the general call flow illustrated in Figure 6.71. Due to application packages not always being continuous, the network may release an RRC connection due to the inactivity of a user. Upon

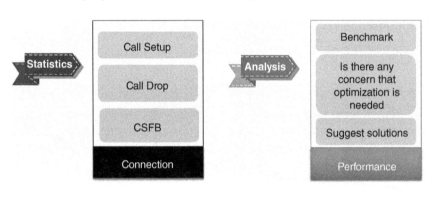

Figure 6.70 Data connection performance.

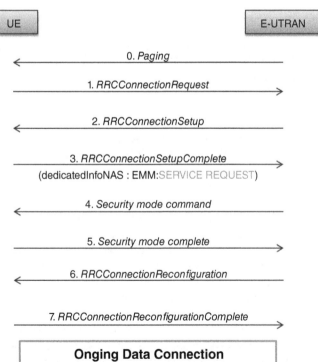

Figure 6.71 General call flow.

receiving an application package or paging for an MT package, the UE will trigger the establishment of a DRB. Steps 1 to 3 in Figure 6.71 indicate the RRC connection setup success rate. Steps 6 to 7 indicate the E-RAB establishment success rate. SRB2 and DRB can be set up after security has been activated. The link between the EPS bearer and the DRB is via eps-BearerIdentity and drb-Identity. This will also contain dedicated radio resource configuration. After Step 7, any drop in connection will indicate abnormal release and counts towards call drops.

6.7.1 RRC Connection Performance

Figure 6.72 provides the RRC connection performance while Figure 6.73 illustrates the number of RRC connections for connected users. It can be observed from Figures 6.72 and 6.73 that the RRC connection establishment phase is stable, while the median RRC connection

establishment success rate is 99.96%; the ratio of RRC connection attempts to the number of connected users is 1:14 (7%) per second. This translates to an average of four RRC connections per user every 1 minute.

6.7.2 E-RAB Performance

Figure 6.74 shows the E-RAB establishment performance while Figure 6.75 illustrates E-RAB call drop rates. It can be observed from Figure 6.74 that the E-RAB establishment phase is stable, as the median E-RAB establishment success rate is 99.97%. Figure 6.75 shows that the E-RAB call drop rate is also stable – one call drops out of every 1300 calls for 50% of the cells (median).

6.7.3 CSFB Performance (LTE Only)

Figures 6.76 and 6.77 show that the CSFB success rate is stable (on the LTE side only, with redirection, it includes

Figure 6.72 RRC connection performance.

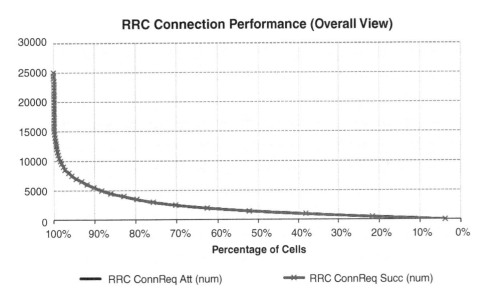

Figure 6.73 Number of RRC connections for connected users.

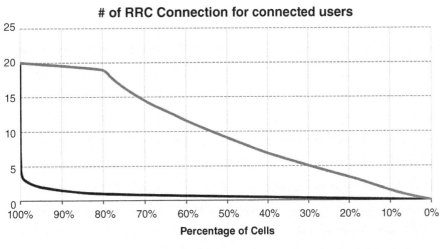

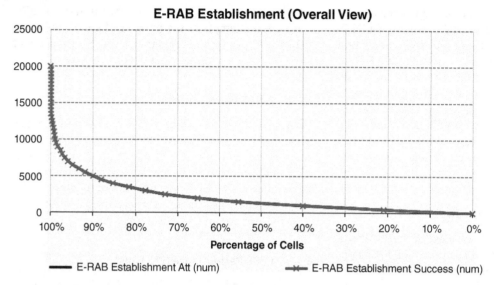

Figure 6.74 E-RAB performance.

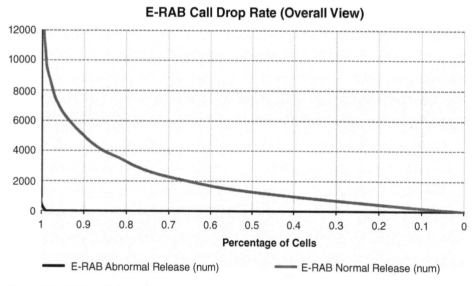

Figure 6.75 E-RAB call drop rate performance.

only the signaling portion). A high success rate in the range of 99% → 55% of CSFB calls are made from Idle mode, as 16% of users move to 3G/2G due to a CSFB call attempt (user impact for data connection due to mRAB in 3G). The median number of CSFB attempts per LTE user is six calls → deploying VoLTE is important to retain users in LTE. On average, 2% of RRC connection request attempts are due to CSFB call attempts from Idle. On average, there is one CSFB call every minute.

6.7.4 Summary of Observations

Smartphone applications create frequent transmission of small packets due to the background activity of the operating-system-related network services and the applications.

There is an average of four RRC connections per user every minute for PS data and an average of one CSFB call per minute.

Packet-data-related signaling is higher than voice-related signaling.

Frequent signaling in LTE can create challenges for terminal power consumption in addition to higher signaling load.

C-DRX is used in the current network, but the user-inactivity timer seems to be conservative (5 seconds). In general, a short inactivity timer, around 5 to 10 seconds, is preferred to minimize smartphone power consumption as well as the number of connected users. The timer can be increased to 10 seconds with C-DRX enabled.

Figure 6.76 CSFB performance.

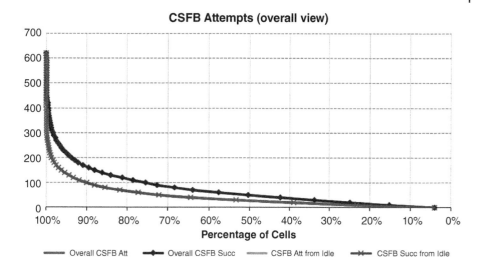

Figure 6.77 CSFB call attempts per user.

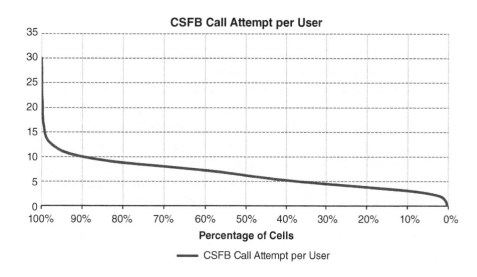

Comparing handover rate with RRC connection setup rate, there is an average of six handover attempts per minute and 20 E-RAB setup attempts per minute. This indicates that signaling due to call setup attempts is much higher than for handover attempts. The reason is that the number of packet calls is very high and the packet calls are very short, and handover is unlikely to occur during a short packet call. Therefore, mobility signaling is no problem in current networks, compared to the call setup. If the cell size becomes considerably smaller in future small-cell networks, mobility signaling may increase.

6.8 Link Reliability Analysis

The statistics relating to link reliability analysis focus on PDCP losses, PDCP delays, and brief QCI analysis, while the performance analysis consists of optimization

concerns and suggested solutions, as demonstrated in Figure 6.78.

Normally, the PDCP SDUs are received consecutively for PDCP entities associated with AM RLC. The received PDCP SDUs are delivered in ascending order to the application layer. Delays and discard rates are important in influencing the application layer performance within the delay/loss budget for the QCI used, as shown in Figure 6.79.

6.8.1 QCI = 9 PDCP Performance

It can be observed from Figures 6.80 and 6.81 that the DL packet discard rate is within the acceptable range for QCI = 9. The overall median value on the DL is 0.002%. The median discard rate over the Uu interface on the DL is 0.0002%, while on the UL it is 0.003%. High UL packet loss compared to the DL is due to backhaul implementation (high priority given to DL traffic over UL), which

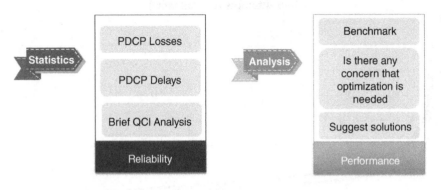

Figure 6.78 Link reliability performance.

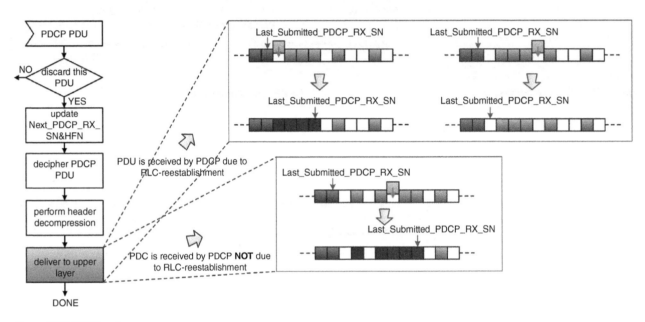

Figure 6.79 PDCP performance.

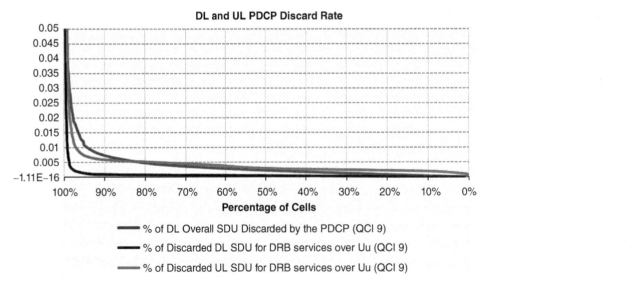

Figure 6.80 DL/UL discard rate.

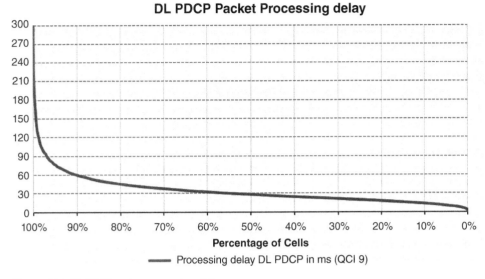

Figure 6.81 DL PDCP packet processing delay.

contributes to 10% of the overall discard rate on the DL. This indicates that the majority of the discard rate comes from backhaul.

6.8.1.1 QCI = 8 versus QCI = 9

On average, and in 50% of the cell, QCI = 8 traffic is present with a median value of 1.3% compared to total traffic on QCI = 9. The current network configures only QCI = 9. This indicates that there are roaming users present. The majority of traffic is observed around hotel areas.

A slightly higher packet discard rate is observed for QCI = 8. For QCI = 8, the discard rate on the Uu interface is 30% higher than the discard rate for QCI = 9. As a result, there is a slight impact on the user's roaming experience.

6.9 Main KPI Comparison for Different Operators

In the following study, a KPI comparison has been made between two different operators that use the same frequency band (band 20) with different bandwidths. Operator 1 uses band 20 with bandwidth 15 MHz, while Operator 2 uses band 20 with bandwidth 10 MHz.

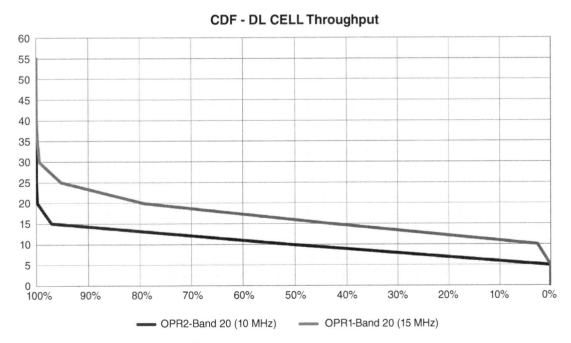

Figure 6.82 DL cell throughput for different operators.

6.9.1 DL Cell Throughput Comparison

As explained earlier, DL cell throughput can be calculated using Equation (6.6). It can be observed from Figure 6.82 that the average DL cell throughput for Operator 1 (with bandwidth 15 MHz) is 16 Mbps, while the average DL cell throughput for Operator 2 (with bandwidth 10 MHz) is 10 Mbps.

6.9.2 DL Resource Block (RB) Utilization

As explained earlier, the DL physical resource block utilization can be calculated from Equation (6.23). It can be

observed from Figure 6.83 that the average DL RB utilization for Operator 1 is about 8%, while the average DL RB utilization for Operator 2 is about 14%.

6.9.3 Mean CQI Comparison

As explained earlier, the mean CQI can be estimated from Equation (6.14). Figure 6.84 shows that the overall mean CQI for Operator 1 is about 11, while the overall mean CQI for Operator 2 is about 9.

For more details on LTE practical performance and LTE capacity planning, refer to [4] and [5].

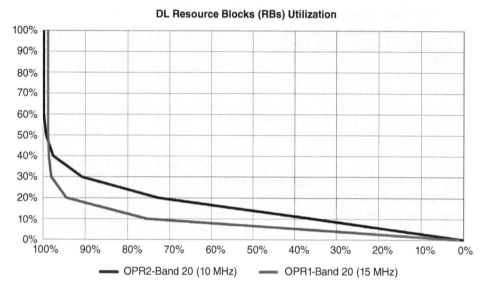

Figure 6.83 DL RB utilization for different operators.

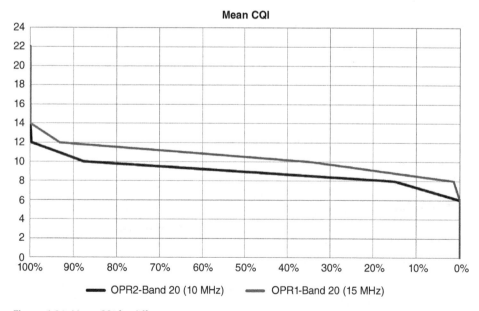

Figure 6.84 Mean CQI for different operators.

References

1 3GPP TS 36.213 V10.8.0: "Evolved Universal Terrestrial Radio Access (E-UTRA); Physical layer procedures," (Release 10).

2 Ball, C., Mullner, R., Lienhart, J. and Winkler, H. (2009) "Performance Analysis of Closed and Open Loop MIMO in LTE," *European Wireless Conference (EW)*, May, pp. 260–265.

3 3GPP TS 36.321 V10.9.0: "Universal Terrestrial Radio Access (E-UTRA); Medium Access Control (MAC) protocol specification," (Release 10).

4 Elnashar, A. and El-Saidny, M. A. (2013) "Looking at LTE in Practice: A Performance Analysis of the LTE System based on Field Test Results," *IEEE Vehicular Technology Magazine*, **8**(3), 81–92, September.

5 Elnashar, A. and Al-saidny, M. (2014) *Design, Deployment, and Performance of 4G-LTE Networks: A Practical Approach*, John Wiley & Sons.

7

IoT Evolution Towards a Super-connected World

7.1 Overview

The Internet of Things (IoT) is one of the key components of digital and digital transformation along with big data and analytics. Together with big data, analytics, and the cloud, the IoT can help enable numerous possibilities and new opportunities. These possibilities will impact our daily lives substantially and open new business models for enterprises; the number of connected IoT devices could go up to 50 billion by 2022.

In this chapter, we will cover the evolution of the Internet of Things from different aspects. We will focus on 3GPP cellular IoT evolution for connectivity, i.e., narrowband IoT (NB-IoT). Prior to a detailed technical description of the NB-IoT, we will summarize the IoT evolution from an end-to-end perspective, including the IoT platform, IoT protocols, connectivity, and the sensors layer. The IoT evolution is different from the regular mobile evolution; the latter is focusing on connectivity only, while the IoT evolution should be addressed from end to end. This is because the IoT connectivity is only 5% to 10% of the IoT value chain and therefore, a service provider should focus on the end-to-end use case. From a customer perspective, the most important factor is the outcome of the IoT use case and the associated business case (BC) behind it. Therefore, the focus on connectivity has been driven by the legacy telecom vendor as the main factor, while this is not the case from a commercial point of view. There are a lot of successful and innovative IoT use cases which have been implemented by small companies. Thus, service providers should adopt a digital transformation strategy aimed at providing end-to-end IoT use cases instead of focusing on connectivity only. The focus on connectivity is in the DNA of the legacy mobile operators, but it is no longer valid for the current digital evolution, and the connectivity portion is just a small fraction of the IoT use case.

The classical telecom vendors are still focusing on dump pipe expansion and they are describing this as an evolution such as 4.5G, but this is just a capacity expansion that will be deployed based on network traffic increase. Telecom service providers should instead adopt a digital transformation strategy and focus on monetizing connectivity and data by offering OTT applications and IoT use cases. Unfortunately, the legacy telecom vendors are struggling to adopt such a digital transformation and they are still promoting connectivity as the key factor, however this is not the case with the current decline in voice and data revenue. Therefore, service providers should change their operating models and address the enterprise market with a different approach. The majority of IoT use cases target enterprise customers and therefore a new model is mandatory to customize the use case to address the enterprise customer requirements. More importantly, a positive BC is a real challenge for the IoT use cases, especially for the enterprise market.

With all that in mind, this chapter will be organized using a different approach. We will cover the IoT evolution from technical and commercial points of view and we will cover 3GPP and non-3GPP technologies to address all IoT use cases. The aim is to provide the reader with an holistic overview of IoT evolution from end to end and a guide to how to build, design, and customize a successful IoT use case from all perspectives including technical and commercial aspects.

7.2 Introduction to the IoT

The IoT is defined as the interconnection of uniquely identifiable, embedded computing devices within the existing Internet infrastructure. Typically, the IoT is expected to offer advanced connectivity of devices, systems, and services that goes beyond machine-to-machine communications (M2M) and covers a variety of protocols, domains, and applications. The IoT topology includes the IoT Application Enablement Platform (AEP), the IoT device management layer, IoT data aggregation, and connectivity and sensors layers. In recent developments, the majority of IoT platforms have offered both AEP and device management, sometimes termed the device cloud.

Practical Guide to LTE-A, VoLTE and IoT: Paving the Way Towards 5G, First Edition. Ayman Elnashar and Mohamed El-saidny.

In this chapter, we will provide detailed descriptions and guiding principles for each of the major layers of the IoT topology, including the IoT platform, IoT connectivity, and differing IoT technologies, with a focus on the Wireless Sensors Network (WSN) and the IoT sensors layer.

The IoT/M2M network will aggregate the information from different IoT devices and sensor networks. These will vary in network architecture (i.e. infrastructure, topology, transport/communication methods, network QoS, bandwidth requirements, transmit intervals, processing capacity, and "north-bound" IP connectivity) based on their functional needs. The WSN will have a key role in this area because of its low-power sensor requirements, and with the scalability to connect a multitude of sensors to the IoT infrastructure or use a gateway-based architecture, where a single gateway (or a cascading gateway solution) collects data from a multitude of different sensor types and technologies.

The IoT covers many areas, as demonstrated in Figure 7.1 [1] ranging from enabling technologies and components to several mechanisms to effectively integrate these low-level components. Software is then a discriminating factor for IoT systems. IoT operating systems are designed to run on small-scale components in the most efficient way possible, while at the same time providing basic functionalities to simplify and support the global IoT system in its objectives and purposes. Middleware, programmability – in terms of Application Programming Interfaces (APIs) – and data management seem to be key factors for building a successful system in the IoT domain. Management capabilities are needed in order to properly handle systems that can potentially grow up to millions of different components. In this context, self-management and self-optimization of each individual component and/or subsystem may be strong requirements. In other words, autonomic behaviors could become the norm in large and complex IoT systems. Data security and privacy will play an important role in IoT deployments. IoT systems will produce and deal with personally identifiable information, therefore, data security and privacy will be critical from the very beginning. Services and applications will be built on top of this powerful and secure platform to satisfy and address business needs. Many applications are envisioned as well as generic and reusable services. This outcome will require new, viable business models for IoT and its related ecosystems of stakeholders. Finally, the IoT will have an impact on people and the society they live in, and so it must be conceived and conducted within the constraints and regulations of each country [1].

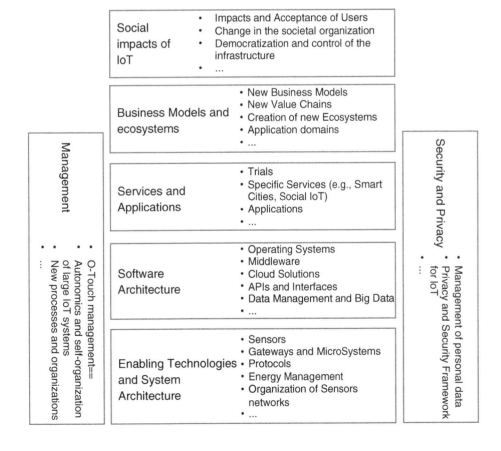

Figure 7.1 Technological and social aspects related to the IoT.

7.3 IoT Standards

Many organizations work on the IoT standardization. Accordingly, we considered IoT definitions from the European Telecommunications Standards Institute (ETSI), ITU, and IEEE. Other organizations include the Internet Engineering Task Force (IETF), the National Institute of Standards and Technology (NIST), the Organization for the Advancement of Structured Information Standards (OASIS), and the World Wide Web Consortium (W3C) [1]. In the following sections, we will summarize the IoT LoRa Alliance and Wi-SUN alliance effort in the three main organizations: IEEE, ETSI, and ITU.

7.3.1 IEEE

IEEE P2413 – Standard for an Architectural Framework for the Internet of Things (IoT) [2] – is currently considering the architecture of the IoT as being three-tiered, with the layers explained in Figure 7.2 [1].

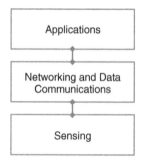

Figure 7.2 IEEE P2413 three-tier architecture of the IoT.

IEEE P2413 also currently suggests the extent of an IoT market and the stakeholders of IoT, as demonstrated in Figure 7.3 [1, 2]. IEEE is also standardizing PHY and MAC layers technologies relevant to IoT, for example: IEEE 802.11 (Wi-Fi Task Group), IEEE 802.15 Task groups (Wi-SUN, VLC).

7.3.2 ETSI

ETSI does not explicitly mention the words "Internet of Things" in its document; it discusses a similar concept under the name "machine-to-machine (M2M) communication" [3]. Accordingly, ETSI defines M2M communication as:

> "Machine-to-Machine (M2M) communications is the communication between two or more entities that do not necessarily need any direct human intervention. M2M services intend to automate decision and communication processes."

From a practical perspective, M2M usually refers to devices that have an embedded SIM card, while the IoT goes beyond classical SIM-based devices to sensors and is sometimes termed the "Internet of Everything (IoE)."

OneM2M is a global partnership developing standards for machine-to-machine (M2M) communications, enabling large-scale implementation of the IoT. OneM2M works in partnership with various standardization organizations, vendors, and service providers like ETSI, IEEE, and others. OneM2M does not offer a precise definition of M2M/IoT systems; instead it provides an exhaustive list of requirements that an M2M/IoT system fulfills [1].

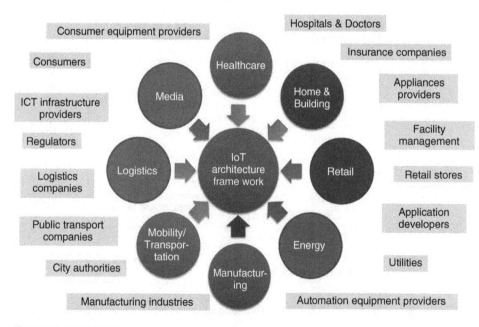

Figure 7.3 IEEE P2413 IoT markets and stakeholders.

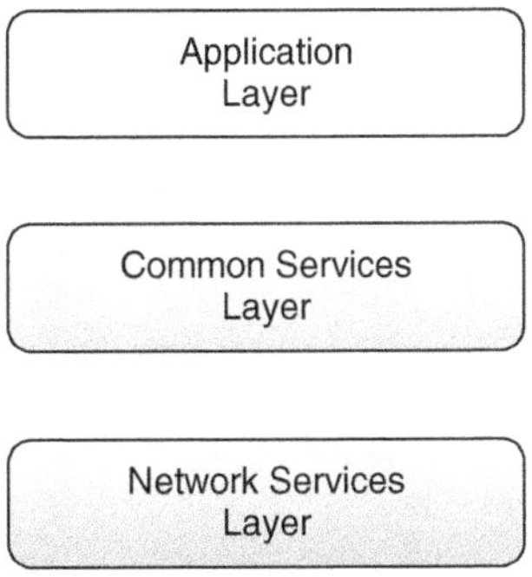

Application
Layer

Common Services
Layer

Network Services
Layer

Figure 7.4 OneM2M layered architecture.

OneM2M provides a detailed standard for M2M/ IoT in relation to architecture, interfaces, security, communication protocols and so on. OneM2M has a three-layer model, which is illustrated in Figure 7.4.

The three layers of the oneM2M model are described as follows:

Application layer: Comprises oneM2M applications and related business and operational logic.
Common services layer: Consists of oneM2M service functions that enable oneM2M applications (e.g., management, discovery, and policy enforcement).
Network services layer: Provides transport, connectivity, and service functions.

Therefore, the IoT can be simplified into mainly three horizontal layers: the IoT platform layer, a connectivity layer, and a sensors layer. In the context of the smart city, the IoT system is part of the infrastructure layer in a smart city platform. Figure 7.5 illustrates a smart city platform. The upper layers of the smart city platform include a data orchestration layer, a service enablement layer, and an application layer. The infrastructure layer includes data sources such as IoT sensors, service provider networks, government platforms, social media,

etc. The IoT platform in the smart city platform can be used for device management only, while applications can be created in the application layer. Alternatively, applications can be created on top of the IoT platform to offload the smart city platform.

7.3.3 ITU

The ITU defines the IoT as a network that is "Available anywhere, anytime, by anything and anyone." In this context, consumer products might be tracked using tiny radio transmitters or tagged with embedded hyperlinks and sensors. As illustrated in Figure 7.6, showing the ITU definition of the IoT, connectivity will take on an entirely new dimension. Today, users can connect at any time and at any location. Tomorrow's global network will not only consist of humans and electronic devices, but all sorts of inanimate things as well (hence, the term Internet of Everything, IoE). These things will be able to communicate with other things, for example, fridges with grocery stores, laundry machines with clothing, implanted tags with medical equipment, and vehicles with stationary and moving objects. However, device-to-device or thing-to-thing communication will usually be through the IoT platform, which acts as the brain for the IoT use case. In certain cases, an edge device may be equipped with some intelligence to reduce the load on the IoT platform and for immediate response. In fact, the current trend is to move from a centralized IoT solution to distributed or edge computing IoT solutions. The same concept is being evolved in cloud computing, where the concept of core cloud and edge cloud is being adopted to reduce latency and increase reliability.

The ITU has announced the formation of a new ITU-T study group to address standardization requirements of IoT technologies. The group, ITU-T SG20, will focus on IoT standards as related to smart cities. It will be responsible for international standards to enable the

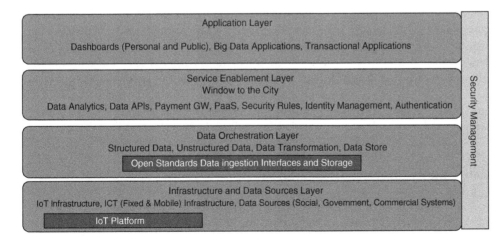

Figure 7.5 Smart city platform.

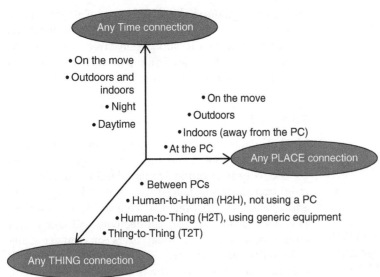

Figure 7.6 ITU definition of the IoT.

coordinated development of IoT technologies, including machine-to-machine communications and ubiquitous sensor networks [4].

Several innovations were developed recently to enable the extension of Internet technologies to constrained devices to avoid proprietary architectures and protocols. Most of these efforts focused on the networking layer: IPv6 over Low-Power Wireless Personal Area Networks (RFC 4919), Transmission of IPv6 Packets over IEEE 802.15.4 Networks (RFC 4944), IETF routing over low-power and lossy networks or the ZigBee adoption of Internet Protocol Version 6 (IPv6). These new standards enable the realization of an Internet of Things, where end-to-end IP-based network connectivity with tiny objects such as sensors and actuators becomes possible [37].

It has now become possible to deploy a self-organizing sensor network, to interconnect it with IPv6 Internet, and to build applications that interact with these networks using embedded web service technology. Application developers using high-level programming languages can count on these standardized technologies in the realization of a Semantic Web of Things or Sensor Web, which enables data producers and users to publish and access sensor information via web- and standards-based interfaces [37].

To ensure wide adoption, new solutions have to be interoperable with the most widely used protocols in the Internet, initially IP and, in a later stage, HTTP. To address these needs, the IETF — responsible for the development of high-quality Internet standards — has formed several working groups: IPv6 over Low Power WPAN (6LoWPAN), Routing Over Low Power and Lossy Networks (ROLL), and Constrained Restful Environments (CORE). The 6LoWPAN group tackles the transmission of IPv6 packets over IEEE 802.15.4

networks, the ROLL group develops IPv6 routing solutions for Low Power and Lossy Networks (LLNs), and the CoRE group aims at providing a framework for resource-oriented applications intended to run on constrained IP networks. Together, these protocols allow the IP-based integration of constrained devices into the Internet in a standardized way [37].

7.4 IoT Platform

The IoT platform consists of two layers: the application enablement platform and the device management layer. The IoT platform should support open standard protocols to communicate with different sensors, devices, and gateways from different vendors. The IoT platform selection is based on several factors including, but not limited to, the following:

Device-agnostic network interface protocols: Seamless connectivity of IoT devices, sensors, concentrators, gateways, and access points using interface bus adaptors.

Pre-integration of the maximum number of gateways and IoT devices from differing vendors and different technologies including long- and short-range technologies such as 2G, 3G, LTE, LPWA networks, Wi-Fi, Zigbee, 6LoWPAN, 802.15.4, NB-IoT, and LTE-MTC.

A rich set of supported communication protocols: CoAP, MQTT, AMQP, TCP, UDP, HTTP(s), XMPP, FTP, SMTP, etc.

Pre-configured integration of the widest possible set of IoT technology platforms and sensors.

Device boarding and device certification processes: The widest range of supported interfaces, the widest range of supported devices, support for a devices model library, and support for a device certification process or program.

Support for multi-tenancy and multi-applications per tenant.

Full device management: Device registration, Firmware Over The Air (FOTA) upgrades, OTP, OTPA, provisioning, configuration, commissioning, and invocation.

Message brokers and message queueing: AMMP, AMQP.

Security and authentication: SSL, TLS, and AES128, AES256 encryptions, FIPS and white/black listing via ACL.

Data storage and database: Data collection, data encryption, visualization, and analysis.

Device administration and monitoring.

Application enablement platform: AEP including IDEs and SDKs (C, Java, .NET) for developing complementary mobile apps and Web apps.

API exposure: REST and HTTP support for third-party application development and integration with other platforms or as part of a smart city platform.

Open Web service interfaces: Integration with other Web services.

In a nutshell, the IoT platform should support open standard protocols as well as proprietary protocols if needed to communicate with different sensors, devices, and gateways from different vendors. The IoT platform should be designed to be a fully distributed, fault-tolerant, and geo-redundant platform that supports edge computing (at the device and/or gateway level) and cloud-ready back-end computing (on premises or the cloud) to provide fast, local, and semi-connected operations that pure-cloud-based solutions simply cannot provide.

7.4.1 Horizontal IoT Topology

A horizontal IoT platform can manage multiple verticals and multiple technologies based on many standard interfaces. A typical horizontal IoT platform is illustrated in Figure 7.7 [3, 5]. The unique approach of a horizontal IoT platform is the core of the IoT ecosystem to overcome the silo approach and fragmentation in the IoT landscape. The horizontal IoT platform can manage different devices and different technologies and should be pre-integrated with hundreds of devices and tens of protocols. This will reduce the time to market for new use case development, provide integration of new devices, unify the IoT domain, provide a central management platform for different government entities to deploy, and manage and create applications for their sensors/things. The IoT platform has proven capability and is ready to manage millions of things.

ETSI [3, 5] is working towards "horizontalizing" the pipes (M2M vertical use cases). Vertical and horizontal pipe standardization scenarios are shown in Figure 7.7. A vertical pipe scenario is one in which there is one application, one network, and one (or a few) type(s) of device(s). On the other hand, a "horizontal pipe scenario" can also describe a model where applications share common infrastructure, environments, and network elements. ETSI oneM2M topology is depicted in Figure 7.8. The components of the oneM2M topology are summarized as follows:

OneM2M architecture entities:

AE: Application entity, containing the application logic of the M2M solution like home management functions, fleet management, blood sugar monitoring.

CSE: Common service entity, containing a set of common service functions (CSFs) that are common to a

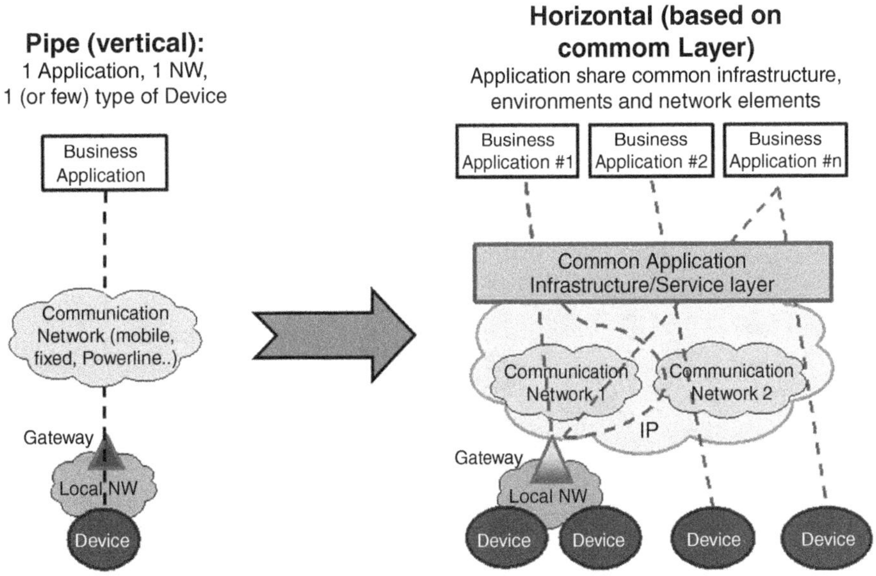

Figure 7.7 Standards-based horizontal IoT/M2M service layer.

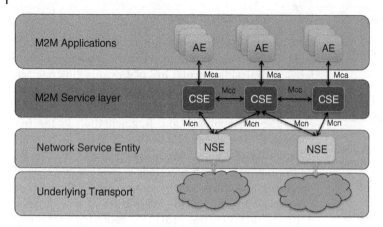

Figure 7.8 OneM2M simplified architecture.

broad range of M2M environments (verticals). This is the main part of the oneM2M specification.

CSF: Common service functions included in a CSE. CSFs can be mandatory or optional; a CSF can contain sub-functions (mandatory or optional).

NSE: Network service entity, provides network services to the CSE, like device triggering, device management support, and location services. These services are related to the underlying network capabilities.

OneM2M architecture reference points:

Mca reference points: The interface points between the AE and the CSE, the Mca point provides the M2M applications with access to the common services included in the CSE. The AE and CSE may be collocated in the same physical entity or in different entities.

Mcc reference points: This is the reference point between two CSEs. The Mcc reference point allows a CSE to use the services of another CSE in order to fulfill necessary functionality. Accordingly, the Mcc reference point between two CSEs is supported over different M2M physical entities. The services offered

via the Mcc reference point are dependent on the functionality supported by the CSEs.

Mcn reference points: This is the reference point between a CSE and the underlying network services entity. The Mcn reference point allows a CSE to use the services (other than transport and connectivity services) provided by the underlying network services entity in order to fulfill the necessary functionality.

Mcc reference points: The interface between two M2M service providers; as similar as possible to the Mcc reference point. However, due to the nature of inter-M2M service provider communications, some differences are anticipated.

A simplified typical IoT topology that aligns with ETSI/IEEE and other standardization bodies is illustrated in Figure 7.9. The proposed IoT platform will support existing technologies including 2G/3G/LTE/Wi-Fi/PLC/ Fixed and low-power, wide-area (LPWA) networks such as long-range (LoRaWAN) and Sigfox as well as 3GPP standardized technologies such as LTE-MTC (Release 13) and 3GPP NB-IoT (Release 13). The horizontal IoT platform architecture approach ensures an

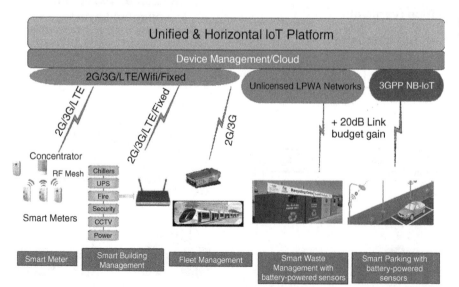

Figure 7.9 Horizontal IoT topology/architecture.

inter-operability framework for the diverse set of gateways and sensors across different sets of applications and network connectivities.

The horizontal IoT platform will enable customers to build, evolve, and operate scalable and secure IoT solutions *rapidly*, for any thing, any data, any network, and any application. A purpose-built and horizontal IoT platform enables users to build connected applications; these applications can be fully integrated into the network and business processes. Such a solution is secure, scalable, and flexible. Enterprise customers can enjoy enhanced IoT services without the additional burden of managing complex IT infrastructures, applications, and connectivity issues. The IoT platform is a horizontal platform, which factorizes re-usable services across different domains, such as device management, data abstraction, protocol translation, data storage and queries, business rules, multi-tenancy, and many others. The IoT platform connects with any device through standard protocols, or through an embedded agent, or using adapters. Once the device is connected and on-board, the platform starts to collect data, process and transform these data, and feed applications and integration connectors. The Intel reference architecture IoT platform is demonstrated in Figure 7.10.

Figure 7.10 IoT platform reference architecture. Source: Intel.

The IoT platform architecture enables the rapid creation of IoT solutions to reduce customers' development costs and deployment times. The horizontal multi-layered

IoT platform makes it easy to create solutions, greatly reducing the amount of custom code required to achieve results. In addition to the AEP layer, the IoT platform offers a device cloud for the management of connected "things". All lead to the following benefits:

Simplified solution creation
Reduced development costs
Minimized deployment time
Suitability for all verticals, including smart cities and other use cases
Data that are interchangeable between different use cases and that create cross-use cases.

In addition, the IoT platform should be device/gateway agnostic; it may be hosted within premises or on a public or private cloud. It can be integrated with a choice of M2M connectivity platform for mobile SIM provisioning and management, tariff management, and connectivity management. However, it should be noted that the IoT platform is completely different from the M2M platform designed mainly for SIM and connectivity management. The IoT platform enables a broad range of public and privately managed IoT services, including provisions for everything from dynamic security, healthcare, hospitality, education, retail, transportation, and infrastructure to financial applications. Delivery of critical real-time data from diverse urban assets to decision makers, business stakeholders, and citizens enables significant new operational efficiencies, more effective services, and reduced operating costs for cities and enterprises.

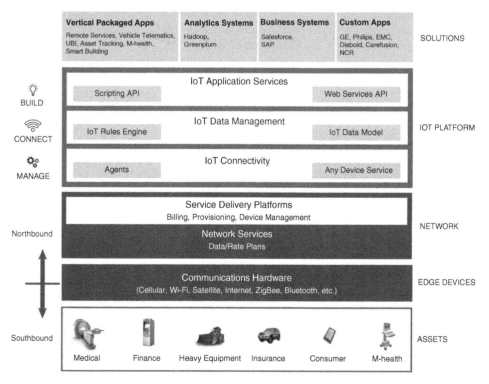

Figure 7.10 IoT platform reference architecture. Reproduced by permission of Intel.

The IoT platform should be scalable and based on a pay-as-you-grow model. The IoT platform license model is an OPEX model based on the number of connected things. The things can be classified into light things (such as environmental sensors), medium things (such as fleet management), and heavy things (such as a smart building GW with multiple inputs such as CCTV, fire alarm panel, HAVC and so on). For example, let us assume an IoT platform that will manage 1M light sensors, 500k medium things, and 100k heavy things. Assuming 0.5 USD for a light thing license, 1 USD for medium things, and 3 USD for heavy things, the total annual license will be $1M \times 0.5 + 500k \times 1 + 3 \times 100k = 1.6M$ USD per annum. These prices are for illustration purposes only. A unified global license per thing is also a recommended option to avoid capacity bottlenecks in the future. A one-time installation and integration fee may be required as well.

7.5 IoT Gateways, Devices, and "Things" Management

In an IoT system, we can have different topologies to connect the sensors to the horizontal IoT platform. There are three main scenarios, as shown in Figure 7.11. The three scenarios can be described as follows:

Scenario 1: Legacy sensors/devices that already exist with their own proprietary/legacy platforms such as SCADA systems. This is the case for utility organizations that deployed proprietary platforms for the management of existing sensors and use cases. The integration with a horizontal IoT platform will allow multiple utilities and different organizations under city municipalities to monitor and control multiple verticals and produce a common dashboard city-wide. Also, this will offer a unified window for the city.

Moreover, the integration of a smart city platform will enrich the generated dashboards, improve efficiency across multiple use cases, and allow the seamless creation and introduction of new use cases.

Scenario 2: Sensors and network devices connected to concentrators or gateways that will support bidirectional communication with the horizontal IoT platform. These concentrators and gateways can be "dummy" or "smart", depending on their built-in capabilities. A "dummy" concentrator has only data aggregation capabilities. A "smart" concentrator has data aggregation + processing capabilities. It often comes with a "light" processor or microcontroller, and some small-to-medium on-board memory where a Real Time Operating System (RTOS) can run and manage any application-specific remote software agents (often called clients). In such a scenario, the horizontal IoT platform not only receives data from the concentrators or gateways but is also able to interact with devices for maintenance and operational tasks, such as, for instance, remote firmware upgrades, configuration updates, device restarts, etc. In addition, we can conduct analytics and computing at the edge, i.e., at the GW or the concentrators. Pre-integration with different GWs from different vendors based on different protocols is a key differentiator for IoT platform selection. The on-boarding of a new GW or new device is another key topic that needs to be considered. A typical SLA for on-boarding of a new device based on a standard protocol should be within one week.

This scenario suits short-range sensors such as Bluetooth, ZigBee, Wi-Fi, and Z-Wave devices that can be integrated with a GW and then the GW can be integrated with the IoT platform through the existing 2G/3G/LTE or fixed network. Typical use cases for this scenario are the smart home, smart meters, and smart buildings.

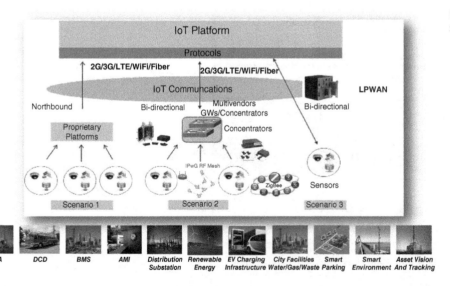

Figure 7.11 IoT deployment scenarios.

Scenario 3: Sensors/devices will be connected directly to the horizontal IoT platform using a wireless sensor network or mobile network. The connectivity can be through existing legacy 2G/3G/LTE networks or via the recently introduced LPWA networks designed for low data rates, long ranges, and low-power sensors.

Therefore, the horizontal IoT platform will eliminate the fragmentation in IoT technologies and will overcome such challenges, as the IoT platform will be able to support, integrate, and interoperate different technologies, different protocols, and different GWs into a unified platform. In addition, sensors/things from the same use case or different use cases can be seamlessly integrated on different networks thanks to the horizontal IoT platform.

In Scenario 2, where it is necessary to deploy concentrators to connect the sensors, the GW may meet the following requirements based on different verticals and different use cases:

Hardened/rugged switches for outdoor/harsh environments.

PoE support to reduce cabling requirements.

Copper or fiber connectors.

Ring networking solutions (REP) to reduce cabling requirements.

A mobile wireless router on cellular networks (3G/LTE) to extend coverage and provide mobility.

Wireless mesh technologies.

Advanced routing with RPL.

Extensive backhaul options: Wi-Fi, wired, 2G/3G/4G, Ethernet, and Fiber.

Extensive access technologies: wired, Wi-Fi, 802.15.x, PLC, LPWAN, Ethernet, and Serial.

Extensive QoS support.

Smart ports with automatic configuration.

Location service for wired and wireless devices.

Secure networking.

Easy deployment with auto-configuration, zero-touch provisioning, etc.

7.6 Edge and Fog Computing

Gateways are powerful elements in the IoT technology solution stack. As secure devices for IoT solution deployment, gateways also allow unprecedented flexibility in the types of end point device – SCADA devices, 6LoW-PAN sensors, manufacturing PLCs, LPWAN-based sensors, and others – that are available to an organization deploying an IoT solution.

Computations can be made at the edge as soon as there is a microcontroller in the GW, similar to the Cisco fog computing edge GW demonstrated in Figure 7.12. If we take a car maintenance use case, instead of sending all the data of vibration measurements directly to the IoT platform, we can only send the result of a computation or

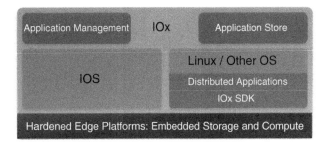

Figure 7.12 Edge computing topology. Reproduced by permission of Cisco.

aggregation (e.g. the harmonics from the Fourier transform of the vibration data) to the horizontal IoT platform. This will allow a reduction in data to offload the IoT platform and promote distributed, or fog, computing versus centralized cloud computing. In addition, we can also access all the logic from the IoT platform, to use it at the edge when required.

Fog computing is a paradigm that extends cloud computing and services to the edge of the IoT network. Similar to the cloud, fog provides data, computation, storage, and application services to end users. The distinguishing characteristics of fog are its proximity to end users, its dense geographical distribution, and its support for mobility. Services are hosted at the network edge or even end devices such as gateways. By doing so, fog reduces service latency and improves QoS, resulting in superior user experience. Fog computing supports emerging IT/OT convergence, which demands real-time/predictable latency (industrial automation, transportation, networks of sensors and actuators). Figure 7.13 demonstrates the Cisco GW that hosts different connectivity modules and supports fog computing.

Figure 7.14 demonstrates the IoT architecture migration from legacy to fog/edge computing topology.

Figure 7.15 illustrates the IoT topology with edge/fog computing capability.

The defining characteristics of fog computing are:

Edge location, low latency, and location and context awareness.

Widespread geographic distribution.

Very large number of nodes.

Predominant role of wireless access.

Real-time analytics and control close to source.

Heterogeneity – different form facts, different environments.

Extends the cloud computing paradigm to the network edge.

Enables a new breed of applications and services.

Provides distributed computation, storage, and network services.

Figure 7.16 illustrates the disruption in analytics between edge and cloud.

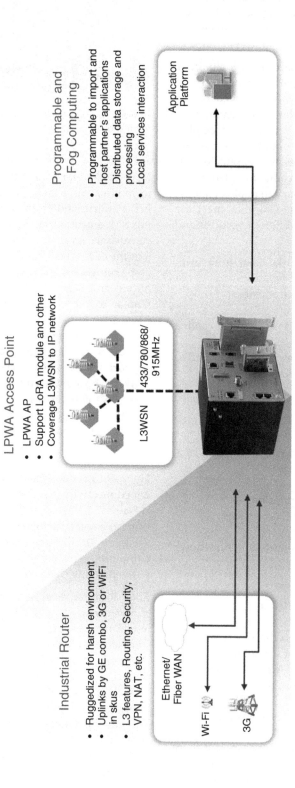

Figure 7.13 Edge/fog computing GW. Reproduced by permission of Cisco.

Figure 7.14 Edge/fog computing architecture migration. Reproduced by permission of Cisco.

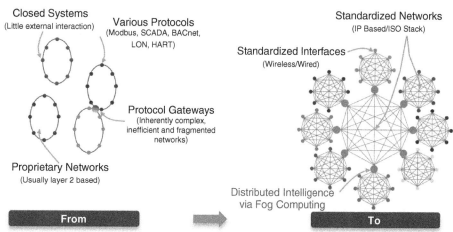

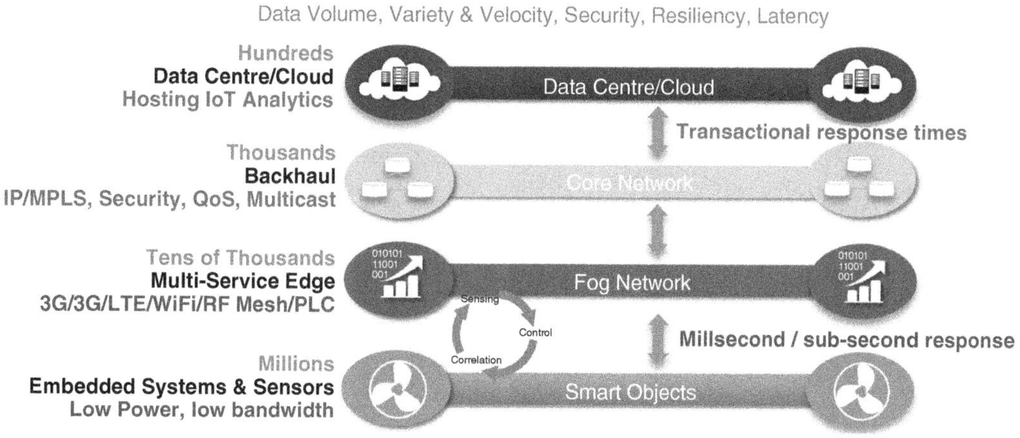

Figure 7.15 Edge/fog computing topology. Reproduced by permission of Cisco.

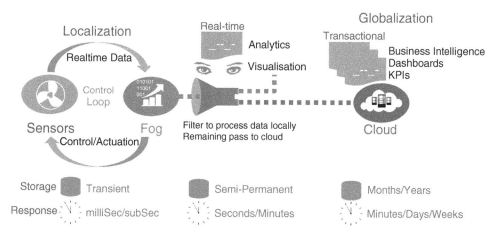

Figure 7.16 Edge/fog computing distribution. Reproduced by permission of Cisco.

This is achieved by a ubiquitous device connectivity layer focused on edge connectivity. The device connectivity layer or device cloud provides scalable, secure, embeddable, and easily deployable communications designed for connecting sensors, devices, and equipment across any network topology and any communication scenario.

The IoT gateway interacts with the devices/things (sensors and actuators), integrating the variety of formats, protocols, and technologies that can be found in this environment and converting them into a common language. The IoT platform maintains a pluggable architecture, which enables any new type of device to be integrated easily by plugging in an adapter for the device.

This IoT gateway enables connectivity to a multitude of devices in a highly scalable manner with support for existing, legacy, and upcoming communication protocols. The device cloud part of the IoT platform is pre-integrated with hundreds of IoT GWs and already supports tens of protocols over different technologies, as explained above.

7.7 IoT Sensors

In today's competitive market, and with the continued urgency to comply with governmental, social, and environmental regulations to meet citizens' daily life expectations and corporate financial objectives, it is deemed that smart and intelligent, low-cost automation is required at several levels to improve people's, institutions', and governmental efficiencies, productivity, and overall seamless collaboration. Such a vision can be achieved via the deployment and proliferation of Wireless Sensors Networks (WSNs). Wireless sensors

typically have a leading advantage over traditional legacy wired sensors due to their possession of inherent intelligence-processing, flexibility, ease of deployment, long lifetime, small form factor and footprint, and intelligent self-organizing and self-healing. In the coming years, data generated by sensors and M2M or IoT-connected "things" will fill the storage systems of data centers across a wide range of industries. In fact, IDC forecasts that machine-generated data will increase to 42% of all data by 2020.

Advances in wireless networking, micro-fabrication, and integration (for example, sensors and actuators manufactured using micro-electro mechanical system technology, or MEMS), and embedded microprocessors have enabled a new generation of massive-scale sensor networks suitable for a range of commercial and military applications. The technology promises to revolutionize the way we live, work, and interact with the physical environment.

Figure 7.17 illustrates different areas in which sensors are applied to improve the quality of life of citizens as

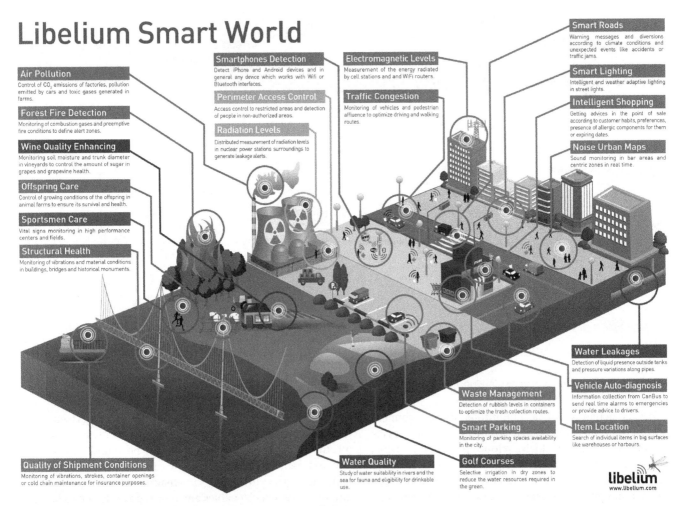

Figure 7.17 Sensor ecosystem. Reproduced by permission of Libelium.

well as to make business processes more efficient. A key and important factor in sensor technology, besides its undisputed core function, is the ease of integration into the IoT platform (plug and play), its reachability across the network, associated ICT requirements, and, of course, the lifetime. While selecting an IoT platform, we must make sure that we select from the best breed of horizontal platforms with well-constructed, device-agnostic *device management modules* with support for, and proven field deployments of, low-power, long-range wireless sensors such as NB-IoT, LTE-M, LoRa, LPWAN sensors and gateways, Sigfox, etc.

Figures 7.18 and 7.19 provide typical lists of practical IoT sensors from Libelium for different IoT use cases.

7.8 IoT Protocols

One of the biggest challenges in IoT space technology is the protocol selection – this challenge stems from the fact that currently, standardization in the IoT protocols is virtually non-existent [6]. Several standardization bodies have developed standards for the IoT, as depicted in Figure 7.20 [6]. The root of the problem with standardization is the constrained environment of IoT, characterized by low memory availability, low power, low bandwidth requirements, and high packet loss – combined, these do not allow TCP/IP stack and Web technologies to be used easily for the IoT. However, to solve this challenge, there are hundreds of proprietary protocols in the IoT, M2M, and Home Automation space, such as ZigBee and Z-Wave. Although these protocols are supported by an alliance of product vendors, they are not standardized like TCP, IP, HTTP, or SMTP [6]. A set of open, standardized protocols has started to emerge. Most bodies, such as IEEE, IETF, and W3C, have standardized protocols such as 6LowPAN and CoAP.

Rather than trying to cover all the IoT protocols on top of existing architecture models like the OSI model, the protocols have been broken in [7] into the following layers to provide some level of organization:

Infrastructure (e.g.: 6LowPAN, IPv4/IPv6, RPL, IEEE 802.15.4)
Identification (e.g.: EPC, uCode, IPv6, URIs)
Comms/Transport (e.g.: Wi-Fi, Bluetooth, LPWAN)
Discovery (e.g.: Physical Web, mDNS, DNS-SD)
Data Protocols (e.g.: MQTT, CoAP, AMQP, Websocket, Node)
Device Management (e.g.: TR-069, OMA-DM)
Semantic (e.g.: JSON-LD, Web Thing Model)
Multi-layer Frameworks (e.g.: Alljoyn, IoTivity, Weave, Homekit).

Another representation of the IoT stack is presented in Figure 7.21. As indicated in this figure, the IoT landscape is a mess – too many protocols, too many standards, too many revolutions. The IoT layers demonstrated in Figure 7.21 are summarized in [8].

In this section, we will provide a summary of the main IoT-relevant protocols. More details on the IoT protocols and use cases for each protocol can be found in [7] and references therein. The IoT-relevant protocols are summarized as follows [7]:

IPv6: An Internet layer protocol for packet-switched internetworking, providing end-to-end datagram transmission across multiple IP networks.

6LoWPAN: An acronym for IPv6 over Low-Power Wireless Personal Area Networks. It is an adaptation layer for IPv6 over IEEE 802.15.4 links. This protocol operates only in the 2.4 GHz frequency range with a 250 kbps transfer rate.

IETF 6Lo WG focuses on the work that facilitates IPv6 connectivity over constrained node networks with the characteristics of:

- limited power, memory, and processing resources
- hard upper bounds on state, code space, and processing cycles
- optimization of energy and network bandwidth usage
- lack of some layer 2 services like complete device connectivity and
- broadcast/multicast.

UDP (User Datagram Protocol): A simple OSI transport layer protocol for client/server network applications based on the Internet Protocol (IP). UDP is the main alternative to TCP and one of the oldest network protocols in existence, introduced in 1980. UDP is often used in applications especially tuned for real-time performance.

uIP: The uIP is an open-source TCP/IP stack capable of being used with tiny 8- and 16-bit microcontrollers.

DTLS (Datagram Transport Layer): The DTLS protocol provides communications privacy for datagram protocols. The protocol allows client/server applications to communicate in a way that is designed to prevent eavesdropping, tampering, or message forgery. The DTLS protocol is based on the Transport Layer Security (TLS) protocol and provides equivalent security guarantees.

Time-Synchronized Mesh Protocol (TSMP): A communications protocol for self-organizing networks of wireless devices called motes. TSMP devices stay synchronized with each other and communicate in time slots, like other TDM (time-division multiplexing) systems.

mDNS (multicast Domain Name System): Resolves host names to IP addresses within small networks that do not include a local name server.

Physical Web: The Physical Web enables you to see a list of URLs being broadcast by objects in the environment

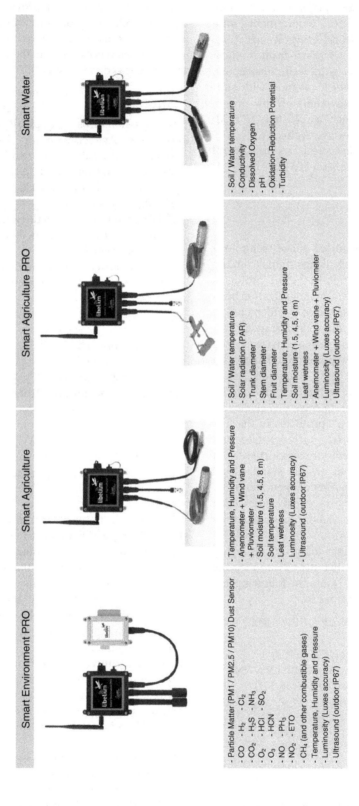

Figure 7.18 IoT sensor samples. Reproduced by permission of Libelium.

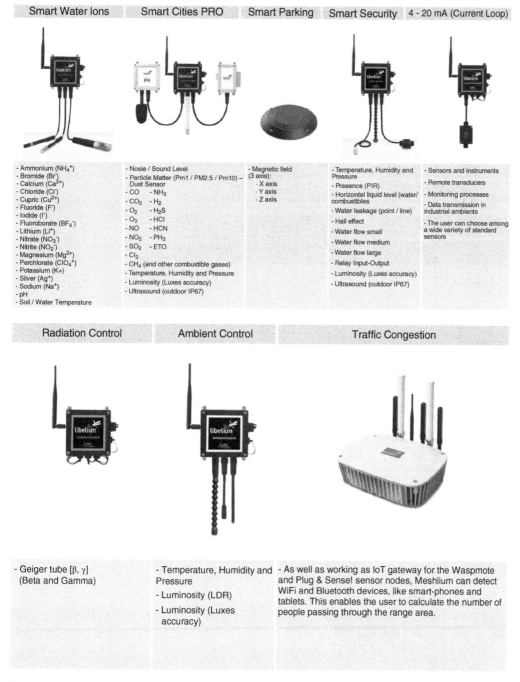

Smart Water Ions	Smart Cities PRO	Smart Parking	Smart Security	4 - 20 mA (Current Loop)

- Ammonium (NH_4^+)
- Bromide (Br^-)
- Calcium (Ca^{2+})
- Chloride (Cl^-)
- Cupric (Cu^{2+})
- Fluoride (F^-)
- Iodide (I^-)
- Fluoroborate (BF_4^-)
- Lithium (Li^+)
- Nitrate (NO_3^-)
- Nitrite (NO_2^-)
- Magnesium (Mg^{2+})
- Perchlorate (ClO_4^+)
- Potassium ($K+$)
- Silver (Ag^+)
- Sodium (Na^+)
- pH
- Soil / Water Temperature

- Noise / Sound Level
- Particle Matter (Pm1 / PM2.5 / Pm10) – Dust Sensor
- CO   - NH_3
- CO_2 - H_2
- O_2 - H_2S
- O_3 - HCl
- NO   - HCN
- NO_2 - PH_3
- SO_2 - ETO
- Cl_2
- CH_4 (and other combustible gases)
- Temperature, Humidity and Pressure
- Luminosity (Luxes accuracy)
- Ultrasound (outdoor IP67)

- Magnetic field (3 axis):
  - X axis
  - Y axis
  - Z axis

- Temperature, Humidity and Pressure
- Presence (PIR)
- Horizontal liquid level (water/ combustibles
- Water leakage (point / line)
- Hall effect
- Water flow small
- Water flow medium
- Water flow large
- Relay Input-Output
- Luminosity (Luxes accuracy)
- Ultrasound (outdoor IP67)

- Sensors and instruments
- Remote transducers
- Monitoring processes
- Data transmission in industrial ambients
- The user can choose among a wide variety of standard sensors

Radiation Control	Ambient Control	Traffic Congestion

- Geiger tube [β, γ] (Beta and Gamma)

- Temperature, Humidity and Pressure
- Luminosity (LDR)
- Luminosity (Luxes accuracy)

- As well as working as IoT gateway for the Waspmote and Plug & Sense! sensor nodes, Meshlium can detect WiFi and Bluetooth devices, like smart-phones and tablets. This enables the user to calculate the number of people passing through the range area.

Figure 7.19 IoT sensor samples. Reproduced by permission of Libelium.

around you with a Bluetooth Low Energy (BLE) beacon.

HyperCat: An open, lightweight JSON-based hypermedia catalog format for exposing collections of URIs.

UPnP (Universal Plug and Play): Now managed by the Open Connectivity Foundation, this is a set of networking protocols that permits networked devices to discover each other's presence seamlessly on the network and establish functional network services for data sharing, communications, and entertainment.

MQTT (Message Queuing Telemetry Transport): The MQTT protocol enables a publish/subscribe messaging model in an extremely lightweight way. It is useful for connections with remote locations where a small code footprint is required and/or network bandwidth is at a premium.

MQTT-SN (MQTT for Sensor Networks): An open and lightweight publish/subscribe protocol designed specifically for machine-to-machine and mobile applications.

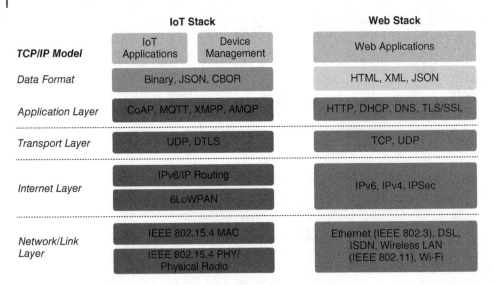

Figure 7.20 Standardized IoT protocol stack.

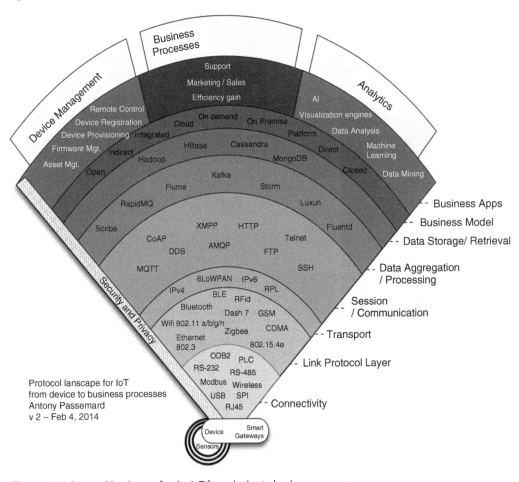

Figure 7.21 Protocol landscape for the IoT from device to business process.

Mosquitto: An open-source MQTT v3.1 broker.

CoAP (Constrained Application Protocol): An application layer protocol that is intended for use in resource-constrained Internet devices, such as WSN nodes. CoAP is designed to translate easily to HTTP

for simplified integration with the Web, while also meeting specialized requirements such as multicast support, very low overhead, and simplicity. The CoRE group has proposed the following features for CoAP: RESTful protocol design minimizing the complexity of mapping with HTTP, low header overhead and parsing complexity, URI and content-type support, support for the discovery of resources provided by known CoAP services, simple subscription for a resource and resulting push notifications, simple caching based on max-age.

XMPP (Extensible Messaging and Presence Protocol): An open technology for real-time communication, which powers a wide range of applications including instant messaging, presence, multi-party chat, voice and video calls, collaboration, lightweight middleware, content syndication, and generalized routing of XML data.

XMPP-IoT: In the same manner as XMPP, has silently created people-to-people interoperable communication: "We are aiming to make communication machine to people and machine to machine interoperable."

Mihini/M3DA: The Mihini agent is a software component that acts as a mediator between an M2M server and the applications running on an embedded gateway. M3DA is a protocol optimized for the transport of binary M2M data. It is made available in the Mihini project both for device management, by easing the manipulation and synchronization of a device's data model, and for asset management, by allowing user applications to exchange typed data/commands back and forth with an M2M server in a way that optimizes the use of bandwidth.

AMQP (Advanced Message Queuing Protocol): An open standard application layer protocol for message-oriented middleware. The defining features of AMQP are message orientation, queuing, routing (including point-to-point and publish-and-subscribe), reliability, and security.

DDS (Data-Distribution Service for Real-Time Systems): The first open international middleware standard directly addressing publish–subscribe communications for real-time and embedded systems.

LLAP (Lightweight Local Automation Protocol): A simple, short message that is sent between intelligent objects using normal text; it's not like TCP/IP, Bluetooth, ZigBee, 6LowPAN, Wi-Fi, etc., which achieve, at a low level, "how" to move data around. LLAP can run over any communication medium. The three strengths of LLAP are: it will run on anything now, anything in the future, and it's easily understandable by humans.

LWM2M (Lightweight M2M): A system standard in the Open Mobile Alliance. It includes DTLS, CoAP, Block, Observe, SenML, and Resource Directory and weaves them into a device–server interface along with an object structure.

SSI (Simple Sensor Interface): A simple communications protocol designed for data transfer between computers or user terminals and smart sensors.

REST (Representational State Transfer) and RESTful HTTP: Additional resources in the context of the IoT.

HTTP/2: Enables a more efficient use of network resources and a reduced perception of latency by introducing header field compression and allowing multiple concurrent exchanges on the same connection.

SOAP (Simple Object Access Protocol): JSON/XML, WebHooks, Jelastic, MongoDB.

WebSocket: Developed as part of the HTML5 initiative, this introduced the WebSocket JavaScript interface, which defines a full-duplex, single-socket connection over which messages can be sent between client and server. The WebSocket standard simplifies much of the complexity around bidirectional Web communication and connection management.

7.9 IoT Networks

IoT deployments currently use a broad variety of communication technologies depending on the use case, requirements, availability of networks, etc. The most important factor is the end-to-end use case rather than the connectivity, as highlighted earlier. Figure 7.22 demonstrates the typical IoT value chain. As depicted in Figure 7.22, the connectivity part is not the main revenue stream and therefore this model drives telecom operators to change their strategy from being legacy connectivity providers to end-to-end digital service providers. However, building a positive BC for IoT use cases is the main challenge that may hamper their delivery. To overcome this challenge, service providers may bundle the IoT use cases with other services for the enterprise. In addition, building a nationwide standard IoT network such as NB-IoT based on the 3GPP standard may reduce the OPEX and CAPEX compared to using proprietary protocols or connectivity. However, the service providers should expect a touch of competition from vertical vendors that build and customize a single use case. The enterprises or the IoT customers may prefer to build their own, controlled use cases rather than relying on service providers' global networks. Adopting a horizontal IoT platform will reduce the TCO but will need software development expertise to develop customized use cases. The multinational operators have better opportunities in this regard, since their investment can be justified by deploying the same use cases in different countries or even having different instances of the same IoT platform based on a global deal. Amazon and software giants such as Microsoft and IBM deployed their IoT platforms on their public clouds, and this is another thread for service providers. In this section, we will analyze the different IoT connectivity networks and provide the reader with a strategy for IoT connectivity based on the use case requirements and other relevant factors.

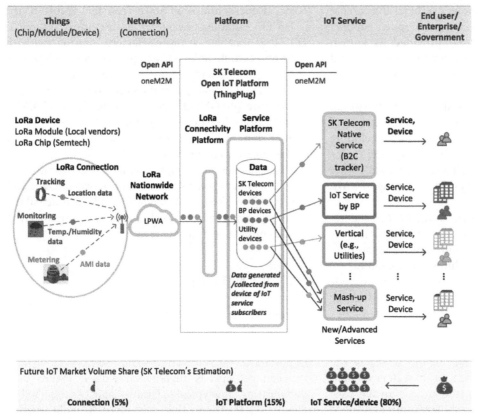

SK Telecom's E2E IoST Architecture (Source: Netmanias.com)

Figure 7.22 SK Telecom's E2E IoT architecture. Reproduced by permission of Netmanias.

We can classify the IoT/M2M connectivity into three classes based on the use case requirements in terms of throughput, as indicated in Figure 7.23. The first and second classes can be covered by existing 2G/3G/LTE/Wi-Fi networks. The third class of IoT connectivity should consider new, smart approaches to cater for low power, low throughput, a very high number of connections, and a very low cost for the end unit or modem. As indicated in Figure 7.22, there is a need for new, low-power, wide-area networks (LPWANs) to meet the requirements of WSNs, which need massive numbers of sensors, low power consumption, batteries that can last for five to ten years, low-cost modems/modules, and long ranges for wide area coverage. In this section, we will focus on the last class of connectivity, which addresses the LPWAN and similar technologies for wireless sensors. This is the current evolving domain of the IoT and many technologies are available in licensed and unlicensed bands.

Over the next decade, the LPWAN will play an important role in connecting a range of IoT devices that have low mobility, low power consumption, long range, low cost, and are secured. The LPWAN connections are expected to serve a diverse range of vertical industries and cover a range of smart city applications and deployment scenarios in which existing mobile network technology may not be best placed to provide such connectivity. Analyses Mason, in its study for the GSMA published in April 2015 [9], identified seven categories of LPWAN applications: agriculture and environment, consumer, industrial, logistics, smart building, smart city, and utilities, as illustrated in Figure 7.24.

Although the average revenue per LPWAN connection is likely to be relatively low, this new technology will enable the mobile industry to add substantial value to the IoT. Analyses Mason forecasts LPWA technologies will generate $970 million globally in connectivity revenue in 2018, rising to $7.5 billion in 2022 [9]. By that time, Strategy Analytics estimates network operators could be generating more than $13 billion from LPWA connectivity, as well as significant additional revenues from value-added services, such as data analytics and security [10]. It is worth considering the points listed earlier in this section that may impact these forecasts.

LPWA connectivity is particularly well-suited to IoT applications, such as environmental sensors, parking sensors, energy meters, logistics tracking and animal and crop monitoring, that require large numbers of widely dispersed devices to send occasional status updates. It can also be used to activate devices remotely, such as sprinklers, lights, and air conditioning. As many sensors

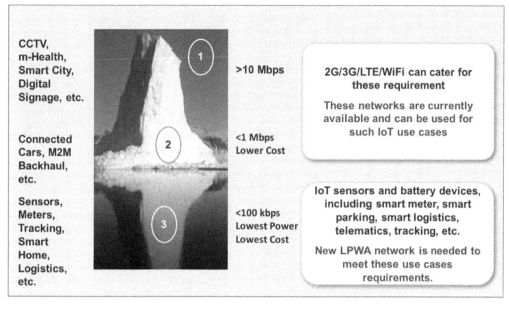

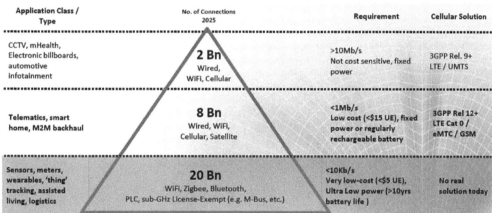

Figure 7.23 IoT connectivity classification.

are scattered, they need to be run on batteries for years. As well as enabling many more devices to be connected, LPWA will be used to provide backup connectivity, or more robust connectivity, for some applications, such as intruder alarms or vehicle accident alerts [10].

The LPWA technologies will have the following characteristics [10]:

Very low power consumption – a battery life of more than 10 years for some use cases such as smart parking, smart bins, smart environment, etc.

Optimized for brief messages – about the length of an SMS.

Very low device unit cost – the connectivity module will eventually cost a few dollars.

Good coverage outdoors and indoors – enabling connectivity in rural and underground locations and building basements.

Easy network installation – re-using existing cellular infrastructure wherever possible.

Scalable – able to support large numbers of devices over a wide geographic area.

Secure connectivity and support for authentication appropriate to the IoT application.

The ability to be integrated into a unified, horizontal IoT platform.

The main LPWAN technologies can be summarized as follows:

3GPP-based, licensed technologies:

NB-IoT (Narrowband IoT): A technology being standardized by the 3GPP standards body.

:LTE-MTC (LTE-Machine-Type Communication): Standards-based family of technologies supporting several technology categories, such as Category 1 and Category M1, suitable for the IoT.

EC-GSM-IoT (Extended Coverage-GSM-IoT): Enables new capabilities for existing cellular networks with LPWA (low-power, wide-area)

UTILITIES

→ Gas and water metering, including smart meter consumption tracking and pipeline monitoring
→ Mocrogeneration (monitoring status of generation equipment (solar, wind, thermal) with sensors
→ Smart Grid (energy infrastructure monitoring)

LOGISTICS

→ Asset tracking (e.g. industrial assets, containers, vehicles; location and status update)

SMART CITY

→ Parking sensors (to monitor and report availability of parking spaces)
→ Waste management (monitoring status of waste containers to optimise the collection of waste)
→ Smart lighting (remove management of street lights, e.g. remotely turning lights on or off or adjusting the strength of the light)

AGRICULTURE & ENVIRONMENT

→ Agricultural applications - live stock tracking (fishing, cattle and wild animal tracking and monitoring)
→ Agricultural - stationary tracking and monitoring of soil, temperature, weather conditions
→ Environmental monitoring - near real-time (e.g. fire hydrant network monitoring, manhole cover status reporting, alerts for chemical emission levels)
→ Environmental monitoring - data collection (Pollution, noise, rain, wind, river flow speed, health hazard, bore hole, etc)

INDUSTRIAL

→ Industrial: tank process/safety monitoring, (petrochemical and waste storage monitoring, hazardous fluids tank monitoring, High RF interference environments)
→ Industrial: machinery control (equipment status, factory control, process and safety monitoring)
→ Industrial: propane tank monitoring (monitoring and reporting fill level of propane tanks for rural and suburban, residential and light industry customers)
→ Vending machines - general (item filling level report, device diagnosis report, raising alarm on burglary attempt)
→ Vending machines - privacy / data verification (e.g. age) for cashless payments

CONSUMER & MEDICAL

→ Wearables (e.g. fashionable low-end leisure devices such as smart watches
→ VIP/Pet tracking (e.g. reporting location of persons or pets, paging when queried)
→ White goods/appliances (e.g. refrigerators, washing machines)
Smart bicycles (tracking location)
→ Assisted Living / Clinical remote monitoring (e.g. temperature sensors, alert buttons)

SMART BUILDINGS

→ Smoke detectors (regular auto-test, battery check, real-time alerts to the relevant parties in case of fire)
→ Alarm systems, atuators
→ Home automation (e.g. values for temperature, humidity, control of garage doors, blinds)

Figure 7.24 Low-power, wide-area network (LPWAN) applications. Reproduced by permission of GSMA.

IoT applications. EC-GSM-IoT can be activated through new software deployed over a very large GSM footprint, adding even more coverage to serve IoT devices.

Non-3GPP-based, unlicensed technologies:

LoRaWAN: Network protocol intended for wireless battery-operated things in regional, national, or global networks. LoRaWAN is promoted by the LoRa Alliance.

Sigfox: A single global network based on proprietary technology from Sigfox. Sigfox is among the early promoters of LPWA networks and the company partners with local service providers to build a global IoT network. However, they may face resistance from operators due to the proprietary technology and also from local authorities, as they have a global IoT platform although they recently

Table 7.1 Comparison between 3GPP IoT licensed technologies.

Technology	LTE MTC Cat 0	LTE eMTC Cat M	EC-GPRS	NB-IoT
Governing body/Standards	3GPP Release 12	3GPP Release 13	3GPP Release 13	3GPP Release 13
Frequency (MHz)	Uses LTE technology and frequency bands		GSM bands	In-band LTE carrier, or within LTE guard bands, or standalone in re-farmed GSM spectrum
DL bandwidth	20 MHz	1.4 MHz	200 kHz	180 kHz (15 kHz subcarrier spacing)
UL bandwidth				Single-tone: 180 kHz (3.75 kHz or 15 kHz spacing) Or Multi-tone: 180 kHz (15 kHz subcarrier spacing)
Multiple access DL	OFDMA		TDMA	OFDMA
Multiple access UL	SC-FDMA			Single-tone FDMA or multi-toned SC-FDMA
Modulation DL	QPSK, 16QAM, 64QAM		GMSK, optional 8PSK	BPSK, QPSK, optional 16QAM
Modulation UL	QPSK, 16QAM			BPSK, QPSK, 8PSK optional 16QAM
Peak data rate	1 Mbps		10 kbps to 240 kbps	DL up to 250 kbps; UL single tone up to 20 kbps, multi-toned up to 250 kbps
Maximum range/Coverage (link budget)	−41 dB	−156 dB	−164 dB	−164 dB
Description	LTE enhancements for MTC with new power-saving mode	Further LTE enhancements for MTC building on the work started in Category 0, long battery life, low device cost, extended coverage compared to LTE MTC Category 0	Low device cost, long battery life and extended coverage compared to GPRS/GSM devices	Clean-state technology added to the LTE platform optimized for the low end of the IoT market; provides lower cost, extended coverage and long battery life compared to eMTC Category M
Comments		Also, referred to as Category M1		Also, referred to as Category M2

became open for local deployment based on the market size.

RPMA (Random Phase Multiple Access): A technology communication system employing Direct-Sequence Spread Spectrum (DSSS) with multiple access. It is now offered by Ingenu company.

Weightless: A proposed proprietary open wireless technology standard for exchanging data between a base station and thousands of machines around it (using wavelength radio transmissions in unoccupied TV transmission channels) with high levels of security.

A comparison between 3GPP-based technologies for IoT networks is provided in Table 7.1. A summary of unlicensed LPWAN is provided in Table 7.2.

The Gartner Hype Cycle: Figure 7.25 demonstrates generic Gartner's Hype Cycle. Gartner's July 2017 Hype Cycle for IoT standards and Protocols, 2017 can be downloaded by Gartner clients or subscribers from [11].

Because IoT is a combination of multiple technologies, there are a plethora of standards available, and a considerable amount of consolidation and attrition is to be expected. Each standard is seeking to de ne the way that devices are designed, and how they

Table 7.2 Comparison between unlicensed LPWA technologies.

Technology	LoRaWAN	Sigfox	OnRamp/Ingenu
Governing body/Standards	LoRa Alliance	Sigfox	Ingenu (formerly OnRamp)
Frequency (MHz)	Regional sub-GHz bands		Global band: 2.4 GHz
DL bandwidth	125, 500 kHz	Base station listening bandwidth: 200 kHz, 100 Hz UL channels; 600 Hz DL channels	1 MHz
UL bandwidth	125, 250, 500 kHz		
Multiple access DL	Proprietary Chirp Spread Spectrum (CSS)	Proprietary UNB/FHSS	Random Phase Multiple Access (RPMA)
Multiple access UL			
Modulation DL	LoRa; (G)FSK	GFSK	BPSK, OQPSK, FSK, GFSK, P-FSK, P-GFSK
Modulation UL		DBPSK	
Peak data rate	0.3 kbps − 50 kbps	100 bps up, 600 bps down	20 kbps
Maximum range/Coverage (link budget)	− 150 − 157 dB[3]	− 146 − 162 dB[3]	∼ 4 km (urban)
Description	Long-range, low-power connectivity	Good fit for any application that needs to send smack, infrequent burst of data such as alarm systems or simple metering	Long range connectivity for smart grid, intelligent lighting, advanced metering infrastructure, oil and gas automation plus more
Comments	LoRa physical layer LoRaWAN is the MAC layer. The intellectual property (IP) for LoRaWAN PHY layer is owned by Semtech While LoRaWAN MAC layer is defined by LoRa Alliance.	Sigfox is an LPWAN dedicated to IoT	Solution includes machine network, management platform and chipset/modules; OnRamp wireless rebranded to Ingenu in September 2015

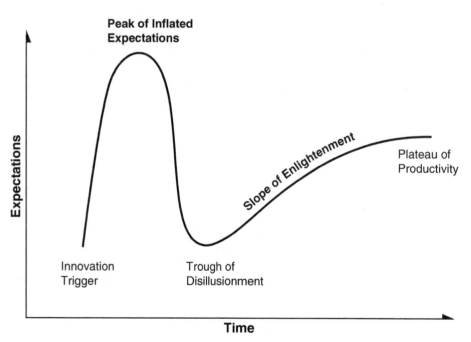

Figure 7.25 Generic Gartner's Hype Cycle. Reproduced by permission of Gartner. *Source*: Gartner Methodologies, Gartner Hype Cycle, http://www.gartner.com/technology/research/methodologies/hype-cycle.jsp

communicate with each other, with the cloud, and with other products.

In many cases, the standards overlap or compete, especially in areas with large commercial potential. There are also areas of the IoT in which standards are incomplete (such as security). Consequently, new standards will likely emerge in the coming years. Most IoT projects are currently custom projects, so integration of technologies plays a vital role. This means consultants and system integrators often play a vital role. Enterprises can therefore use the Hype Cycle to get an overall perspective and examine how certain standards are faring in adoption compared with others.

The Hype Cycle for IoT Standards and Protocols, 2017, includes standards that impact the design and development of IoT hardware products — the basic building blocks of an IoT hardware device. The Hype Cycle includes only standards that are supported across manufacturers, even if some of them have a single patent or license holder.

Several standards on the Hype Cycle appear to be tightly packed just after exiting the Peak of Inflated Expectations. We believe this reflects the way that the IoT itself continues to rise in hype and mind share. We feel technologies near the Peak of Inflated Expectations are not necessarily more important than the other technologies, they are simply enjoying a greater degree of hype, market expectation and press interest than other technologies.

The Hype Cycle for IoT Standards and Protocols, 2017 provides a glimpse into tomorrow's technology standards and represents Gartner's view of the progress of some of the most interesting and significant standards and protocols relating to IoT. Most of the technologies on this Hype Cycle should be viewed as underlying and enabling technologies that facilitate and support a wide range of IoT products and solutions. We believe their adoption rates will vary across different industries.

The Priority Matrix: Gartner's Priority Matrix is demonstrated in Figure 7.26 (figure 2 of the report [11]). The Priority Matrix "maps the time to mainstream adoption of a technology against its benefit rating. Five technology standards have been designated with a low benefit rating in this Hype Cycle,

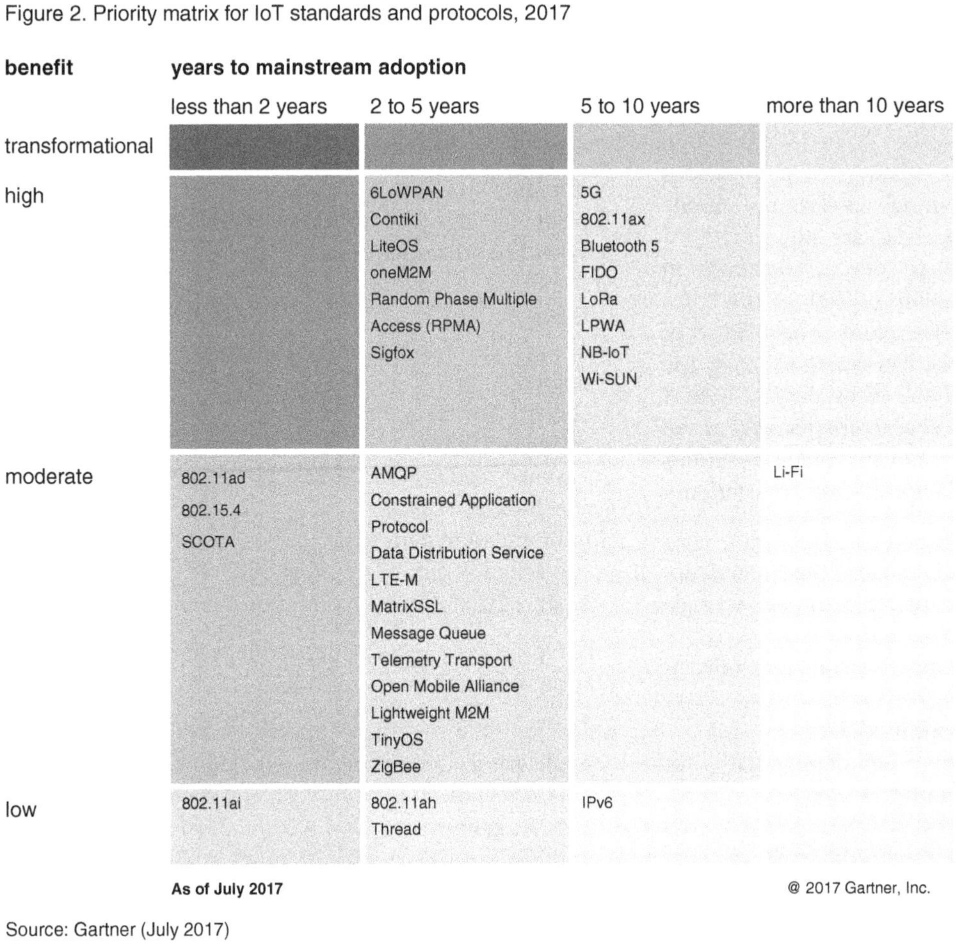

Figure 2. Priority matrix for IoT standards and protocols, 2017

benefit	years to mainstream adoption			
	less than 2 years	2 to 5 years	5 to 10 years	more than 10 years
transformational				
high		6LoWPAN Contiki LiteOS oneM2M Random Phase Multiple Access (RPMA) Sigfox	5G 802.11ax Bluetooth 5 FIDO LoRa LPWA NB-IoT Wi-SUN	
moderate	802.11ad 802.15.4 SCOTA	AMQP Constrained Application Protocol Data Distribution Service LTE-M MatrixSSL Message Queue Telemetry Transport Open Mobile Alliance Lightweight M2M TinyOS ZigBee		Li-Fi
low	802.11ai	802.11ah Thread	IPv6	

As of July 2017 @ 2017 Gartner, Inc.

Source: Gartner (July 2017)

Figure 7.26 Priority matrix for IoT standards and protocols. Reproduced by permission of Gartner.

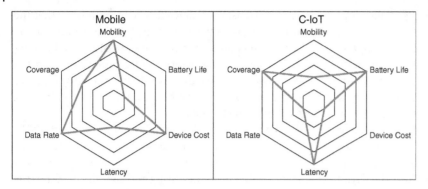

Figure 7.27 New paradigm for cellular IoT.

some because their commercial success is far from certain (e.g., Thread), while some are minor updates that smooth the experience rather than shift the paradigm (e.g., 802.11ai)."

We believe no technology standard has been designated as having transformational benefit, which intuitively fits with the reality of standards development. However, 15 of the 30 technology standards profiled in this Hype Cycle have been marked to deliver high business benefit, and seven of those are expected to reach mainstream adoption within the next two to five years. The chart shows clearly how early we are in the evolution of the IoT — in five years, another 21 standards should be well on their way to the Plateau of Productivity, making the whole IoT more productive and increasing the digitization of the real world.

To summarize, most of the existing IoT applications today are supported via short-range wireless and mobile WAN technologies, however these are not suited for some application areas due to power consumption levels, the total cost of ownership (TCO), or the complexity of the infrastructure. Therefore, a new breed of purpose-built LPWA networks has emerged to fill the gap between all of these LPWA technologies. LPWA technologies possess several characteristics that make them particularly attractive for devices and applications that have low mobility, low levels of data transfer, and long battery lifetimes:

Low power consumption (to the range of the Nano Amp) that enable devices to last for ten years on a single charge.

Optimized data transfer (supports small, intermittent blocks of data).

Low device unit cost (sub-$5 per module).

Simplified network topology and deployment, for example, via software upgrades.

Optimized for low throughput.

Improved outdoor and indoor penetration coverage over existing wide-area technologies, such as GSM/GPRS.

Secured connectivity and adequate level of authentication.

Integrated into a unified/horizontal IoT/M2M platform. Network scalability for capacity upgrades.

The characteristics of the LPWA market opportunity dictate the adoption of new approaches. As depicted in Figure 7.27, the current traditional mobile technologies are not able to meet the LPWA requirements cost-efficiently. Figure 7.27 illustrates the need for new networks to address the IoT requirements. Also, each application or use case may have a different profile, as demonstrated in Figure 7.28. Therefore, the selection of connectivity technology will rely on several factors and there is no single standard that will meet all IoT use case requirements. The classifications of the IoT market along with appropriate technologies are demonstrated in Figure 7.29.

The profiling of the LPWA IoT use cases depends on many factors, including, but not limited to:

Battery life

Type of coverage (indoor/outdoor/mixed)

Bandwidth (downlink/uplink)

SLA

Latency

Security (encryption, channel, environment requirements)

Mobility (stationary, mobile)

Service availability

Location-based services (LBS)

Critical IoT use cases

Communication patterns:
    Pre-scheduled or event-triggered
    Features that require bidirectional communication.

A typical service provider strategy for IoT networks may be summarized as follows:

Offer multiple/hybrid technologies for the use case/application requirements, including: throughput, coverage, power, latency, cost, and spectrum.

The existing networks (2G/3G/LTE/Wi-Fi) will meet the applications that need long range and high data rates (**10% of IoT market volume**).

Short-range technologies such as ZigBee, RF Mesh (802.15.4), PLC, Wi-Fi, etc. will be used for short-range

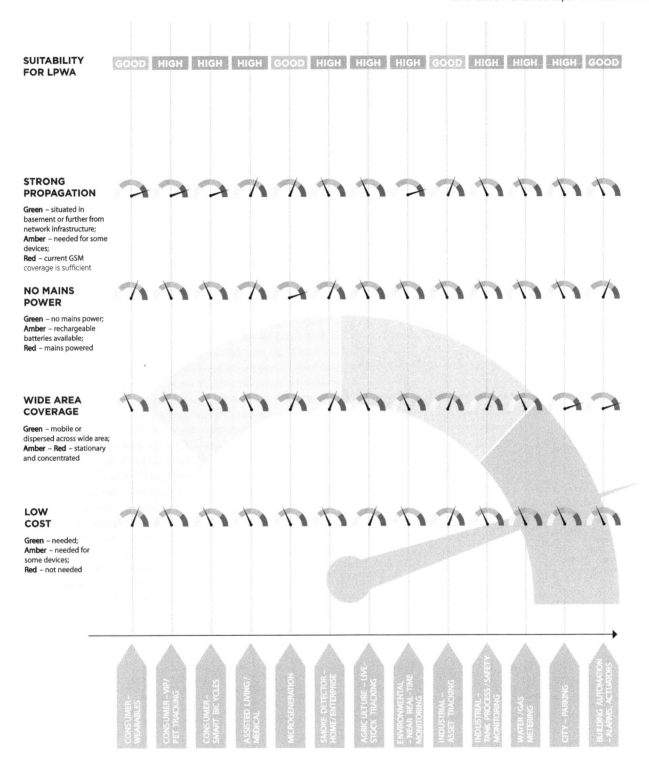

Figure 7.28 LPWA IoT application profiling. Reproduced by permission of GSMA.

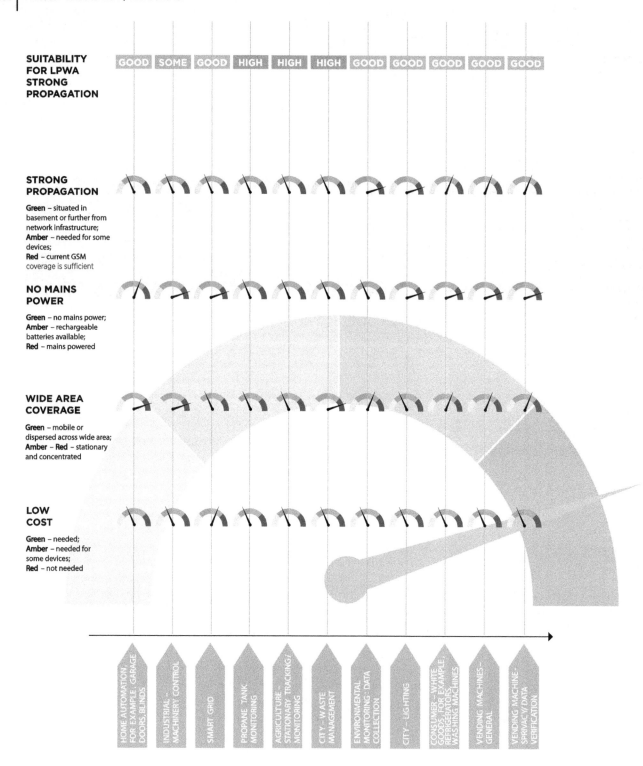

Figure 7.28 *(Continued)*

Figure 7.29 IoT market classifications.

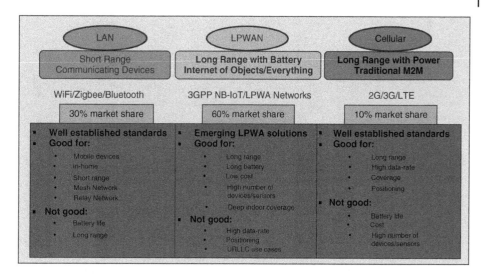

applications such as smart meters, smart homes, smart parking, etc. (**30% of IoT market volume**).

Introduce NB-IoT as mainstream technology for LPWA network use cases. This will be the mainstream connectivity for critical IoT application (**60% of IoT market volume**).

Adopt a hybrid approach for IoT network connectivity short-range, long-range, and LPWA networks. This will allow operators to address the major part of the IoT/M2M verticals and compete with niche deployment in the unlicensed bands.

A unified/horizontal IoT/M2M platform will be able to manage all technologies and devices from different networks, different GWs, and different standards and protocols.

7.10   3GPP Standards for IoT

In this section, we will focus on 3GPP-based IoT evolutions. 3GPP Release 13 has made a major effort to address the IoT market. The portfolio of technologies that 3GPP operators can now use to address their different market requirements includes:

eMTC: Further LTE enhancements for Machine-Type Communications (MTC), building on the work started in Release 12 (UE Category 0, new power-saving mode: PSM).

NB-IoT: New radio added to the LTE platform optimized for the low end of the market.

EC-GSM-IoT: EGPRS enhancements, which, in combination with PSM, prepare GSM/EDGE markets for IoT.

A freeze on the protocol specifications for NB-IoT was completed by the end of Q2-16. The above agreement was concluded in February 2016 after long debate inside 3GPP on which technology to adopt for LPWA

networks. Figure 7.30 summarizes the 3GPP roadmap for IoT technologies [13]. The debate inside 3GPP has delayed the deployment of the standardized LPWA networks and provided an opportunity for proprietary solutions to be widely adopted and deployed. Unlike the cellular technology, the IoT use cases are mainly derived by enterprise organizations and therefore, proprietary or local deployment is preferred for reasons such as security, lack of license fees, less dependency on service providers, a CAPEX model rather than an OPEX model, and ease of deployment. Therefore, the IoT landscape became a mess in terms of connectivity. The major part of this mess is due to the lack of a global standard and the different requirements of the IoT use cases, as explained earlier in the chapter. In addition, the service providers' focus on connectivity rather than e2e use cases provided opportunities for other, proprietary technologies to be inserted within ready-to-use cases. Delays by 3GPP have also contributed, as a result of the technology debate and multiple IoT technologies, as demonstrated in Figure 7.30. On the other hand, this wide range of technologies, either licensed or unlicensed, has led to innovation and competition. Moreover, this aggressive competition in the IoT market has led to a transformation in service providers' strategy to offer end-to-end IoT use cases rather than focusing on connectivity only.

The convergence to a horizontal IoT approach, as demonstrated earlier, is a key advantage in tackling the IoT silos. The transformation from vertical silos to a unified horizontal IoT approach is not an easy transformation, as it needs software expertise to develop the use cases on top of the application enablement layer of the IoT platform. The existing customized IoT verticals will resist such an approach to keep their use cases.

Looking at 3GPP technologies, we can see an overlap, and three technologies have been proposed to cater for LPWA IoT use cases in addition to the LTE for MBB,

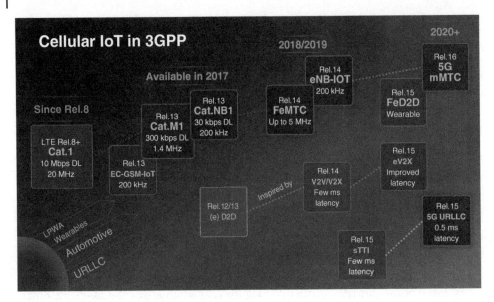

Figure 7.30 3GPP roadmap for IoT technologies.

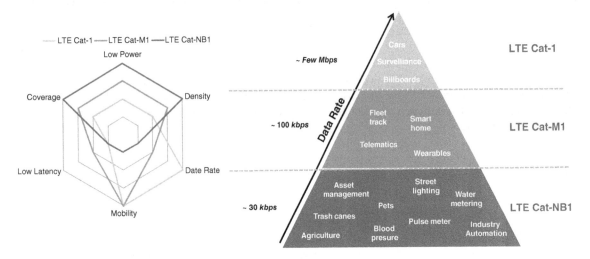

Figure 7.31 3GPP IoT application profiling.

as demonstrated in Figure 7.31 by GSMA [14]. The NB-IoT is the main stream for LPWA networks and has mainly been developed to compete with unlicensed technologies such as LoRa and Sigfox. Table 7.3 provides a comparison between NB-IoT and unlicensed LPWA networks. The main difference between NB-IoT and unlicensed LPWAN is the use of existing mobile networks in licensed bands versus unlicensed bands, such as ISM bands, and therefore, the latter can be deployed by enterprises and small organizations without relying on service providers' networks. However, some operators have adopted unlicensed technologies such as LoRa and Sigfox as they are mature, have a variety of use cases, and are easy to deploy, and hence have a faster time to market.

There are three deployment scenarios for NB-IoT, as demonstrated in Figure 7.32 [13]. NB-IoT supports three different modes of operation, as follows:

Standalone operation:

Utilizing, for example, the spectrum currently being used by GERAN systems as a replacement for one or more GSM carriers as well as scattered spectrum for potential IoT deployment.

No use of LTE resources by NB-IoT.

Additional spectrum used by NB-IoT.

Typically, in sub-1GHz spectrum like GSM; needs to co-exist with other RATs deployed in the vicinity.

Guard band operation:

Utilizing the unused subcarriers within an LTE carrier's guard band.

No use of LTE resources by NB-IoT.

No additional spectrum used by NB-IoT.

NB-IoT channels in guard band limited.

Deployed in LTE guard bands, e.g. for 20 MHz two guard bands 1MHz wide and for 5 MHz two guard

Table 7.3 Comparison between NB-IoT and unlicensed LPWA networks.

Criteria	NB-IoT	Unlicensed LPWA Networks
Standardization body	3GPP	Proprietary or industry alliance.
Frequency type	Licensed	Unlicensed
Exact frequency band	In-band, guard band, and standalone.	ISM bands such as 868 MHz and 915 MHz
Network type	Public LTE network with NB-IoT upgrade.	Public/Private LPWA network.
Hosting	Only sensors are deployed at enterprise and other network elements, and IoT platform is hosted and managed by the service provider.	Can be fully hosted at enterprise including sensors, GWs, and IoT platform.
HW/SW	SW upgrade to LTE network for most of the infra-vendors. Some vendors will need an upgrade to baseband.	New network to be deployed, however small GW with omni antennas and any available backhauling such as LTE/Wi-Fi/Fiber/MW.
Status	Standardization was completed in June 2016 and commercial deployment was expected by Q1 2017, although only POCs materialized.	Deployed for several LPWA use cases.

bands 250 kHz wide; co-existence of 3.75 kHz subcarrier spacing in UL and LTE system to be studied.

Up to 6 dB boost in Tx power compared to LTE PRBs.

In-band operation:

Utilizing resource blocks within a normal LTE carrier.

Trade NB-IoT carriers versus LTE capacity.

No additional spectrum used by NB-IoT.

Close to 100 kHz grid single PRB locations within LTE transmission BW, DL supposed to be orthogonal; co-existence of 3.75 kHz subcarrier spacing in UL and LTE system to be studied.

Up to 6 dB boost in Tx power compared to other LTE PRBs.

The 100 kHz carrier allocation grid is maintained from legacy LTE for NB-IoT. The distance of the PRB center from the 100 kHz grid varies with PRB instance and LTE channel bandwidth. Hence, only a subset of PRBs is eligible for NB-IoT anchor carrier:

For 10 MHz and 20 MHz channels, the NB-IoT anchor PRB shall only be 2.5 kHz off the nearest 100 kHz grid point.

For 3 MHz, 5 MHz, and 15 MHz channels, the NB-IoT anchor PRB shall only be 7.5 kHz off the nearest 100 kHz grid point.

The central 6 PRBs cannot be assigned an NB-IoT anchor carrier as they carry the legacy PSS/SSS/PBCH channels.

For example, the in-band scenario with LTE-800 MHz network to provide deep indoor coverage is an optimal solution for operators that have an LTE network in this band. Another option is to use one GSM carrier at 900 MHz, but the issue here is the RF head of the GSM 900 MHz needs to support NB-IoT.

The NB-IoT is introduced with the promise of a smooth upgrade for existing networks and it is mainly based on the LTE standard. Some vendors claim it is only a software upgrade to an LTE base station, while others need a hardware upgrade to the base band unit to support the NB-IoT processing. The service provider needs to benchmark the CAPEX of NB-IoT versus the unlicensed LPWA networks. The latter offer very attractive, low-cost deployment with a variety of backhauling options such as Wi-Fi, LTE modem, and Ethernet. In addition, since the unlicensed LPWA networks were developed from scratch, a lite core network is needed or the GW will be directly connected to the IoT platform. On the other hand, the core of NB-IoT will need an upgrade to existing LTE evolved packet switch (EPS) or an additional slice will need to be added for the NB-IoT core. The latter is the preferred option to avoid major interruption and upgrades to the mobile broadband EPC core, which is not designed to accommodate such a high volume of connectivity. In addition, the license model of NB-IoT connectivity is different from the MBB license and it should be a fraction of the MBB license. Figure 7.33 provides a typical network topology for the EPC core based on a virtualized environment with different slices for different services. Since the NB-IoT is being introduced while most operators are exploring network virtualization evolution, it makes sense to deploy the NB-IoT as a virtualized node from day one. So, even if the legacy MBB EPC is not virtualized, the NB-IoT core can be introduced as a separate, virtualized slice, as indicated in Figure 7.33. (Please refer to Chapter 8, where network slicing evolution is addressed in detail.)

A summary of 3GPP LTE-based technologies for the IoT is provided in Table 7.4 and the corresponding UE categories are summarized in Table 7.5. EC-GSM is not included as it is intended for LTE evolutions only. In addition, there are no commercial plans for EC-GSM so

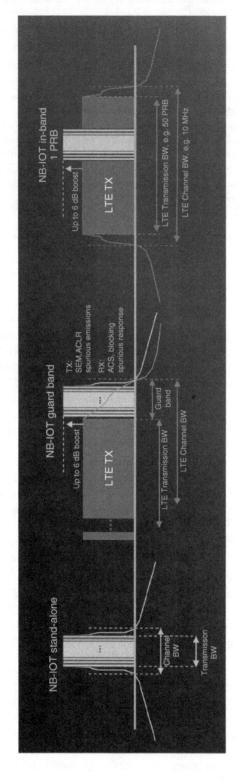

Figure 7.32 NB-IoT deployment scenarios.

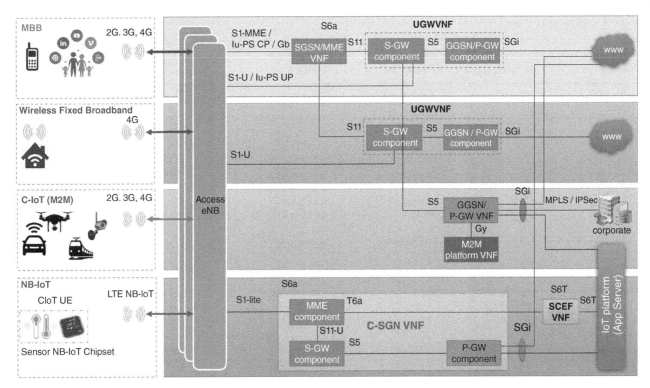

Figure 7.33 Core network slicing to support NB-IoT.

far and it is expected that NB-IoT will be the mainstream for LPWA networks for mobile operators. Several use cases require higher data rates (e.g. audio streaming), broadcast support (e.g. for software updates), positioning (e.g. asset tracking, fleet management), and support for voice (e.g. wearables); FeMTC (Release 14) is being developed for this purpose while guaranteeing low power consumption, low complexity, and extended coverage.

NB-IoT is designed with three main factors in mind: low cost, low power consumption for battery-operated sensors, and extended coverage. Table 7.6 provides a summary of technologies explored by 3GPP to meet these requirements. The uplink and downlink parameters of NB-IoT are summarized in Table 7.7. As indicated, NB-IoT will accommodate latency up to 10 seconds, and therefore it is not designed for critical IoT applications. In addition, only Idle mode reselection for mobility is adopted. Most importantly, the link budget is improved by 20 dB to support deep indoor coverage. Therefore, the planning of NB-IoT sites overlaid on normal LTE sites will be different, and fewer sites will be needed thanks to the 20 dB improvement in the link budget.

7.11 3GPP NB-IoT

Based on the above discussion for LPWA networks, the following standard specific requirements for NB-IoT are derived [15]:

Minimize the signaling overhead, especially over the radio interface.

Provide appropriate security for the complete system, including the core network.

Improve battery life.

Support delivery of IP and non-IP data [16, 17].

Support for SMS as a deployment option [18].

To fulfill these requirements, many advanced and even basic features of LTE Release 8/9 are not supported [5]. The main mobility feature, which is the handover for UEs in the Connected state, is not included in NB-IoT. Only cell reselection in the Idle state is supported within the NB-IoT with no inter-RAT handover and without relevant features. Examples are the lack of LTE-WLAN interworking, interference avoidance for in-device co-existence, and measurements to monitor the channel quality. Most LTE-Advanced features are not supported either, such as carrier aggregation (CA), dual connectivity, and device-to-device communications. In addition, there is no Quality of Service (QoS) concept, because NB-IoT is not used for delay-sensitive data packets. Consequently, all services requiring a guaranteed bit rate, like real-time IMS, are not offered in NB-IoT. Therefore, the air interface for small, non-delay-sensitive data packets is split off and optimized separately. UEs which support working on NB-IoT technology are tagged with the new UE category, Category NB1 [15].

3GPP NB-IoT evaluation provides the following benefits to service providers:

Table 7.4 Comparison between 3GPP IoT technologies for LPWAN.

Criterion	Category 1 (Release 8+)	Category M1 (Release 13)	Category NB1 (Release 13)	FeMTC (Release 14)	eNB-IoT (Release 14)
Bandwidth	20 MHz	1.4 MHz	180 kHz	Up to 5 MHz (CE Mode A and B for PDSCH and A only for PUSCH)	180 kHz
Deployments/HD-FDD	LTE channel/No HD-FDD	Standalone, in LTE channel/HD-FDD preferred	Standalone, in LTE channel, LTE guard bands, HD-FDD	Standalone, in LTE channel/HD-FDD, FD-FDD, TDD	Standalone, in LTE channel, LTE guard bands, HD-FDD preferred
MOP	23 dBm	23 dBm/20 dBm	23 dBm/20 dBm	23 dBm/20 dBm	23 dBm/20 dBm/14 dBm
Rx ant/layers	2/1/	1/1	1/1	1/1	1/1
Coverage, MCL	145.4 dB DL, 140.7 dB UL (20 kbps, FDD]	155.7 dB	Deep coverage: 164 dB +3	155.7 dB (at 23 dBm)	Deep coverage: 164 dB
Data rates (peak)	DL: 10 Mbps, UL: 5 Mbps	~800 kbps (FD-FDD) 300/375 kbps DL/UL (HD-FDD)	30 kbps (HD-FDD)	DL/UL: 4 Mbps FD-FDD@5 MHz	TBS in 80/105 kbps 1352/1800 peak rates t.b.d.
Latency	Legacy LTE: < 1s	~ 5s at 155 dB	< 10s at 164 dB	At least the same as Category M1 legacy LTE (normal MCL)	At least the same as Category NB1, some improvements are FFS
Mobility	Legacy support	Legacy support	Cell selection, reselection only	Legacy support	More mobility compared to Category NB1
Positioning	Legacy support	Partial support	Partial support	OTDA with legacy PRS and frequency hopping	50 m H target, new PRS introduced, details FSS. UTDOA under study
Voice	Yes (possible]	No	No	Yes	No
Optimizations	N/A	MPDCCH structure, frequency hopping, repetitions	NPDCCH, NPSS/NSSS, NPDSCH, NPUSCH, NPRACH etc., frequency hopping, repetitions, MCO	Higher bandwidth will be DCI or RRC configured, multi-cast, e.g. SC-PTM	Multi-cast, e.g. SC-PTM
Power saving	DRX	eDRX, PSM	eDRX, PSM	eDRX, PSM	eDRX, PSM
UE complexity BB	100%	−45%	< 25%	~ 55%	~ 25%

Table 7.5 Summary of 3GPP IoT module categories.

Terminal	3GPP Release	Availability	Details
LTE Category 1	Release 8	Available	Cat 1 devices have been launched for both data applications and VoLTE/voice applications.
			Cat 1 will continue for VoLTE/voice application devices, i.e., not replaced by NB-IoT/Cat 0/Cat M.
			Cat 1 single Rx also launched for certain applications..
LTE Category 0	Release 12	2H16/1H17	Cat 0/power save mode/half duplex/extended ACB/1Rx.
			Cat 0 is used for replacing data application devices of Cat 1, but cannot replace voice/VoLTE.
			Cat 0 is the interim solution prior to Cat M rollout, i.e., a short lifecycle. It may be dropped to go direct to Cat M.
Category M	Release 13	Q1 2017	1.4 MHz BW, reduced power, extended DRX, coverage enhancements.
			Cat M is used for replacing data application devices of Cat 1/Cat 0, but cannot support VoLTE either for sure.
NB-IoT	Release 13	Q1 2017	Overlaps LTE, 1 PRB is used, reduced power, coverage enhancements, 200 kbps.
			The differentiators of Cat M and NB-IoT are "mobility" and throughput. Cat M is still needed unless the evolution of NB-IoT will provide sufficient mobility support.

Table 7.6 Summary of NB-IoT exploited technologies.

Criteria	Low Cost	Low Power	Coverage Extension
Target Solutions	< $1USD	10+years battery life	20 dB enhanced coverage
	HD-FDD	Extended DRX	Repetition
	Single-Rx antenna	Power-saving mode	PSD boosting
	Reduced BW (200 kHz)	Reduced BW (200 kHz)	

Maximum spectrum utilization: NB-IoT supports standalone deployment, LTE guard-band deployment, and LTE in-band deployment, which fully utilize operators' spectrum resources and increase spectrum efficiency.

Support for many low-rate users: The low-rate and low-activity machine-to-machine (M2M) application service model supports many users.

Deep coverage: The coverage gains provided by NB-IoT are 20 dB greater than those provided by GSM with the help of time-domain repetition and power spectrum density (PSD) increase.

Low power consumption of UEs: NB-IoT optimizes the protocol stack, streamlines signaling processing and signaling interaction, and reduces the power consumption of UEs.

7.12 NB-IoT DL Specifications

For the DL, three physical channels are defined, as follows:

NPBCH, the narrowband physical broadcast channel.
NPDCCH, the narrowband physical downlink control channel.
NPDSCH, the narrowband physical downlink shared channel.

Two physical signals are defined, as follows:

NRS, the narrowband reference signal,
NPSS and NSSS, the primary and secondary synchronization signals.

Table 7.7 UL and DL parameters of NB-IoT.

Parameter	DL	UL
RF bandwidth	180 kHz	
Duplex mode	HD-FDD only	
Multiple access	OFDMA 15 kHz	SC-FDMA 3.75 kHz and 15 kHz
Modulation	QPSK	Pi/2 BPSK, Pi/4 QPSK for single tone
		QPSK for multi-tone
HARQ process number	1 HARQ process	1 HARQ process
Max TBS	680 bits	1000 bits
Channel coding	TBCC	Turbo for PUSCH Format 1
		Repetition for PUSCH Format 2 (ACK/NACK)
Peak data rate	In-band: 26 kbps	Single-tone: 16.6 kbps
	Standalone/Guard band: 27 kbps	Multi-tone: 58.6 kbps
Coverage	164 dB	164 dB
Mobility	Idle mode cell reselection	
Battery life	Up to 10 years	
Latency	< 10 seconds	

There are fewer channels compared to LTE; the physical multicast channel, PMCH, is not included because there is no MBMS service for NB-IoT. Figure 7.34 demonstrates the connection between the transport channels and the physical channels [15].

MIB information is always transmitted over the NPBCH, the remaining signaling information and data over the NPDSCH. The NPDCCH controls the data transfer between the UE and eNB. The physical DL channels are always QPSK modulated. NB-IoT supports the operation with either one or two antenna ports, AP0 and AP1. For the latter case, Space Frequency Block Coding (SFBC) is applied. Once selected, the same transmission scheme applies to NPBCH, NPDCCH, and NPDSCH. Like in LTE, each cell has an assigned physical cell ID (PCI), the narrowband physical cell ID (NCellID). In total, 504 different values for NCellID are defined. Its value is provided by the secondary synchronization signal (NSSS) [15].

The NB-IoT DL and UL characteristics are illustrated in Figure 7.35 and can be summarized as follows:

NB-IoT DL:
OFDMA on the downlink
Subcarrier bandwidth = 15 kHz
Usable subcarriers = 12 = 180 kHz
Guard bandwidth at either end = 10 kHz.
Only applicable to standalone.
Time-domain frame/slot structure
Same as LTE
1 slot = 0.5 ms

1 subframe = 2 slots
1 frame = 10 subframes
CP length for DL
Only normal CP is supported for both downlink and uplink in Release 13.

NB-IoT UL:
Single-tone and multi-tone transmissions are both supported
Single-tone transmission: Supports 15 kHz and 3.75 kHz tone spacing
Multi-tone transmission: Supports 15 kHz only
For 15 kHz tone spacing
OFDM/SC-FDMA symbol boundary is no change from LTE
For 3.75 kHz tone spacing
1 NB-slot = 2 ms = 7 symbols.

NB-IoT supports the half-duplex FDD (HD-FDD) operation, as described in Figure 7.36, which can be summarized as follows:

The HD-FDD guard period is Type B. In Release 13, NB-IoT supports FDD only.
When an NB-IoT UE is transmitting, the UE is not expected to monitor or receive any downlink channels.
For Type B half-duplex FDD operation:
Guard periods, each referred to as a half-duplex guard subframe, are created by the UE by:
Not receiving a downlink subframe immediately preceding an uplink subframe from the same UE, and

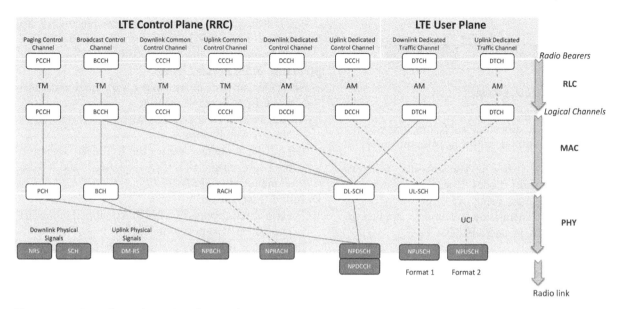

Figure 7.34 NB-IoT Channel Mapping of Protocol Layers.

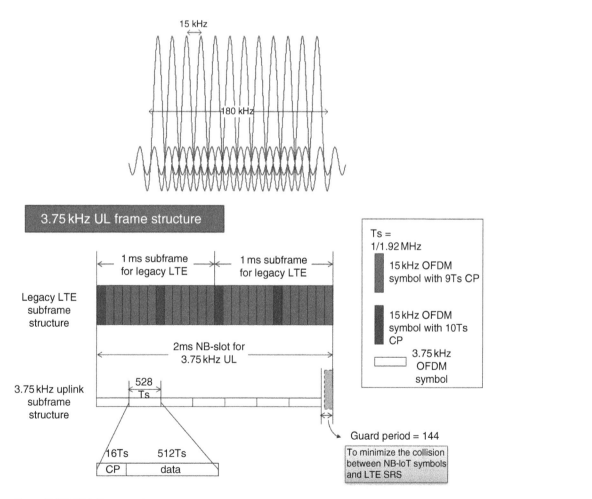

Figure 7.35 NB-IoT numerology.

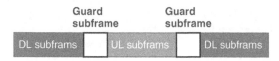

Figure 7.36 NB-IoT HD-FDD operation.

Not receiving a downlink subframe immediately following an uplink subframe from the same UE.

NB-IoT supports two DL transmission schemes defined in all operation modes: a single antenna port and two antenna ports, using transmit diversity, i.e. SFBC. The same transmission scheme is applied to NB-PDCCH, NB-PBCH, and NB-PDSCH.

7.12.1 NB-IoT Reference Signals (NB-RS)

The concept of NB-RS is introduced for NB-IoT, as indicated in Figure 7.37, and can be summarized as follows:

Presence of NB-RS

NB-RS is always present and is used for single-antenna-port and two-antenna-port transmission schemes.

NB-RS is present without the condition for NB-PDCCH/NB-PDSCH in-band; only NB-RS in standalone or guard band.

In cell-specific valid DL PRB pairs, NB-RS is present; in invalid DL PRB pairs, there is no NB-RS.

No NB-RS in NB-PSS and NB-SSS.

In an in-band NB-IoT carrier, without cell-specific valid DL subframes the NB-IoT device expects NB-RS in subframes 0 and 4 and in subframe 9 if it does not contain NSSS.

In a guard band or standalone NB-IoT carrier, assume NB-RS in all subframes except for NPSS/NSSS.

Not precluded: LTE CRS for DL demodulation or measurements if the number of APs for LTE CRS and NRS is the same and either 1 or 2.

NB-RS and LTE CRS

When the same PCI indicator is set to true, the cell ID is identical for NB-IoT and LTE, the number of antenna ports is the same for LTE CRS as for

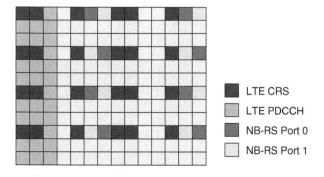

■ LTE CRS
□ LTE PDCCH
■ NB-RS Port 0
□ NB-RS Port 1

Figure 7.37 NB-IoT reference signals (NB-RS).

NB-RS, channel estimation is possible on NB-RS and on LTE CRS, LTE CRS is available in NB-IoT PRB wherever NB-RS is available.

Structure of PRB with NB-RS

NB-RS for antenna ports 0 and 1 is mapped to the last two OFDM symbols of a slot.

NB-RS uses a cell-specific frequency shift derived as NB-IoT Cell ID mod 6.

The NB-RS sequence re-uses the LTE CRS sequence; the center of the LTE CRS sequence is the NB-RS sequence for all PRBs.

Detailed Description

The two antenna ports of NB-RS are mapped to the last two OFDM symbols of the slot.

The number of NB-RS ports (1 or 2) is indicated by NB-PBCH CRC masking, as explained in the next section. All 0s for one port and all 1s for two ports, as in the current specification for LTE CRS.

Cell-specific cyclic shifting with physical-layer cell-identity groups: v_shift = NB-IoT PCID mod 6. NB-IoT PCID from the LTE PCID indicates the same LTE CRS position.

For the NB-RS sequence, the LTE CRS sequence is re-used.

The center of the LTE CRS sequence is used as the NB-RS sequence for all NB-IoT carriers.

In cell-specific valid DL PRB pairs, an NB-IoT UE may assume that NB-RS is present.

NB-RS is not present in cell-specific invalid DL PRB pairs.

"cell-specific valid DL PRB pairs" are configured in SIB1.

In the PRB pair to carry NB-PSS and NB-SSS, an NB-IoT UE should not expect NB-RS.

For all operation modes, in an NB-IoT carrier, a UE without a valid configuration of the cell-specific valid DL subframes may assume NB-RS is transmitted in subframes 0 and 4 and in subframe 9 if it does not contain NB-SSS, before getting "valid DL subframes configuration" from SIB1 and "after cell search and before SIB1 acquisition", which happens in the anchor PRB. Table 7.8 summarizes the locations of NB-RS.

7.12.2 NB-PBCH Resource Allocation

The NB-PBCH resource allocation is illustrated in Figure 7.38 and can be summarized as follows:

Transmission of Master Information Block (MIB-NB)
Demodulation based on narrowband RS (NRS)
Uses subframe 0 of every radio frame
  MIB-NB TTI is 640 ms
  NPBCH consists of eight independently decodable 80 ms blocks

Table 7.8 Reference signals (NB-RS) presence.

Anchor	Subframe Index	Phy Channel/Signals	NB-RS Assumption
Anchor PRB	0	NB-PBCH; every radio frame	Always present
	4	NB-PDSCH of SIB1; not every radio frame	Always present
		Others : Valid or invalid depending on SIB1	(*)
	5	NB-PSS; every radio frame	Not present
	9	NB-SSS; every other radio frame	Not present
		Others: Valid or invalid depending on SIB1	(**)
	1/2/3/6/7/8	Valid or invalid depending on SIB1	Valid: present
			Invalid: not present
Non-anchor PRB	0 ~ 9	Valid or invalid depending on dedicated configuration	Valid: present
			Invalid: not present

			NB-RS Assumption
(*) Before getting valid subframe configuration in SIB1			Always present
(**) After getting valid subframe configuration in SIB1			Valid: present
			Invalid: not present

It makes more sense if the network always sets subframe 4 and subframe 9 as valid subframes in the anchor PRB. However, there is no guarantee.

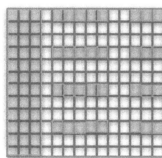

Figure 7.38 NB-PBCH resource allocation.

80 ms boundary is identified from NSSS
Subframe pattern
 In FDD mode, NB-PBCH is transmitted in subframe 0
  in every radio frame
Used REs
 A unified RE usage for all deployment modes
 For rate-matching purposes for NB-PBCH, the num-
  ber of NB-RS ports is based on two.
 Reserved for potential LTE usage
  three-OFDM symbol PDCCH region
  four-port CRS
   The PCID from NB-SSS and the LTE PCID indi-
    cate the same LTE CRS position.

The coding chain of the NB-PBCH is demonstrated in Figure 7.39. The NB-PBCH re-uses similar functionalities to the LTE PBCH, as shown in Figure 7.39.

7.12.3 NB-PDCCH Resource Mapping

NB-PDCCH resource mapping is illustrated in Figure 7.40 and it can be summarized as follows:

DL transmission:
 Single AP (port 0) and two APs (ports 0 and 1) with
  transmit diversity (SFBC).
 Modulation scheme: QPSK.
NB-PDCCH structure, follows LTE PDSCH, as below:
 Two NB-IoT Control Channel Elements (NCCEs) per
  PRB pair: upper 6 SC allocated to one NCCE, lower
  6 SC to the other.
 No resource element groups (REGs).
 Legacy PDCCH avoided only in the in-band case.
 LTE CRS and NRS rate-matched.
 Frequency duplex of NB-PDCCH is only supported for
  $R = 1$.
 No support for frequency duplex with NB-PDSCH.
 Maximum aggregation level (L): 2, both NCCEs in the
  same subframe.
CCE definition:
 A CCE for the NB-PDCCH is composed of resources
  within a subframe.
  Within a PRB pair, two CCEs are defined.

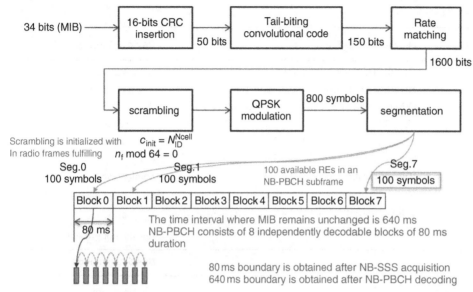

Figure 7.39 shows:

34 bits (MIB) → 16-bits CRC insertion → 50 bits → Tail-biting convolutional code → 150 bits → Rate matching → 1600 bits → scrambling → QPSK modulation → 800 symbols → segmentation

Scrambling is initialized with $c_{init} = N_{ID}^{Ncell}$
In radio frames fulfilling $n_f \bmod 64 = 0$

Seg.0 100 symbols   Seg.1 100 symbols   100 available REs in an NB-PBCH subframe   Seg.7 100 symbols

Block 0 | Block 1 | Block 2 | Block 3 | Block 4 | Block 5 | Block 6 | Block 7

80 ms

The time interval where MIB remains unchanged is 640 ms
NB-PBCH consists of 8 independently decodable blocks of 80 ms duration

80 ms boundary is obtained after NB-SSS acquisition
640 ms boundary is obtained after NB-PBCH decoding

Repeat 8 times within each 80 ms NB-PBCH block

Figure 7.39 NB-PBCH coding chain.

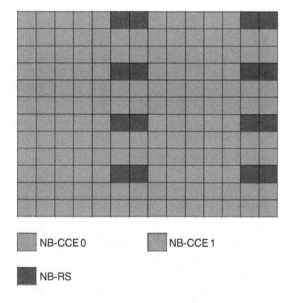

NB-CCE 0   NB-CCE 1

NB-RS

Figure 7.40 NB-PDCCH resource mapping.

Upper 6 subcarriers are allocated to one CCE and lower 6 subcarriers are allocated to the other CCE within a PRB pair.

Resource mapping details:
In in-band, the first few OFDM symbols are not used for NB-PDCCH.
CFI is signaled by NB-SIB1.
In standalone and guard band, all OFDM symbols are assumed to be available for the NB-PDCCH.

Transmission scheme:
Within a CCE, an SFBC pair is transmitted consecutively on two available REs within an OFDM symbol.

REs within the SFBC pair can be separated at most by one tone.
LTE CRS (if present) is rate-matched.
NB-RS is rate-matched.
NB-PDCCH is punctured on REs used for CSI-RS in the in-band case.
No signaling of LTE CSI-RS nor PRS configurations is provided to NB-IoT UEs; i.e., the UE has no idea of the existence of LTE CSI-RS and PRS.

The NB-PDCCH encoding chain re-uses most LTE functionalities, as demonstrated in Table 7.9. An NPDCCH block of subframes is scrambled in the same way as for PDCCH.

7.12.4 NB-IoT Downlink Control Information (NB-DCI)

The NB-DCI format is described in Tables 7.10 and 7.11 and may be summarized as follows:

DCI Format N0:
NB-PUSCH assignment
Will be described in the NB UL section.

Table 7.9 NB-PDCCH versus LTE PDCCH.

Function	Identical to PDCCH of LTE?	Note
CRC	Yes	16 bits
Channel coding		TBCC
Modulation	Yes	QPSK

Table 7.10 Summary of NB-DCI Format N0/N1.

	NB-DCI Format N0/N1			
DCI format	N0 (NPUSCH assignment)		N1 (NPDSCH assignment/NPDCCH order)	
Sub-type	UL (15 kHz)	UL (3.75 kHz)	DL (15 kHz)	NPDCCH order
N0/N1 Flag	1 bit			
Sub-type indication	– (15 kHz/3.75 kHz indicated in RAR)		1 bit (NPDCCH order or not)	
Scheduling delay	2 bits		3 bits	2 bits for #Rep of NPRACH;
Subcarrier allocation	6 bits (15 kHz: 5 bits are used) (3.75 kHz: 6 bits are used)		—	6 bits for subcarrier indication of NPRACH; Others set to "1, …, 1"
N_{PRB} (or N_{RU})	3 bits			
I_{TBS}	4 bits			
New data indicator	1 bit			
RV	1 bit (RV0 or RV2)		—	
# NB-PDCCH repetitions	2 bits			
Repetition number	3 bits		4 bits	
HARQ-ACK resource	—		2 + 2 for 15 kHz 3 + 1 for 3.75 kHz	
Sum above	23 bits			
	After 16 bits CRC attachment, total size is 39 bits			

Table 7.11 Summary of NB-DCI Format N2.

NB-DCI Format N2

Fields	Paging	Direct indication
Flag	1 bit (1 for paging; 0 for direct indication)	
N_{PRB} (or N_{RU})	3 bits	
I_{TBS}	4 bits	N/A
# NB-PDCCH repetitions	3 bits	
Repetition number	4 bits	
Direct indication information	N/A	8 bits
Reserved bits		6 bits
Sum above		15 bits
CRC		16 bits
Total size		31 bits

DCI Format N1:
NB-PDSCH assignment (not for paging)
or NB-PDCCH order.

DCI Format N2:
NB-PDSCH assignment for paging
or direct indication of system information update and
other fields.

7.12.5 NB-PDSCH

The NB-PDSCH re-uses most of the LTE functionalities, such as channel coding, and RM follows LTE functionalities for control channels, as highlighted in Table 7.12.

DL transmission:
Single AP (port 0) and two APs (ports 0 and 1) with transmit diversity (SFBC).
Modulation scheme: QPSK.
Structure (follows LTE PDSCH, as summarized below):
Narrowband RS (NRS) for demodulation.
Maximum Transport Block Size (TBS): 680.
Codeword across 1–6, 8, or 10 subframes.
Separate I_TBS (4 bits), N_SF (3 bits) indication in DCI.
Tail-biting Convolutional Code (TBCC).

Table 7.12 NB-PDSCH parameters.

Parameter	Value
Max TBS	680
CRC bit no.	24
Channel coding	TBCC
RV	Not supported
Modulation	QPSK only
HARQ	Single, adaptive, asynchronous

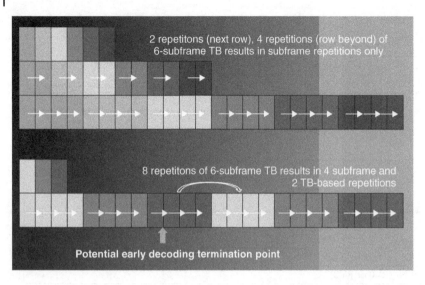

Figure 7.41 NB-PDSCH.

Table 7.13 PRB indexes for different BWs.

LTE system bandwidth	3MHz	5MHz	10MHz	15MHz	20MHz
PRB indices for NB-PSS/SSS transmission	2, 12	2, 7, 17, 22	4, 9, 14, 19, 30, 35, 40, 45	2, 7, 12, 17, 22, 27, 32, 42, 47, 52, 57, 62, 67, 72	4, 9, 14, 19, 24, 29, 34, 39, 44, 55, 60, 65, 70, 75, 80, 85, 90, 95

24-bit CRC.

2112 soft channel bits in Category NB1.

Scrambling sequence and scrambling initialization as in TS36.211.

Repetition:

Repetition cycle is the basic unit in which a full transport block (TB) is repeated.

Hybrid scheme of cyclic (subframe-based) and TB-based repetitions, as shown in Figure 7.41 [13].

Maximum subframe accumulation inside TB is four.

HARQ:

One HARQ process supported.

Adaptive and asynchronous HARQ.

No Redundancy Version (RV) scheme supported in NB-PDSCH HARQ.

7.12.6 Channel Raster

LTE's carrier center frequency must be an integer multiple of 100 kHz. NB-IoT's carrier center frequency depends on the deployment mode, as follows:

Standalone deployment:

Follow LTE's principle: Must be an integer multiple of 100 kHz.

In-band/Guard band: For both, two rules are to be satisfied

Rule 1: Maintain orthogonality of LTE system:

One complete PRB is used for in-band

At least 15k subcarrier spacing shall be maintained.

Rule 2: Within ± 7.5 kHz of an integer multiple of 100 kHz.

For the NB-IoT UL, frequency carrier determination for all deployment scenarios is as follows:

For initial access, the NB-IoT DL/UL frequency separation is configured by higher layers (SIBx) and is cell-specific.

After the success of the initial random access procedure, there can also be a UE-specific configuration for the NB-IoT DL/UL frequency separation.

For *in-band mode*, the PRB indexes for allocating the anchor NB-IoT carrier are listed in Table 7.13. The offset between the raster and the center of the anchor NB-IoT carrier ≤ 7.5 kHz. An example for a 3 MHz carrier is demonstrated in Figure 7.42.

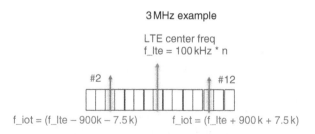

3 MHz example

LTE center freq
$f\_lte = 100\,kHz * n$

#2   #12

$f\_iot = (f\_lte - 900k - 7.5k)$ $f\_iot = (f\_lte + 900\,k + 7.5\,k)$

Figure 7.42 Example of NB-IoT PRB for a 3 MHz carrier.

Figure 7.43 NB-IoT carriers for a 5 MHz LTE carrier in guard band.

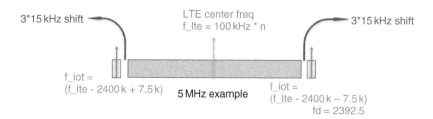

For *guard-band mode*, the center frequency of an NB-IoT carrier which transmits NB-PSS/SSS is *fd* kHz away from the LTE center. Each *fd* is such that the NB-IoT carrier is within the guard band, and the center frequency has a frequency separation from the LTE center which is at most 7.5 kHz offset from the 100 kHz channel raster. Note that the 15 kHz numerology is still assumed in the guard band. Figure 7.43 demonstrates NB-IoT carriers for a 5 MHz LTE carrier in guard band.

7.12.7 NPSS/NSSS: Principle

NPSS/NSSS is illustrated in Figure 7.44 and Table 7.14. It supports 504 PCIDs, indicated by NSSS. The unified resource mapping rules for the three operational modes are:

Common synchronization signal for all deployment modes.
The first three LTE OFDM symbols are not used by NPSS/NSSS.
NPSS/NSSS are punctured by LTE CRS (if a collision exists).
NPSS/NSSS occupy a fixed number (11 for normal CP) of OFDM symbols in each synchronization subframe.
NPSS: Base sequence length-11 Zadoff–Chu (ZC) sequence with root index 5, no cyclic shift; code cover: [1 1 1 1 − 1 − 1 1 1 1 − 1 1].

NSSS: Length-131 Zadoff–Chu sequence with time-domain cyclic shifts and binary scrambling sequence (Hadamard).

Within a subframe, NPSS/NSSS use the same antenna port. Similar to LTE, there is no transmit diversity scheme specified. NB-PDSCH/PDCCH on a given NB-IoT carrier are not mapped to the subframes containing NPSS/NSSS/PBCH on that carrier. Any transmission in such a subframe is postponed and there is no use of the first three OFDM symbols for all three modes.

7.12.8 Multiple NB-IoT Carrier Operation

NB-IoT supports multiple-carrier operation. Multiple NB-IoT carrier operation is demonstrated in Figure 7.45 and may be summarized as follows:

Multi-carrier operation (MCO):
NB-IoT supports operation with multiple NB-IoT carriers for in-band, guard-band, and standalone modes of operation.
One NB-IoT PRB with NB-PSS/NB-SSS/NB-PBCH is defined as the anchor carrier.
Flexible time-domain scheduling:
Support for dynamic indication of DL and UL scheduling delays for easier time-domain resource multiplexing.

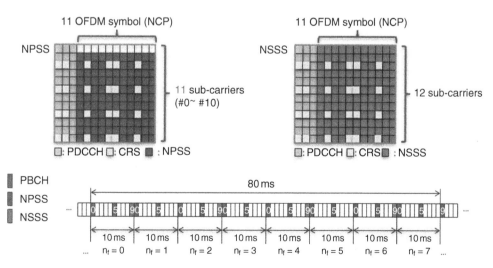

Figure 7.44 Unified resource mapping rules for NPSS/NSSS.

Table 7.14 NPSS/NSSS description.

	NPSS	NSSS
Subframe index in frame	#5	#9
Periodicity	Every frame	20 ms (when (n_f mod 2) = 0)
OFDM symbol in subframe	The last 11 OFDM symbols	
Used REs per OFDM symbol	11	12
	Subcarrier #0 − 10	
Sequence design	Base sequence : ZC 11 with root index 5	ZC 131 and four Hadamard code sequence:
	Code cover :	to indicate 504 PCell IDs
	[1 1 1 1 − 1 − 1 1 1 1 − 1 1]	four time-domain cyclic shifts:
		to indicate the 80 ms boundary
LTE CRS collision	Puncture	
NB-RS	Not present	

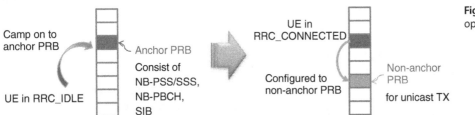

Figure 7.45 Multiple NB-IoT carrier operation.

UL compensation gaps (UCG):

Introduced in NB-IoT in consideration of long, continuous UL transmissions by HD-FDD UEs.

Co-existence considerations:

Introduction of the bit-map-based valid subframes concept (termed "NB-IoT DL/UL subframes").

For in-band/guard-band/standalone modes, the following are applied:

The UE in RRC Idle camps on the NB-IoT carrier on which the UE has received NB-PSS/SSS, NB-PBCH, and SIB transmissions.

The UE in RRC Connected can be configured, via UE-specific RRC signaling, to a non-anchor PRB for all unicast transmissions.

The UE is not expected to receive NB-PBCH, NB-PSS/SSS, or any transmissions other than unicast transmissions in the non-anchor PRB.

The allowed multi-PRB combinations are:

In-band + in-band, in-band + guard band, guard band + in-band, and guard band + guard band.
 Within the same LTE donor cell, from the same eNB, synchronized.
Standalone + standalone.
 Synchronized.

The non-allowed multi-PRB combinations are: Standalone + in-band, in-band + standalone, standalone + guard band, guard band + standalone.

7.13 NB-IoT UL Specifications

The NB-IoT frame structures for 15 kHz and 3.75 kHz subcarrier spacing are illustrated in Figure 7.46 and Figure 7.47, respectively, and are summarized in Table 7.15. Subcarrier spacing is indicated in RAR and all the UL transmissions follow it. For 3.75 kHz subcarrier spacing, besides the seven symbols located from the beginning of the 2 ms period, the remaining time (2304 Ts) is used as a guard period to minimize collision between NB-IoT symbols and LTE SRS.

In the uplink, both 3.75 and 15 kHz carrier spacing are supported, eNB decides which one to use. For the 3.75 kHz carrier spacing, number of sub-carriers expand to 48 within 180 kHz bandwidth. The symbol duration of 3.75 kHz is four times than 15 kHz so that duration of slot is 2 ms. Uplink physical channel resources are in units of timeslots in the time domain. Each timeslot lasts 2 ms for 3.75 kHz sub-carrier spacing and lasts 0.5 ms for 15 kHz sub-carrier spacing. Uplink physical channel resources are scheduled in units of resource units (RUs), whose duration varies depending on scenarios.

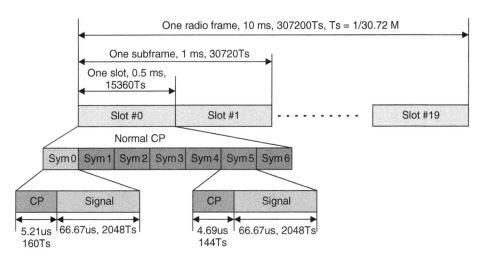

Figure 7.46 NB-IoT frame structure for 15 kHz.

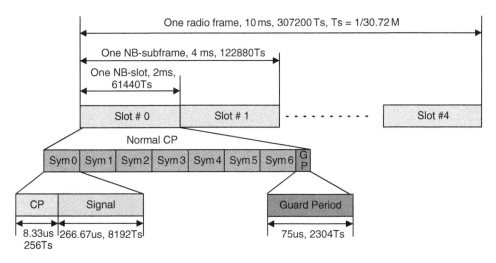

Figure 7.47 NB-IoT frame structure for 3.75 kHz.

Table 7.15 NB-IoT UL frame structure.

Subcarrier Spacing	Number of Subcarrier N_{SC}^{UL}	CP Length	Number of Symbols per Slot	Length of Slot
3.75 kHz	48	256 Ts	7	61440 Ts
15 kHz	12	160 Ts for 1st symbol	7	15360 Ts
		144 Ts for 2nd–7th symbols		

Why needs 3.75 kHz carrier spacing?

- Smaller bandwidth leads to larger power spectrum density with the same transmission power which improves the coverage.
- 6dB coverage gain by 3.75 kHz carrier spacing than 15 kHz carrier spacing.

7.13.1 NB-PUSCH Specifications

The NB-PUSCH can be summarized as follows:

UL transmission:
 SC-FDMA for 15 kHz multi-tone allocations (optional in Release 13).

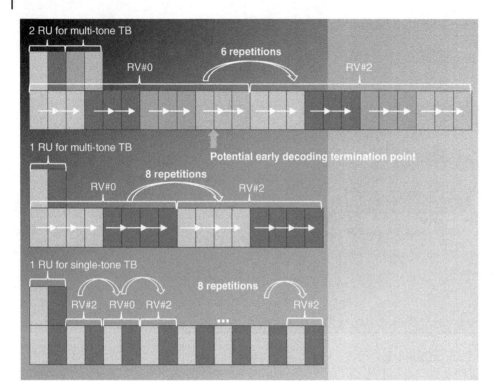

Figure 7.48 NB-PUSCH repetitions.

15 kHz multi-tone and single-tone allocations {12, 6, 3, 1}.

3.75 kHz single-tone allocation:

CP of 8.33 μs and symbol duration of 528 T_s = 275 μs

After seven OFDM symbols, 75 μs guard period up to 2 ms slot (also enables collision avoidance with legacy LTE SRS).

Modulation scheme, coding, scrambling:

Multi-tone: QPSK.

Single-tone: p/2 BPSK and p/4 QPSK for minimization of PAPR.

Phase-rotated BPSK/QPSK applied to data and DMRS, new UL DRS for single-tone and sub-PRB multi-tone.

NB-PUSCH Format 1 UL data:

Convolutional Turbo Coding (CTC), two RVs (RV#0, RV#2).

A single Transport Block (TB) can be scheduled over multiple Resource Units (RUs).

Enhanced scrambling and cyclic repetition patterns for NB-PUSCH Format 1.

NB-PUSCH Format 2 "NB-PUCCH" and UCI:

Single-tone only with p/2-BPSK.

Repetition coding.

Frequency and time locations with respect to baseline resource indicated via DL assignment (DCI Format N1).

UCI = 1 − bit A/N feedback only, no CSI feedback, no SR.

Repetition:

Redundancy version cyclic repetition is described in Figure 7.48 [13].

In each cycle of an RV, subframes are repeated Z times: Z = min{4, repetitions/2} for multi-tone.

Z = 1 for single-tone.

After cycling within the first RV, the next RV is treated analogously.

The set of possible repetitions for NPUSCH: {1, 2, 4, 8, 16, 32, 64, 128}.

HARQ:

One HARQ process supported.

Adaptive and asynchronous HARQ.

PHICH not supported for NPUSCH, UL HARQ-ACK indicated by NPDCCH carrying UL grant (DCI Format N0) and (non-)toggled NDI.

The resource units (RUs) and modulation are described in Table 7.16 and may be summarized as follows:

Resource units:

For data

Single tone, 8 ms for 15 kHz subcarrier spacing, 32 ms for 3.75 kHz subcarrier spacing (112 REs).

Three tones, 4 ms (168 REs).

Six tones, 2 ms (168 REs).

Twelve tones, 1 ms (168 REs).

Table 7.16 NB-IoT resource and modulation.

Physical Channel	Δf	N_{SC}^{RU}	N_{slots}^{UL}	N_{Symb}^{UL}	Modulation
NB-PUSCH Format 1 (Data)	3.75 kHz	1	16		Pi/2 BPSK
	15 kHz	1	16	7	Pi/4 QPSK
		3	8		QPSK
		6	4		
		12	2		
NB-PUSCH Format 2 (UCI)	3.75 kHz	1	4		Pi/2 BPSK
	15 kHz	1	4		

Table 7.17 MCS/TBS mapping table for multi-tone (four bits for I_{TBS}).

I_{mcs}	Modulation	I_{tbs}
0	QPSK	0
1	QPSK	1
2	QPSK	2
3	QPSK	3
4	QPSK	4
5	QPSK	5
6	QPSK	6
7	QPSK	7
8	QPSK	8
9	QPSK	9
10	QPSK	10
11	QPSK	11
12	QPSK	12

Table 7.18 MCS/TBS mapping table for single-tone (four bits for I_{TBS})

I_{MCS}	Modulation	I_{tbs}
0	Pi/2-BPSK	0
1	Pi/2-BPSK	2
2	Pi/4-QPSK	1
3	Pi/4-QPSK	3
4	Pi/4-QPSK	4
5	Pi/4-QPSK	5
6	Pi/4-QPSK	6
7	Pi/4-QPSK	7
8	Pi/4-QPSK	8
9	Pi/4-QPSK	9
10	Pi/4-QPSK	10

For UCI
   Single tone (28 REs).
     2 ms for 15 kHz
     8 ms for 3.75 kHz.
Modulation:
   Data
     Single tone: Pi/2 BPSK; Pi/4 QPSK.
     Multi-tone: QPSK.
   UCI
     Single tone: Pi/2 BPSK.

The MCS/TBS mapping tables for multi-tone and single-tone NB-IoT are provided in Table 7.17 and Table 7.18, respectively. Adaptive HARQ is supported for the uplink. The HARQ retransmissions in the uplink are asynchronous. PHICH is not supported for NB-PUSCH.

The existing timing advance procedure is re-used for NB-IoT. The timing advance update does not impact the phase settings. For example, phase is determined in the same way with or without timing advance. The adjustment of the uplink transmission timing is applied from the start of the first NPUSCH transmission which

starts at least 12 ms after the end of the corresponding timing advance command transmission.

7.13.2 NB-IoT Physical Random Access Channel (NPRACH)

The NPRACH is introduced in NB-IoT and the physical transmission scheme is as follows:

Single-tone NPRACH preamble with 3.75 kHz subcarrier spacing.
Coverage extension by NPRACH symbol group repetitions (NPRACH symbol = 266.7 µs).
NPRACH transmissions identified by their starting subcarrier location (i.e., FDM instead of CDM).
Multi-level NB-PRACH frequency hopping:
   Single-subcarrier hopping between the first and second and between the third and fourth symbol groups.
   Six-subcarrier hopping between the second and thirdsymbol groups.
   Pseudo-random hopping used every four symbol groups.
UE support for single-/multi-tone Msg3 indicated via NB-PRACH resource partitioning.
Up to three NB-PRACH resource configurations in a cell.

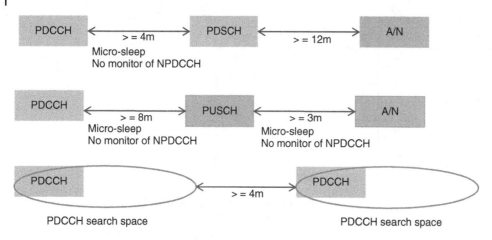

Figure 7.49 Timing relationship and micro-sleep.

Random access procedure and related NB-PDCCH CSS design follows Release 13 eMTC.

Based on single-tone frequency hopping symbol groups.

Using 3.75 kHz subcarrier spacing.

Two CP lengths for different cell sizes:

Format 0: 66.67 μs CP (10 km cell radius).

Format 1: 266.67 μs CP (40 km cell radius).

Frequency hopping between four groups of symbols:

A group is defined as one CP and five identical symbols:

Format 0: $5 * 266.67 + 66.67 = 1400$ μs

Format 1: $5 * 266.67 + 266.67 = 1600$ μs

Four identical symbol groups following two levels of frequency hopping:

First level of frequency hopping interval is 3.75 kHz (one subcarrier).

Second level of frequency hopping interval is 22.5 kHz (six subcarriers).

7.13.3 Timing Relationship and Micro-sleep

An NB-IoT UE that has received a grant from an NB-PDCCH is not required to monitor the NB-PDCCH for any further DL or UL grant during the period between the end of the NB-PDCCH that schedules the grant and the start of the corresponding NB-PDSCH or NB-PUSCH transmission. Figure 7.49 shows the timing relationship and micro-sleep.

7.13.4 UL Transmission Gap

NB-IoT introduced uplink transmission gaps for long uplink (i.e. NPUSCH/NPRACH) transmissions to enable the UE to switch to the DL and perform time/frequency synchronization, and this has been adopted for both NPUSCH and NPRACH, as shown in Figure 7.50. During uplink transmission gaps, the UE may switch to the DL and perform time/frequency synchronization. The uplink transmission gap is defined by a period of 256 ms and a gap length of 40 ms for NPUSCH. For NPRACH, $X = 64 \times$ (preamble duration), $Y = 40$ ms. This means that, for normal CP, $X = 5.6 \times 64 = 358.4$ ms, for extended CP, $X = 6.4 \times 64 = 409.6$ ms. Figure 7.50 demonstrates the UL transmission gap.

In addition, a DL gap is also introduced, as indicated in Figure 7.51, to prevent long transmissions from the UE and to block short transmissions from other UEs; this has been adopted for both NPDCCH and NPDSCH.

A comparison between NB-IoT and LTE physical layer channels is provided in Table 7.19 [13]. Finally, a summary of NB-IoT higher layer features is provided in Table 7.20 [13].

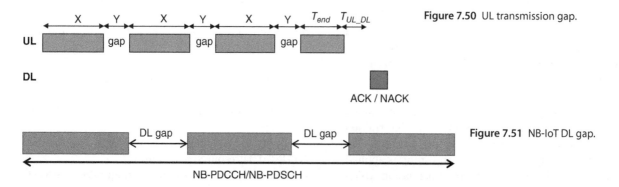

Figure 7.50 UL transmission gap.

Figure 7.51 NB-IoT DL gap.

Table 7.19 Comparison between NB-IoT and LTE physical layer channels.

	Channel	NB-IoT	Legacy LTE
DL	NPSS	• New ZC sequence for single PRB fit • All cells share one NPSS	• LTE PSS part of six center PRBs • Three different LTE PSSs
	NSSS	• New ZC sequence for single PRB fit • NSSS provides three least significant bits of SFN	• LTE NSS part of six center PRBs • N/A for LTE NSS
	NPBCH	• 640 ms TTI (decodable 80 ms blocks)	• 40 ms TTI
	NPDCCH	• Uses single PRB on multiple subframes in TD	• Uses a single subframe with multiple PRBs in FD
	NPDSCH	• QPSK, TBCC, only one redundancy version, single-layer transmission, max TBS 680 bits	• Up to 64QAM [256QAM], CTC, multiple RVs, multi-layer transmission, max TBS per layer > 70,000 bits
UL	NPRACH	• New preamble format based on 3.75 kHz single-tone frequency hopping • Coverage extension by NPRACH symbol group repetitions (NPRACH symbol = 266.7 μs)	• LTE PRACH occupies six PRBs using multi-tone format based on 1.25 kHz SCS
	NPUSCH Format 1 Format 2	• Min allocation: single-tone, max TBS 1000 bits, 15 kHz, 3.75 kHz (single-tone), Pi/2 BPSK, Pi/4 QPSK, single-layer transmission • Single-tone, repetition coding, 1-bit A/N	• Min allocation: 1 PRB, max TBS per layer > 70,000 bits, 15 kHz SCS only, QPSK up-to 64QAM [256QAM], multi-layer transmission • N/A
	[PUCCH]	• N/A	• LTE PUCCH dedicated channel for A/N, SR, CSI FB

Table 7.20 Summary of NB-IoT of higher layer features.

Feature	Description
Power savings	PSM, DRX in Connected mode, DRX in Idle mode
	Max PSM time: 310 hours (> 12 days)
	Release 13 extended C-DRX and I-DRX operation
	Connected mode (C-eDRX): Extended DRX cycles of 5.12 s and 10.24 s are supported
	Idle mode (I-eDRX): Extended DRX cycles up to ~ 44 min for Category M1, up to ~ 3 hrs for Category NB1
Limited mobility support	Intra-frequency and inter-frequency cell reselection.
	Support for dedicated priorities.
	MFBI (Multi-frequency Band Information) not supported:
	Network-controlled HO, inter-RAT cell reselection and mobility in Connected mode, speed-dependent scaling of mobility parameters and mobility
Limited positioning support	UE and network implementation to support positioning for NB-IoT.
	Subject to the final NB-IoT design, all methods and protocols (LPP, LPPa) can be used where applicable, e.g. the RAT-independent methods such as GNSS, WLAN, BT, MBS, and RAT-dependent methods such as E-CID.
	eNB-assisted methods such as E-CID using the TA Type 2 method can be supported if dedicated RACH for positioning purposes is supported.
	On the other hand, E-CID using the TA Type 1 method cannot be supported, as in Release 13 no UE support for RAT-measurements (such as UE Rx–Tx time difference measurement) was assumed.
Not supported	PWS, ETWS, CMAS, CSG, relaying, RS services, MBMS, CS services, CSFB, limited service state and emergency call.
	VoLTE, dual connectivity, IDC, RAN-assisted WLAN interworking, D2D/ProSe, MDT.

7.14 Release 13 Machine-type Communications Overview

3GPP Release 13 introduced a new low-complexity, low-power, wide-range category called Category M1 with the following characteristics:

Low complexity:

A bandwidth-reduced, low-complexity (BL) UE can operate in any LTE system bandwidth but with a limited channel bandwidth of 6 PRBs (1.4 MHz) in DL and UL.

Supports: FDD (FD-FDD and HD-FDD UEs) and TDD:

Optimized support for half-duplex (HD)-FDD operation using a single local oscillator → Type B HD-FDD operation is assumed.

A guard-time of 1 subframe (1 ms) is provisioned for DL-to-UL and UL-to-DL switching.

Reduced maximum DL and UL transport block (TB) sizes of 1000 bits:

No simultaneous reception/transmission of multiple TBs.

Single Rx antenna reception is assumed to limit the complexity.

Low power:

eDRX or PSM (see power reduction techniques).

Wide range:

Support for coverage enhancement feature targeting 155 dB Maximum Coupling Loss (MCL).

For the DL and UL, the LTE system bandwidth is divided into a set of non-overlapping narrow bands (NBs). A narrow band comprises six contiguous PRBs and therefore the total number of NBs for the DL and UL is the number of available resource blocks divided by 6. The remaining RBs are divided evenly at both ends of the system bandwidth. The extra PRB for odd system bandwidths (e.g. 3, 5, and 15 MHz) is located at the center of the system bandwidth, as indicated in Figure 7.52. The PSS/SSS/PBCH are mapped to the central 72 subcarriers as in LTE – the location of PSS/SSS/PBCH is independent of the narrow bands defined. Due to reduced bandwidth support, UEs need to perform re-tuning as they switch narrow bands within the larger LTE system bandwidth. For eMTC, when referring to 1.4 MHz RB(s), the specification also uses the term "narrowband (NB)"; however, this term should not be confused with the 180 kHz NB(s) term used for NB-IoT.

3GPP eMTC introduced coverage enhancement due to low bandwidth of 1.4 MHz with 6 PRBs, with a target of 155.7 dB maximum coupling loss for both UL and DL. Two modes are configurable via RRC:

Mode A: For no repetitions and a small number of repetitions.

Mode B: For large numbers of repetitions where a UE supporting Mode B also supports Mode A.

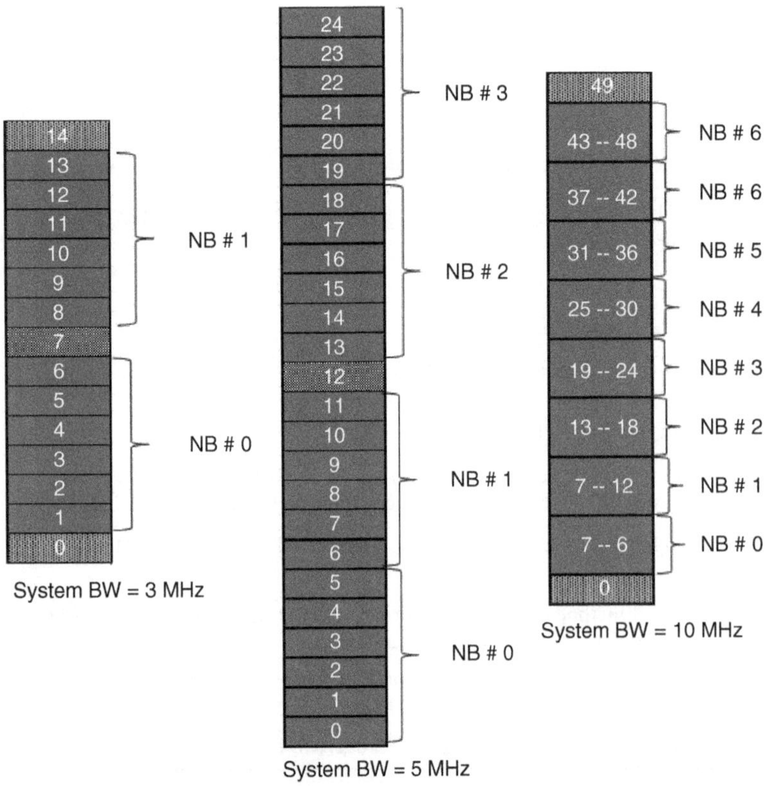

Figure 7.52 eMTC narrow band definition.

Table 7.21 Link budgets for eMTC channels and LTE Category 1+.

Technology	LTE Cat 1+	Cat M1	LTE Cat 1+	Cat M1	LTE CAT 1+	Cat M1
Physical channel name	PDSCH	PDSCH 6 PRBs	PUSCH	PUSCH 1 PRB	PDCCH (1A)	MPDCCH
Data rate/payload	20 kbps	4.8 kbps	20 kbps	1 kbps	36 bits	34 bits
Repetitions		192		16		64
Transmitter						
Max Tx power (dBm)	46	46	23	23	46	46
(1) Actual Tx power (dBm)	32	36.8	23	23	42.8	36.8
Receiver						
(2) Thermal noise density (dBm/Hz)	−174	−174	−174	−174	−174	−174
(3) Receiver noise figure (dB)	9	9	5	5	9	9
(4) Interference margin (dB)	0	0	0	0	0	0
(5) Occupied channel bandwidth (Hz)	360000	1080000	360000	180000	4320000	1080000
(6) Effective noise power $= (2) + (3) + (4) + 10 \log((5))$ (dBm)	−109.4	−104.7	−113.4	−116.4	−98.6	−104.7
(7) Required SINR (dB)	−4	−14.2	−4.3	−16.3	−4.7	−14.2
(8) Receiver sensitivity $= (6) + (7)$ **(dBm)**	−113.4	−118.9	−117.7	−132.7	−103.34	−118.9
(9) **MCL** = (1) − (8) **(dB)**	145.4	155.7	140.7	155.7	146.1	155.7

7.15 Link Budget Analysis

The link budgets for eMTC channels PDSCH, PUSCH, and MPDCCH are shown in Table 7.21. In addition, the corresponding channels for LTE Category 1+ are provided for comparative analysis. In the downlink, the best strategy is to map the TBS over the largest bandwidth to maximize the data rate. In the uplink, the best strategy is to concentrate the power over as narrow a band allocation as possible (1 PRB) while for larger bandwidths, such as 3 PRBs, very low SNR is required to meet the target MCL, which is challenging. In PUSCH, 16 repetitions are enough to achieve 8 dB SNR improvement, as shown in Table 7.21. For MPDCCH, 64 repetitions are enough to achieve the required MCL.

The link budgets for NB-IoT with different subcarrier spacing are shown in Table 7.22. The maximum coupling loss (MCL) is 164 dB, which has been achieved with the assumptions given in Table 7.23: assuming ∼ 55000 devices per cell for urban clutter with deep in-building penetration and rural clutter with long-range coverage (10 − 15 km). The high-level MCL estimation is provided in Table 7.22 for 15 kHz and 3.75 kHz subcarrier spacing. The NB-IoT deployment parameters are provided in Table 7.23. Coverage analysis for NB-IoT PUSCH (NPUSCH) is provided in Table 7.24. The SINR of −11.8 dB achieves an MCL of 164 dB.

Comparative coverage analysis is provided in [19] for NB-IoT, eMTC, and LTE. The simulation parameters are summarized in Table 7.25. The coupling loss distribution for the outdoor, road, and indoor users with varying penetration loss is demonstrated in Figure 7.53 [19, 20]. The vertical lines indicate the minimum supported MCL for LTE, LTE-M, and NB-IoT. LTE provides coverage for almost 99% of the outdoor and road users, while only

Table 7.22 Maximum coupling losses for NB-IoT.

Subcarrier spacing (kHz)	15	3.75
(1) Transmit power (dBm)	23.0	23.0
(2) Thermal noise density (dBm/Hz)	−174	−174
(3) Receiver noise figure (dB)	3	3
(4) Occupied channel bandwidth (Hz)	15000	3750
(5) Effective noise power $= (2) + (3) + 10 * \log((4))$ (dBm)	−129.2	−135.3
(6) Required SINR (dB)	−11.8	−5.7
(7) Receiver sensitivity $= (5) + (6)$ (dBm)	−141.0	−141.0
(8) Maximum coupling loss $= (1) − (7)$ (dB)	164.0	164.0

Table 7.23 NB-IoT deployment parameters.

	NB-IoT
Deployment	In-band and guard-band LTE, standalone
Coverage (MCL)	164 dB
Downlink	OFDMA, 15 kHz tone spacing, TBCC, 1 Rx
Uplink	Single tone: 15 kHz and 3.75 kHz spacing, SC-FDMA: 15 kHz tone spacing, turbo code
Bandwidth	180 kHz
Highest modulation	QPSK
Link peak rate (DL/UL)	DL: ~ 30 kbps UL: ~ 60 kbps
Duplexing	HD-FDD
Duty cycle	Up to 100%, no channel access restrictions
MTU	Maximum PDCP SDU size 1600 B
Power saving	PSM, extended Idle mode DRX with up to 3 h cycle, Connected mode DRX with up to 10.24 s cycle
UE power class	23 dBm or 20 dBm

Table 7.24 Coverage analysis for NB-IoT PUSCH.

	Extreme	Robust	Normal
Subcarrier spacing (kHz)	15	15	15
Data rate (kbps)	0.332	3.03	21.25
Burst duration (ms)	2048	224	32
Number of subcarriers in a burst	1	3	12
Modulation	BPSK	QPSK	QPSK
Transmitter			
(1) Tx power (dBm)	23	23	23
Receiver			
(2) Thermal noise density (dBm/Hz)	−174	−174	−174
(3) Receiver noise figure (dB)	3	3	3
(4) Interference margin (dB)	0	0	0
(5) Occupied channel bandwidth (Hz)	15,000	45,000	180,000
(6) Effective noise power = (2) + (3) + (4) + 10 log ((5)) (dBm)	−129.2	−124.5	−118.4
(7) Required SINR (dB)	−11.8	−6.6	−3.4
(8) Receiver sensitivity = (6) + (7) (dBm)	−141	−131.1	−121.8
(9) Rx processing gain	0	0	0
(10) MCL = (1) (8) + (9) (dB)	164	154.1	144.8

supporting one third of the deep indoor users, experiencing an additional 30 dB loss. The LTE-M (eMTC) provides coverage for almost 99.9% of the light indoor users, while only about 80% of the deep indoor users can be served. Therefore, NB-IoT is an important update to 3GPP's MTC portfolio, but, as the results in Figure 7.53 indicate, almost 5% of the deep indoor users cannot obtain the target data rates.

It is worth comparing the performance of the 3GPP-based technologies with unlicensed LPWAN technologies such as LoRa and Sigfox. Table 7.26 summarizes LoRa and Sigfox parameters [21, 22] in the ISM 868 MHz band. The 868 MHz EU ISM band enables two basic mechanisms for sharing the spectrum: duty cycle restrictions and Listen Before Talk (LBT). Both Sigfox and LoRaWAN use duty cycle restrictions for access in

Table 7.25 Simulation parameters.

Parameter	Variable
Scenario	71 sectors operating in LTE band 20 (800 MHz)
Shadow fading	$\sigma = 8.7$dB, sector correlation = 1, site correlation = 0.5
Power model	According to [7], 40% power amplifier efficiency
Antenna configuration	BS: 1 Tx, 2 Rx; UE: 1 Tx, 1 Rx
BS receive diversity	6 dB gain (using maximum ratio combining)
Indoor loss	10, 20, 30 dB in addition to the outdoor path loss
Carrier load	60%
Available resources	1 PRB per device (180 kHz)
System bandwidth	6 PRBs for LTE-M, 1 PRB for NB-IoT
Application 1	4 UL payload, 4 DL payload
Application 2	1 UL payload, 1 DL acknowledgment (29 bytes)
Payload	128 or 256 bytes
Application sessions	1, 2, 4, or 8 per day
RACH opportunities	300 per second per preamble

Figure 7.53 Coverage performance for LTE, LTE-M, and NB-IoT.

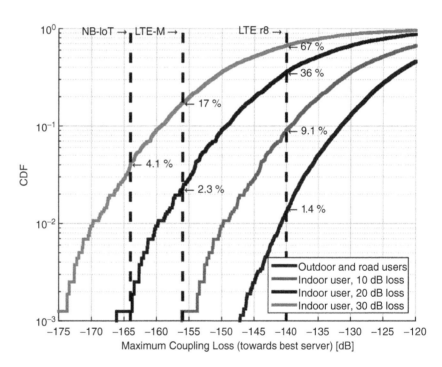

the EU ISM band. The duty cycle restriction varies within the ISM band from 0.1% to 10%, where the latter is only available for the 250 kHz band in 869.4 − 869.65 MHz, as illustrated in Figure 7.54. Certain parts of the ISM band are pre-allocated to specific use cases such as alarms and voice systems, which are limited to a maximum radiated power of 10 dBm. The remaining parts of the ISM band allow a maximum radiated power of 14 dBm, while the aforementioned 250 kHz band may use 27 dBm [23].

Sigfox

The Sigfox network [24] relies on Ultra-Narrow Band (UNB) modulation using DBPSK at 100 bps. Sigfox is based on a simple access scheme where the device initiates a transmission by sending three uplink packages containing the same data, in sequence on three random carrier frequencies. The base station will successfully receive the package even if two of the transmissions are lost due to, for

Table 7.26 LoRa and Sigfox parameters.

	LoRa	Sigfox	
	UL and DL	UL	DL
Spectrum (MHz)	863 – 870	868.1 – 868.3	869.425 – 869.625
Tx power (dBm)	14 – 27	14	27
Modulation	Chirp spread spectrum	DBPSK	GFSK
Bandwidth (kHz)	125	0.1	0.6
Maximum payload (bytes)	51	12	8
Scheduling	Uplink initiated	Uplink initiated	

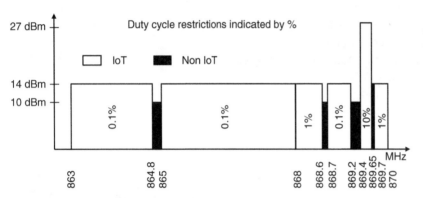

Figure 7.54 868 MHz EU ISM band power and duty cycle restrictions.

example, collision with other devices or interference from other systems using the same frequency. Sigfox uses 868.1 – 868.3 MHz for uplink with a maximum radiated power of 14 dBm under the EU ISM band regulations [23].

The duty cycle of this frequency band is a maximum of 1%, allowing the Sigfox device to transmit only 36 seconds every hour. With a time on air of 2 s per transmission, that is 6 s in total for a single Sigfox package; this allows a maximum of six messages per hour with a payload of 4, 8, or 12 bytes [25].

Sigfox was initially designed without a downlink channel, but it has been added in the recent Sigfox standards [24]. The downlink channel uses Gaussian Frequency-Shift Keying (GFSK) with 600 bps in the frequency band 869.425– 869.625 MHz. This ISM band allows a maximum radiated power of 27 dBm and a duty cycle of 10% or 360 seconds per hour. The downlink payload is always 8 bytes and, depending on the Sigfox subscription, up to four downlink messages per day are allowed.

LoRa

The LoRa LPWA solution consists of two major components: the LoRa physical layer specifications and the LoRaWAN, which is the network protocol. The LoRa physical layer is based on chirp spread spectrum with GFSK modulation and high bandwidth-time product (BT > 1) to protect against in-band and out-band interference. LoRa provides six different spreading

factors from 6 to 12. This enables multiplexing of different devices without causing performance degradation and reducing time on air. LoRa can operate in the entire EU ISM band but has three mandatory channels at 868.1, 868.3, and 868.5 MHz. The maximum LoRaWAN payload depends on the spreading factor, and for the best protected channels it is limited to 51 bytes.

Both the Sigfox and LoRaWAN LPWA networks are based on a typical star protocol where each device communicates with a base station that relays the information to and from a central server via an IP-based protocol. Each end-user device can transmit at any time, and the LoRaWAN devices use any data rate unless instructed otherwise by the base station. Finally, each end device needs to track the time spent for each transmission to observe the local spectrum regulations.

The MCL for LoRa and Sigfox is calculated as shown in Table 7.27. Below we will compare the simulated MCL with the claimed MCL of both technologies and the impact on the interference in this band on the coverage and capacity.

The simulation in [21] is based on the path loss estimates for the urban areas of northern Denmark, combined with the MCL and time on air of Sigfox and LoRaWAN. The path loss calculations are made for outdoor positions, but additional losses of 10, 20, and 30 dB are added to account for outdoor-to-indoor penetration losses in buildings [21]. Finally, the modeled

Table 7.27 Link budget for Sigfox and LoRaWAN.

Technology	LoRa		Sigfox	
UL/DL	UL	DL	UL	DL
Transmitter	14	14	14	27
(1) Tx power (dbm)				
Receiver				
(2) Thermal noise density (dBm/Hz)	−174	−174	−174	−174
(3) Receiver noise figure (dB)	3	5	3	5
(4) Occupied channel bandwidth (kHz)	125	125	0.1	0.6
(5) Effective noise power $= (2) + (3) + (4) + 10 \log((4))$ (dBm)	−120	−118	−118	−141
(6) Required SINR (dB)	−20	−20	7	7
(7) Receiver sensitivity $= (5) + (6)$ (dBm)	−140	−138	−144	−134
(8) MCL = (l) − (7) (dB)	154	152	158	161

interference is included to determine the impact on coverage and capacity.

Coverage analysis [21]

The estimated Sigfox and LoRaWAN uplink coverage in the urban areas of northern Denmark is illustrated in Figure 7.55. Both technologies provide more than 99% coverage for up to 20 dB indoor penetration loss when interference is not included, as illustrated by the solid lines. For deep indoor coverage (penetration loss of 30 dB), Sigfox provides 96% coverage and LoRaWAN 90% coverage under ideal conditions without interference. Considering the interference from external sources, as modeled in [21], reduces the coverage area, as shown by the dashed lines in Figure 7.55. Sigfox only covers 90% of the outdoor area while LoRaWAN provides 95% coverage. According to the modeling, Sigfox is more sensitive to interference than LoRaWAN, as LoRaWAN uses spread-spectrum techniques. Even though Sigfox transmits three times on random frequencies, each Sigfox message is likely to collide with interference units during the transmission [21]. For indoor coverage (20 dB penetration loss), the impact from interference is even worse, as the link budget is reduced and LoRaWAN has 78% coverage and Sigfox less than 50% coverage.

Figure 7.56 shows the coverage of Sigfox and LoRaWAN in downlink. The observations are like those made for the uplink in Figure 7.55, but

Sigfox performs slightly better in the downlink due to the higher transmit power of the base station (27 dBm versus 14 dBm according to Table 7.27).

Capacity analysis [21]

The uplink failure rate is shown in Figure 7.57 for both LoRaWAN and Sigfox. Without interference (solid lines) both Sigfox and LoRaWAN are able to provide one 10-byte message every hour per device for both indoor (20 dB indoor penetration loss) and outdoor locations with 95 percentile uplink failure (combination of collision and lack of coverage) less than 1%.

Considering the external interference (dashed lines), the LoRaWAN outdoor devices have a 95-percentile uplink failure of about 7% while Sigfox has an uplink failure rate of 17%. For indoor deployment, LoRaWAN has 95-percentile uplink failure rate of 50% while Sigfox has an uplink failure rate of more than 60%.

Figure 7.58 shows the capacity of Sigfox and LoRaWAN in the downlink. Generally, the downlink failure rate is like the uplink in Figure 7.57, but again, Sigfox performs slightly better in the downlink due to the higher transmit power.

The analysis in this section indicates a major gap in coverage and reliability between 3GPP-based technologies and unlicensed LPWA technologies, mainly due to the interference in the unlicensed bands. Therefore, coverage and capacity cannot be guaranteed in ISM bands due to external interference, and therefore such drawbacks need to be considered while adopting use cases in these bands. However, it is a viable option to use the ISM bands for specific use cases such as inside a campus or within an enterprise HQ where the interference can be controlled. In addition, low-priority use cases can be considered in such a band, or when the customer is asking for a fully controlled IoT network. Since the 3GPP-based technologies are based on the evolution of the cellular network, the business model will be different and it will be based on an OPEX model. On the other hand, the LPWA networks in unlicensed bands such as the ISM band can be based on a CAPEX model and there is no need to rely on service provider networks. More importantly, LPWAN technologies have rich, customized use cases which will take some time before being realized for 3GPP-based technologies. Also, the simplified topology of the LPWA networks allows many small organizations to jump into this boat and to offer quick, user-friendly use cases, which is the main differentiator for IoT use cases.

Although the NB-IoT standardization was concluded in June 2016, the availability of its use cases is still limited. In other words, the IoT use cases have been overestimated and many use cases have failed due to negative business cases. The next few years until 2020 will allow us to better evaluate the early forecast of the IoT market size.

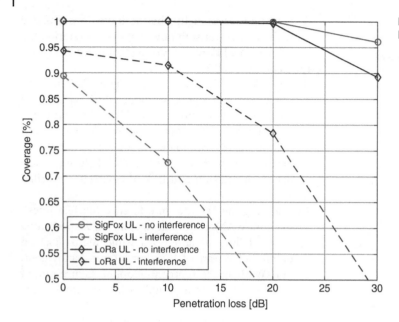

Figure 7.55 Uplink coverage relative to penetration loss with and without interference.

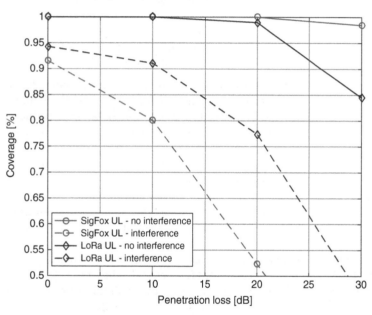

Figure 7.56 Downlink coverage relative to penetration loss with and without interference.

7.16 NB-IoT Network Topology

Figure 7.59 illustrates the end-to-end network topology of NB-IoT and relevant protocols. The NB-IoT architecture can be summarized as follows:

NB-IoT UE communicates with the eNodeB over the air interface.

The eNB performs functions such as air interface access processing and cell management. The eNB communicates with the IoT core through the S1-lite interface and sends Non-Access Stratum (NAS) data to the IoT core for processing. The IoT core could be an upgraded version of the LTE EPC or a new, overlaid virtual EPC (vEPC).

The IoT core interacts with the NAS of the UE and forwards IoT data to the IoT platform for processing.

The IoT platform converges IoT data from access networks and forwards different data to corresponding application servers based on protocol and use case requirements.

The application server receives and processes IoT data. The use case could use different protocols. In addition, the application server or Application Enablement Platform (AEP) could be part of the horizontal IoT platform.

The IoT platform will be connected to the NB-IoT core and to the unified PS core (2G/3G/LTE) and therefore multiple use cases over 2G/3G/LTE/NB-IoT can be connected to the same horizontal IoT platform.

Figure 7.57 CDF of the total uplink failure from random access collision and coverage limitations. The penetration loss is 20 dB for indoor devices.

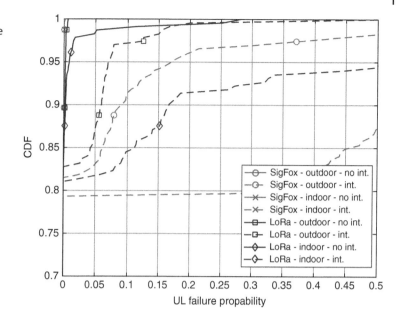

Figure 7.58 CDF of the total downlink failure from blocking and coverage limitations. The penetration loss is 20 dB for indoor devices.

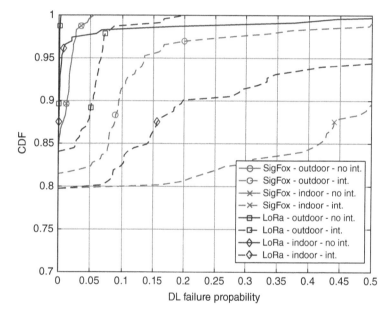

The NB-IoT core network has been optimized as follows [26]:

Dedicated core network: Overlaid core to serve and fulfill requirements of IoT devices. A dedicated core network can be deployed to serve IoT devices in an optimized manner. Typically, the dedicated core network contains the P-GW and HSS but may also contain the MME and control elements. IoT-specific features are supported in the dedicated core network. Cloud deployment is beneficial for IoT, as it brings scalable capacity, elasticity, and efficient use of resources.

Subscription optimization: Optimizations for subscriber data storage and retrieval. For a large number of IoT devices sharing the same subscriber attributes, the subscriber data storage can be optimized in the HSS as well as the signaling to retrieve subscriber data.

Signaling reduction/overload control: Prevention of overload situations caused by signaling storms. Signaling can be reduced by guiding IoT devices to perform periodic location updates less frequently and by optimizing paging. Reducing signaling destined for overloaded elements can help avoid overload situations.

Small data transmission: Transmission of small data from/to IoT devices. Network resource utilization can be optimized for IoT devices sending small data transmissions infrequently using methods such as

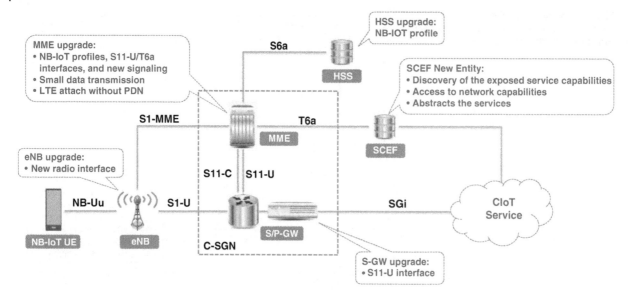

Figure 7.59 Network architecture of NB-IoT.

bearerless solutions with non-IP data over NAS signaling, as explained in the next section.

Resource optimization: Optimizing resource usage both in IoT devices and in the network. The battery lifetime of IoT devices can be extended by, for example, allowing the IoT devices to switch off functionality and go to sleep.

Monitoring: Reporting the connectivity status of IoT devices to IoT applications. Monitoring can allow IoT applications to obtain information on the connectivity status of IoT devices.

7.16.1 Service Capability Exposure Function (SCEF)

The Service Capability Exposure Function (SCEF) is a new node designed especially for machine-type data. It is used for the delivery of non-IP data over the control plane and provides an abstract interface for network services (authentication and authorization, discovery and access network capabilities) [15, 18]. The SCEF is the key entity within the 3GPP architecture for service capability exposure that provides a means to expose the services and capabilities provided by 3GPP network interfaces securely to external entities to support IoT applications. To help meet the requirement of low power, the power-hungry protocol to establish IP data bearers has been replaced by extending the NAS protocol to allow small amounts of data to be transferred over the control plane. The IP stack is not necessary, so this type of transfer has been named Non-IP Data Delivery (NIDD). The path for NIDD between the UE and the AS is defined by 3GPP to traverse the MME (or SGSN, or a new node that has been named C-SGN) and the SCEF.

Initially, the exposing of capabilities was to facilitate operators being able to charge for use of those capabilities. From the perspective of operators, SCEF was regarded as a nice-to-have opportunity. Now, operators need the SCEF so that LPWA technologies like NB-IoT can be deployed. Therefore, SCEF has become a requirement for operators to support NIDD for billions of devices. Figure 7.60 illustrates a typical IoT core network, which could be provided either by an MVNO or by the MNO itself. The SCEF may reside at the edge of the IoT domain, as shown in Figure 7.60, or the SCEF may lie completely within the IoT domain interfacing with an external API management platform at the edge [27].

In order to send data to an application, two optimizations for the cellular IoT (CIoT) in the Evolved Packet System (EPS) were defined: user-plane CIoT EPS optimization and control-plane CIoT EPS optimization, as demonstrated in Figure 7.61. The control-plane CIoT EPS optimization is indicated in dark gray, while the user-plane CIoT EPS optimization is indicated in mid gray. Both optimizations may be used but are not limited to NB-IoT devices. In control-plane CIoT EPS optimization, UL data are transferred from the eNB (CIoT RAN) to the MME. From there, they may either be transferred via the Serving Gateway (S-GW) to the Packet Data Network Gateway (P-GW), or to the SCEF, which is only possible for non-IP data packets. From these nodes, they are finally forwarded to the application server (CIoT services). DL data are transmitted over the same paths in the reverse direction. In this solution, there is no data radio bearer setup, data packets are sent on the signaling radio bearer instead. Consequently, this solution is most

Figure 7.60 Typical IoT core network including SCEF.

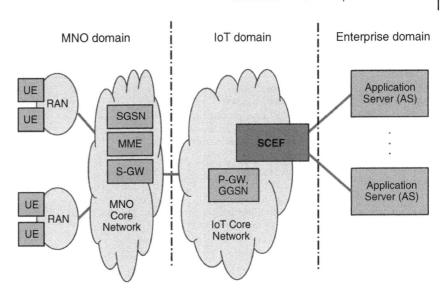

Figure 7.61 Network for NB-IoT data transmission and reception.

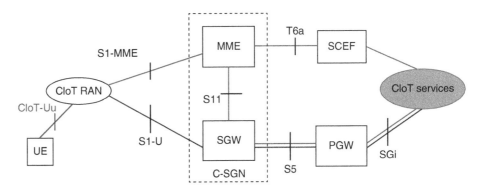

appropriate for the transmission of infrequent and small data packets [15].

With user-plane CIoT EPS optimization, data are transferred in the same way as the conventional data traffic, i.e. over radio bearers via the S-GW and the P-GW to the application server. Thus, it creates some overhead on building the connection; however, it facilitates a sequence of data packets to be sent. This path supports both IP and non-IP data delivery [15].

In Release 13, 3GPP defined the SCEF to be the interface for small data transfers and control messaging between enterprises and operators' core networks. The SCEF provides APIs to the enterprises for small data transfers and control messages, and uses 3GPP-defined interfaces with the network elements in the operators' core networks in the performance of its functions. The major features of the SCEF are as follows [27]:

APIs and AAA
External ID
NIDD
Device trigger request
Exposing capabilities for new revenue.

The SCEF could expose the APIs to an API gateway of the operator, and from there, it can be combined with other APIs and exposed to enterprises or other external customers.

7.17 Architecture Enhancement for CIoT

The following assumptions are considered for CIoT, as provided in [28]:

The user plane data rate requirements of CIoT on the core network are very low compared to those for an LTE core network. Traffic models for CIoT are summarized in Table 7.28.

Control plane efficiency is important for the CIoT system. Even a low data rate due to a large number of devices will result in a high number of RRC connections being established and released.

Applications expected to be supported on CIoT are generally expected to be delay-tolerant. For certain applications requiring strict delay profiles, a maximum delay of 10 seconds is considered.

Table 7.28 Traffic models for cellular IoT.

Category	Application Example	UL Data Size	DL Data Size	Frequency
Mobile Autonomous Reporting (MAR) exception reports	Smoke alarm detectors, power failure notifications from smart meters, tamper notifications etc.	20 bytes	0 ACK payload size is assumed to be 0 bytes	Every few months; Every year
Mobile Autonomous Reporting (MAR) periodic reports	Smart utility (gas/water/electric) metering reports, smart agriculture, smart environment etc.	20 bytes with a cut off of 200 bytes, i.e. payloads higher than 200 bytes are assumed to be 200 bytes.	50% of UL data size ACK payload size is assumed to be 0 bytes	1 day (40%), 2 hours (40%), 1 hour (15%), and 30 minutes (5%)
Network command	Switch on/off, device trigger to send uplink report, request for meter reading	0–20 bytes 50% of cases require UL response.	20 bytes	1 day (40%), 2 hours (40%), 1 hour (15%), and 30 minutes (5%)
Software update/reconfiguration model	Software patches/updates	200 bytes with a cut off of 2000 bytes, i.e. payloads higher than 2000 bytes are assumed to be 2000 bytes.	200 bytes with a cut off of 2000 bytes, i.e. payloads higher than 2000 bytes are assumed to be 2000 bytes.	180 days

Support for inter-RAT mobility and intra-RAT network-controlled handover is not required.

Support for CS services is not required.

Traffic models for cellular IoT are defined and summarized in Table 7.28.

7.17.1 Key Issues to be Addressed in IoT Optimized Core Network

The following issues need to be addressed in the CIoT core network [29].

7.17.1.1 Key Issue 1: Architecture Reference Model for Cellular IoT

This key issue aims to describe an architecture reference model that is appropriate to fulfill the requirements of cellular IoT for ultra-low-complexity and low-throughput "Internet of Things" devices that may also be constrained, for example with regard to processing power, memory, battery capacity, etc. We will address a high-level architecture reference model, network elements, and reference point for CIoT.

7.17.1.2 Key Issue 2: Efficient Support for Infrequent Small Data Transmissions for Cellular IoT

This key issue aims to provide a solution to support highly efficient handling of infrequent, small data transmissions for ultra-low-complexity, power-constrained, and low-data-rate "Internet of Things" devices, called

CIoT devices. It is expected that the number of such devices will increase exponentially but the data size per device will remain small. Infrequent, small data traffic characteristics for MTC applications (as described in Annex E of TR 45.820 [28]) may lead to inefficient use of resources in the 3GPP system. It should be noted that the existing E-UTRAN/EPC (i.e. S1-based) architecture might not be best optimized for infrequent, small data transfer, and therefore would require optimization to support data handling for CIoT devices.

The following architecture requirements will be supported:

Minimize system signaling load, especially over radio interface.

Appropriate security for the EPS system.

Improve battery life.

Support delivery of IP data.

Support delivery of non-IP data.

Support for SMS (as a deployment option).

7.17.1.3 Key Issue 3: Efficient Support for Tracking Devices Using Small Data Transmissions for Cellular IoT

This key issue aims to provide a solution to support highly efficient handling of tracking devices using small data transmissions for ultra-low-complexity, power-constrained, and low-data-rate "Internet of Things" devices, called CIoT devices. It is expected that the number of such devices will increase exponentially but the data size per device will remain small. Small data traffic characteristics for tracking MTC applications may lead to inefficient use of resources in the 3GPP system. It should be noted that the existing E-UTRAN/EPC (i.e. S1-based) architecture might not be best optimized for tracking devices using small data transfer, and therefore would require optimization to support data handling for CIoT devices. Support for tracking devices is not expected to fit within the TR 45.820 [28] traffic models for the inter-arrival time periodicity. Periodic uplink reporting is expected to be common for tracking cellular IoT applications in different sectors, e.g. transportation (automotive industry, fleet management, railroad), health and medical (patient monitoring, sports), and asset management (tracking/location management).

7.17.1.4   Key Issue 4: Support for Efficient Paging Area Management for Cellular IoT

3GPP SA2 has been discussing the paging optimization issue due to the scarcity of radio interface resources and core network interface resources in regard to no/low-mobility devices. Accordingly, the idea of making paging available only to the last-used eNB/cell rather than whole cells in tracking areas of the UE in order to save paging resources both for the radio interface and the core network interface has been discussed.

The "scarcity of paging resource" issue is more serious for cellular IoT due to the following reasons:

The number of CIoT devices is much higher than the number of legacy cellular devices in a given area.

A CIoT RAT operating in narrowband spectrum may not supply enough paging resources and the number of UE identities (S-TMSI, IMSI) included in a single paging message would be very limited due to the small message size compared to the legacy access system (e.g. E-UTRAN).

Coverage enhancement is a mandatory requirement, so each paging message may occupy a long period of time (e.g. repetition of the same paging message).

The majority of cellular IoT devices would have no/low-mobility characteristics. Therefore, limiting the paging area to cell(s) rather than paging all eNB(s)/cell(s) in tracking area(s) is more appropriate and beneficial in terms of saving paging resources. However, the last-known cell information, given to the MME when the UE is entering Idle mode, may not be accurate, even for a stationary CIoT device. The serving cell can change while the UE is not moving due to various reasons, such as a change in radio load conditions or a change in neighboring conditions (i.e. blocked by a new building). In addition to no-mobility UEs, limiting paging areas is also beneficial and necessary for low-mobility UEs, since a low-mobility UE will not move across a wide area in a given time. Therefore, paging area management

in order to limit the paging area is required for paging optimization in cellular IoT devices.

The following paging area management requirements need to be met:

The system needs to support an efficient paging area management procedure for no/low-mobility UEs.

The system needs to consider dynamic environment radio condition changes, even for no-mobility UEs.

The system should consider the CIoT UE not sending a measurement report to CIoT RAT.

The system should consider avoiding frequent signaling exchange.

7.17.1.5   Key Issue 5: Selection of CIoT–EPC Dedicated Core Network

There may be one or more EPC dedicated core networks supporting CIoT (CIoT–EPC). Depending on RAN work, a CIoT–EPC DCN may be connected to both the E-UTRAN and CIoT RAN. This key issue is to determine how to select a CIoT–EPC DCN for IoT devices. Redirection of CIoT devices is assumed to use the existing DECOR functionality. Redirection of CIoT devices between DCNs should be reduced.

7.17.1.6   Key Issue 6: Support for Non-IP Data

It is not uncommon to find applications in the machine-to-machine world which utilize non-IP data, for example, 6LowPAN, MQTT-SN, etc. When such applications are deployed in the mobile domain (over CIoT), non-IP data need to be transferred between application/service capability servers and CIoT devices. Therefore, the means of facilitating the transfer of such non-IP data needs to be investigated.

7.17.1.7   Key Issue 7: Support for SMS

In the scope of this key issue, we need to identify how to support SMS efficiently and at least address these topics:

Whether the combined attach is used to obtain an SMS transfer service or just a PS attach.

Whether CIoT devices will exist which do not need any PDN connection and just operate using SMS-based data transfer, and identify related solutions.

7.17.1.8   Key Issue 8: Control of Small Data Misuse

This key issue aims to provide a solution to controlling small data misuse. For CIoT in the scope of this key issue, we need to identify how to control the misuse of small data as a result of misuse in relation to the service level agreement.

7.17.1.9   Key Issue 9: Optimized Support for SMS Transmission

The current Short Message Service standard, as defined in TS 23.060 [30] uses protocols that are quite "talkative" over the radio (for example, the RP/CP protocols, as defined in TS 24.011 [31]) and have quite an excessive number of layers in the protocol stack. In CIoT there is an opportunity to optimize the way SMSs are sent. In the scope of this key issue, we need to identify how to optimize support for SMS and at least address these topics:

Reduce the "talk" over the radio.

Optimizations are needed for devices that mainly originate short messages only, so that the network does not attempt to deliver pending short messages.

7.17.1.10 Key Issue 10: Authorization of Use of Coverage Enhancement

The use of coverage enhancement may require extensive resources from the network, i.e. it should be possible to authorize usage of the coverage enhancement functionality to ensure that only specific subscribers are able to benefit from the feature. The system should support procedures to authorize the use of coverage enhancement.

7.17.1.11 Key Issue 11: Header Compression Enhancements for CIoT

This key issue aims to provide an efficient header compression scheme to support the CIoT traffic models. Based on the traffic model in Table 7.19, most of the categories make use of small data size and send data infrequently. For example, periodic MAR and network command are using the data size of 20 – 200 bytes every 30 minutes or longer. Considering the IP and transport layer headers, for example, 20 B for IPv4 and 40 B for IPv6, 8 B for UDP, 20 B for TCP, and 12 B for RTP, the overhead is significant. Therefore, header compressing is necessary for the efficient support of a large number of devices for the NB-IoT. For instance, the traffic/user data for the NB-IoT typically include (though not exclusively) status information, measurement data, and alarm data from M2M applications. Due to infrequent data sending and mobility, the header compression context may be kept in the eNB and the UE may be reset if it enters Idle mode or changes eNB. This leads to packets with full-size headers, or even with additional overhead, sent frequently. In addition, there could also be non-IP traffic for CIoT use. In this case, the header compression scheme must be able to turn the scheme off when non-IP traffic is present.

It is therefore desirable to develop a header compression mechanism to support the efficient transport of CIoT traffic.

For data delivery over the control plane, the header compression function for CIoT traffic needs to be supported to allow Connected–Idle mode transition and mobility. The header compression function for CIoT needs to be able to handle non-IP traffic when the UE requests non-IP support.

7.17.2 Optimized EPS Architecture Options for CIoT

In this section, we will briefly describe some 3GPP solutions to address the above issues and meet the CIoT requirements.

7.17.2.1 Solution 1: Lightweight CN Architecture for CIoT

NB CIoT RAT [28] and LTE eMTC [32] include the following properties:

Ultra-low UE power consumption
A large number of devices per cell
Applied in narrowband spectrum
Increased coverage.

Figures 7.62 and 7.63 illustrate the simplified topology for non-roaming and roaming CIoT, respectively. Considering the typical use cases for CIoT devices, the CIoT system only supports reduced and necessary functionalities compared with the existing MME, S-GW, and P-GW, as follows:

Support for efficient small data procedures
Simplified NAS signaling

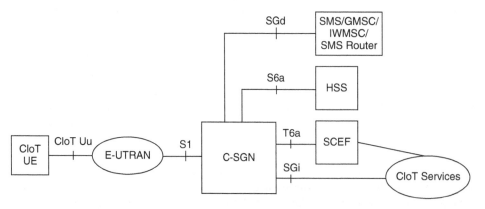

Figure 7.62 Optimized EPS architecture option for CIoT – non-roaming architecture.

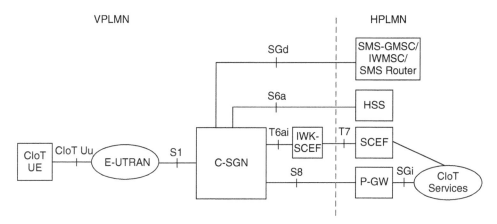

Figure 7.63 Optimized EPS architecture option for CIoT – roaming architecture.

USIM-based security
Authentication and authorization
NAS signaling security
Simplified mobility management
Attach, TAU, Detach
Tracking area list management
UE reachability in Idle state (including paging)
Simplified session management
Only one PDN connection and one bearer for IP (or non-IP) data (no need for DRB establishment)
Lawful interception
Paging for coverage enhancements
Charging and accounting.

The P-GW functions can be separated from the C-SGN as an implementation option. In this case, the S5 interface is used between C-SGN and P-GW. As mentioned above, CIoT traffic patterns and UE properties have different characteristics than "normal" UEs. From the CN perspective, it is expected that they will differ from "normal" UEs in the following areas:

For session management, only the procedures required for infrequent, small data procedures will be supported.

From a mobility management perspective, it is expected that a large number of devices will not be in an ECM-Connected state and less MM signaling will be performed due to low mobility/extended pTAU timer.

The C-SGN, given that it is largely based on S1 with optimizations, can be the CN node of use for UEs supporting the CIoT RAT or LTE eMTC. RAN (eNB or CIoT BS) allocates all CIoT UEs to a dedicated CN when the UE indicates that only IP (or non-IP) infrequent, small data transmission is required to be supported. No dedicated bearers with different QoS, no Connected mode mobility, and no CS procedures are supported by the dedicated CN.

C-SGN The C-SGN (CIoT Serving Gateway Node) is a combined node EPC implementation option that minimizes the number of physical entities by collocating EPS entities in the control and user plane paths (for example, MME, S-GW, P-GW), which may be preferred in CIoT deployments. The external interfaces of the C-SGN implementation option are the interfaces of the respective EPC entity supported by the C-SGN, such as the MME, S-GW, and P-GW.

A C-SGN supports a subset of necessary functionalities compared with the existing EPS core network elements, and also supports at least some of the following CIoT optimizations:

Control plane CIoT EPS optimization for small data transmission.

User plane CIoT EPS optimization for small data transmission.

Necessary security procedures for efficient small data transmission.

SMS without combined attach for NB-IoT-only UEs.

Paging optimizations for coverage enhancements.

Support for non-IP data transmission via SGi tunneling and/or SCEF.

Support for attach without PDN connectivity.

S1-lite S1-lite is an optimized version of S1-C. The protocol stack (see Figure 7.64) is based on S1-C with only necessary components for the support of efficient small data. From S1-C only, the necessary S1-AP messages and IEs are supported for the related CIoT procedures and only the optimized security procedures are supported. The user plane data are carried in S1-AP in order to support efficient small data handling.

Protocol Stack The CIoT-RAN supports the selection of C-SGN for CIoT UE. Both the CIoT-RAN and C-SGN support the S1-lite interface to transmit small data

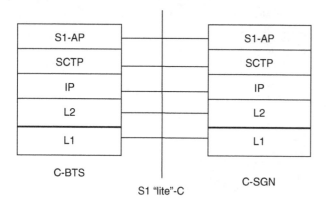

Figure 7.64 S1-lite protocol stack.

efficiently. The protocol stacks for CIoT small data transmission for non-roaming and roaming scenarios are illustrated in Figure 7.65 and Figure 7.66, respectively.

Note that, as an implementation option, a P-GW may terminate the SGi. In this case, the S5 interface is used between the C-SGN and the P-GW.

This solution simplifies the existing EPC architecture in order to address the key issues identified in the previous section. It bases small data transmission on a simplified architecture that is oriented towards transporting small data over NAS signaling messages. A dedicated core network node for the CIoT profile provides combined C-plane and U-plane functions, for example, aggregating some of the functions traditionally residing in the MME and S-GW (and in some instances, the P-GW).

The solution supports certain modifications (for example, native support of SMS in the PS domain, attach without PDN connection, simplified NAS, etc.) being applicable only to CIoT UEs without requiring procedures to maintain backward compatibility with other UEs. These solutions may also be applicable to

other architectures, not necessarily assuming the C-SGN functional grouping. The reduced number of messages over the radio interface can potentially allow for a lower power consumption profile.

7.17.2.2 Solution 2: Infrequent Small Data Transmission Using Pre-established NAS Security

This solution corresponds to Key Issue 2 – "Efficient Support for Infrequent, Small Data Transmissions for Cellular IoT" and Key Issue 6 – "Support for Non-IP Data." The solution is based on the lightweight CN architecture for CIoT described in Solution 1. Figure 7.67 demonstrates this solution. The P-GW is only used to support the roaming case. For the non-roaming case, the SGi interface terminates on C-SGN and C-SGN can send/receive data on SGi directly.

The benefits of this solution are:

No establishment of DRBs and S1-U bearers.
Reduced signaling procedures in the core network and radio interface.
The solution provides the same efficiency to both stationary and mobile devices.
The same transport mechanism can cover a wide range of applications sending/receiving infrequent small data (for example, IP, non-IP, and SMS).

This solution piggybacks uplink small data onto the initial NAS uplink message that is extended, and it uses an additional NAS message for carrying downlink response small data. The effort of setting up the user plane, i.e. related RRC messages and AS security setup, can be avoided. This solution does not preclude delivery of very infrequent large data (for example, software updates/software patches). The C-SGN capacity needs to be suitable for this traffic. Data transfer on the radio interface depends on the CIoT RAN design.

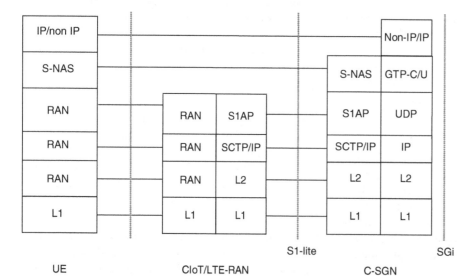

Figure 7.65 Protocol stack for CIoT small data transmission – non-roaming.

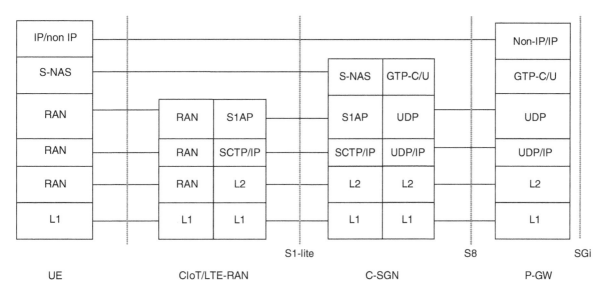

Figure 7.66 Protocol stack for CIoT small data transmission – roaming.

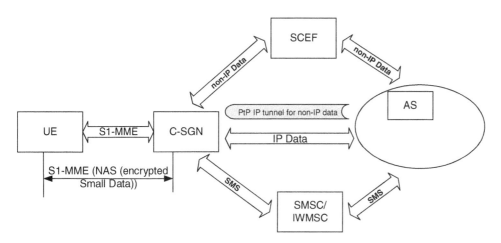

Figure 7.67 E2E small data flow.

7.17.2.3 Solution 3: Non-IP Small Data Transmission via MTC-IWF

This solution corresponds to Key Issue 6 – "Support for Non-IP Data." This solution is based on the lightweight CN architecture for CIoT defined in Solution 1. Figure 7.68 shows the cellular IoT network architecture proposed for efficient, non-IP small data transmission based on the T5* interface. The MTC-IWF protocol entity may be a standalone entity or may be combined with the MME in non-roaming cases. A pre-requisite for small data transfer is that the UE is attached to the network, and the capabilities for the small data service have been exchanged between the UE and the C-SGN during EPS Attach or TAU.

The MTC-IWF has similar functions to those defined in Release 12 for T4 small data transmission.

This solution also works for the existing EPC architecture, in which case, the T5* reference point will terminate in the MME.

This solution has the following advantages:

Re-use of existing data flows and messages to encapsulate small data payloads.
Re-use of existing security procedures with no changes.

Figure 7.68 CIoT architecture for non-IP small data transmission via MTC-IWF.

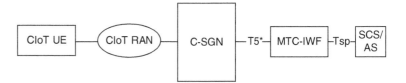

Signaling-only small data transfer requires neither user-plane resources nor combined attach.

Flexible SD message payload size is constrained only by RAN design.

Sessionless one-shot messages are an efficient way to provide small data transport.

The existing, always-on paradigm is not affected, as this small data solution works with or without always-on.

The solution has the following disadvantage:

Complete re-use of existing implementations is not possible, but the implementation of the affected nodes needs to change for CIoT.

7.17.2.4 Solution 4: Non-IP Small Data Transmission via the SCEF

This solution corresponds to Key Issue 6 – "Support for Non-IP Data." The solution is based on the lightweight CN architecture optimized for CIoT. Figure 7.69 shows the cellular IoT network architecture proposed for efficient, non-IP small data transmission via the SCEF.

This solution also works for the existing EPC architecture, in which case, the T6a* reference point will terminate at the MME.

7.17.2.5 Solution 5: Non-IP Small Data Transmission via the SCEF with Minimized Load to HSS

This solution corresponds to Key Issue 6 – "Support for Non-IP Data." The solution is based on the lightweight CN architecture optimized for CIoT. Figure 7.70 shows the cellular IoT network architecture proposed for efficient, non-IP small data transmission via the SCEF.

This solution also works for the existing EPC architecture, in which case, the T6a reference point terminates between the MME and the SCEF.

7.18 Sample IoT Use Cases

In this section, we will present some practical IoT use cases. We will present the end-to-end use case from sensors all the way to the application layer. The IoT verticals presented in this section may or may not follow the horizontal IoT approach that we presented earlier in this chapter. More than 50 IoT practical use cases are listed in [33].

7.18.1 Smart Mobile Site

This IoT solution is based on the Cisco Asset Management (CAM) [34] system along with Azeti BTS monitoring software [35]. Figure 7.71 illustrates a typical practical topology of this IoT use case. The system has been deployed at two mobile sites to monitor, control, and improve the efficiency of the site assets. The following data points were monitored from the mobile sites:

Voltage (40)
Temperature (36)
Power (26)
Battery and site current (19)
Consumption (4)
Energy (4)
Codes (2)
Site access (2)
AC frequency (2)
Volume (2)
Duration (2)
Humidity (2)
Fuel flow rate (1)
CCTV.

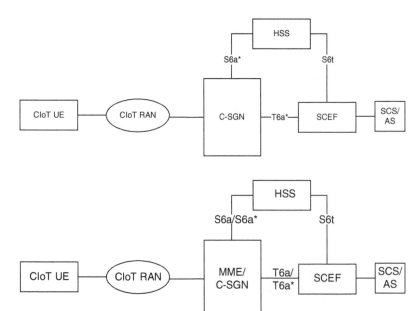

Figure 7.69 CIoT architecture for non-IP small data transmission via the SCEF.

Figure 7.70 CIoT architecture for non-IP small data transmission via the SCEF with minimized load to HSS.

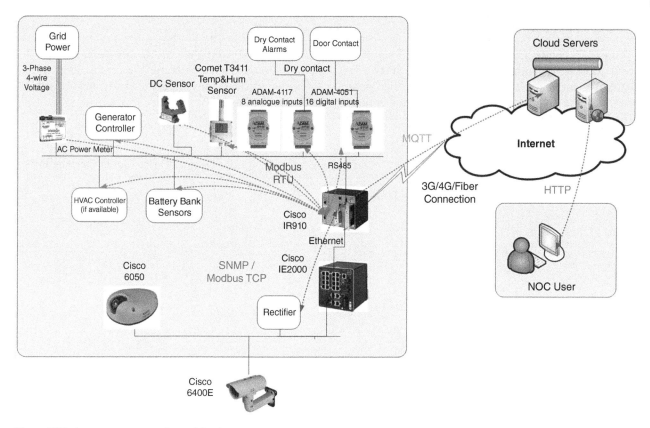

Figure 7.71 Asset management for mobile sites.

Cisco Asset Management (CAM) software empowers you to manage connected assets remotely across your organization. It provides real-time asset identification, location, tracking, monitoring, analysis of asset consumption and availability, and physical security features from a single, consolidated view. By visualizing asset energy consumption, you can optimize energy use and reduce operational expenses [34].

BTS monitoring systems provide solutions for telecoms operators and service providers. Most monitoring applications check Quality of Service and detect nearly 100% of mobile service errors or glitches. It is usually not limited to Quality of Service (QoS) monitoring. It also includes monitoring passive infrastructure such as ultrasonic fuel levels, battery cells, power metering, physical security of a cell site, and other environmental parameters such as temperature, humidity, or any kind of leak. Some monitoring solutions are SNMP-based and can be integrated easily with the NOC [35].

For this specific use case, every time a motion is detected inside or outside the mobile site, a picture is taken by the camera. With this feature, you know who accessed the site and when. CAM reporting allows extensive analytics of the above-gathered data points, which can be used to optimize the site. Setting alerts is a powerful feature that allows you to be notified (in real time) when a sensor value changes and is not within the limits of your defined thresholds. Adding new alerts in CAM is easy and ensures you are aware of what is happening on your sites.

The system allows monitoring and, accordingly, proposes efficient techniques for asset management, such as the cases described below.

7.18.2 Smart Meter

The smart meter system is replacing legacy meters by connecting digital meters to an Automated Meter Reading (AMR) platform using Advanced Metering Infrastructure (AMI). Figure 7.75 provides a schematic diagram for an AMR system. As indicated in Figure 7.75, we have two scenarios for meter connectivity: either through a GW or concentrator or directly by installing a GSM/3G/LTE module inside the meter. The latter scenario is used in main stations and for scattered villas. The GW or concentrator scenario can be deployed in multi-story buildings to aggregate smart meters on different floors. In addition, water and gas meters can be aggregated to the same GW. Several short-range options are available to connect meters on different floors to the concentrators, such as power-line carriers (PLCs), proprietary RF at 868 MHz, RS485, Wi-Fi, and even hybrid solutions are supported for redundancy.

Case 1: Temperature and Humidity Range

Figure 7.72 demonstrates the temperature for the two different sites. As demonstrated in the figure, the two sites have different ranges of temperature, which could potentially increase A/C consumption or degrade the batteries. The site temperature should be set within the range 20 to 24 °C. We should investigate what causes the fluctuations in humidity (probably temperature) and configure/optimize A/C systems to keep the temperature and humidity within the optimal values.

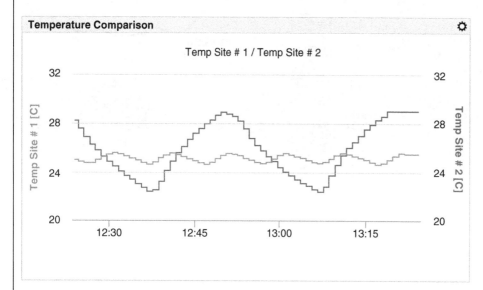

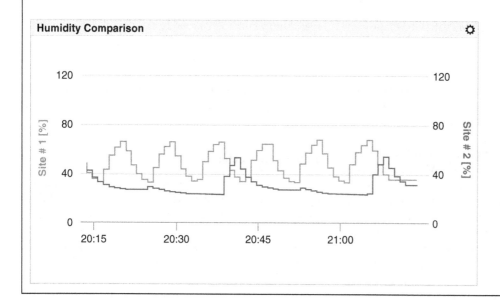

Figure 7.72 Temperature and humidity.

The introduction of NB-IoT could replace both scenarios and eliminate the need for the concentrators. However, deep indoor coverage is needed for such a case. In addition, the latency of the NB-IoT could be a bottleneck for power meters, as action may be needed immediately.

The AMI system will offer the following benefits to utility companies:

OPEX reduction: A traditional meter cannot support remote meter reading, and the billing relies on manual meter reading and fee collection. In order to cover wide areas, the power companies often need lots of manpower. AMI supports remote meter reading, pre-paid payments, and remote control.

Power consumption reduction: Smart meters have been designed based on electronic components, so

Case 2: Power Consumption Comparison

Figure 7.73 demonstrates the power for the two different sites. As illustrated in the figure, the two sites have different patterns, as the cycle of A/C is different. The average is different due to the different configurations of the sites, but site # 2864 has less peak time due to the cycle of the A/C. Relaxing the A/C cycle of the other site could potentially reduce the average power consumption.

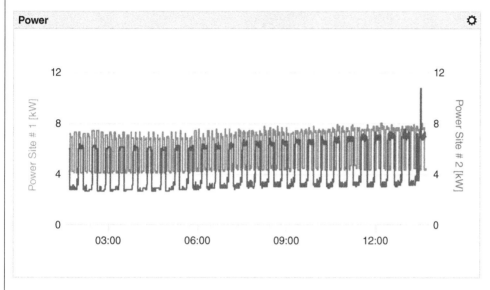

Figure 7.73 Power consumption.

Case 3: Phase Voltage

Figure 7.74 demonstrates the three-phase voltage. The difference in phase voltage is called "voltage imbalance". The difference between the highest voltage phase and the lowest voltage phase cannot be more than 2% of the lowest voltage, otherwise defects in equipment can occur.

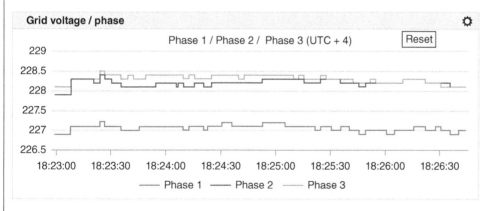

Figure 7.74 Three-phase voltage from the grid.

the average power consumption is 0.6–0.7 Watts. But traditional induction-type meters will be 1.7–2 Watts. Therefore, the power consumption from the meter will be reduced greatly by smart meters. For example, there are 500,000 meters in the energy network, so the average saved power consumption will be 500,000 Watts.

Reading accuracy: Normally, a meter of level 2 means the measurement error is ±2% within the 5% to 400% current range; smart meters are normally level 1.

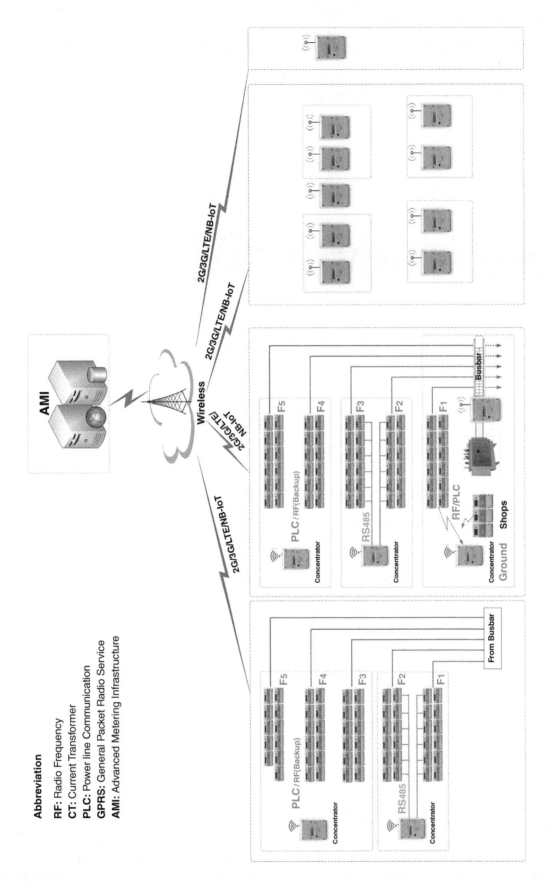

Abbreviation

RF: Radio Frequency
CT: Current Transformer
PLC: Power line Communication
GPRS: General Packet Radio Service
AMI: Advanced Metering Infrastructure

Figure 7.75 Smart meter system.

The tolerance of the induction-type meter is 0.86% to −5.7% The insurmountable defects in terms of mechanical wear lead to slower induction-type meters and, ultimately, increased error. Substantial evidence indicates that for more than 50% of meters after five years, the error exceeds the allowable range, which causes additional revenue loss to the energy company.

Extended overload and over frequency range: Smart meters have a wide work range which can bear up to 6 ~ 8 times overload; some smart meters can bear up to 8 ~ 10 times, and some even up to 20 times. For frequency, smart meters can work within 40 Hz ~ 100 Hz normally. For an induction-type meter, the overload is four times and the operating frequency range is only 45 ~55 Hz. So, in terms of voltage instability, smart meters can guarantee a long working life and can avoid overload damage.

Functions: AMI is based on smart meters which involve electronic technology. As a result, it can support the relevant communication protocols to connect with a computer. Therefore, AMI supports remote control (remote reading, remote connect or disconnect), multi-tariffs, recognition of malignant loads, anti-tampering, and pre-paid functionality. These functions, for traditional, induction-type meters, are too difficult to implement.

Line loss analysis and control: For a traditional meter, meter reading relies on manual reading and cannot ensure consistent reading times and accurate data, so line loss analysis is difficult. For AMI, the above problems will be resolved, so the system can undertake entire line loss analysis and raise alarms, to help the power companies improve and deal with line loss. Substantial evidence indicates that line losses are more than 30% in most African countries, but the figure can still be about 6% in most developed countries. So AMI can benefit all power companies greatly.

7.18.3 Smart Building

A smart building solution offers efficient management for the resources within a multi-story building, campus, company HQ, etc. Figure 7.76 illustrates a typical schematic diagram for a smart building system. As illustrated in the figure, all systems within the building are connected to a rigid industrial IoT GW including a fire system, chillers, security, CCTV, power, and UPS. An IoT platform is needed to manage the resources of the building. In addition, power management software could be installed to optimize the power consumption. Cisco Energy Management (CEM) is an example of such software. Figure 7.77 provides typical dashboards extracted from the CEM platform [36].

The Cisco Energy Management Suite automatically discovers all IT assets connected to your network and dynamically calculates energy use on a per-device basis. You can view usage by physical location, business unit, data center row, rack, slot, device, virtual machine, and more. You can measure, prioritize, and report on energy usage and carbon reduction efforts. Even more critically, policy and automation features simplify energy-saving management and control, such as powering off devices when they are not in use. In addition, alerting capabilities ensure that you are informed when you reach or exceed power, utilization, and temperature thresholds [36].

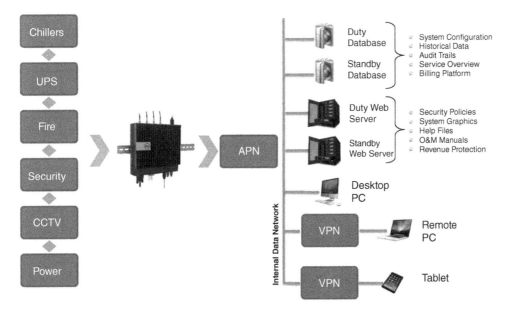

Figure 7.76 Smart building system.

Figure 7.77 CEM dashboard example.

References

1 Minerva, R., Biru, A. and Rotondi, D. (2015) "Towards a definition of the Internet of Things (IoT)," May.Online: https://www.scribd.com/document/364863585/IEEE-IoT-Towards-Definition-Internet-of-Things-Issue1-14MAY15

2 IEEE P2413: "Standard for an Architectural Framework for the Internet of Things (IoT)." Online: https://standards.ieee.org/develop/project/2413.html

3 ETSI TS 102 689 V1.1.1 (2010 I08): "Machine-to-Machine Communications (M2M); M2M Service Requirements," technical specification.

4 ITU-T SG20: "IoT and its applications including smart cities and communities (SC&C)." Online: http://www.itu.int/en/ITU-T/studygroups/2013-2016/20/Pages/default.aspx

5 ETSI: "oneM2M Requirements Technical Specification." oneM2METSI0002IVI 0.6.2. 2013I10I17. Online: http://www.ttc.or.jp/jp/document_list/pdf/j/TS/TSIM2MI 0002v0.6.2.pdf.

6 Chakrabarti, A. (2015) *"Emerging Open and Standard Protocol Stack for IoT,"* White Paper, December. Online: https://www.linkedin.com/pulse/emerging-open-standard-protocol-stack-iot-aniruddha-chakrabarti

7 http://postscapes2.webhook.org/internet-of-things-resources/

8 https://entrepreneurshiptalk.wordpress.com/2014/01/29/the-internet-of-thing-protocol-stack-from-sensors-to-business-value/

9 Rebbeck, T., Mackenzie, M. and Afonso, N. (2014) *"Low-powered wireless solutions have the potential to increase the M2M market by over 3 billion connections,"* Analyses Mason, September.

10 GSMA (2016) *"Mobile Internet of Things Low Power Wide Area Connectivity,"* Industry Paper, March.

11 Gartner, (2017) *"Hype Cycle for IoT Standard and Protocols,"* Bill Ray, Ganesh Ramamoorthy, July 18, 2017 G00314546.

12 GSMA White Paper, "3GPP Low Power wide Area Technologies," Online: https://www.gsma.com/iot/3gpp-low-power-wide-area-technologies-white-paper/

13 Chatterjee, D., Drewes, C., Hausner, J., Roessel, S., Sesia, S. and Tarradell, M. (2016) "Cellular Internet-of-Things – A Deep Dive into Technology and Devices", *Intel Corporation Industry Seminar at Globecom 2016*. Online: http://globecom2016.ieee-globecom.org/sites/globecom2016.ieee-globecom.org/files/u14/Cellular%20IoT.pdf

14 GSMA: "Emerging Mobile IoT Technologies: Use Cases, Business and Security Requirements," GSMA ETSI M2M Workshop, 9 December 2015. Online: https://docbox.etsi.org/Workshop/2015/201512_M2MWORKSHOP/S04_WirelessTechnoforIoTandSecurityChallenges/GSMA_GRANT.pdf

15 Rohde & Schwarz (2016) "Narrowband Internet of Things," White Paper, August. Online: www.rohde-schwarz.com/appnote/ 1MA266.

16 3GPP TS 23.272 V13.3.0 (2016-03) "Circuit Switched (CS) Fallback in Evolved Packet System (EPS)."

17 3GPP TS 23.401 V13.6.1 (2016-03) "General Packet Radio Service (GPRS) Enhancements for Evolved Universal Terrestrial Radio Access Network."

18 3GPP TS 23.682 V13.5.0 (2016-03) "Architecture enhancements to facilitate communications with packet data networks and applications."

19 Lauridsen, M. et al. (2016) "Coverage and Capacity Analysis of LTE-M and NB-IoT in a Rural Area," *84th IEEE Vehicular Technology Conference (VTC-Fall).*

20 Lauridsen, M. et al. (2017) "From LTE to 5G for Connected Mobility," *IEEE Communications Magazine,* **55**(3), March.

21 Vejlgaard, B. et al. (2017) "Interference Impact on Coverage and Capacity for Low Power Wide Area IoT Networks," *IEEE Wireless Communications and Networking Conference (WCNC).*

22 Lauridsen, M. et al. (2017) "Interference Measurements in the European 868 MHz ISM Band with Focus on LoRa and Sigfox," *IEEE Wireless Communications and Networking Conference (WCNC).*

23 ETSI EN 300 220-1 V2.4.1 (2012-01) "Electromagnetic Compatibility and Radio Spectrum Matters; Short Range Devices; Radio equipment to be used in the 25 MHz to 1 000 MHz frequency range with power levels ranging up to 500 mW; Part 1."

24 Sigfox: accessed January 4 2017. Online: https://www.sigfox.com/

25 Libelium Comunicaciones Distribuidas (2015) "Waspmote Sigfox Networking Guide." Online: v4.1.

26 3GPP: "SCEF: "Primer: Definition Networks. Online: http://definitionnetworks.com/3gpp-scef-primer/

27 Nokia: "LTE evolution for IoT connectivity," White Paper.

28 3GPP TR 45.820 (2015-11) "Cellular System Support for Ultra Low Complexity and Low Throughput Internet of Things."

29 3GPP TR 23.720 V13 (2016-03) "Study on architecture enhancements for Cellular Internet of Things."

30 3GPP TS 23.060 (2017-09) "General Packet Radio Service (GPRS); Service description; Stage 2."

31 3GPP TS 24.011 (2017-06) "Point-to-Point (PP) Short Message Service (SMS) support on mobile radio interface."

32 3GPP TR 36.888 (2013-06) "Study on provision of low-cost Machine-Type Communications (MTC) User Equipment (UEs) based on LTE."

33 Libelium (2017) "*50 real IoT success stories after ten years of experience in the market,*" White Paper.

34 http://www.cisco.com/c/en/us/products/switches/asset-management-suite/index.html

35 http://www.azeti.net/bts-monitoring.html

36 http://www.cisco.com/c/en/us/products/switches/energy-management-technology/index.html

37 Ishaq, I. et al. (2013) "IETF Standardization in the Field of the Internet of Things (IoT): A Survey," *J. Sens. Actuator Netw.* **2**, pp. 235–287.

8

5G Evolution Towards a Super-connected World

8.1 Overview

Mobile communication technologies have been evolving for many years, with each generation transforming the way we experience new services. The smartphone market has expanded significantly in recent years and is expected to grow further in the years to come; therefore, the network must continue to evolve to keep pace with users' demands, even beyond the common usage connectivity. The envisioned market space for the next generation technology is being driven by requirements to enhance mobile broadband connectivity, reach a massive range of machine-type communication (MTC), and target services with ultra-reliable and low-latency (URLLC) communications. To deliver these requirements, 5G must be designed with scalability and diversity across many components from the spectrum, core network, radio access, and devices.

8.2 Introduction

The envisioned market space for 5G technology involves a design that has the capability to unify the system needed by various use cases. New use cases keep evolving according to the need for higher peak data rates, reduced end-to-end service latency, and increased network capacity in terms of user traffic and density. In addition, use cases suitable for the commercial 4G LTE network capabilities keep appearing beyond today's usage, and potentially even beyond what the network and devices were designed for.

A wide range of new applications and use cases requires advanced connectivity capabilities that can be both intense and diverse in their requirements, including high capacity and data rate connectivity (3D video and virtual reality connectivity), real-time communications with low latencies (interactive video, automotive, critical type control, and tactile Internet), and massive Internet-of-Things connectivity (sensor networks, smart metering).

8.2.1 Definition and Use Cases for 5G

At the moment, whenever any potential market space opens up, it requires significant design changes to the existing technologies, and in some cases, a totally new radio access and core network. For example, when the M2M-type applications for different types of industries started to materialize, it was realized that a totally new, Internet-of-Things technological design was needed, which did not help in accelerating the deployment of new services. A revamped radio access called Narrowband Internet of Things (NB-IoT) had to be designed by 3GPP to address the emerging adjacent industries as presented in Chapter 7.

All these factors require a new system designed to handle any new use case at any time without the need to re-dimension, re-design, or even invest heavily in a totally new network and technology components for each single use case. 5G is promising to deliver a unified system that can be considered all in one: more than just developing new radio technology, it is for new possibilities and use cases. It aims to accelerate new business cases that keep promoting continuous, costly, and non-trivial changes into the existing ecosystem, from new cellular towers to investing in new cloud computation and core networks, and devices that users can utilize efficiently in their everyday lives.

In order to enable services for a wide range of users and industries with new requirements, the capabilities of 5G must extend beyond those of previous wireless access generations. These capabilities will include enhanced mobile broadband connectivity, massive system capacity for machine-type communications, very high data rates everywhere, very low latency, ultra-high reliability and availability, and low device cost, as Figure 8.1 shows.

8.2.2 LTE Evolution to Enable 5G Use Cases

The major differences expected in 5G compared to the legacy 4G generation will not only be at the level of combining old and new radio access technologies; 5G will also enable new use cases and requirements

Practical Guide to LTE-A, VoLTE and IoT: Paving the Way Towards 5G, First Edition. Ayman Elnashar and Mohamed El-saidny.
© 2018 John Wiley & Sons Ltd. Published 2018 by John Wiley & Sons Ltd.

Figure 8.1 5G and the industry driving force: an all-in-one system.

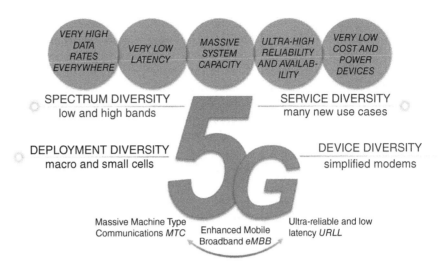

of mobile communication beyond 4G systems. It will be an integration of existing cellular standards and technologies, including new disruptive technologies like millimeter-wave (mmWave) and spectrum sharing among other new concepts like network slicing. These concepts will facilitate the integration of vertical industries into the mobile ecosystem, whilst opening up new business models and revenue streams for operators. Mobile broadband access and service availability with low latencies are key use cases driving the requirements for 5G, especially at the initial phase of deployment. The key enablers for any new design in 5G will be the spectrum and the means by which to utilize concepts from previous technologies in order to accelerate the migration to 5G. In order to meet the 5G requirements and bring this visionary system to reality, the future 5G network will be one that is built upon the small-cell backbone either in standalone or non-standalone deployment. As spectrum suitable for mobile communication becomes more and more scarce, densification is the only way to meet the area traffic-capacity demand. Even for the millimeter-wave band, where spectrum is abundant,

the channel's propagation characteristics will likely limit its range for mobile access to that of a small cell, at least in the early phase of 5G before device technology matures [1]. 5G design also targets bringing the radio access point closer to the end device, thereby shortening the last and most challenging segment of an end-to-end communication link and consequently reducing latency and increasing reliability. Many of the massive number of machine-type communications can also benefit from the extended battery life resulting from a shorter uplink distance.

Figure 8.2 shows an example of traffic distribution from both smartphone-centric and router-centric LTE networks. This commercial network deployment example shows that the disproportionate traffic across cells may require the re-design of network concepts in future evolution. It shows that in a smartphone-centric LTE network, 62% of users are camping on 38% of cells and generating 57% of traffic, with an average user throughput of less than 6 Mbps; one third of cells generate ~ 50% of traffic to handle 60% of users. On the other hand, the router-centric LTE network has a traffic

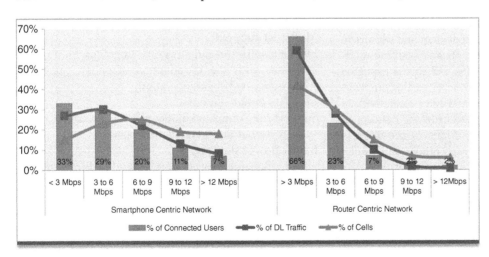

Figure 8.2 LTE-A traffic and throughput correlation.

distribution where 66% of users are camping on 42% of cells and generating 59% of traffic, with an average user throughput of less than 3 Mbps.

In the current LTE network deployment, the available use cases for connectivity lead to concentrated traffic in a small area, which impacts the end-user experience. Moving towards small cells is important in order to have a contiguous network with diversity of traffic and utilize uniform network resources rather than deploying different networks to handle different types of traffic. Additionally, smartphone-type traffic requires the network to be dimensioned differently, so that more cells are needed to ensure coverage everywhere while the cell resources are less utilized because of the bursty nature of the smartphone data connectivity. In this example, the overall downlink bit rate per resource block (RB = 180 kHz) is 0.22 Mbps only. The maximum theoretical bit rate per RB is 1.42 Mbps in LTE. This means that only 15.5% of a single RB throughput is utilized on the downlink across cells in such a network, leading to a waste of OFDM capacity.

To facilitate the deployment of a future mobile network that has the small cell as its primary traffic-bearing workhorse, the current radio access architecture needs to undergo some major revamps and new technologies need to be introduced. Initially, the common layer could use the LTE waveform (OFDM) but the dedicated layer could use the new 5G waveform. Therefore, 5G devices could be served by LTE for locations where the 5G service was not available. Over time, 5G coverage will improve and 5G UEs will become popular; therefore, the bandwidth of the LTE-based network can be reduced, or even replaced by 5G capabilities. This is the expected migration plan from a non-standalone deployment to a standalone one. We discuss this issue in the next section.

8.2.3 Massive Machine-type Communication (mMTC)

The Internet of Things (IoT) defines the way in which intelligently connected devices and systems can leverage and exchange data between small devices and sensors in machines and objects. IoT concepts and working models have started to spread rapidly, and this is expected to provide a new dimension for services that improve the quality of consumers' lives and the productivity of enterprises. The IoT effort started from the concept of Machine-to-Machine (M2M) solutions to use wireless networks to connect devices to each other and through the Internet, in order to deliver services that meet the needs of a wide range of industries. Next to eMBB radio access, 5G will incorporate systems that enable massive machine-type communications (MTC). In 3GPP Release 13, NB-IoT was defined to operate within a 200 kHz bandwidth. In the work on 5G specifications, this is expected to be optimized further toward a high number

of supported devices, low device cost, and ultra-low power consumption. In order to support different types of deployments, NB-IoT targets three different modes of operation; one mode involves utilizing the spectrum currently being used by GERAN systems as a replacement for one or more GSM carriers (standalone operation). The second mode utilizes the unused resource blocks within an LTE carrier's guard band (guard-band operation). The third mode utilizes resource blocks within a normal LTE carrier (in-band operation).

The NB-IoT will support the following main objectives:

OFDMA on the downlink with 15 kHz subcarrier spacing.

SC-FDMA on the uplink with single and multi-tone of 3.75 kHz and 15 kHz subcarrier spacing.

A single synchronization signal design for the different modes of operation, including techniques to handle overlap with legacy LTE signals while reducing the power consumption and latencies.

Utilization of the existing LTE procedures and protocols and relevant optimizations to support the selected physical layer and core network interfaces targeting signaling reduction for small data transmissions.

The deployment bands: 1, 2, 3, 5, 8, 12, 13, 17, 18, 19, 20, 26, 28, and 66. Other bands not supported in Release 13 are being studied for inclusion in Release 14 for NB-IoT.

3GPP Release 14 will introduce further enhancements to the NB-IoT network and device capabilities in order to extend the solution to more use cases and applications [2]:

Positioning enhancements:
Support for Observed Time Difference Of Arrival (OTDOA), or Uplink-Time Difference of Arrival (UTDOA) for better positioning accuracy, without adding significant device complexity or power consumption impact.

Multicast support:
Efficient software and firmware upgrades for massive devices with the introduction of enhancements to support narrowband operation.

Mobility enhancements:
Support for Connected mode mobility for service continuity for both user and control plane.

Lower power support:
Lower transmit power class (e.g. 14 dBm) to support lower current consumption suitable for small form-factor batteries (e.g. for wearable devices).

Higher data rate support:
Multiple HARQ process support and higher Transport Block Size (TBS) to increase the downlink data rate from ∼ 28 kbps to ∼ 112 kbps.

8.3   5G New Radio (NR) and Air Interface

The evolution of LTE in Release 14 is expected to offer a first step towards 5G by enabling wireless access for frequency bands below 6 GHz (sub-6). Hence, LTE-Advanced Pro (LTE-A Pro) might be considered a special case of 5G in those frequency bands. For higher bands, a new radio-access technology (RAT) and inherent support and integration solutions will be introduced. Therefore, the 5G architecture will be an integration of multi-RAT, supporting the simultaneous operation of multiple heterogeneous technologies.

It is therefore expected that 5G will initially work on three areas of improvement, taking the LTE-A Pro design concepts into consideration:

Radio access that provides service multiplexing for eMBB, mMTC, and URLLC. This requirement is expected to provide scalable numerology and a flexible time–frequency grid. The target is to design a waveform that is flexible for different subsystems within the same carrier, where the same spectrum resource is used to deploy new services.

Radio access that is designed to have lossless physical layer transmission to reduce the pilot overhead and utilize connectionless transmission to reduce the control channel overhead, similar to that in NB-IoT. This requirement is expected to provide a reduction in overhead and achieve high efficiency in 5G networks. In the current LTE deployment, the overhead of control and pilot channel in 20 MHz bandwidth deployment can reach up to 28%. Therefore, 5G may aim to reduce the transmission overhead of the control/pilot channels, restricting the necessary overhead in the narrow band for initial access and dynamically configuring a wider bandwidth operation for eMBB.

Radio access that is capable of low-latency transmission. This can be achieved by having shorter OFDM symbol length, shorter Transmission Time Interval (TTI), contention-based uplink, and modified carrier spacing in order to meet latency requirements of < 10 ms for eMBB services.

Therefore, 5G is a portfolio of access and connectivity solutions addressing the demands and requirements of mobile communication for a wide range of services and applications. Current and future mobile networks have to overcome several challenges:

How to manage highly diverse deployment strategies and topologies.

How to maintain a consistent user experience across all network layers and locations in dense interference environments.

How to reduce cost per bit, increase the capacity, and at the same time maximize the return on investment.

Therefore, the options for 5G architecture may be summarized as follows:

Two radio technology migration plans have to be considered in the 5G discussions:

LTE (in its Release 15 version), known for now as "eLTE". LTE is still capable of attracting capacity and coverage layers.

Next Generation Radio (or New Radio – NR).

Two core network concepts have to be considered in the 5G discussions:

EPC (with potential evolutions).

Next Generation Core Network (NGCN).

Standalone and non-standalone NR operation: Initially, both starting in conjunction and running together:

Target use cases: Enhanced Mobile Broadband (eMBB) as well as low latency and high reliability to enable some Ultra-Reliable and Low-Latency Communications (URLLC).

Frequency ranges below 6 GHz and above 6 GHz.

Figure 8.3 illustrates the NR air interface concepts, whereas Figure 8.4 demonstrates co-existence with the legacy LTE waveform by keeping the OFDM structure where LTE will occupy most low-band spectrum in the next decade. The 5G NR will be flexible in terms of supporting different subsystems/use cases within the same carrier by re-using the same spectrum resource to deploy new services as, demonstrated in Figure 8.3.

In order to accelerate next-generation mobile technology that is capable of meeting these challenges, a network that co-exists with LTE-A Pro (eLTE in its Release 15 version) is expected to be a key enabler for 5G. There is strong industry interest in completing the non-standalone (NSA) version of the 5G new radio

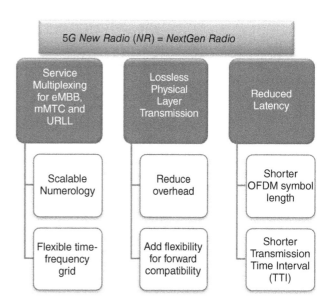

Figure 8.3 5G NR design concepts.

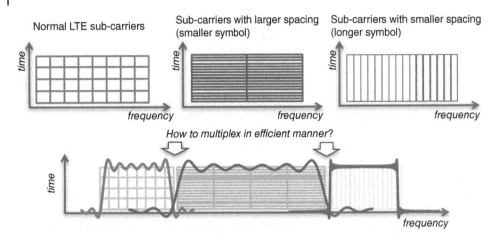

Figure 8.4 Efficient multiplexing of different services on the same carrier.

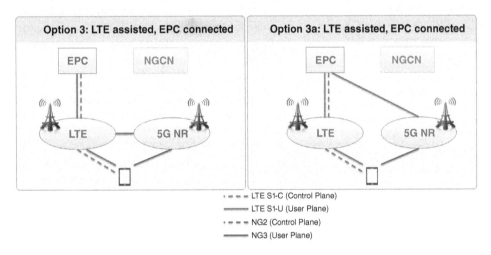

Figure 8.5 5G migration scenarios in 3GPP: non-standalone (NSA) options.

specifications on the basis of the legacy LTE architecture (EPS) before March 2018. Several options are under discussion, including the ones shown in Figure 8.5. The deployment strategy (among 11 possible strategies) suggests that the 5G design will consist of Next Generation Radio (or New Radio – NR) at the access side and Next Generation Core (NGCN) at the core network side. These two entities may be deployed as standalone entities or may be combined with the LTE radio and core network. 3GPP is taking the direction of independent radio and core migrations, whereby NR and NGCN do not necessarily come together. The key aspect of NGCN will be slicing. Therefore, some operators may want slicing to accommodate new businesses without introducing a new radio (i.e. they may use the existing LTE radio but with NGCN).

The NSA architecture will have a radio access that is LTE-assisted (LTE as an anchor layer) with 5G radio in a dual-connectivity mode, while the core network will remain on the top of the legacy Evolved Packet Core (EPC), as in LTE [3].

On the other hand, there is also strong industry interest in completing the standalone (SA) option 2 and option 4/4a/5/7/7a by the agreed deadline of June 2018. Figure 8.6 shows the possible deployment options under further study now. Both cases will initially deal with eMBB- and URLLC-related use cases [3].

8.4 What is Next for LTE-A Pro Evolution?

From now until 5G realizes real deployment traction in 2020, much of the available mobile network coverage will continue to be provided by LTE. Because 5G will most likely co-exist with LTE and other technologies, such as Wi-Fi access, it becomes important that operators with deployed 4G networks have the opportunity to manage the existing network efficiently and provide a good underlying access layer into 5G, especially for NSA-type 5G deployment. The evolution of LTE to LTE-A introduced three main categories: carrier aggregation above

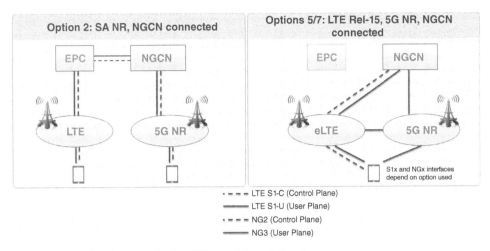

Figure 8.6 5G migration scenarios in 3GPP: standalone (SA) options.

20 MHz bandwidth, higher-order modulation beyond 64QAM, and higher-order MIMO (Multiple-Input, Multiple-Output) beyond 2 × 2. These features can be deployed in order to improve the peak data rates and spectral efficiency.

The theoretical peak data rates increased from 300 Mbps in the downlink and 75 Mbps in the uplink (Release 8) to 3 Gbps in the downlink and 1.5 Gbps in the uplink (Release 12). The most important feature LTE-A introduced to meet these requirements was carrier aggregation to enable peak data rates above 150 Mbps on the downlink and above 50 Mbps on the uplink. The current common deployment uses up to three component carriers (CCs) in the downlink and two CCs in the uplink with up to 450 Mbps and 100 Mbps. Furthermore, 3GPP specifies MIMO extensions to 4 × 4 in the downlink, and also adds higher-order modulation with 256QAM on the downlink and 64QAM on the uplink. LTE-Advanced Pro maximum downlink data rates are expected to exceed 600 Mbps when combining these features, as in Category 15 and 16 deployments shown in various scenarios in Figure 8.7.

Packet latency is another performance metric used by mobile network operators and end users to measure end-to-end Quality of Service. Many existing applications would benefit from reduced latency by improving perceived quality of experience, including real-time applications like Voice over LTE (VoLTE), video telephony, and gaming. Furthermore, the number of delay-critical applications will increase: we will see remote control and autonomous driving of vehicles, augmented reality applications, and specific machine communications requiring low latency as well as highly reliable communications. 3GPP has specified work items in order to improve latencies in the LTE network with concepts that are expected to be carried over to 5G developments at a later stage. All these improvements in LTE-A deployments will reflect positively on the evolution to 5G. It is therefore good to summarize the current gaps in the 4G LTE network and detail how 5G will bridge them over the next few years. Figure 8.8 shows the gaps between LTE and 5G.

8.5 5G Spectrum View

One of the significant design concept changes coming into 5G is enabling cellular transmission with all types of spectrum and bands to support a wide range of new services with different deployment requirements. In order to meet the demand of the increasing traffic capacity and enable the transmission bandwidths needed to support very high data rates at one end and a diversity of use cases at the other end, the 5G design will extend the range of frequencies used for cellular bands. This includes utilizing new and existing spectrum below 6 GHz (sub-6), as well as defining new spectrum for cellular use in higher frequency bands (above 6 GHz).

The standardization and regulation bodies worldwide are defining 5G roadmaps with input from 5G Americas, 5G Forum Korea, 5GMF Japan, 5G-PPP Europe, and the IMT-2020 5G Promotion Group China. The typical alignment mostly takes place between ITU-R (the International Telecommunication Union, Radio communication sector) and 3GPP. In early 2012, ITU-R initiated a program to develop an International Mobile Telecommunication (IMT) system for 2020 and beyond (IMT-2020), thereby officially kicking off the global race towards a yet-to-be-defined 5G mobile network. The vision of this next-generation system began taking shape with ITU-R WP5D proposing a work plan on spectrum and technology timelines [4].

The key ITU-R IMT-2020 roadmap is shown in Figure 8.9. ITU-R will open the evaluation criteria as of October 2017, open the window of proposal submissions as of June 2019, and finally set the specification

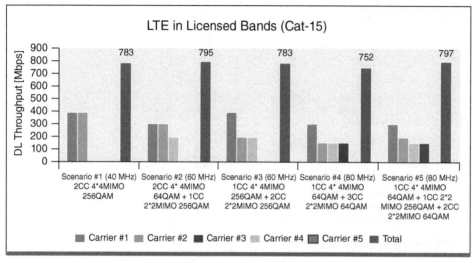

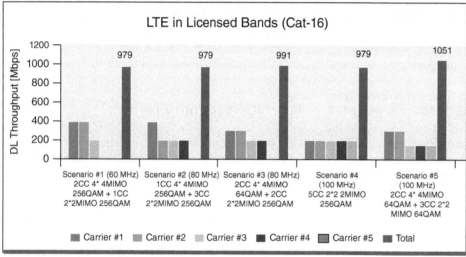

Figure 8.7 LTE-Advanced Pro peak throughput deployment scenarios.

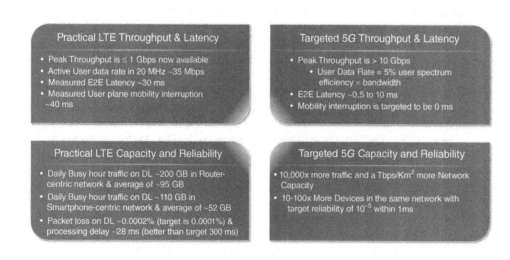

Figure 8.8 5G and LTE gap analysis.

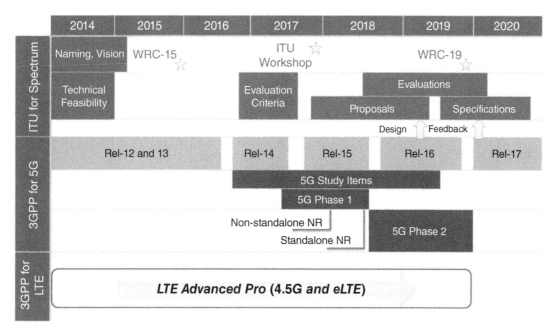

Figure 8.9 Timeline for IMT-2020 (5G) development.

details as of October 2020. In the meantime, 3GPP set up its roadmap to address 5G in two phases, with the final phase in 2019 having its specifications ready for submission to ITU-R as part of IMT-2020 in February 2020. 3GPP started work on 5G in September 2015. The 3GPP specifications for 5G will come in two phases; Figure 8.10 highlights the release timing for 5G Phase 1 and Phase 2. 5G standardization for Phase

1 and Phase 2 will be in Releases 15 and 16, respectively, based on the outcome of the Release 14 study items.

The two 5G phases may be summarized as follows:

5G Phase 1 to be completed towards the end of 2018 (3GPP Release 15) and set up the priorities for:
Spectrum and waveforms up to 40 GHz for both eMBB and low-latency use cases.

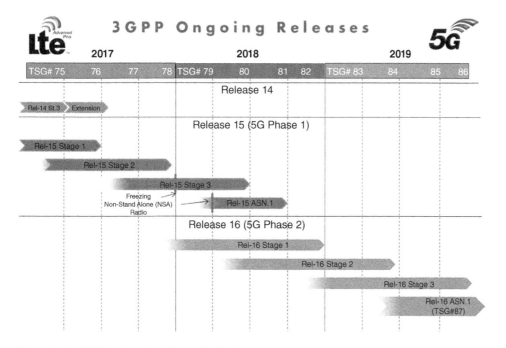

Figure 8.10 3GPP timelines for 5G standardization.

Phase 1

🔊 Network Slicing Support

🔊 QoS Framework

🔊 UE/Mobility Management

🔊 Data Session Management

🔊 Data Session Continuity

🔊 Efficient User Plane path

🔊 Network Function Interaction

🔊 Policy/Charging Control, Security

🔊 Interworking & Migration from 4G

🔊 Support IMS in providing voice, ...

🔊 Network discovery/selection 3GPP

🔊 Network Capability Exposure

Subsequent Phases

🔊 Broadcast/Multicast Capabilities

🔊 Proximity Services

🔊 Communications via Relays

🔊 Off-Network communication

🔊 Netw. discovery/selection non3GPP

🔊 Traffic steering/switching between 3GPP and non3GPP accesses

🔊 Extremely rural deployments

🔊 ...

Figure 8.11 5G features for Phase 1 and Phase 2.

Radio migration to/from LTE, as discussed in the previous section, with non-standalone being the feasible operation, network slicing, mobility session management, basic core network policies and security, and IMS (voice/SMS).

5G Phase 2 to be completed towards the beginning of 2020 (3GPP Release 16) and set up the priorities for:

Spectrum and waveforms above 40 GHz and adding mMTC, URLLC for vehicle-to-vehicle/everything (V2V and V2X) use cases.

Shared and unlicensed spectrum including interworking with other cellular systems.

Additional use cases and services including proximity services, multimedia broadcast services, public warning/emergency alerts, satellite communication, etc.

Figure 8.11 summarizes Phase 1 and Phase 2 features.

The key requirement for NR (New Radio of 5G) design is that it should be forward compatible at its core, so that features can be added in later releases in an optimal way. Therefore, the new RAT should be inherently forward compatible. The new RAT will consider frequency ranges up to 100 GHz. From now until the first phase of 3GPP is completed, the industry needs to address the available spectrum and technical capabilities to address the initial stage of deployment. Therefore, the ITU World Radio Communication Conference 2015 (WRC-15) addressed an extra level of harmonized spectrum for different industries including mobile and wireless communications. One of the key achievements in this context was the allocation of an additional IMT spectrum within 470 MHz to 6 GHz. For example, WRC-15 defined the largest contiguous range of 200 MHz between 3400 and 3600 MHz, known as C-Band. WRC-19 is expected to deal with the range of bands above 6 GHz for IMT-2020, from 24.25–86 GHz, as shown in Figure 8.12.

8.6 5G Design Considerations

The latest plenary meeting of the 3GPP Technical Specifications Groups (TSG #72) has agreed on a detailed work plan for Release 15, the first release of 5G specifications. The plan includes a set of intermediate tasks and checkpoints to guide the ongoing studies in the Working Groups.

The 3GPP TSG RAN further agreed that the target NR scope for Release 15 includes support for the following:

Standalone and non-standalone NR operation (with work for both starting in conjunction and running together):

Non-standalone NR in this context implies using LTE as the control plane anchor. Standalone NR implies full control plane capability for NR.

5G target use cases: Enhanced Mobile Broadband (eMBB) as well as low latency and high reliability to enable some Ultra-Reliable and Low-Latency Communications (URLLC) use cases.

Frequency ranges below 6 GHz and above 6 GHz.

Three usage scenarios defined by ITU-R IMT for 2020 and beyond [4] are envisaged to expand and support diverse families of usage scenarios and applications that will continue beyond the current IMT. Furthermore, a broad variety of capabilities will be coupled tightly with these different usage scenarios and applications intended for IMT for 2020 and beyond. The families of usage scenarios for IMT for 2020 and beyond include:

eMBB (enhanced mobile broadband)

mMTC (massive machine-type communications)

URLLC (ultra-reliable and low-latency communications).

Figure 8.13 illustrates these use cases along with applications.

In order to meet the ITU use cases and applications, Figure 8.14 summarizes the 5G targets. The latency

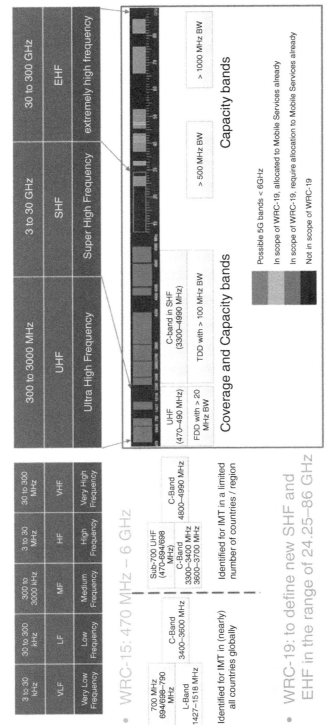

Figure 8.12 WRC spectrum definitions.

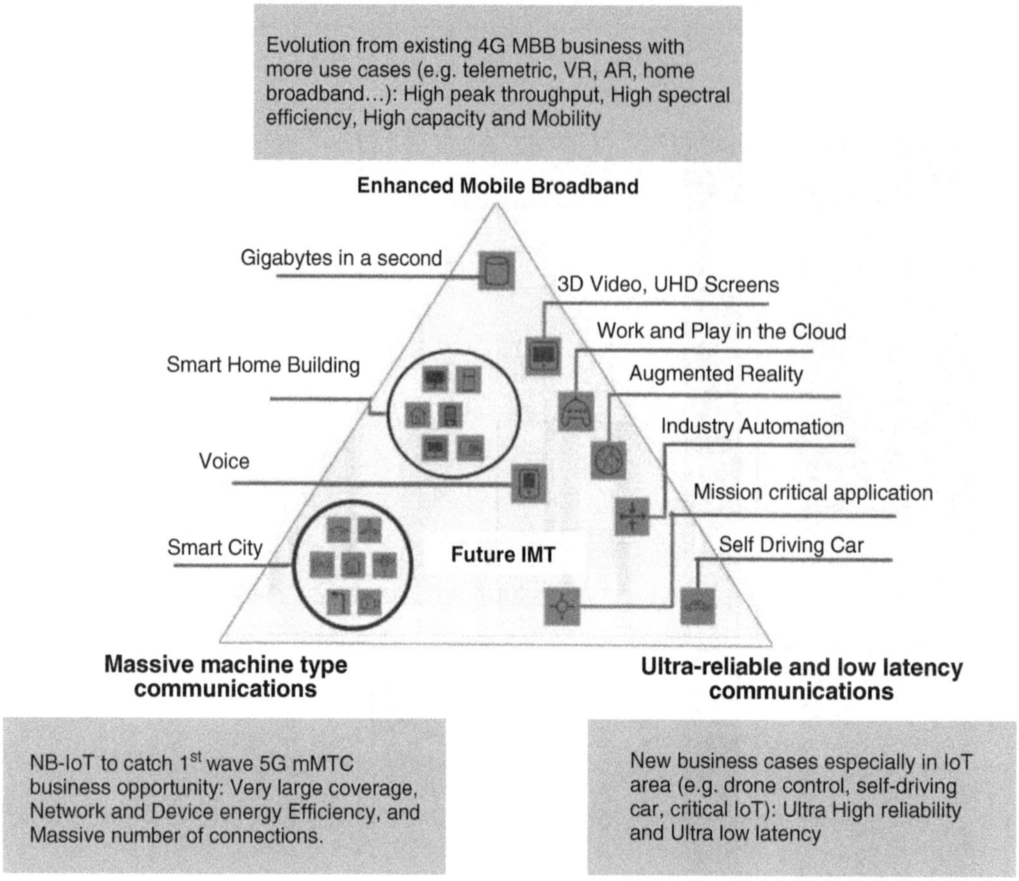

Evolution from existing 4G MBB business with more use cases (e.g. telemetric, VR, AR, home broadband...): High peak throughput, High spectral efficiency, High capacity and Mobility
Enhanced Mobile Broadband
Gigabytes in a second
3D Video, UHD Screens
Work and Play in the Cloud
Smart Home Building
Augmented Reality
Industry Automation
Voice
Mission critical application
Smart City
Future IMT
Self Driving Car
Massive machine type communications
Ultra-reliable and low latency communications
NB-IoT to catch 1st wave 5G mMTC business opportunity: Very large coverage, Network and Device energy Efficiency, and Massive number of connections.
New business cases especially in IoT area (e.g. drone control, self-driving car, critical IoT): Ultra High reliability and Ultra low latency

Figure 8.13 5G use cases and applications.

Peak Data Rate (20 Gbps)
Edge Throughput (100 Mbps)
20
100
IMT-2020
10
1
Area Traffic Capacity (Mbit/s/m2)
10
1
0.1
Spectrum Efficiency
3x
1x
1x
10x
100x
Network Energy Efficiency
350
400
500
Mobility (km/h)
IMT-Advanced
10^5
10^6
10
1
Connection Density (devices/km2)
Latency (1 ms)

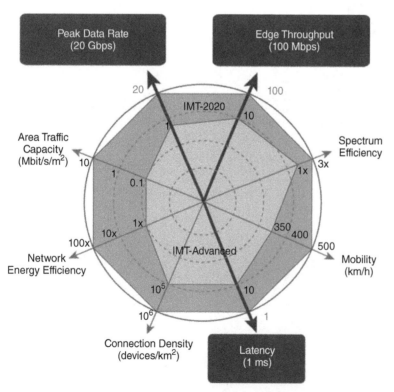

Figure 8.14 5G targets compared to IMT-Advanced

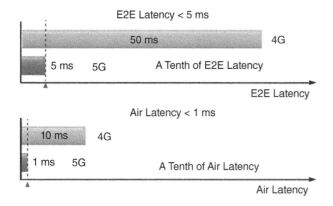

Figure 8.15 5G latency versus 4G.

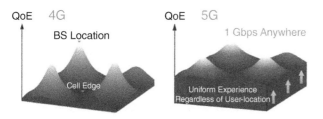

Figure 8.16 Cell-edge throughput comparison.

Whatever the numerology used, a resource block consists of 12 sub-carriers, but for each numerology a minimum and maximum number of resource blocks is defined, and with the knowledge of the corresponding sub-carrier spacing the minimum and maximum bandwidth for the channel could be calculated, as shown in Table 8.1.

There is a mixture of numerology formats in the same deployment to support converged scenarios. Different OFDM formats become mutually non-orthogonal, leading to interference between them. To maintain acceptable performance for any allocation, good scheduling strategy and interference suppression techniques should be used.

The following may be adopted to reduce latency:

A reduction in the length of an OFDM symbol (or an increase in subcarrier spacing) with $\geq$ 60 kHz subcarrier spacing.

A reduction in the number of OFDM symbols within a TTI with 15 kHz subcarrier spacing.

Shorter Transmission Time Interval (TTI).

In order to meet the eMBB requirement, 5G needs to unlock the mmWave bands. Table 8.2 provides a comparison between lower bands and higher bands from different aspects. The sub-6 GHz band will focus on massive MIMO for capacity boosting, while the mmWave bands will focus on beam-steering for coverage gain. The mmWave bands can offer high bandwidth and lower modulation for robustness, while the sub-6 GHz band will go to 100 MHz or a few hundreds due to lack of spectrum resources. Digital beamforming techniques will be implemented in sub-6 GHz based on baseband processing, while the mmWave bands will adopt hybrid architecture based on digital and analog beamforming due to the cost and complexity of RF components. Figure 8.18 provides a comparison between digital beamforming and hybrid beamforming from the antenna

expectation and comparison with 4G are demonstrated in Figure 8.15. Finally, a cell-edge throughput comparison is shown in Figure 8.16 [5–7].

The air interface for 5G should be flexible and assign resources dynamically to meet different use case requirements. Figure 8.17 demonstrate 5G frame structure.

It was agreed that max channel bandwidth is 100 MHz for sub-6 and 400 MHz for mm Wave. Compared to LTE, 5GNR is designed to have higher bandwidth efficiency, reaching 99% (compared to 90% in LTE, where 100 RB covered only 18 MHz in a 20 MHz Bandwidth carrier).

Another difference with LTE is that there is no explicit DC sub-carrier reserved both for downlink and uplink.

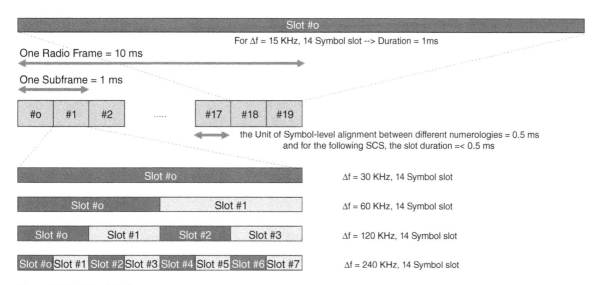

Figure 8.17 5G flexible frame structure concept.

Table 8.1 Scalable and flexible 5G design

μ	$\Delta f = 15 \times 2^\mu$	Cyclic prefix	N^{slot}_{symb}	$N^{frame,\mu}_{slot}$	$N^{subframe,\mu}_{slot}$	min RB	Max RB	Frequency BW min (MHz)	Frequency BW max (MHz)
					Slot Configuration				
0	15	Normal	14	10	1	24	275	4.32	49.5
1	30	Normal	14	20	2	24	275	8.64	99
2	60	Normal	14	40	4	24	275	17.28	198
		Extended	12	40	4				
3	120	Normal	14	80	8	24	275	34.56	396
4	240	Normal	14	160	16	24	138	69.12	397.44

Table 8.2 Comparison between sub-6 GHz and mmWave bands.

3 GHz →	28 GHz →	70 GHz →
Bandwidth limited	↔	Huge bandwidth
Interference limited	↔	Noise limited
Emphasis on spectral efficiency	↔	Emphasis on gain
Pre-antenna channel knowledge	↔	Pre-beam channel knowledge
Baseband architecture	↔	Hybrid/RF architecture
Small-scale array: SU-MIMO sufficient. Large-scale array: High-order MU-MIMO.	↔	Large-scale arrays are required, with an initial emphasis on SU-MIMO.

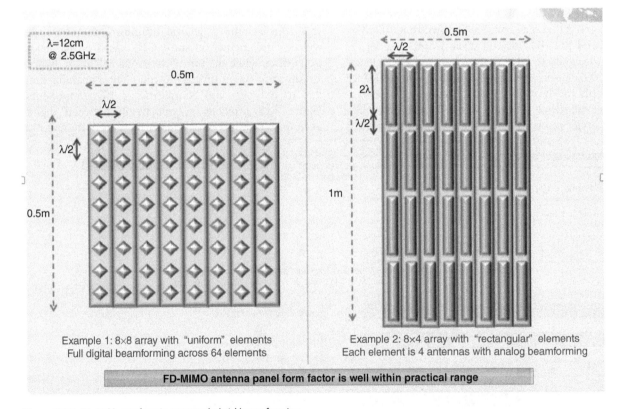

Figure 8.18 Digital beamforming versus hybrid beamforming.

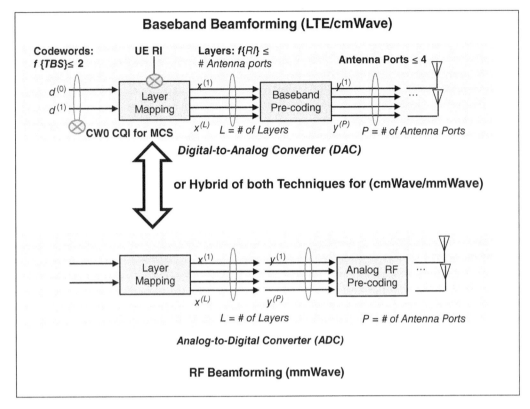

Figure 8.19 Comparison between the RF beamforming in 4G LTE (cmWave) and 5G (mmWave).

array point of view. Figure 8.19 provides an architecture comparison between beamforming techniques.

Analog processing in the RF domain with variable phase shifters becomes available; there is no need to support each antenna with a dedicated RF chain. Hybrid baseband and RF processing have to strike a balance between system complexity and capacity.

The lower operating frequencies (sub-6 GHz) are more interference-limited:

LTE is already designed for high spectral efficiency (< 8 antenna ports).
Low bandwidth leads to a need for capacity-enhancing solutions.
Emphasis on improved spectrum efficiency (e.g. with multi-user spatial multiplexing).
Full digital array.
Extend Release 13/14 full-dimension MIMO framework to 64, 128, 256 total physical elements.
Full support for SU-MIMO and MU-MIMO.

The higher operating frequencies (mmWave) have poor path-loss conditions:

A single carrier waveform for higher frequencies leads to the need for coverage-enhancing solutions.
Emphasis on cell range extension by beamforming gain.
RF array is most likely to be considered.

Baseband architectures are not viable initially for very high bandwidths.
DAC/ADC high power consumption at high bandwidth.
Can support RF or a hybrid structure as well.
Support for SU-MIMO.

The massive MIMO gain at lower bands is expected to boost the site capacity. However, it is important to note that this capacity boosting is not linear compared to the array size. The maximum number of special streams is expected to be eight and this will happen only in rich, multi-path environments. This is why the move to mmWave band is mandatory for high-density capacity gain, such as small-cell and pico-cell scenarios. The expected gain from massive MIMO is shown in Figure 8.20.

As is evident from Figure 8.20, a threefold capacity gain can be attained with a 64-antenna element array. Potential 5G technologies are summarized in Table 8.3.

Figure 8.21 illustrates the three potential bands for 5G and relevant design criteria. TDD is the main duplexing scheme for cmWave and mmWave due to channel reciprocity, which is more suitable for beam-steering algorithms. The characteristics of cm/mmWave may be summarized as follows:

5G small cells in 6–30 GHz band (cmWave) with 500 MHz carrier bandwidth can provide hundreds of Gbps / km^2.

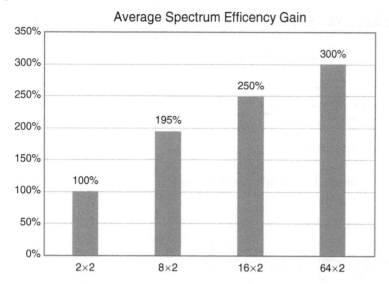

Figure 8.20 Massive MIMO gain.

Table 8.3 5G potential technologies.

Solution	Benefit	Solution	Benefit
Usage of cm and mm waves	10x. 100x more capacity	Enhanced interference coordination	Higher efficiency
UE agnostic MIMO and beamforming	Network-based massive MIMO evolution	Aggregation of LTE + 5G carriers	Higher data rate with smooth migration
Lean carrier design	Low power consumption, less interference	Wireless backhaul with full duplex	Improved performance
Flexible frame structure	Low latency, high efficiency	Flexible connectivity, mobility, and sessions	Optimized end-to-end for any services
Dynamic TDD	Improved performance	New waveforms	Multi-service flexibility

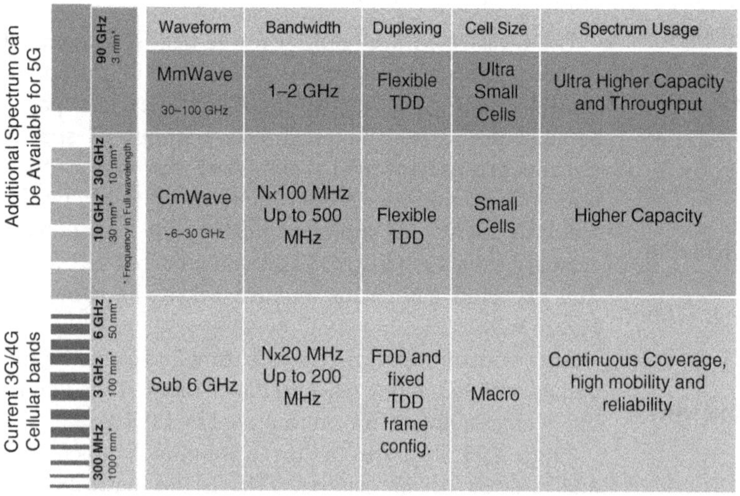

Figure 8.21 5G potential bands and relevant design criteria.

	Waveform	Bandwidth	Duplexing	Cell Size	Spectrum Usage
90 GHz 3 mm*	MmWave 30–100 GHz	1–2 GHz	Flexible TDD	Ultra Small Cells	Ultra Higher Capacity and Throughput
10 GHz / 30 GHz / 10 mm* / 30 mm*	CmWave ~6–30 GHz	Nx100 MHz Up to 500 MHz	Flexible TDD	Small Cells	Higher Capacity
6 GHz / 3 GHz / 300 MHz / 50 mm* / 100 mm* / 1000 mm*	Sub 6 GHz	Nx20 MHz Up to 200 MHz	FDD and fixed TDD frame config.	Macro	Continuous Coverage, high mobility and reliability

Additional Spectrum can be Available for 5G

Current 3G/4G Cellular bands

* Frequency in Full wavelength

5G small cells in up to 100 GHz band (mmWave) with 2 GHz carrier bandwidth can provide a Tbps / km². Both can further provide backhaul to the small cells. Challenges can be found in antennas, RF, beamforming, and new spectrum to be secured (ITU WRC-18/19).

Higher bandwidths have greater propagation losses. Massive MIMO can compensate path loss for cmWave by using high beamforming gains and many MIMO streams. Massive MIMO may not be required for mmWave, as the small wavelengths enable the use of a large number of

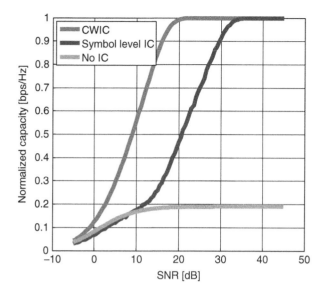

Figure 8.22 Example MU capacity performance by different UE IC capability.

antennas that can be used to generate narrow beams used for spatial multiplexing with high energy.

Massive MIMO at sub-6 GHz with MU-MIMO will need Interference Cancellation (IC) at the UE side for the following reasons: Multi-user (MU) transmission via the same channel may create non-orthogonality; it is difficult to separate signals in the spatial domain due to the air interface being crowded with many users. Therefore, MU interference limits the capacity of the MU-MIMO channel.

Sub-6 GHz massive MIMO for small cells is important, because greater than 6 GHz spectrum may not be available for 2020 deployment.

Wider beamwidths from the smaller, 6 GHz massive MIMO antenna result in high interference on the downlink.

Figure 8.22 demonstrates the capacity gain of interference cancellation with MU-MIMO. To cancel MU interference, two techniques are demonstrated in Figure 8.22, as follows [8]:

Symbol-level IC (SLIC): In this technique, the interfering modulation symbols are detected without decoding.

Codeword IC (CWIC): In this technique, the interfering data are detected and decoded before deploying the IC technique.

As indicated in Figure 8.22, the CWIC provides better performance at the cost of high receiver complexity.

8.6.1 IEEE WiGig Evolution

Similar to the 3GPP 5G effort, the IEEE Wi-Fi 802.11xx standard has also been evolved to the mmWave band at unlicensed 60 GHz (57 GHz to 66 GHz), i.e., 802.11ad (a.k.a. WiGig) [9–11]. Table 8.4 provides a comparison between LTE, 802.11ad, and the potential 5G standard at eBand. The unlicensed frequency allocations at around 60 GHz in each region do not match exactly, but there is substantial overlap: at least 3.5 GHz of contiguous spectrum is available in all regions that have allocated spectrum. Unlike the 2.4 GHz and 5 GHz unlicensed bands, the 60 GHz area is also relatively uncongested. Transmission at 60 GHz covers less distance for a given power, mainly due to the increased free space path loss (loss over 1 m at 60 GHz is 68 dB, which is 21.6 dB worse than for the 5 GHz band), compounded by propagation losses through materials and human body shadowing (losses from a few dB to 30 dB+).

Figure 8.23 demonstrates the atmospheric absorption at 60 GHz [12]. This chart shows the average atmospheric absorption of millimeter-wave signals, under some nominal assumptions for environmental moisture, altitude,

Table 8.4 Comparison between LTE at sub-6 GHz and mmWave technologies.

	LTE		802.11ad		Potential 5G (MMW)
Frequency band	< 6 GHz		60 GHz		70 GHz
Supported bandwidths	TBD		2160 MHz		2000 MHz
Maximum QAM	64	16	64		64
Modulation	OFDM	SC	OFDM		NullCP-SC
Channel spacing (B)	20 MHz	2.16 GHz	2.16 GHz		2 GHz
FFT size	2048	512	512		1024
Subcarrier spacing	15 kHz	4.2 MHz	5.1 MHz		1.5 MHz
Sampling frequency	3.072 MHz	1.76 GHz	2.46 GHz		1.54 GHz
T sampling	32.6 ns	5.68 ps	406 fs		651 fs
T symbol	66.7 μs	245 ns	198 ns		666.7 ns
T guard	4.7 μs	36.4 ns	52 ns		10.4 ns
T	71.4 μs	291 ns	250 ns		666.7 ns

Figure 8.23 Atmospheric absorption at 60 GHz.

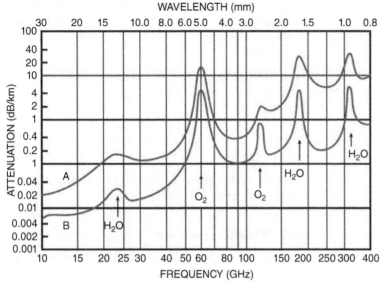

Average atmospheric absorption of MMWs: (A) Sea Level: T = 20 °C,
P = 760 mm, H_2O = 7.5 g/m^3 (B) 4 km altitude: T = 0 °C H_2O = 1 g/m^3

etc. The chart only goes down to 10 GHz, but it can be assumed that lower-frequency signals have absorption less than or equal to that shown in this chart. We can see that absorption generally goes up as frequency increases. There are a few places where there is an absorption peak, such as the oxygen peak at 60 GHz, but significant peaks such as these do not occur below 10 GHz. The substantial RF absorption peak in the 60 GHz band is due to a resonance of atmospheric oxygen molecules, and this is often cited as a limitation on range in this band. However, this absorption effect only starts to become significant at > 100 m range, which is fine for the WiGig application. Thus, low-power transmissions will not propagate very far, but this is considered an advantage; it reduces the likelihood of co-channel interference and increases the possible frequency re-use density. Another perceived advantage of limited range is the reduced opportunity for "theft" of protected content by eavesdropping on nearby transmissions. (In fact, 60 GHz was first proposed for battlefield communications for this very reason.)

High path loss can be mitigated by increasing antenna gain. The small geometries required at 60 GHz permit the use of fabrication techniques such as Low-Temperature Co-fired Ceramic (LTCC) and thermoplastic substrates to create suitable, physically small, high-gain directional antennas. Multiple-antenna configurations using beam-steering are an optional feature of the WiGig specifications. Beam-steering can be employed to circumnavigate minor obstacles like people moving around a room or a piece of furniture blocking line-of-sight transmission, but longer free-space distances (e.g. > 10 m) and more substantial obstructions (e.g. walls, doors, etc.) will prevent transmission. It would be unlikely, for example, for a media server in one room

to be able to transmit HD video directly and reliably to a display in another room. The 60 GHz spectrum allocation is summarized in Figure 8.24 [9]. The ITU-R's recommended channelization comprises four channels, each 2.16 GHz wide, centered on 58.32 GHz, 60.48 GHz, 62.64 GHz, and 64.80 GHz. As Figure 8.24 illustrates, not all channels are available in all countries. Channel 2, which is globally available, is therefore the default channel for equipment operating in this frequency band [9].

Table 8.4 shows a comparison between LTE at sub-6 GHz, IEEE 802.11ad at 60 GHz, and a B4GMMW (Beyond 4G mmWave) prototype developed by Nokia and National Instrument at eBand, i.e. 70 GHz.

The spectrum mask of 802.11ad is shown in Figure 8.25 and is expressed in decibels relative to the signal level at the band center (dBr). This mask is significantly different to the masks specified at lower frequencies. The breakpoints at −20 dBr have been pushed out slightly to accommodate both OFDM and Single-Carrier (SC) modulations, and the adjacent channel requirements have been considerably relaxed, specifically to ease the circuit design challenges at 60 GHz by permitting higher levels of out-of-band distortion. The maximum permitted transmitter power varies by country, but, in general, +10 dBm can be taken as a practical limit [9–11].

The IEEE 802.11ad-2012 Directional Multi-Gigabit (DMG) physical layer (PHY) supports three distinct modulation methods [9–11]:

Spread-spectrum modulation: the Control PHY.
Single-Carrier (SC) modulation: the Single-Carrier PHY and the Low-Power, Single-Carrier PHY.
Orthogonal Frequency Division Multiplex (OFDM) modulation: the OFDM PHY.

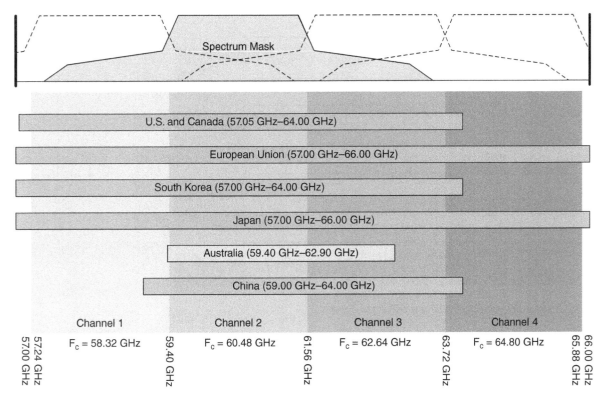

Figure 8.24 Channel and region allocation of unlicensed 60GHz band ITU-R WP 5A.

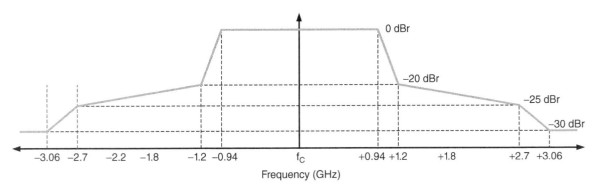

Figure 8.25 IEEE 802.11ad spectrum mask.

Each PHY type has a distinct purpose and packet structure, shown in Figure 8.26, but care has been taken to align the packet structures, and in particular the preambles, to simplify signal acquisition, processing, and PHY-type identification in the receiver.

The specification tabulates 32 different modulation and coding schemes [9–11]. Thus, we can quickly simplify the picture by dividing the MCS list into four basic classifications, as shown in Table 8.5. It is clearly important for the reliable establishment and maintenance of connectivity that the control channel should be as robust as possible. The purpose of the control PHY and the reasons for its emphasis on reliability over raw speed are considered evident. It is perhaps less clear why so many

MCSs are required. Given the anticipated diversity of device type that will want to support 802.11ad, there are persuasive arguments for and against both OFDM- and SC-based modulations, and for seriously constrained devices, there is a further argument in favor of trading the strength of LDPC-based error correction for further power savings. Within each of the SC, OFDM, and LPSC categories, the specific MCS selects a different pairing of error-protection coding and modulation depth, which, taken together, provide the user with a logical progression of link quality versus throughput operating points.

We can summarize the MCS of 802.11ad [9–11] as follows:

Very robust 27.5 Mbps control channel.

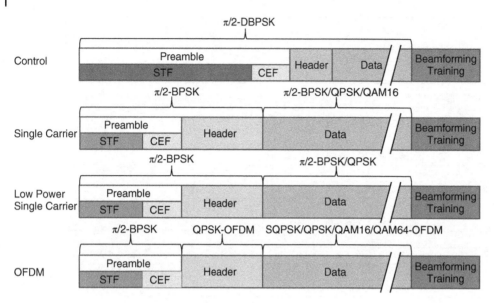

Figure 8.26 Packet structures for each of the three modulation types.

Table 8.5 Modulation and coding schemes of P802.11ad ITU-R WP 5A.

Control (CPHY)			
MCS	Coding	Modulation	Raw Bit Rate
0	1/2 LDPC, 32x spreading	p/2-DBPSK	27.5 Mbps
Single Carrier (SC-PHY)			
MCS	Coding	Modulation	Raw Bit Rate
1–12	1/2 LDPC, 2x repetition 1/2 LDPC, 5/8 LDPC, 3/4 LDPC, 13/16 LDPC	p/2-BPSK, p/2-QPSK, p/2-16QAM	385 Mbps to 4620 Mbps
Orthogonal Frequency Division Multiplex (OFDM-PHY)			
MCS	Coding	Modulation	Raw Bit Rate
13–24	1/2 LDPC, 5/8 LDPC, 3/4 LDPC, 13/16 LDPC	OFDM-SQPSK, OFDM-QPSK, OFDM-16QAM, OFDM-64QAM	693 Mbps to 6756.75 Mbps
Low-Power Single-Carrier (LPSC-PHY)			
MCS	Coding	Modulation	Raw Bit Rate
25–31	RS (224,208) + Block Code (16/12/9/8,8)	p/2-BPSK, p/2-QPSK	625.6 Mbps to 2503 Mbps

Variable error protection.

Variable modulation complexity.

Therefore, the Error Vector Magnitude (EVM) is specified from −6 dB to −25 dB.

Variable data rates from 385 Mbps (MCS1) to 6756.75 Mbps (MCS24).

Mandatory modes ensure all 802.11ad devices are capable of at least 1 Gbps.

MCS0–4 mandatory.

MCS13–16, if OFDM invoked.

8.7 5G Deployment Scenarios for Mobile Applications

Deployment scenarios for eMBB, mMTC, and URLLC are described in this section. Other deployment scenarios related to eV2X (enhanced Vehicle-to-Everything) services are also described. However, some of the eMBB deployment scenarios may possibly be re-used to evaluate mMTC and URLLC, or some specific evaluation tests (e.g., link-level simulation) can be developed to check whether the requirements can be achieved. High-level descriptions of deployment scenarios including carrier frequency, aggregated system bandwidth, network layout/ISD, BS/UE antenna elements, UE distribution/speed and service profile are proposed. It is assumed that more detailed attributes and simulation parameters, for example, the channel model, BS/UE Tx power, the number of antenna ports, etc., will be defined in the 3GPP study item [13]. Table 8.6 provides four deployment scenarios; these are the legacy mobile site deployments.

The indoor hotspot deployment scenario focuses on small coverage per site/TRPx (transmission and reception point) and high user throughput or user density in buildings. The key characteristics of this deployment scenario are high capacity, high user density, and a consistent user experience indoors.

The dense urban microcellular deployment scenario focuses on macro TRPxs with or without micro TRPxs and high user densities and traffic loads in city centers and dense urban areas. The key characteristics of this deployment scenario are high traffic loads, outdoor and outdoor-to-indoor coverage. This scenario will be interference-limited, using macro TRPxs with or without micro TRPxs. A continuous cellular layout and the associated interference are assumed.

The urban macro deployment scenario focuses on large cells and continuous coverage. The key characteristics of this scenario are continuous and ubiquitous coverage in urban areas. This scenario will be interference-limited using macro TRPxs (i.e. radio access points above rooftop level).

Finally, the rural deployment scenario focuses on larger and continuous coverage. The key characteristic of this scenario is continuous wide-area coverage supporting high-speed vehicles. This scenario will be noise-limited and/or interference-limited using macro TRPxs.

The high-speed deployment scenario focuses on continuous coverage along tracks in high-speed trains. The key characteristics of this scenario are consistent passenger user experience and critical train communication reliability with very high mobility. In this deployment scenario, dedicated linear deployment along railway lines and deployments including Single Frequency Network (SFN) scenarios are considered, and passenger UEs are located in train carriages. For the passenger UEs, if the antenna of the relay node for eNB-to-Relay is located on top of one carriage of the train, the antenna of the relay node for Relay-to-UE could be distributed to all carriages. Recommended parameters for high-speed train deployment are provided in Table 8.7. Two high-speed train deployment scenarios are illustrated in Figure 8.27 and Figure 8.28 for sub-6 and mmWave deployment, respectively.

Three other specific deployment scenarios (highway, extreme rural, and urban grid for connected car) are

summarized in Table 8.8. The highway deployment scenario focuses on vehicles on highways with high speeds. The main KPIs evaluated under this scenario are reliability/availability under high speeds/mobility (and thus frequent handover operations). The extreme long range deployment scenario is defined to allow for the provision of services for very large areas with a low density of users, whether they are humans or machines (e.g. low ARPU regions, wilderness, areas where only highways are located, etc.). The key characteristics of this scenario are macro cells with very large area coverage supporting basic data speeds and voice services, with low to moderate user throughput and low user density. The urban macro deployment scenario focuses on high numbers of densely deployed vehicles in an urban area. It could cover a scenario where freeways lead through an urban grid. The main KPIs evaluated under this scenario are reliability/availability/latency in high network load and high UE density scenarios.

Finally, the urban coverage for massive connection scenario focuses on large cells and continuous coverage to provide mMTC. The key characteristics of this scenario are continuous and ubiquitous coverage in urban areas, with very high connection density of mMTC devices. This deployment scenario is for the evaluation of the KPI of connection density. Recommended parameters for this scenario are listed in Table 8.9.

8.8 Air-to-Ground and Satellite Scenarios

Two air-to-ground communications deployment scenarios are described in Table 8.10. The commercial air-to-ground deployment scenario is defined to allow for the provision of services for commercial aircraft, to enable both humans and machines aboard the aircraft to initiate and receive mobile services. It is not for the establishment of airborne-based base stations. The key characteristics of this scenario are upward-pointing macro cells with very large area coverage supporting basic data and voice services with moderate user throughput that are optimized for high-altitude users traveling at very high speeds. The commercial airlines' aircraft are likely to be equipped with an aggregation point (e.g. relay). The light aircraft scenario is defined to allow for the provision of services for general aviation aircraft to enable both humans and machines aboard helicopters and small airplanes to initiate and receive mobile services. It is not for the establishment of airborne-based base stations. The key characteristics of this scenario are upward-pointing macro cells with very large area coverage supporting basic data and voice services with moderate user throughput and low user density that are optimized for moderate-altitude users

Table 8.6 5G deployment scenarios for legacy mobile clusters.

Attributes	Indoor Hotspot	Dense Urban	Urban Macro	Rural
Carrier frequency	Around 30 GHz or around 70 GHz or around 4 GHz	Around 4GHz+ around 30 GHz (two layers)	Around 2 GHz or around 4 GHz or around 30 GHz	Around 700 MHz or around 4 GHz (for ISD 1) Around 700 MHz and around 2 GHz combined (for ISD 2)
Aggregated system bandwidth	Around 30 GHz or around 70 GHz: Up to 1 GHz (DL + UL) Around 4 GHz: Up to 200 MHz (DL + UL)	Around 30 GHz: Up to 1 GHz (DL + UL) Around 4 GHz: Up to 200 MHz (DL + UL)	Around 4 GHz: Up to 200 MHz (DL + UL) Around 30 GHz: Up to 1 GHz (DL + UL)	Around 700 MHz: Up to 20 MHz (DL + UL) Around 4 GHz: Up to 200 MHz (DL + UL)
Layout	Single layer: – Indoor floor (Open office)	Two layers: – Macro layer: Hex. Grid – Micro layer: Random drop Step 1 : Around 4 GHz in Macro layer Step 2: Both around 4 GHz and around 30 GHz may be available in Macro and Micro layers (including 1 macro layer, macro cell only)	Single layer: – Hex. Grid	Single layer: – Hex. Grid
ISD	20 m (Equivalent to 12 TRPxs per 120 m × 50 m)	Macro layer: 200 m Micro layer: 3 micro TRPxs per macro TRPx All micro TRPxs are outdoor	500 m	ISD 1: 1732 m ISD 2: 5000 m
BS antenna elements	Around 30 GHz or around 70 GHz: Up to 256 Tx and Rx antenna elements Around 4 GHz: Up to 256 Tx and Rx antenna elements	Around 30 GHz: Up to 256 Tx and Rx antenna elements Around 4 GHz: Up to 256 Tx and Rx antenna elements	Around 30 GHz: Up to 256 Tx and Rx antenna elements Around 4 GHz or around 2 GHz: Up to 256 Tx and Rx antenna elements	Around 4 GHz: Up to 256 Tx and Rx antenna elements Around 700 MHz: Up to 64 Tx and Rx antenna elements
UE antenna elements	Around 30 GHz or around 70 GHz: Up to 32 Tx and Rx antenna elements Around 4 GHz: Up to 8 Tx and Rx antenna elements	Around 30 GHz: Up to 32 Tx and Rx antenna elements Around 4 GHz: Up to 8 Tx and Rx antenna elements	Around 30 GHz: Up to 32 Tx and Rx antenna elements Around 4 GHz: Up to 8 Tx and Rx antenna elements	Around 4 GHz: Up to 8 Tx and Rx antenna elements Around 700 MHz: Up to 4 Tx and Rx antenna elements
User distribution and UE speed	100% indoor, 3 km/h, 10 users per TRPx	Step1: Uniform/macro TRPx, 10 users per TRPx Step2: Uniform/macro TRPx+ Clustered/micro TRPx, 10 users per TRPx 80% indoor (3 km/h), 20% outdoor (30 km/h)	20% outdoor in cars: 30 km/h, 80% indoor in houses: 3 km/h 10 users per TRPx	50% outdoor vehicles (120 km/h) and 50% indoor (3 km/h), 10 users per TRPx
Service profile	Whether to use full-buffer traffic or non-full-buffer traffic is FFS. For certain KPIs, full-buffer traffic is desirable to enable comparison with IMT-Advanced values.	Whether to use full-buffer traffic or non-full-buffer traffic is FFS. For certain KPIs, full-buffer traffic is desirable to enable comparison with IMT-Advanced values.	Whether to use full-buffer traffic or non-full-buffer traffic is FFS. For certain KPIs, full-buffer traffic is desirable to enable comparison with IMT-Advanced values.	Whether to use full-buffer traffic or non-full-buffer traffic is FFS. For certain KPIs, full-buffer traffic is desirable to enable comparison with IMT-Advanced values.

Table 8.7 5G deployment scenario for high-speed trains.

Parameter	Recommendations for High-speed Train Deployment
Carrier frequency	Macro only: Around 4 GHz Macro +relay nodes: 1) For BS to relay: Around 4 GHz For relay to UE: Around 30 GHz or around 70 GHz or around 4 GHz 2) For BS to relay: Around 30 GHz For relay to UE: Around 30 GHz or around 70 GHz or around 4 GHz.
Aggregated system bandwidth	Around 4 GHz: Up to 200 MHz (DL + UL) Around 30 GHz or around 70 GHz: Up to 1GHz (DL + UL)
Layout	Macro only: Around 4 GHz: Dedicated linear deployment along the railway line, as in Figure 8.27. RRH site to railway track distance: 100 m. Macro + relay nodes: Around 4 GHz: Dedicated linear deployment along the railway line, as in Figure 8.27. RRH site to railway track distance: 100 m. Around 30 GHz: Dedicated linear deployment along the railway line, as in Figure 8.28. RRH site to railway track distance: 5 m.
ISD	Around 4 GHz: ISD 1732 m between RRH sites, two TRPxs per RRH site. See Figure 8.27. Around 30 GHz: 1732 m between BBU sites, 3 RRH sites connected to 1 BBU, one TRPx per RRH site, inter-RRH site distance (580 m, 580 m, 572 m). See Figure 8.28. Small cell within carriages: ISD = 25 m.
BS antenna elements	Around 30 GHz: Up to 256 Tx and Rx antenna elements Around 4 GHz: Up to 256 Tx and Rx antenna elements
UE antenna elements	Relay Tx: Up to 256 antenna elements Relay Rx: Up to 256 antenna elements Around 30 GHz: Up to 32 Tx and Rx antenna elements Around 4 GHz: Up to 8 Tx and Rx antenna elements
User distribution and UE speed	100% of users in train For non-full buffer, 300 UEs per macro cell (assuming 1000 passengers per high-speed train and at least 10% activity ratio) Maximum mobility speed: 500 km/h
Service profile	Option 1: Full buffer Option 2: FTP model 1/2/3 with packet size 0.5 Mbytes, 0.1 Mbytes (another value is not precluded) Other traffic models are not precluded, e.g., for critical train communications.

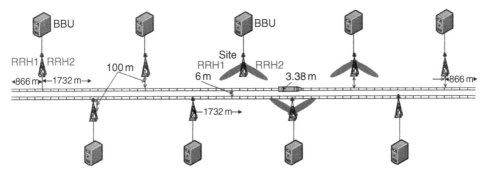

Figure 8.27 4 GHz deployment.

who might be traveling at high speeds. General regime aircraft are not equipped with relays.

The satellite extension to the terrestrial deployment scenario (see Table 8.11) is defined to allow for the provision of services for those areas where the terrestrial service is not available and also for those services that can be more efficiently supported by satellite systems, such as the broadcasting service. Satellite acts as a fill-in, especially on roadways and in rural areas where the terrestrial service is not available. The supported services via the satellite system are not limited to just data and voice, but also include others such as machine-type communications, broadcasting, and other delay-tolerant services. "Bent pipe" refers to the architecture where the satellite

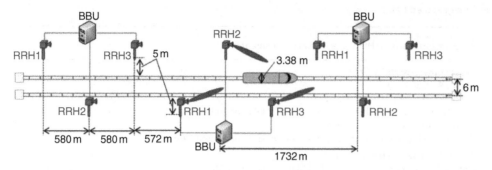

Figure 8.28 30 GHz deployment.

Table 8.8 Highway, extreme rural, and urban grid for connected car deployment scenarios.

Parameter	Highway Scenario	Extreme Rural	Urban Grid for Connected Car
Carrier frequency	Macro only: Below 6 GHz (around 6 GHz) Macro + RSUs: 1) For BS to RSU: Below 6 GHz (around 6 GHz) 2) RSU to vehicles or among vehicles: Below 6 GHz	Below 3 GHz With a priority on bands below 1 GHz Around 700 MHz	Macro only: Below 6 GHz (around 6 GHz) Macro + RSUs: 1) For BS to RSU: Below 6 GHz (around 6 GHz) 2) RSU to vehicles or among vehicles/pedestrians: Below 6 GHz
Aggregated system bandwidth	Up to 200MHz (DL + UL) Up to 100 MHz (SL)	40 MHz (DL + UL)	Up to 200 MHz (DL + UL) Up to 100 MHz (SL)
Layout	Option 1: Macro only Option 2: Macro + RSUs	Single layer: Isolated macro cells	Option 1: Macro only Option 2: Macro + RSUs
ISD/Cell range	Macro cell: ISD = 1732m, 500 m (Optional) Inter-RSU distance = 50 m or 100 m	100 km range (isolated cell) to be evaluated through system-level simulations. Feasibility of higher range will be evaluated through link-level evaluation (for example, in some scenarios ranges up to 150–300 km may be required).	Macro cell: ISD = 500 m RSU at each intersection for Option 2. Other values (50 m and 100 m) should also be considered for Option 2
BS antenna elements	Tx: Up to 256 Tx Rx: Up to 256 Rx		Tx: Up to 256 Tx Rx: Up to 256 Rx
UE antenna elements	RSU Tx: Up to 8 Tx RSU Rx: Up to 8 Rx Vehicle Tx: Up to 8 Tx Vehicle Rx: Up to 8 Rx		RSU Tx: Up to 8 Tx RSU Rx: Up to 8 Rx Vehicle Tx: Up to 8 Tx Vehicle Rx: Up to 8 Rx Pedestrian/bicycle Tx: Up to 8 Tx Pedestrian/bicycle Rx: Up to 8 Rx
User distribution and UE speed	100% in vehicles Average inter-vehicle distance (between two vehicles' center) in the same lane is 0.5 seconds or 1 second * average vehicle speed (average speed: 100–300 km/h)	User density: Speed up to 160 km/h	Urban grid model (car lanes and pedestrian/bicycle sidewalks are placed around a road block. Two lanes in each direction, four lanes in total, one sidewalk, one block size: 433 m × 250 m) Average inter-vehicle distance (between two vehicles' center) in the same lane is 1 second * average vehicle speed (average speed 15–120 km/h) Pedestrian/bicycle dropping: Average distance between UEs is 20 m
Traffic model	50 messages per 1 second with absolute average speed of either 100–250 km/h (relative speed: 200–500 km/h) or −30 km/h	Average data throughput at busy hours/user: 30 kbps User experienced data rate: Up to 2 Mbps DL while stationary and 384 kbps DL while moving	50 messages per 1 second with 60 km/h, 10 messages per 1 second with 15 km/h

Table 8.9 Parameters for the urban coverage for massive connection deployment scenario.

Attribute	Values or Assumptions
Carrier frequency	700 MHz, 2100 MHz as an option
Network deployment including ISD	Macro only, ISD = 1732 m, 500 m
Device deployment	Indoor and outdoor in-car devices
Maximum mobility speed	20% of users are outdoor in cars (100 km/h) or 20% of users are outdoors (3 km/h)
	80% of users are indoors (3 km/h)
	Users dropped uniformly in entire cell
Service profile	Non-full buffer with small packets
BS antenna elements	2 and 4 Rx ports (8 Rx ports as optional)
UE antenna elements	1 Tx

satellites are deployed in a wide range of frequency bands including L band (1–2 GHz), S band (2–4 GHz), C band (3.4–6.725 GHz), Ku band (10.7–14.8 GHz), Ka band (17.3–21.2 GHz, 27.0–31.0 GHz), and Q/V bands (37.5–43.5 GHz, 47.2–50.2 GHz, and 50.4–51.4 GHz) and more.

8.9   5G Evaluation KPIs

Some of the 5G key performance indicators that are mostly used for benchmarking and evaluation are listed below:

Peak data rate: The target for peak data rate should be 20 Gbps for downlink and 10 Gbps for uplink.
Peak spectral efficiency: The target for peak spectral efficiency should be 30 bps/Hz for downlink and 15 bps/Hz for uplink.
Control plane latency: Control plane latency refers to the time to move from a battery-efficient state (e.g., Idle) to the start of continuous data transfer (e.g., Active).
The target for control plane latency should be 10 ms.
For a satellite communications link, the control plane should be able to support RTT of up to 600 ms in the case of GEO and HEO, up to 180 ms in the case of MEO, and up to 50 ms in the case of LEO satellite systems.
User plane latency: The time it takes to deliver an application layer packet/message successfully from the radio protocol layer 2/3 SDU ingress point to the

transponders are transparent – they only amplify and change frequency but preserve the waveform. On-board processing satellite transponders incorporate regeneration – including modulating and coding the waveform. A "mobile" constitutes both handhelds and other moving platform receivers such as automobiles, ships, planes, etc. Currently, handhelds are limited to L and S bands but research is ongoing to support higher bands. The carrier frequencies noted here are for evaluation purposes only;

Table 8.10 Air-to-ground deployment scenarios

Parameter	Commercial Air-to-Ground Deployment Scenario	Light Aircraft Scenario
Carrier frequency	Macro + relay: For BS to relay: Below ITU-R report M.2135 [4] GHz, for relay to UE: [TBD] GHz	Macro only: Below [4 GHz]
System bandwidth	[40] MHz (DL + UL)	[40 MHz] (DL + UL)
Layout	Macro [layout including number of base stations is FFS] + relay nodes	Single layer: Macro cell [layout including number of base stations is FFS]
Cell range	Macro cell: [100] km range to be evaluated through system-level simulations. Feasibility of higher range will be evaluated through link-level evaluation.	[100 km] range to be evaluated through system-level simulations. Feasibility of higher range will be evaluated through link-level evaluation.
	Relay: up to [80] m	
User density and UE speed	End user density per macro:	End user density per aircraft: Up to [6 users]
	UE speed: Up to [1000] km/h	UE speed: Up to [370 km/h]
	Altitude: Up to 3GPP TS 22.268 [14] km	Altitude: Up to [3 km]
Traffic model	Average data throughput at busy hours/user: [TBD] kbps	Average data throughput at busy hours/user: [TBD] kbps
	End user experienced data rate: [384 kbps] DL.	End user experienced data rate: [384 kbps] DL.

Table 8.11 Examples for satellite deployment.

Attribute	Deployment-1	Deployment-2	Deployment-3
Carrier frequency	Around 1.5 or 2 GHz for both DL and UL	Around 20 GHz for DL Around 30 GHz for UL	Around 40/50 GHz
Duplexing	FDD	FDD	FDD
Satellite architecture	Bent pipe	Bent pipe, On-board processing	Bent pipe, On-board processing
Typical satellite system positioning in the 5G architecture	Access network	Backhaul network	Backhaul network
System bandwidth (DL + UL)	Up to 2 * 10 MHz	Up to 2 * 250 MHz	Up to 2 * 1000 MHz
Satellite orbit	GEO, LEO	LEO, MEO, GEO	LEO, MEO, GEO
UE distribution	100% outdoors	100% outdoors	100% outdoors
UE mobility	Fixed, Portable, Mobile	Fixed, Portable, Mobile	Fixed, Portable, Mobile

radio protocol layer 2/3 SDU egress point via the radio interface in both uplink and downlink directions, where neither device nor base station reception is restricted by DRX.

For URLLC, the target for user plane latency should be 0.5 ms for UL and 0.5 ms for DL. Furthermore, if possible, the latency should also be low enough to support the use of the next-generation access technologies as a wireless transport technology that can be used within the next-generation access architecture.

For eMBB, the target for user plane latency should be 4 ms for UL and 4 ms for DL.

For the satellite case, the evaluation needs to consider the maximum RTT that is associated with GEO satellite systems.

When a satellite link is involved in the communication with a UE, the target for user plane RTT can be as high as 600 ms for GEO satellite systems, up to 180 ms for MEO satellite systems, and up to 50 ms for LEO satellite systems.

Latency for infrequent small packets: For infrequent application layer small packet/message transfer, this is the time it takes to deliver an application layer packet/message successfully from the radio protocol layer 2/3 SDU ingress point at the mobile device to the radio protocol layer 2/3 SDU egress point in the RAN, when the mobile device starts from its most "battery-efficient" state. The latency should be no worse than 10 seconds on the uplink for a 20-byte application packet (with uncompressed IP header corresponding to 105-byte physical layer) measured at the maximum coupling loss (MCL) of 164 dB.

Mobility interruption time: Mobility interruption time means the shortest time duration supported by the system during which a user terminal cannot exchange user plane packets with any base station during transitions. The target for mobility interruption time should be 0 ms. This KPI is for both intra-frequency and inter-frequency mobility for intra-NR mobility. Mobility support can be relaxed for extreme rural scenarios for the provision of minimal services for very-low-ARPU areas: Inter-RAT mobility functions can be removed. Intra-RAT mobility functions can be simplified if it helps to decrease the cost of infrastructure and devices. Basic Idle-mode mobility shall be supported as a minimum.

Reliability: Reliability can be evaluated by the success probability of transmitting X bytes within a certain delay, which is the time it takes to deliver a small data packet from the radio protocol layer 2/3 SDU ingress point to the radio protocol layer 2/3 SDU egress point of the radio interface, at a certain channel quality (e.g., coverage-edge).

A general URLLC reliability requirement for one transmission of a packet is $1-10-5$ for 32 bytes with a user plane latency of 1 ms.

For eV2X [15], for communication availability and resilience and user plane latency of delivery of a packet of size [300 bytes], the requirements are as follows:

Reliability = $1 - 10 - 5$, and user plane latency = [3 − 10 ms], for direct communication via sidelink and communication range of, e.g., a few meters.

Reliability = $1 - 10 - 5$, and user plane latency = [2] ms, when the packet is relayed via BS.

Coverage: MCL in uplink and downlink between device and base station site (antenna connector(s)) for a data rate of 160 bps, where the data rate is observed at the egress/ingress point of the radio protocol stack in uplink and downlink. The target for coverage should be 164 dB.

Extreme coverage:

For a basic MBB service characterized by a downlink data rate of 2 Mbps and an uplink data rate of 60 kbps for stationary users, the target maximum coupling loss is 140 dB. For mobile users, a downlink data rate of 384 kbps is acceptable.

For a basic MBB service characterized by a downlink data rate of 1 Mbps and an uplink data rate of 30 kbps for stationary users, the target maximum coupling loss is 143 dB. At this coupling loss, relevant downlink and uplink control channels should also perform adequately.

UE battery life: UE battery life can be evaluated by the battery life of the UE without recharge. For mMTC, UE battery life in extreme coverage shall be based on the activity of mobile-originated data transfer consisting of 200 bytes UL per day followed by 20 bytes DL from MaxCL of 164 dB, assuming a stored energy capacity of 5 Wh.

The target for UE battery life for mMTC should be beyond 10 years; 15 years is desirable.

Connection density: Connection density refers to the total number of devices fulfilling a target QoS per unit area (per km^2), where the target QoS is to ensure a system packet drop rate less than 1% under given packet arrival rate l and packet size S. Packet drop rate = (Number of packets in outage)/(number of generated packets), where a packet is in outage if this packet failed to be received successfully by the destination receiver beyond the packet-dropping timer.

The target for connection density should be 1 000 000 devices/km^2 in an urban environment.

Mobility: Mobility means the maximum user speed at which a defined QoS can be achieved (in km/h).

The target for mobility should be 500 km/h.

8.10 Next-generation Radio Access Requirements

The RAN design for the next-generation radio access technologies should be designed to fulfill the following requirements [13]:

The RAN architecture should support tight interworking between the new RAT and LTE.

Consider high-performing inter-RAT mobility and aggregation of data flows via at least dual connectivity between LTE and the new RAT. This should be supported for both collocated and non-collocated site deployments.

The RAN architecture should support connectivity through multiple transmission points, either collocated or non-collocated.

The RAN architecture should enable a separation of control plane signaling and user plane data from different sites.

The RAN architecture should support interfaces supporting effective inter-site scheduling coordination.

Different options and flexibility for splitting the RAN architecture should be allowed.

The RAN architecture should allow for deployment flexibility, e.g. to host relevant RAN, CN, and application functions close together at the edges of the network when needed, for example, to enable context-aware service delivery, low-latency services, etc.

The RAN architecture should allow for C-plane/U-plane separation.

The RAN architecture should allow deployments using network function virtualization.

The RAN architecture should allow for the RAN and the CN to evolve independently.

The RAN architecture should allow for the operation of network slicing, 3GPP TR 23.799 [16].

The RAN architecture should support sharing of the RAN between multiple operators.

The design of the RAN architecture should allow the deployment of new services rapidly and efficiently.

The design of the RAN architecture should allow the support of 3GPP-defined service classes (e.g. interactive, background, streaming, and conversational).

The design of the RAN architecture should enable lower CAPEX/OPEX with respect to current networks to achieve the same level of services.

RAN–CN interfaces and RAN internal interfaces (both between new RAT logical nodes/functions and between new RAT and LTE logical nodes/functions) should be open for multi-vendor interoperability.

The RAN architecture should support operator-controlled sidelink (device-to-device) operation, both in coverage and out of coverage.

The study item (SI) in [17] aims to develop an NR access technology to meet a broad range of use cases including enhanced mobile broadband, massive MTC, critical MTC, and additional requirements defined during the RAN requirements study. The new RAT will consider frequency ranges up to 100 GHz [13].

Detailed objectives of the SI item are:

Target a single technical framework addressing all usage scenarios, requirements, and deployment scenarios defined in TR38.913 including:

Enhanced mobile broadband

Massive machine-type communications

Ultra-reliable and low-latency communications.

The new RAT shall be inherently forward compatible:

It is assumed that the normative specification will occur in two phases: Phase 1 (to be completed in June 2018) and Phase 2 (to be completed in December 2019).

Phase 1 specification of the new RAT must be forward compatible (in terms of efficient co-cell/site/carrier operation) with Phase 2 specification and beyond, and backward compatibility to LTE is not required.

Phase 2 specification of the new RAT builds on the foundation of Phase 1 specification, and meets all the set requirements for the new RAT.

Smooth, future evolution beyond Phase 2 needs to be ensured to support later advanced features and to enable support for service requirements identified later than Phase 2 specification.

Initial work on the study item should allocate high priority to gaining a common understanding of what is required in terms of radio protocol structure and architecture to fulfill Objectives 1 and 2, with a focus on progressing in the following areas:

Fundamental physical layer signal structure for the new RAT:

Waveform based on OFDM, with potential support for non-orthogonal waveform and multiple access.

FFS: Other waveforms if they demonstrate justifiable gain.

Basic frame structure(s).

Channel coding scheme(s).

Radio interface protocol architecture and procedures.

Radio access network architecture, interface protocols, and procedures.

Study on the above two points should at least cover:

The feasibility of different options of splitting the architecture into a "central unit" and a "distributed unit", with a potential interface in between, including transport, configuration, and other required functional interactions between these nodes:

Study the alternative solutions with regard to signaling, orchestration, …, and OAM, where applicable;

An outline of the RAN–CN interface and functional split;

Identification of the basic structure and operation of realization of RAN network functions (NFs). Study to what extent it is feasible to standardize RAN NFs, the interfaces of RAN NFs and their interdependency;

Identification of specification impacts of enabling the realization of network slicing;

Identification of additional architecture requirements, e.g. support for QoS concept, SON, support for side-link for D2D.

Fundamental RF aspects – especially where they may impact decisions on the above.

Study and identify the aspects related to the testability of RF and performance requirements.

Study and identify the technical features necessary to enable the new radio access to meet Objectives 1 and 2, also including:

Tight interworking between the new RAT and LTE.

Interworking with non-3GPP systems.

Operation in licensed bands (paired and unpaired), and licensed assisted operations in unlicensed bands. [Standalone operation in unlicensed bands is FFS].

Efficient multiplexing of traffic for different services and use cases on the same contiguous block of spectrum.

Standalone operation in licensed bands.

Provide performance evaluation of the technologies identified for the new RAT and analysis of the expected specification work.

Identify relevant RF parameters to be used for sharing and co-existence studies.

Study and identify technical solutions that enable support for wireless relay.

The NR should be able to support highly reliable (i.e. with a low ratio of erroneous packets) and low-latency services. Support for high-reliability services is subject to the following requirements and assumptions [18]:

High reliability: High reliability is about providing a high likelihood of delivering error-free packets through the 3GPP system within a bounded latency. A performance metric for high reliability is the ratio of successfully delivered error-free packets within a delay bound over the total number of packets. The required ratio and latency bound may be different for different URLLC use cases.

High availability: High availability is related to a communication path through the 3GPP system providing reliable services. This communication path between the communication end points is made up of radio links as well as transport links and different HW and SW functions. The NR should provide high availability in addition to deploying redundant components and links for radio, transport, and HW/SW.

Low latency: According to [13], the NR should support latencies down to 0.5 ms UL/DL for URLLC.

It is FFS whether a new function is needed at the NR to support high-reliability, low-latency services. It is FFS whether and how end-to-end delays will be considered.

The new RAN consists of the following logical nodes [18]:

gNBs providing the NR U-plane and C-plane protocol terminations towards the UE; and/or

eLTE eNBs providing the E-UTRA U-plane and C-plane protocol terminations towards the UE.

The logical nodes in the new RAN are interconnected by means of the Xn interface. The logical nodes in the new RAN are connected to the NGC by means of the NG

Figure 8.29 New RAN architecture.

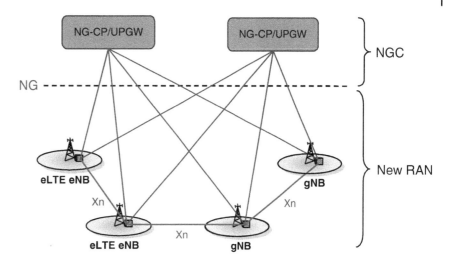

interface. The NG interface supports a many-to-many relationship between NG-CP/UPGWs and the logical nodes in the new RAN. The new RAN architecture is illustrated in Figure 8.29 [18].

8.10.1 Deployment Scenarios for 5G NR

The following example scenarios should be considered for support by the new RAN architecture. Although it is not always explicitly specified, it should be assumed that an inter-BS interface may be supported between a gNB and other gNBs or (e)LTE eNBs. A heterogeneous deployment comprises two or more deployments in the same geographical area, as defined in the following sections [18].

8.10.1.1 Non-centralized Deployment
In this scenario, shown in Figure 8.30, the full protocol stack is supported at the gNB, e.g. in a macro deployment or indoor hotspot environment (this could be public or enterprise). The gNB can be connected to "any" transport. It is assumed that the gNB is able to connect to other gNBs or (e)LTE eNBs via a RAN interface.

8.10.1.2 Co-sited Deployment with E-UTRA
In this scenario (Figure 8.31), the NR functionality is co-sited with E-UTRA functionality, either as part of the same base station or as multiple base stations at the same site. Co-sited deployment is applicable in all NR deployment scenarios, for example, urban macro. In this scenario, it is desirable to utilize fully all spectrum resources assigned to both RATs by means of load balancing or connectivity via multiple RATs (for example, utilizing lower frequencies as coverage layer for users on cell edges).

8.10.1.3 Centralized Deployment
The NR should support centralization of the upper layers of the NR radio stacks. Different protocol split options between the central unit and lower layers of gNB nodes

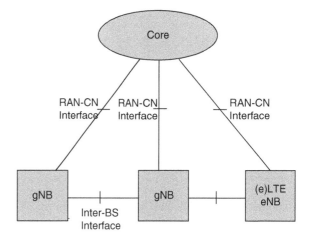

Figure 8.30 Non-centralized deployment.

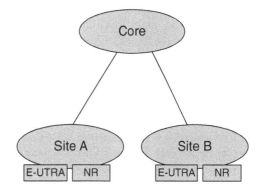

Figure 8.31 Co-sited deployment with E-UTRA.

may be possible. The functional split between the central unit and lower layers of gNB nodes may depend on the transport layer. High-performance transport between the central unit and lower layers of gNB nodes, e.g. optical networks, can enable advanced CoMP schemes and scheduling optimization, which could be useful in high-capacity scenarios, or scenarios where cross-cell

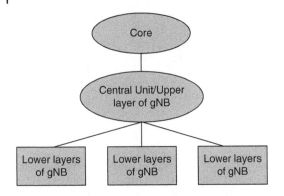

Figure 8.32 Centralized deployment.

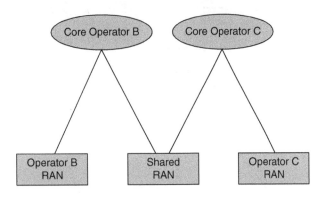

Figure 8.33 Shared RAN deployment.

coordination is beneficial. Low-performance transport between the central unit and lower layers of gNB nodes can enable the higher protocol layers of the NR radio stacks to be supported in the central unit, since the higher protocol layers have lower performance requirements on the transport layer in terms of bandwidth, delay, synchronization, and jitter. Both non-co-sited deployment and co-sited deployment with E-UTRA can be considered for this scenario, as shown in Figure 8.32.

8.10.1.4 Shared RAN Deployment

The NR should support shared RAN deployments, supporting multiple hosted core operators, as shown in Figure 8.33. The shared RAN could cover large geographical areas, as is the case for national or regional network sharing. The shared RAN coverage could also be heterogeneous, i.e. limited to a few or many smaller areas, for example in the case of shared in-building RANs. A shared RAN should be able to interoperate efficiently with a non-shared RAN. Each core operator may have its own non-shared RAN serving areas adjacent to the shared RAN. Mobility between the non-shared RAN and the shared RAN should be supported in a manner at least as good as for LTE. The shared RAN may (as for the case of LTE) operate either on shared spectrum or on the spectrum of each hosted operator.

8.10.2 RAN Deployment Scenarios

In terms of the cell layout served by the NR [19], the following scenarios are assumed:

Homogeneous deployment where all cells provide similar coverage, e.g. macro or small cell only.
Heterogeneous deployment where cells of different size are overlapped, e.g. macro and small cells.

Figure 8.34 shows deployment scenarios in terms of cell layout and Node B location, where both NR and LTE coverage exist in the geographical area. The left side of Figure 8.34 shows a scenario where both LTE and NR cells are overlaid and collocated, providing similar coverage. Both LTE and NR cells are macro or small cells. The right side of Figure 8.34 shows another scenario where LTE and NR cells are overlaid and collocated or not collocated, providing different coverage. In this figure, LTE serves macro cells and NR serves small cells. The opposite scenario is also considered. A collocated cell refers to a small cell together with a macro cell for which their eNBs are installed at the same location. A non-collocated cell refers to a small cell together with a macro cell for which their eNBs are installed at different locations.

The deployment scenarios in terms of CN–RAN connection are classified into the following cases:

NR gNB is a master node
LTE eNB is a master node
eNB connected to NextGen Core is a master node
Inter-RAT handover between NR gNB and (e)LTE eNB.

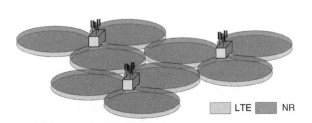
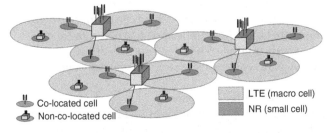

Figure 8.34 Cell layout where NR and LTE coverage co-exist.

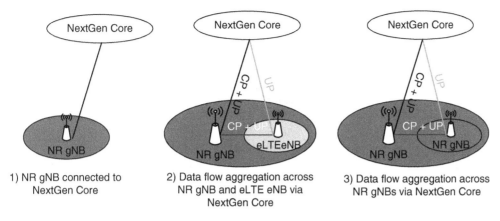

Figure 8.35 CN–RAN deployment scenarios where NR gNB is a master node.

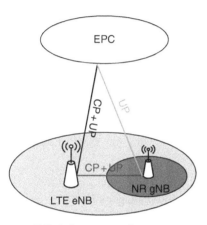

Figure 8.36 CN–RAN deployment scenarios where LTE eNB is a master node.

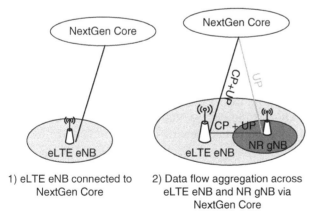

Figure 8.37 CN–RAN deployment scenarios where eNB connected to NextGen Core is a master node.

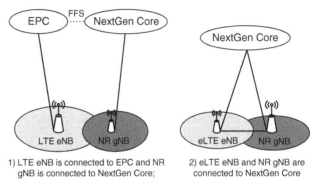

Figure 8.38 CN–RAN connection for inter-RAT mobility between NR gNB and (e)LTE eNB.

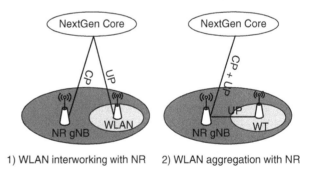

Figure 8.39 CN–RAN connection for WLAN integration with NR.

Figure 8.35 to Figure 8.38 demonstrate the four deployment scenarios, respectively. For more details refer to 3GPP TR 38.804 V1.0.0 [19].

5G NextGen Core integration with WLAN will be considered. Figure 8.39 illustrates the following deployment scenarios in terms of CN–RAN connection assumed for WLAN integration with NR:

WLAN interworking with NR via NextGen Core. WLAN aggregation with NR via NextGen Core.

8.10.3 Overall Topology of NG-RAN

The NG-RAN consists of gNBs, providing the NG-RA user plane (new AS sublayer/PDCP/RLC/MAC/PHY) and control plane (RRC) protocol terminations towards the UE. The gNBs are interconnected by means of the Xn interface. The gNBs are also connected by means of the

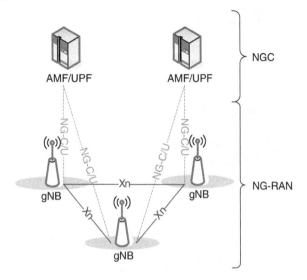

Figure 8.40 Overall architecture.

NG interface to the NGC, more specifically to the AMF (Access and Mobility Management Function) by means of the N2 interface and to the UPF (User Plane Function) by means of the N3 interface.

The NG-RAN architecture is illustrated in Figure 8.40. The gNB hosts the following functions:

Functions for radio resource management: Radio bearer control, radio admission control, connection mobility control, dynamic allocation of resources to UEs in both uplink and downlink (scheduling).

IP header compression and encryption of user data stream.

Selection of an AMF at UE attachment when no routing to an AMF can be determined from the information provided by the UE.

Routing of user plane data towards UPF(s).

Scheduling and transmission of paging messages (originated from the AMF).

Scheduling and transmission of system broadcast information (originated from the AMF or O&M).

Measurement and measurement reporting configuration for mobility and scheduling.

The AMF hosts the following main functions:

NAS signaling termination.

NAS signaling security.

AS security control.

Inter–CN node signaling for mobility between 3GPP access networks.

Idle mode UE reachability (including control and execution of paging retransmission).

Tracking area list management (for UE in Idle and Active modes).

AMF selection for handovers with AMF change.

Access authentication.

Access authorization including check of roaming rights.

The UPF hosts the following main functions (see 3GPP TS 23.501):

Anchor point for intra-/inter-RAT mobility (when applicable).

External PDU session point of interconnect to data network.

Packet routing and forwarding.

Packet inspection and user plane part of policy rule enforcement.

Traffic usage reporting.

Uplink classifier to support routing traffic flows to a data network.

Branching point to support multi-homed PDU session.

QoS handling for user plane, e.g. packet filtering, gating, UL/DL rate enforcement.

Uplink traffic verification (SDF to QoS flow mapping).

Transport-level packet marking in the uplink and downlink.

Downlink packet buffering and downlink data notification triggering.

The Session Management Function (SMF) hosts the following main functions:

Session management.

UE IP address allocation and management.

Selection and control of UP function.

Configuration of traffic steering at UPF to route traffic to proper destination.

Control of part of policy enforcement and QoS.

Downlink data notification.

This is summarized in Figure 8.41, where shaded boxes depict the logical nodes and white boxes depict the main functions.

8.10.4 E-UTRA with NextGen Core

An eNB providing E-UTRA access may connect to the NextGen Core via the NG interfaces, as also described in the deployment scenarios in this section. The overall architecture for E-UTRA with a NextGen Core is illustrated in Figure 8.40. The functions hosted by each logical entity are the same as described earlier.

A number of deployment options for 5G NR, LTE and 5G Core, represented in Figure 8.42. It should be noted that the new 5G System specified by 3GPP includes a new access network (5G-AN) allowing both 3GPP access (NG-RAN) via 5G NR or LTE and Non-3GPP Access (N3IWF) via e.g. fixed access or WiFi and a new core network (i.e. 5G Core) with a single set of interfaces whether 3GPP or N3GPP access is used.

For the user plane of E-UTRA with a NextGen Core, the LTE UP should be used as a baseline and some

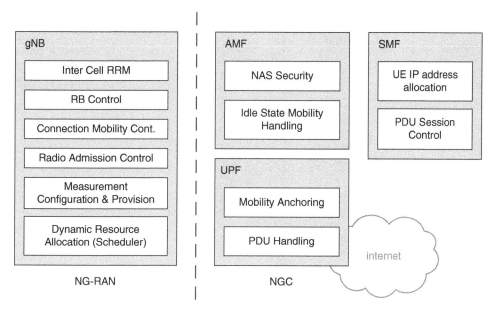

Figure 8.41 Functional split between NG-RAN and NGC.

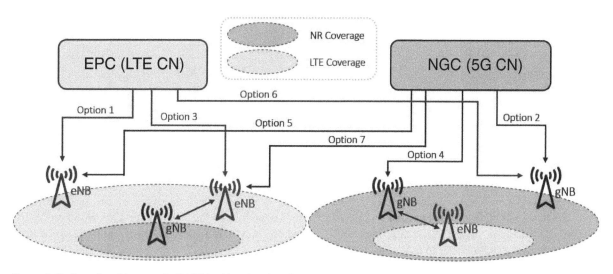

Figure 8.42 Overall architecture for E-UTRA with a NextGen Core.

enhancements (e.g. new QoS-related UP operation) introduced to support the NextGen Core. In particular, the new user plane AS protocol layer above PDCP, accommodating all the functions introduced in AS for the new QoS framework, will also be applicable for E-UTRA with a NextGen Core.

The eNB with connection to the NextGen Core can also have a connection to the EPC, and the LTE cell can support both UEs connected to EPC and the UEs connected to the NextGen Core. In order to support both UEs connected to EPC and UEs connected to the NextGen Core in an LTE cell simultaneously, both the LTE NAS-specific parameters and NextGen NAS-specific parameters should be broadcast in the system information. It should be possible for the eNB

to identify, at the latest by message 5 (which contains the initial NAS message), whether the UE is connecting to EPC or the NextGen Core. Commonality between LTE/NR tight interworking with LTE connected to EPC and LTE/NR tight interworking with LTE connected to the NextGen Core should be maximized. E-UTRA with a NextGen Core supports network slicing functionalities, as described in Section 8.11.

8.10.5   QoS Architecture in NR and NextGen Core

The QoS architecture in NR and NextGen Core is depicted in Figure 8.43 and may be described as follows:

For each UE, the NextGen Core establishes one or more PDU sessions.

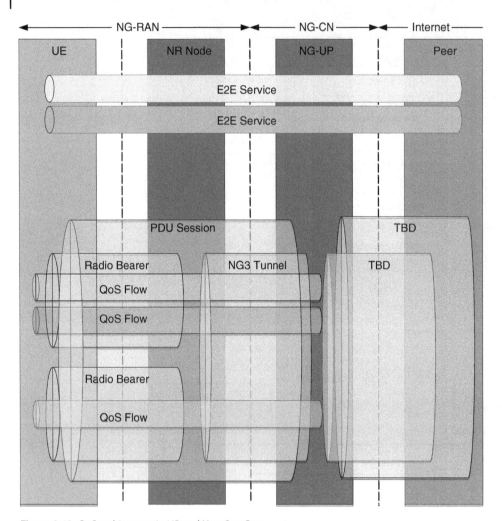

Figure 8.43 QoS architecture in NR and NextGen Core.

For each UE, the RAN establishes one or more data radio bearers per PDU session. The RAN maps packets belonging to different PDU sessions to different DRBs. Hence, the RAN establishes at least one default DRB for each PDU session indicated by the CN upon PDU session establishment.

NAS-level packet filters in the UE and in the NextGen Core associate UL and DL packets with QoS flows.

AS-level mapping in the UE and in the RAN associates UL and DL QoS flows with data radio bearers (DRBs).

The NextGen Core and RAN ensure quality of service (e.g. reliability and target delay) by mapping packets to appropriate QoS flows and DRBs. Hence, there is a two-step mapping of IP flows to QoS flows (NAS) and from QoS flows to DRBs (access stratum).

In NR, the data radio bearer (DRB) defines the packet treatment on the radio interface (Uu).

A DRB serves packets with the same packet-forwarding treatment. Separate DRBs may be established for QoS flows requiring different packet-forwarding treatment.

In the downlink, the RAN maps QoS flows to DRBs based on NG3 marking (QoS Flow ID) and the associated QoS profiles.

In the uplink, the UE marks uplink packets over Uu with the QoS flow ID for the purposes of marking forwarded packets to the CN.

In the uplink, the RAN may control the mapping of QoS flows to DRBs in two different ways:

Reflective mapping: For each DRB, the UE monitors the QoS flow ID(s) of the downlink packets and applies the same mapping in the uplink; that is, for a DRB, the UE maps the uplink packets belonging to the QoS flows(s) corresponding to the QoS flow ID(s) and PDU session observed in the downlink packets for that DRB. To enable this reflective mapping, the RAN marks downlink packets over Uu with QoS flow ID.

Explicit configuration: Besides reflective mapping, the RAN may configure by RRC an uplink "QoS Flow to DRB mapping".

If an incoming UL packet matches neither an RRC-configured mapping nor a reflective "QoS Flow ID to DRB mapping", the UE shall map that packet to the default DRB of the PDU session.

Within each PDU session, it is up to the RAN to decide how to map multiple QoS flows to a DRB. The RAN may map a GBR flow and a non-GBR flow, or more than one GBR flow to the same DRB, but mechanisms to optimize these cases are not within the scope of standardization. The timing of establishing non-default DRB(s) between the RAN and the UE for a QoS flow configured during the establishment of a PDU session can be different from the time when the PDU session is established. It is up to the RAN when non-default DRBs are established. The QoS functions of the current 3GPP architecture are distributed between the UE, the RAN, and the CN [16]. This solution describes an overall QoS solution for the NextGen system, describing how the QoS functionality is distributed between the CN, the RAN, and the UE; see Figure 8.44 for a high-level view of such a functional split.

Table 8.12 lists the QoS functional split corresponding to Figure 8.44. The split may be described as follows:

Subscription (including default QoS profile): The subscription contains information about which QoS parameters are included in the subscription terms. The subscription QoS is an input for the network when authorizing the QoS for a PDU session and a non-service-specific PDU flow in the QoS operator control function.

QoS operator control: With the input from the subscription, operator policies and application requirements input from the service layer, the QoS parameters for PDU sessions and PDU flows are authorized in the QoS operator control function. The QoS operator control function is also responsible for distributing the authorized QoS parameters in the network. In the case where PDU connectivity services are provided via network sharing and/or roaming across, the QoS operator control function also allows the QoS offered by the network providing the access to be limited.

Access and admission control (AN): The access control in the AN regulates the conditions on which the UE establishes a request for connection in the random access channel based on the QoS parameters applied for the session and flows. The admission control function controls which PDU flows shall be admitted on the access network when resources are scarce, based on the QoS parameters applied for the session and flows. Admission control also includes sacrificing already-admitted flows to allow flows with higher priority.

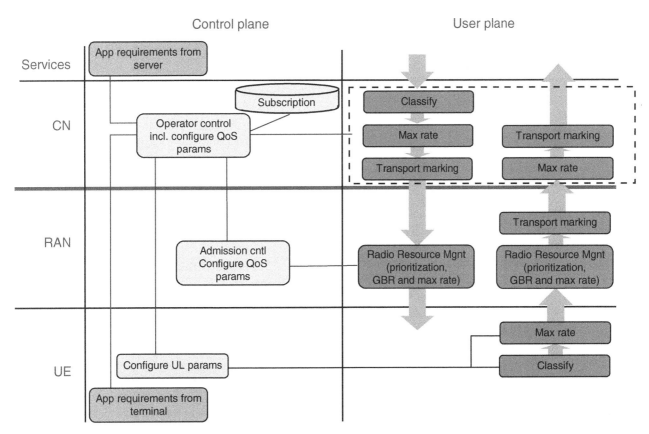

Figure 8.44 QoS functional split including 3GPP RAN.

Table 8.12 QoS functional split between the UE, AN, CN, and the SL.

Function	Distribution	Comment
Subscription (incl default QoS profile)	CN	
QoS operator control	CN	
Access and admission control	AN	Access and admission to AN resources
Configuration of QoS parameters	UE, AN, CN	
Application requirements input	From CN/SL to CN	The application requirements input may be sent from either the server or the client.
	From UE/SL to CN	
	From UE/SL to CN/SL	
Classification:	CN (DL), UE (UL)	Provides classification of packets for QoS purposes
Maximum rate control	CN (DL, UL)	
	AN (DL, UL)	
	UE (UL)	
Transport marking	AN (UL), CN (UL, DL)	
Resource management	AN	Packet scheduling with regards to resource utilization and availability (RRM)
		Resource management is also performed in the transport domain

Configuration of QoS parameters: Each network element in the end-to-end solution is configured with expected behavior with respect to QoS, i.e. how the QoS parameters received from the QoS operator control function will be handled and applied to the PDUs.

Application requirements input: To know the requirements for service data flows transmitted through the network, the network may be informed by the service layer about service behavior and service requirements. The application requirements input is used by the QoS operator control function when authorizing the QoS parameters for PDU sessions and PDU flows.

Classification: Indicates which service data flow and PDU flow each packet belongs to. The classification is used to select which authorized QoS parameters to apply to each PDU in the CN-UP, AN-UP, and UE-UP. Deducible SDFs may be classified based on TFT filters in the DL and UL. Non-deducible SDFs may be classified in the DL based on packet inspection. CN-CP may also provide application-layer information (for example, application identity) to the UE for UL traffic classification. UE-reflective QoS, equivalent to what is described in TS 24.139 for a fixed broadband network and packet inspection in CN-UP, may be used for classification of non-deducible IP or non-IP flows in the UL.

Maximum rate control: The maximum rate control function ensures that the maximum bit rate in the authorized QoS parameters is maintained.

Transport marking: The transport marking function indicates the expected treatment in IP networks as well as in non-3GPP accesses with a stateless QoS mechanism, for example, routers between the network elements. Each PDU is marked by the CN_UP (DL) or AN (UL) based on the QoS associated with the PDU flow.

Resource management: The resource management function is responsible for how the resources are distributed in the access network based on the authorized QoS parameters from the QoS operator control function and the monitoring of the fulfillment of the QoS targets. The resource management function can be different in 3GPP and non-3GPP ANs with regard to the possibilities for controlling resource utilization and availability. Resource management is also done in the transport network.

8.11   5G NextGen Core Network Architecture

Figure 8.45 depicts a reference model of a potential architecture, including potential functional entities and potential reference points, with the intention that the nomenclature, in particular for the reference points, may be used in individual solution proposals for better understanding and comparison. This reference model does not make any assumption with regard to the actual target architecture, i.e. the target architecture may not have all of the depicted reference points or functional

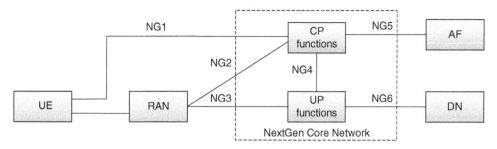

Figure 8.45 Reference point nomenclature.

entities, or may have additional/other reference points or functional entities [16].

The control-plane functions and the user-plane functions of the NextGen Core are depicted as single boxes (CP functions and UP functions, respectively). Individual solution proposals may further split or replicate CP or UP functions. In that case, the naming of additional reference points could add an index to the depicted reference points (e.g. NG4.1, NG4.2). The RAN here refers to a radio access network based on the 5G RAT or evolved E-UTRA (eLTE) that connects to the NextGen Core network.

The following reference points are illustrated in Figure 8.45 [16]:

NG1: Reference point between the UE and the CP functions.

NG2: Reference point between the RAN and the CP functions.

NG3: Reference point between the RAN and the UP functions.

NG4: Reference point between the CP functions and the UP functions.

NG5: Reference point between the CP functions and an application function.

NG6: Reference point between the UP functions and a data network (DN).

8.11.1 Core Deployment Scenarios

As discussed earlier, two radio technologies have to be considered in the 5G discussions:

LTE (in its Release 15 version) (a.k.a. eLTE).
Next-generation Radio (NR).

Similarly, two core network concepts have to be considered in the 5G discussions:

EPC (with potential evolutions)
Next-Generation Core (NGCN).

Accordingly, we have twelve options, some of which were discussed earlier in Section 8.3. See TR 38.801 [18] for further details of the NextGen RAN architecture.

In this section, we provide high-level deployment scenarios corresponding to the options listed in

SP-160464/RP-16266 [19]. Architecture Option 1 is EPC with E-UTRAN, and is not captured here because its support is already specified. Options 6 and 8 SP-160464/RP-16266 [20] are not reflected in this section either. Some of the reference point names and definitions between the UE and the NGC and between the NG RAN and NGC are given above. Reference points for the EPC and their definitions are provided in TS 23.401 [20]. In the scenarios below, the RAN is represented by a box with the RAN type (E-UTRAN or NG RAN) followed by radio access technology in brackets (NR, E-UTRA, or evolved E-UTRA). The interface names between RAN nodes and between the UE and the RAN node are left unnamed until specification is completed by 3GPP [16].

8.11.1.1 Deployment Option 2: Standalone NR in a NextGen System

This deployment option is for NG RAN with the radio access based on NR in a standalone configuration in a NextGen system. Figure 8.46 demonstrates this scenario.

8.11.1.2 Deployment Option 3: Non-standalone NR in EPS

This deployment option is a dual-connectivity deployment with E-UTRA as the anchor RAT and NR as the secondary RAT in a non-standalone configuration in EPS. Figure 8.47 demonstrates this scenario.

8.11.1.3 Deployment Option 4: Non-standalone Evolved E-UTRA in a NextGen System

This deployment option is a dual-connectivity deployment with NR as the anchor RAT and evolved E-UTRA

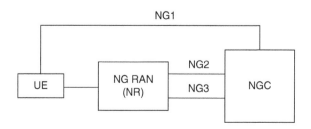

Figure 8.46 Deployment Option 2.

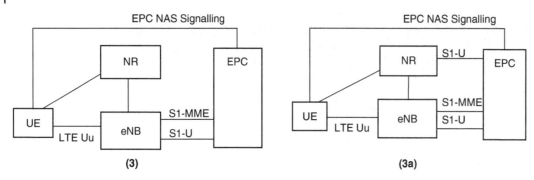

Figure 8.47 Deployment Option 3.

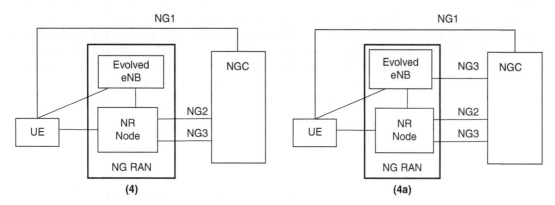

Figure 8.48 Deployment Option 4.

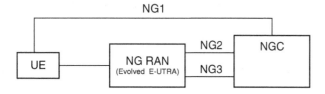

Figure 8.49 Deployment Option 5.

as the secondary RAT in a non-standalone configuration in a NextGen system. Figure 8.48 demonstrates this scenario.

8.11.1.4 Deployment Option 5: Standalone Evolved E-UTRA in a NextGen System

This deployment is for NG RAN with the radio access based on evolved E-UTRA in a standalone configuration in a NextGen system. Figure 8.49 demonstrates this scenario.

8.11.1.5 Deployment Option 7: Non-standalone NR in a NextGen System

This deployment option is for dual-connectivity deployments with evolved E-UTRA as the anchor RAT and NR as the secondary RAT in a non-standalone configuration in a NextGen system. Figure 8.50 demonstrates this scenario.

An overall view of 5G architecture based on Option 5 is shown in Figure 8.51. The architecture involves one PDU session with multiple IP addresses and points of attachment to a DN, and multiple PDU sessions, each with points of attachment to different DNs.

8.11.2 Network Slicing

Network slicing enables the operator to create networks customized to provide optimized solutions for different market scenarios, which gives rise to diverse requirements, for example in the areas of functionality, performance, and isolation. Solutions for this issue will study functionality and capabilities that enable the next-generation system to support network slicing and network slicing roaming requirements, including, but not limited to:

How to achieve isolation/separation between network slice instances and which levels and types of isolation/separation will be required.

How and what type of resource and network function sharing can be used between network slice instances.

How to enable a UE to obtain services simultaneously from one or more specific network slice instances of one operator.

Network slicing creation/composition, modification, and deletion.

Figure 8.50 Deployment Option 7.

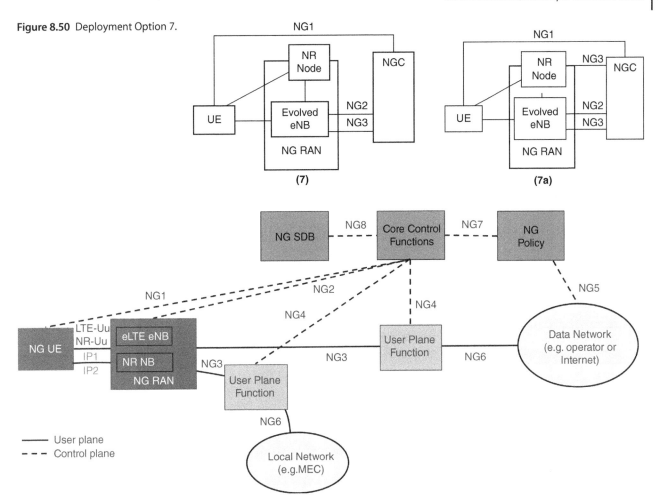

Figure 8.51 5G architecture based on Option 5.

Which network functions may be included in a specific network slice instance, and which network functions are independent of network slices.

The procedure(s) for selection of a particular network slice for a UE.

How to support network slicing roaming scenarios.

How to enable operators to use the network slicing concept to support multiple third parties efficiently (for example, enterprises, service providers, content providers, etc.), when the third parties require similar network characteristics.

Figure 8.52 illustrates the network slicing concept. Eight network-slicing solutions were proposed in [16].

8.11.2.1 Network Slicing Without Slicing the Radio

This section introduces a high-level solution for network slicing without slicing the radio, as demonstrated in Figure 8.52 [16]. In this solution, it is assumed that any slicing of a PLMN is not visible to the UEs at the radio interface. Therefore, in this case, a slice routing and selection function is needed to link the radio access

bearer(s) of a UE with the appropriate core network instance.

The solution does not make any assumption on any potential RAN internal slicing. The main characteristic is that the RAN appears as one RAT + PLMN to the UE and any association with a network instance is performed internally, without the network slices being visible to the UE.

Figure 8.52 depicts the semantics intended by the MDD with UEs that can access a single tenant slice. The case involving UEs accessing multiple tenant NSIs is also possible as long as these NSIs do not require their dedicated CCNFs.

8.11.2.2 Multi-tenant Network Slicing

The service provider may need to offer networking slices per tenant. Figure 8.53 demonstrates a multi-tenant network-slicing topology.

In Figure 8.53, we have a network that is supporting n tenants and up to m network behaviors (slice types) in both the RAN and the core. It is assumed possible for the RAN to dedicate resources to a slice.

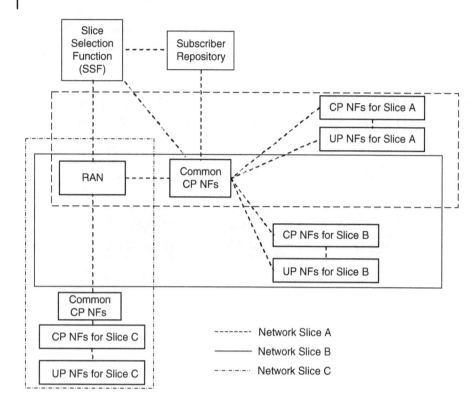

Figure 8.52 Control-plane interfaces for network slicing with common and slice-specific functions.

8.11.2.3 Multiple-slice Connection

To enable a UE to obtain services simultaneously from multiple network slices of one network operator, core network instances can be set up, as depicted in Figure 8.54; that is:

A single set of C-plane functions, common among core network instances, is shared across multiple core network instances.

U-plane functions and other C-plane functions that are not held in common reside in their respective core network instances and are not shared with other core network instances.

C-plane functions common to multiple core network instances include:

Authentication function (AU): AU is responsible for authenticating and authorizing the UE to attach to the operator's network. It also provides security and integrity protection of NAS signaling.

Mobility management function (MM): MM is responsible for UE registration in the operator's network (e.g. storing the UE context) and UE mobility support (e.g. providing a mobility function when the UE is moving across base stations within the operator's network).

Dedicated C-plane functions for each core network instance include:

Session management function (SM): SM is responsible for PDU session establishment, PDU session modification, and PDU session termination.

As a deployment option, all the C-plane functions can be deployed as part of a common C-plane function, as shown in Figure 8.55.

8.11.2.4 Network Slice Selection Based on Usage Class

This solution addresses the support for network slicing and specifically it considers (a) how the UE derives network slice selection information and (b) how the UE can have multiple simultaneous PDU sessions, each associated with a different network slice.

The key aspects of the solution are illustrated in Figure 8.56 and may be described as follows:

Every network slice instance supports one or more usage classes. For example, as shown in Figure 8.56, a network slice instance optimized for massive IoT (mIoT) supports Usage Class x, while a network slice instance optimized for enhanced mobile broadband (eMBB) supports Usage Class y. The usage class of a network slice indicates the use case that the network slice instance is configured to support, i.e. the type of network slice instance.

The UE associates every data flow with a usage class (and thus with a type of network slice). This association is accomplished by using a usage class selection policy provisioned in the UE and containing a list of prioritized rules matching different data flows with different usage classes. For example, the usage class selection policy may contain the following rules:

Rule 1, priority 1: App-id = com.example.app1, Usage Class = x.

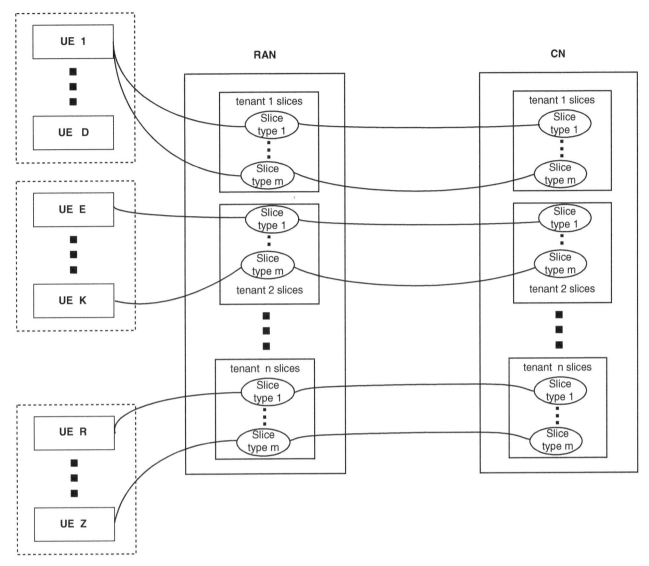

Figure 8.53 Network with *n* tenants and *m* possible slice types (with UEs which can only access a single tenant slice).

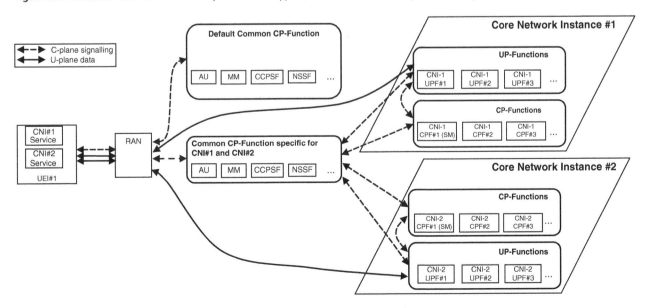

Figure 8.54 Sharing a set of common C-plane functions among multiple core network instances.

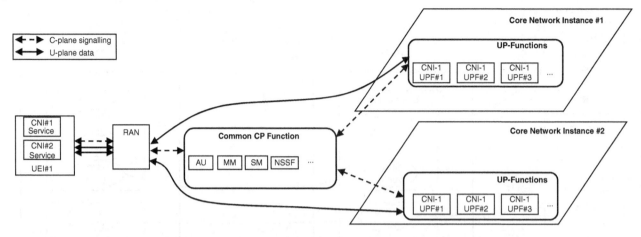

Figure 8.55 Sharing all the C-plane functions in a common CP function for all the core network instances.

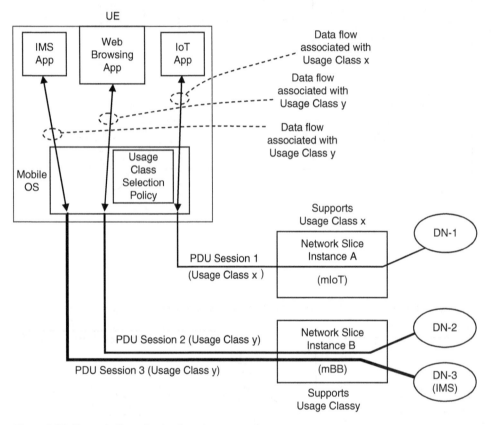

Figure 8.56 Network slice selection based on usage class.

Rule 2, priority 2: App-id = com.example.app2, Usage Class = y.

Default rule: Usage Class = z.

The usage class selection policy may contain only the default rule, in which case, all UE data traffic is associated with the same usage class.

The usage class selection policy can be provided to the UE by the HPLMN and by a VPLMN. In roaming scenarios, the UE applies the policy of the registered PLMN (RPLMN), if it exists.

In the example scenario shown in Figure 8.56, the UE associates the data flow of the IoT app with Usage Class x and the data flows of the Web browsing and IMS apps with Usage Class y.

The usage class selection policy is provided to the UE by the appropriate policy functionality, defined as part of the NextGen policy framework. Details on how this policy is provided to the UE are outside the scope of the present solution.

When the UE initiates a new data flow and this data flow is associated with Usage Class x, the UE routes the data flow via a suitable PDU session that matches this usage class (if it exists). In the example shown in Figure 8.56, the data flow from the IoT app is associated with Usage Class x and it is therefore routed via the PDU session 1 that supports Usage Class x.

If the UE has no suitable PDU session matching the usage class associated with a new data flow, then the UE requests the establishment of a new PDU session with the usage class associated with the data flow.

The UE provides the usage class as "assistance information" for network slice selection: it is used by the network to select a network slice that supports the requested usage class. The UE may provide multiple usage classes as "assistance information" in order to be connected to different network slices.

The UE may be using multiple usage classes simultaneously and it may have multiple PDU sessions, each associated with a usage class. In the example scenario shown in Figure 8.56, the UE is using Usage Class x and Usage Class y simultaneously and has established three different PDU sessions: one with Usage Class x, which is routed via the network slice instance A, and two with Usage Class y, which are both routed via the network slice instance B.

When an app in the UE wants to establish a new PDU session, the app itself may indicate a usage class for the PDU session. If the app does not provide the usage class, the UE uses the default usage class. The default usage class may be determined from the default rule of the usage class selection policy, as shown in Point 2 above.

An example of UE procedures is provided below to clarify the solution further.

The UE powers up and provides the default usage class (determined from the default rule of the usage class selection policy) as "assistance information". The default usage class may depend on the type of UE (e.g. IoT device, smartphone, etc.) and is used by the network to select a network slice instance for the UE. The UE is informed that it is connected to a network slice instance that supports, for example, Usage Class x and Usage Class y.

A Web browser is launched in the UE and initiates a new data flow. By using the usage class selection policy, the UE associates this data flow with Usage Class y. The UE requests the establishment of a PDU session for Usage Class y (the PDU session request message includes Usage Class y). After establishing the PDU session, the UE routes all data flows of the applications associated with Usage Class y (for example, Internet traffic) within this PDU session.

Later, an IoT app is launched in the UE and initiates a new data flow that is associated with Usage Class z. Since the UE is not connected to a network slice that supports Usage Class z, the UE provides to the network Usage Class z as "assistance information" to be connected to another network slice. The network connects the UE to another network slice instance that supports Usage Class z (for example, a network slice optimized for IoT). After that, the UE establishes a PDU session for Usage Class z and routes all data flows of the applications associated with this usage class (for example, all IoT traffic) within the PDU session.

Support for network slicing relies on the principle that traffic for different slices is handled by different PDU sessions. A network can realize different network slices by scheduling and also by providing different L1/L2 configurations. The UE should be able to provide assistance information for network slice selection in an RRC message, if it has been provided by the NAS.

8.12 5G Waveform and Multiple Access Design

Although OFDM has been a great success and still has many advantages, there are many ideas for new 5G waveforms that could bring additional advantages to the new cellular system under certain conditions and circumstances. No single waveform provides all the advantages and answers that are needed. As a result, many anticipate that the final outcome for 5G waveforms may include an adaptive solution – using the optimal waveform for any given situation. OFDM has been an excellent waveform choice for 4G. It provides excellent spectrum efficiency, it can be processed and handled with the processing levels achievable in current mobile handsets, and it operates well with high data rate streams occupying wide bandwidths. It operates well in situations where there is selective fading. However, with the advances in processing capabilities that will be available by 2020 when 5G is expected to have its first launches, other waveforms can be considered. There are several advantages to the use of new waveforms for 5G. OFDM requires the use of a cyclic prefix and this occupies space within the data streams. Other advantages can also be introduced by using one of a variety of new waveforms for 5G [21].

One of the key requirements is the availability of processing power. Although Moore's Law, in its basic form, is running to the limits of device feature sizes and further advances in miniaturization are unlikely for a while, other techniques are being developed, which means the spirit of Moore's Law will continue and processing capability will increase. As such, new 5G waveforms that require additional processing power but can provide additional advantages are still viable. The potential applications for 5G, including high-speed video downloads, gaming, car-to-car/car-to-infrastructure communications, general cellular communications, IoT/M2M communications and the like, all place requirements on the form of 5G waveform scheme that can provide the required performance. Some of the key requirements that need to be supported by the modulation scheme and overall waveform include [21]:

Handling high-data-rate, wide-bandwidth signals.
Providing low-latency transmissions for long and short data bursts, i.e. very short TTIs are needed.
Switching quickly between uplink and downlink for TDD systems that are likely to be used.
Enabling the possibility of energy-efficient communications by minimizing the on-times for low-data-rate devices.
Reducing the peak-to-average power ratio (PAPR).
Increasing spectrum efficiency by eliminating the cyclic prefix (CP).
Supporting different service types by modifying the waveform dynamically.

The limitations of OFDM-based waveforms have been identified as research topics for future 5G waveforms. One aspect is the requirement for much shorter latency to enable new services and applications like autonomous driving, which demands an ultra-low latency and a highly resilient communication link. Another approach is to make the cyclic prefix optional and work with shorter symbol durations. All this has led to several candidate waveforms, such as [22]:

Generalized frequency division multiplex (GFDM): GFDM is a flexible, multi-carrier transmission technique which bears many similarities to OFDM. The main difference is that the carriers are not orthogonal to each other. GFDM provides better control of the out-of-band emissions and reduces the PAPR. These issues are the major drawbacks of OFDM technology [21]. The carriers are not orthogonal to each other, which makes it difficult for the receiver to handle ISI and large guard bands.

Filter bank multi-carrier (FBMC): FBMC has gained a high degree of interest as a potential 5G waveform candidate. This waveform scheme provides many advantages. In many ways, FBMC has many similarities to CP-OFDM (OFDM using a cyclic prefix), which is used as the 4G waveform. Instead of filtering the whole band as in the case of OFDM, FBMC filters each subcarrier individually. FBMC does not have a cyclic prefix and, as a result, it is able to provide a very high level of spectral efficiency. The subcarrier filters are very narrow and require long filter time constants. Typically, the time constant is four times that of the basic multi-carrier symbol length and, as a result, single symbols overlap in time. To achieve orthogonality, offset-QAM is used as the modulation scheme, so FBMC is not orthogonal with respect to the complex plane [21].

Universal filtered multi-carrier (UFMC): This 5G waveform can be considered an enhancement of CP-OFDM. It differs from FBMC in that, instead of filtering each subcarrier individually, UFMC splits the signal into a number of sub-bands, which it then filters. UFMC does not have to use a cyclic prefix, although one can be used to improve the inter-symbol interference protection [21].

Filtered OFDM (f-OFDM): As the name, f-OFDM, indicates, this form of OFDM uses filtering to provide its unique characteristics. Using f-OFDM, the bandwidth available for the channel on which the signal is to be transmitted is split up into several sub-bands. Different types of services are accommodated in different sub-bands with the most suitable waveform and numerology. This enables a much better utilization of the spectrum for the variety of services to be carried [21]. A sub-band digital filter is applied to shape the spectrum, which leads to a corruption of the subcarriers toward band edges. The bandwidth available for the channel on which the signal is to be transmitted is split up into several sub-bands.

The performance of these waveform candidates is currently being analyzed and evaluated. At the same time, new multiple access schemes are also being researched, including Sparse Code Multiple Access (SCMA), Non-Orthogonal Multiple Access (NOMA), and Resource Spread Multiple Access (RSMA) [22]. However, at this point it has not been decided if any of these waveform candidates or multiple access schemes will be utilized in a future 5G system. It is up to the standardization committees to submit their proposal to the IMT-2020 group within the International Telecommunication Union (ITU). They will decide if the proposal is accepted as a technology for a 5G mobile communications system that can use the frequency bands identified by the ITU at the World Radio Conferences (WRC). It is expected that this will take place between now and November 2019.

Some of the new waveforms proposed to address challenges associated with 5G are enhanced flavors of OFDM, such as f-OFDM [23], Universal Filtered-OFDM (UF-OFDM) [24], and Unique Word-OFDM (UW-OFDM) [25]. These enhanced flavors maintain the main structure of the OFDM frame, which makes them backward compatible. Some other proposed waveforms do not have the same frame structure, such as FBMC [26], Zero-tail DFT-Spread-OFDM (ZT-DFTs-OFDM) [27], and GFDM [28]. In 5G, a hybrid kind of waveform may be a promising vision, where different waveforms can co-exist to support a wide variety of applications [29].

Conventional pulse shaping and filtering might not be sufficient to satisfy a wide range of requirements on the transmitted/received signal. Thus, pre- and post-processing refer to any modification of the waveform and data signal at the transmitter and on the receiver side that cannot be accomplished via a linear filtering and pulse-shaping procedure. In order to optimize the waveform design, the signal and its structure can be modified by additional processing on top of grid design, filtering, and pulse shaping [29]. Some recent modifications to OFDM signaling, i.e., Zero-tail [27] and UW [25], have been proposed for 5G systems and can be considered pre/post processes. Both aim to address some of the disadvantages of the hard-coded CP. In CP-OFDM, the CP is used to separate consecutive symbols in time to prevent inter-symbol interference (ISI) and to ease the equalization of the symbols in the frequency domain by transforming the linear convolution of transmitted OFDM symbols with the channel impulse response to a circular convolution. The UW process not only maintains these gains but also adds more to them by placing the same, known sequence between all consecutive symbols, and it can be implemented both in OFDM and Single-Carrier Frequency Domain Equalization (SC-FDE) [25]. Since the sequence is known, it can be used for synchronization and channel estimation purposes at the receiver, eliminating the need for dedicated pilot subcarriers used in CP-OFDM. Another disadvantage of CP-OFDM is the inflexibility of the duration of the hard-coded CP; the CP length can be either longer than required, causing a waste of resources, or shorter than required, causing insufficient mitigation against ISI. By contrast, the Zero-tail procedure generates low-power samples inside the FFT duration, referred to as "tails," whose length is adjusted according to different channel delay spreads. This maintains the same total symbol length even with different guard durations. For 5G heterogeneous networks operating over different frame structures and different symbol lengths, the usage of flexible guard time is a promising feature for the processing in the OFDM structure.

Moreover, pre/post-processing procedures can have other purposes such as easing equalization [27, 30], reducing OOB emission [31], reducing PAPR [32], improving security [33], and even joint aims [34]. That is, either structural or adaptive processing techniques are used to improve different aspects of the communication system that cannot be solved simply by filtering and lattice design. These procedures can be gathered into the pre- and post-processing at the transmitter and receiver, respectively. Each use case envisioned for 5G has different requirements, which have to be taken into account when designing the waveforms. While each use case has special requirements, even the common requirements differ in importance for different applications. A summary of the waveform performance metrics is provided in Table 8.13; refer to [29] and references therein for more details.

8.12.1 Massive Machine-type Communications (mMTC)

The revolutionary part of 5G will be connected world applications that will be utilizing the frequency bands below 6 GHz; this is mainly due to power consumption requirements for IoT combined with the need for long transmission distances for lifeline communications. In addition, mMTC applications will exploit the suitability of lower frequencies for narrowband signaling and make use of the less complex hardware designed for lower frequencies. Mission critical communications are also required to be operating in channels with well-established models to ensure reliability [35].

The Internet of Things can be envisaged with the following scenario: A network that contains a thousand times more devices than the LTE network, consisting mostly of very cheap sensors that transmit bursts of low-rate data with long intervals and have to have ten times the battery life of currently available similar instruments. Firstly, to maximize battery life while minimizing hardware costs, the transmission of low-rate data with long intervals makes relaxing the strict time and

Table 8.13 Waveform requirements in decreasing order of importance per application.

Application	Key Requirements	Waveform/Multiple Access
Internet of Things (IoT)	Non-orthogonality, High spectral efficiency, Low complexity, Low PAPR, Support short data bursts, long device battery life, and deep indoor coverage.	UL: SC-FDE/RSMA
		DL: CP-OFDM with OFDMA
Ultra-reliable and low latency (URLL)	Low complexity, Immunity to Doppler diversity, Low latency, Low packet data loss rate.	UL: SC-FDM/SC-FDMA or RSMA
		DL: GFDM or CP-OFDM with OFDMA
Enhanced mobile broadband (eMBB)	Low PAPR as devices are power-limited, Immunity to frequency dispersions, Higher spectrum efficiency.	UL: SC-FDM/SC-FDMA and CP-OFDM with OFDMA
		DL: GFDM or CP-OFDM with OFDMA

frequency synchronism requirements a viable option. Although not compulsory, to satisfy the aforementioned requirements, such applications are envisioned to work in an asynchronous network [36]. Considering the capabilities and limitations of the IoT devices under the network operations, relaxing the orthogonality of the waveform itself becomes crucial. Secondly, to support such device density while satisfying the aforementioned requirements, the waveform has to have minimal out-of-band emission. Finally, the power consumption and cost requirements also put waveforms with low complexity and PAPR ahead of others [28]. These requirements will be met initially by NB-IoT of 3GPP Release 13 and its evolution in Release 14.

Another set of applications for 5G has reliability and low latency as key requirements. One example of this scenario is self-driving smart vehicles, since it is considered that operational instructions and information about surroundings will be delivered to the vehicle via the network [37]. Firstly, compared to LTE-A, the five-fold increase in end-to-end latency requirement heavily restricts the computational complexity of the waveform to be used [37]. The five-fold increase in mobility support introduces fast, time-varying, doubly dispersive fading and carrier frequency offset to the channel. This makes maintaining the orthogonality between subcarriers of orthogonal multi-carrier waveforms a challenge, and one that needs to be considered when designing the waveform. The newly introduced Doppler diversity makes achieving the reliability requirement a difficult task, even though the increase in reliability requirement is less than 1% compared to LTE [37].

8.12.2 Enhanced Mobile Broadband (eMBB)

eMBB applications are mainly concerned with transmitting huge volumes of data at much higher data rates, both indoors and outdoors [36]. While this scenario does not push special requirements other than supporting high-volume traffic, a high percentage of such applications is considered to operate above 6 GHz at cm/mmWave, which makes certain characteristics of a candidate waveform important. Firstly, the limited linear range of power amplifiers at such frequencies enforces low PAPR of the waveform used [38], making this an important criterion. Shared MIMO architecture [39] also enforces the low PAPR requirement. Secondly, the phase noise and frequency offset problem present in high-frequency oscillators [38] raise interest in waveforms that are immune to frequency-domain dispersions.

Table 8.14 provides a comparison between 5G waveform candidates. Comparisons in Table 8.14 are based on the best possible performance of each waveform, with OFDM with cyclic prefix (CP-OFDM) used as the benchmark [40]. In addition to the above waveforms, Biorthogonal Frequency Division Multiplexing (BFDM) waveforms have been regarded as being suitable for supporting sporadic data traffic and asynchronous signaling. One appealing feature of BFDM is time and spectral localization balancing through iterative interference cancellation, which, in turn, allows control of degradations due to time and frequency offsets [40]. Non-contiguous OFDM (NC-OFDM) is also presented [40].

Another detailed and practical comparative analysis between 5G waveform candidates was conducted by 5GNOW and is summarized in Table 8.15 [41]. For more details refer to [41], in particular the part where some detailed reports are provided along with MatLab and LabVIEW scripts. A practical performance analysis was conducted on an NI PXI system.

8.12.3 Multiple Access Techniques

The 5G small-cell scenarios envision ultra-dense networks handling large numbers of simultaneous transmissions in a small geographical area. This imposes new challenges for the multiple access (MA), where large numbers of users connect to the cloud, device-to-device (D2D) and a massive set of IoT communications takes place. Such massive amounts of communications place great demands on current multiple access schemes, which need to evolve to permit the deployment of IoT. A non-orthogonal design for multiple access can potentially increase system capacity and spectrum efficiency while exploiting the expected future evolution in the receiver processing capability. Relaxed synchronization requirements between users/services as well as a flexible system design are also desirable in order to adapt the amount of overhead and signaling. Considering the already-congested frequency spectrum, especially at sub-6 GHz, reliable communication is possible by using the underutilized white spaces. This needs a frequency-agile front end and MA using such architecture as well as an opportunistic MA for a cognitive architecture [42].

8.12.3.1 Non-orthogonal Multiple Access (NOMA)

For the downlink, non-orthogonal multiple access (NOMA) is proposed, where multiple users are multiplexed in the power domain on the base station side and multi-user signal separation at the UE side is conducted based on successive interference cancellation (SIC). Signaling aspects, multi-user transmit power allocation, MCS selection, and candidate user set selection are key component technologies that need further analysis in NOMA. At the transmitter side, based on the channel gain (SNR) feedback information from users, multi-user power allocation and MCS selection are conducted, and the user set that maximizes multi-user proportional fairness is scheduled. In addition, dynamic switching

Table 8.14 Candidate waveforms regarded as the most promising for 5G with CP-OFDM used as the benchmark. (TBI = to be investigated.)

Figure of Merit		CP-OFDM	NC-OFDM	DFT-s-OFDM	BFDM	FBMC	GFDM	UFMC
Performance	Peak-to-average power ratio	High	High	Reduced	High	High	Reduced	High
	Spectral efficiency	Low	Low	Low	High	High	High	High
	Overhead	High	High	Variable	Low	Low	Variable	Low
	Frequency localization	Good	Good	Very good	Controllable	Excellent	Excellent	Excellent
	OOB emissions	High	Reduced	Reduced	Variable	Negligible	Reduced	Reduced
	Sidelobe attenuation (dB)	13	20–50	40–60	13–60	60	35	40–60
	Bit error rate	Good	Good	Good	Good	Good	Very good	Very good
	Throughput	Low	Low	Low	High	High	High	High
	Time offset resiliency	Poor	Good	Good	Good	Good	Good	Good
	Frequency offset resiliency	Poor	Good	Good	Good	Good	Good	Good
	Computational complexity	Low	Low	Low	High	High	High	High
Feasibility	Implementation	Efficient	Efficient	Efficient	TBI	Efficient	Efficient	Efficient
	Equalization	Simple	Simple	Simple	TBI	Involved	Simple	Involved
	Resource allocation	Dynamic and fine-grained	Dynamic and fine-grained	Dynamic and fine-grained	Possible	Configurable	Configurable and adaptable	Configurable
Support for	Conventional MIMO	Yes	Yes	Yes	TBI	No	Yes	Yes
	High-order modulation	Yes	Yes	Yes	TBI	TBI	Yes	TBI
	Short-burst traffic (MTC)	No	Yes	Yes	Yes	No	Yes	Yes
	Fragmented spectrum	No	Yes	Yes	Yes	Yes	Yes	Yes
	Low latency (tactile Internet)	No	No	No	No	No	Yes	No

Table 8.15 5G candidate waveforms – comparative analyses.

	UFMC	FBMC	GFDM	BFDM
Orthogonality	Complex domain	Real domain	Orthogonal/non-orthogonal/real domain orthogonal.	Bi-orthogonal
QAM/OQAM preferred?	QAM	OQAM	QAM and OQAM	QAM
BER in multipath Rayleigh	Slightly better than OFDM (no energy wasted for CP). Additional advantages in scenarios with ICI.	Slightly better than OFDM (no energy wasted for CP). Additional advantages in scenarios with ICI.	Potential to reduce amount of CP in the signal (energy savings in the order of 7–15%). Non-orthogonal parameterization can cause self-interference and hence increase BER compared to orthogonal schemes in order to provide benefits with other KPIs.	Improved compared to OFDM due to better adaptation to channel profile.
PAPR	Similar to OFDM. Also allows for single-carrier variant: UFMC-based SC-FDMA possible with, e.g., DFT-pre-coding as in OFDM (any technique can be carried over).	Similar to OFDM.	Can be configured for PAPR reduction beyond SC-FDMA and easy integration with any PAPR-reduction technique available for OFDM.	Similar to OFDM: even complex methods can be transferred.
Out-of-band radiation	Arbitrary low side-lobe level floor due to sub-band filters.	Arbitrary low side-lobe level thanks to per subcarrier filtering. Up to −60 dB attenuation at adjacent subcarriers.	Can be configured for a desired OOB suppression at the cost of other KPIs.	Better than OFDM, comparable to GFDM but worse than FBMC due to block cyclic operation.
Throughput	Better than OFDM (smaller guards).	CP removal increases capacity compared to OFDM but filter rise and fall time must be considered. Throughput is then slightly better than OFDM for long frames.	In the order of 25% better than OFDM: Smaller guard bands, less CP.	Better than OFDM particularly for higher Doppler speeds and channel profile adaptation.
Spectral efficiency	Better than OFDM (smaller guards). 10% more usable resource elements.	Can be better than OFDM, depending on the overlapping factor, the number of OFDM symbols, and the size of the CP.	(see above)	Better than OFDM particularly for higher Doppler speeds and channel profile adaptation.

(Continued)

Table 8.15 (Continued)

	UFMC	FBMC	GFDM	BFDM
Complexity: modulator/demodulator	Efficient frequency-domain FFT-based implementation. Slightly higher than OFDM due to filtering.	Efficient frequency-domain IFFT/FFT-based implementation, but still higher complexity than OFDM due to filtering.	Efficient frequency-domain DFT-based implementation available. Complexity within in the order of 2–10 times OFDM complexity due to filtering and optional self-interference cancellation.	Logarithmic complexity growth in block length is preserved.
Complexity: remaining inner receiver until FEC input	As low as OFDM	On the same order of magnitude compared to OFDM.	Same as OFDM in case of no iterative interference cancellation.	Logarithmic complexity growth in block length is preserved.
Training structure and properties	OFDM know-how can be re-used. All degrees of freedom available for complex sequences with multi-user multiplexing.	OFDM know-how can be re-used but needs adaption for channel estimation and MIMO. Burst modes with preambles allow time–frequency synchronization and channel estimation.	OFDM know-how can be reused. OQAM and/or non-orthogonality can be compensated.	"Quasi-OFDM like" operation possible.
Channel estimation	OFDM know-how can be re-used. Low complexity, frequency-domain channel estimation fully supported.	OFDM know-how can be re-used. Low complexity, frequency-domain channel estimation fully supported when interference cancellation methods are applied at the transmitter side (a more complex frame structure is required).	OFDM know-how can be reused. OQAM and/or non-orthogonality can be compensated.	"Quasi-OFDM like" operation possible.
Sensitivity to time–frequency offsets	Performance gains over OFDM.	Performance gains over OFDM.	Performance gains over OFDM can achieve 20 dB.	Major advantage of BFDM compared to OFDM: order of magnitude better.
Multi-user capability	Block-based modulation configurable in time and frequency, handling mix of sync./async. users, as discussed. Full support for multi-user MIMO multiplexing of sounding/small control elements Synergy with IDMA.	Block-based modulation configurable in time and frequency. Support for multi-user MIMO schemes with adaptation, handling mix of sync./async. users.	Easy integration with LTE approaches. Full support MU- MIMO-GFDM.	Major advantage of BFDM compared to OFDM particularly in asynchronous uplink.

Multi-point capability	Performance gains over OFDM.	Performance gains over OFDM: avoid multiple access interference in uplink, less complex than equivalent CP-OFDM with MAI cancellation, better than OFDM for given scenario with impairments (e.g. precision of local oscillators).	Open topic	No specific advantage of BFDM in synchronous operation.
Resource allocation mechanisms/channel adaptive scheduling	As for OFDM	Same as OFDM	Two-dimensional data block structure enables more degrees of freedom regarding resource allocation than OFDM.	"Quasi-OFDM like" operation possible.
Adaptivity potential	Sub-band-specific waveform parameters possible.	Subcarrier/sub-band-specific waveform parameters possible. Flexible parameters: number of symbols per block, number of subcarriers, ...	Extremely flexible waveform with a large number of degrees of freedom allows, on the one hand, consideration of classical OFDM and SC-FDE as corner cases and, on the other hand, exploration of parameter sets optimized for specific scenarios, e.g. low-latency communication.	Better than OFDM, particularly for higher Doppler speeds and channel profile adaptation.
Co-existence with legacy systems	Good, due to low out-of-band radiation. Can re-use LTE parameter settings, signaling, and pilot symbols.	Very good, due to very low out-of-band radiation. Can re-use LTE parameter settings and part of the signaling and pilot symbols.	Can be parameterized for optimal protection of legacy frequency bands in asynchronous operation mode [MLR + 11]. Can be parameterized to re-use the LTE master clock and resource grid.	Similar to GFDM

between NOMA and OMA is introduced such that NOMA is applied only when it provides gains over OMA [43, 44].

8.12.3.2 Performance Evaluation

Performance evaluation of OMA is provided in [42]. The objectives are to assess the system-level gains of NOMA over orthogonal multiple access (OMA) (e.g., OFDMA) and to provide a design for its radio interface.

8.12.3.3 System and Signal Models

There are K users per cell and the total transmit bandwidth is divided into different sub-bands. The transmit signal at every subcarrier of a sub-band is a simple summation of the coded modulation symbol of all users from a scheduled user set, so that their signals are superposed in the power domain. The received signal at a UE in a sub-band is represented by the sum of the contribution of the superposed signals, which is impaired by the channel and a contribution given by noise plus inter-cell interference. In NOMA, users with high channel gain and low channel gain are paired. At the UE with high channel gain, a SIC receiver is applied in order first to cancel interference from the UE with low channel gain (high transmit power allocation). The UE with low channel gain simply treats interference from the signal of the UE with high channel gain (low transmit power allocation) as noise.

8.12.3.4 Main Results

Using computer simulations, for a 1×2 SIMO system with BW = 10 MHz, $K = 10$ UEs, maximum multiplexing order $m = 1$ or 2 or 3 ($m = 1$ corresponds to OMA), and wideband user scheduling, NOMA performance is evaluated. For multi-user power allocation, FTPA (Fractional Transmit Power Allocation) is used. According to Figure 8.57, the performance gains in the overall cell throughput and cell-edge user throughput for

NOMA over OMA for $m = 2$ ($m = 3$) are approximately 27% (28%) and 34% (39%), respectively. In addition, the performance of NOMA was also evaluated for different average cell-edge user throughputs, as shown in Figure 8.57b. NOMA provides much higher gains in terms of cell throughput with higher target average cell-edge user throughput.

8.12.3.5 SCMA Random Access

Sparse Code Multiple Access (SCMA) is a new frequency-domain, non-orthogonal waveform which offers better spectrum efficiency due to shaping gain of multi-dimensional constellations and allows overloaded non-orthogonal multiple access due to the superposition of sparse codewords. As illustrated in Figure 8.58, over each SCMA layer, the SCMA modulator directly maps a block of coded bits to a complex, multi-dimensional codeword selected from a layer-specific SCMA codebook. Different codewords can be assigned to a user or different users. One application scenario of SCMA is grant-free UL random access, where each SCMA layer represents a user. Since the number of multiplexed users can be more than the spreading factor, an SCMA-based, non-orthogonal multiple access scheme can support more connected devices within each scheduling interval. A blind multi-user reception technique is applied to detect users' activities and extract the original data streams from the received signal. The sparsity of SCMA codewords helps to reduce the complexity of joint detection. This technique is applicable to both mobility and stationary scenarios with timing synchronization maintained [42].

Multi-user MIMO (MU-MIMO) is a well-known multiple access technique to share given time–frequency and power resources among multiple users in a downlink wireless access network. The target is to increase the overall downlink throughput through user multiplexing.

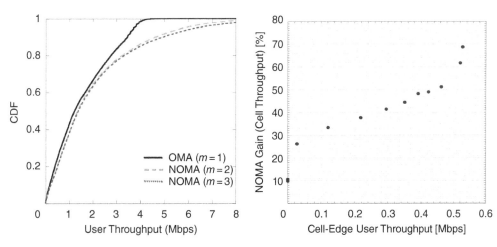

Figure 8.57 (a) CDF of user throughput for OMA ($m = 1$) and NOMA ($m = 2$ or 3); (b) NOMA ($m = 2$) gains in terms of cell throughput versus cell-edge user throughput.

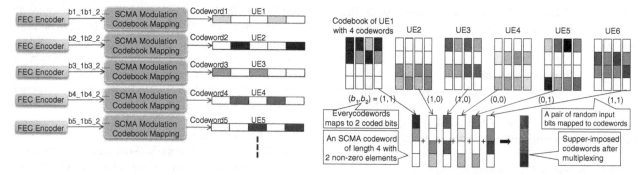

Figure 8.58 SCMA modulation and multiple access using non-orthogonal multiplexing of codewords in the frequency domain. Sparsity facilitates efficient joint detection.

Multiple beams are formed over an array of antennas at a Transmit Point (TP) to serve multiple users distributed within a cell. Every MIMO layer is assigned to a user while layers are orthogonally separated in the space domain, assuming MIMO beamforming pre-coders are properly selected according to the channels of target users. At the receive side, every user can simply match itself to its intended layer while other MIMO layers are seemingly totally muted with no cross-layer interference, provided the pre-coders are properly designed.

Despite the promising throughput gain and the simplicity of detection at user nodes, MU-MIMO as a closed-loop system suffers from some practical difficulties in terms of channel aging and high overhead to feedback channel state information (CSI) of users to a serving TP. CSI is required to form the best set of pre-coders for a selected set of paired users. If CSI is not well estimated, cross-layer interference practically limits the potential performance gain of MU-MIMO [45–50].

Open-loop user multiplexing is a desired approach to avoid the practical limitations of MU-MIMO. Non-orthogonal, code-domain multiple access is an open-loop scheme to pair multiple users over shared time–frequency resources. Sparse Code Multiple Access (SCMA) is a non-orthogonal, codebook-based multiple-access technique with near-optimal spectral efficiency. In SCMA, incoming bits are mapped directly to multi-dimensional complex codewords selected from pre-defined codebook sets. Co-transmitted spread data are carried over superimposed layers [45–50].

SCMA is well matched to user multiplexing, as we can allocate code-domain layers to different users without the need for CSI knowledge of paired users. In this chapter, multi-user SCMA (MU-SCMA) is proposed to improve network throughput. With a very limited need for channel knowledge in terms of channel quality indicators, TP simply pairs users while the transmit downlink power is properly shared among multiplexed layers. Compared to MU-MIMO, this system is more robust against channel variations. In addition, the problem of

CSI feedback is totally removed for this open-loop multiple access scheme. Since layers are not fully separated in a non-orthogonal multiple access system, a nonlinear receiver is required to detect the intended layer of every user. Therefore, further complexity of detection is the cost of non-orthogonal multiple access, especially when a system is heavily overloaded with a large number of multiplexed layers. Sparsity of SCMA codewords lets us take advantage of the low-complexity Message-Passing Algorithm (MPA) detector with ML-like performance. MPA performs well even if the system is overloaded with a large number of layers [45–50].

Low-Density Spreading (LDS) is a special form of SCMA. In LDS, codewords are built by spreading modulated QAM symbols using low-density spreading signatures with a few non-zero elements within a large signature length. Despite the moderate complexity of detection, LDS suffers from poor performance, especially for large constellation sizes above QPSK.

All CDMA schemes, and in particular LDS, can be considered different types of repetition coding in which different variations of a QAM symbol are generated by a spreading signature. Repetition coding is not able to provide desirable spectral efficiency for a wide range of SNR. To overcome this problem, in SCMA the QAM mapper and linear operation of sparse spreading are merged to map incoming bits directly to a complex sparse vector called a codeword. This enables SCMA to benefit from the shaping gain of multi-dimensional constellations as opposed to simple repetition coding of linear sparse sequences. Consequently, SCMA substantially improves the spectral efficiency of linear sparse sequences through multi-dimensional shaping gain of codebooks while still providing other benefits in terms of overloading and moderate complexity of detection [45 –50].

Interference management and robustness of the link quality are concerns in lightly loaded networks. When the traffic demand is low, the resource utilization drops; within the long-term evolution (LTE) context, using an example of an OFDMA system, that is equivalent to muting some resource blocks across the bandwidth of a TP.

In this situation, the interference level at a downlink user changes rapidly within every scheduling interval even if fading channels are quite stable with very slow variation. The interference level at an RB rises if most RBs of neighboring cells are occupied and drops if the corresponding RBs of neighboring cells are empty. This rapid variation in interference level in time and frequency is not predictable in practice, when there is no dynamic cooperation among the neighboring cells. The system has no choice but to adapt itself to the worst-case scenario of channel quality. Poor link adaptation inherently decreases the efficiency of a link [45–50].

Spreading over OFDMA tones can potentially improve the quality of the link-adaptation procedure due to the interference averaging. By using an SCMA spreading technique, interference from different TPs is averaged out over the spread tones. This makes the interference white, which has the advantage of better and more robust link adaptation. In addition, layer multiplexing adds another degree of freedom to the link-adaptation capability of an SCMA system. The number of layers along with codebook sizes, coding rates, and power levels of multiplexed layers are the parameters dictating the rate and quality of a link [45–50].

System-level evaluations are performed in [42] to compare SCMA-based and OFDMA-based UL random access schemes in terms of supported system loading, considering QoE requirements such as maximum latency and average packet drop rate of the overall system. The details of the simulation assumptions and methodology are described in the annex to [42]. In the SCMA case, users share the same time/frequency resources while they are separated in the code domain. Users randomly pick a codebook set and transmit one layer of SCMA signal to the target cell. The instantaneous loading of the system depends on the number of active users in each particular transmission time period. An advanced multi-user detector blindly detects the active users and their corresponding data streams even if users randomly select the same codebook sets. Therefore, the SCMA system is able to tolerate more collisions between users and meanwhile maintain a good quality of detection with affordable complexity, due to the sparsity of colliding codewords. For example, with 6 QPSK user collision, the detection complexity of SCMA can be 150 times less than multi-user detection of multi-carrier CDMA. In the OFDMA random access scenario, a user randomly picks an RB among the available RBs for transmission. A collision happens if two or more users select the same RB for a particular transmission time. A sophisticated MLD receiver is used to separate users even if they collide, but the performance capability of the receiver is limited. In OFDMA, the complexity of multi-user MLD detection grows exponentially with the number of colliding users.

Table 8.16 Supported loading for UL contention-based SCMA and OFDMA.

	Average System Drop Rate: −1%	
	Latency <= 5 ms	Latency <= 10 ms
OFDMA (packets/TTI)	2.222	2.985
SCMA (packets/TTI)	8	9.09
SCMA capacity over OFDMA (times)	3.6	3.0

Some techniques can be applied to reduce the complexity further while maintaining the performance gain and other advantages of SCMA. As shown in Table 8.16, SCMA is able to support around three times more loading compared to OFDMA. The gain is higher with a tighter delay requirement. Further performance analyses for MU-SCMA in the downlink are conducted in [50].

8.13 NFV and SDN

The intersection of telecommunications, the Internet, and IT networking paradigms combined with advances in hardware and software technologies has created an environment that is ripe for rapid innovation and disruption. "Software-defined Networking" (SDN) and "Network Function Virtualization" (NFV) are two promising concepts developed in the telecommunications industry. SDN is based on the separation of control and media planes. The control plane, in turn, lays the foundation for dynamically orchestrating the media flows in real time. NFV, on the other hand, separates software from hardware, enabling flexible network deployment and dynamic operation [51]. Network function virtualization is about implementing network functions in software that run today on proprietary hardware, leveraging (high-volume) standard servers and IT virtualization. NFV supports multi-versioning and multi-tenancy of network functions. It allows the use of a single physical platform for different applications, users, and tenants. Moreover, it enables new ways to implement resilience, service assurance, test and diagnostics, and security surveillance. NFV facilitates innovation towards new network functions and services that are only practical in a pure software network environment. It can be applied to any data plane and control plane functions (fixed or mobile networks). Automation of management and configuration of functions is important in order for NFV to scale. NFV aims ultimately to transform the way network operators design and operate their networks – though change will be incremental.

The reasons behind NFV and SDN are as follows:

Mobility and an explosion in devices and traffic require new ways to expand and automate the network.

The maturity of cloud services and cloud computing over standard IT servers.

High-performance, industry-standard servers shipped in very high volumes.

The convergence of computing, storage, and networks at data centers (DCs). DCs will be converted to pools of NFV resources.

New virtualization technologies that abstract underlying hardware, yielding elasticity, scalability, and automation.

The emergence of software-defined networking techniques.

Huge CAPEX investment to deal with current trends in mobile and fixed networks.

Network operators face an increasing disparity between costs and revenues.

Complexity: Large and increasing variety of proprietary hardware appliances in operators' networks.

Reduced hardware lifecycles.

Lack of flexibility and agility: Inability to move network resources where and when needed.

Launching new services is difficult and takes too long. This often requires yet another proprietary box that needs to be integrated into existing systems.

The benefits of NFV and SDN are as follows:

Flexibility to provision and instantiate new services easily, rapidly, and dynamically in various locations (i.e. no need for the installation of new equipment or physical wiring or on-site configuration).

Reduced time to market as a result of minimizing the typical network operator cycle of innovation. More service differentiation and customization.

Improved operational efficiency as a result of taking advantage of the higher uniformity of the physical network platform and its homogeneity with other support platforms.

Reduced equipment costs through equipment consolidation on high-volume, industry-standard servers, leveraging the economies of scale of the IT industry.

Reduced operational costs: Reduced power, reduced space, improved network monitoring.

Software-oriented innovation (including open source) to prototype and test new services rapidly and generate new revenue streams.

IT-oriented skillset and talent – readily available globally and flexible.

NFV and SDN are highly complementary; they are mutually beneficial but not dependent on each other (NFV can be deployed without SDN and vice versa). SDN can enhance the performance of NFV, simplify compatibility, and facilitate operations. NFV aligns closely with SDN objectives to use software, virtualization, and IT orchestration and management techniques.

Examples of virtual network functions (VNFs) include, but are not limited to:

Switching: BNG, CG-NAT, routers.

Mobile network nodes: HLR/HSS, MME, SGSN, GGSN/PDN-GW, RNC.

Home routers and set-top boxes.

Tunneling gateway elements.

Traffic analysis: DPI.

Signaling: SBCs, IMS.

Network-wide functions: AAA servers, policy control.

Application-level optimization: CDNs, load balancers.

Security functions: Firewalls, intrusion detection systems.

The NFV Infrastructure (NFVI) is required to support the range of use cases and fields of application already identified by the ETSI NFV Industry Specification Group (NFV ISG) while providing a stable platform for the evolution of the VNF ecosystem. The NFVI is the totality of the hardware and software components that build up the environment in which VNFs are deployed. The NFVI provides a multi-tenant infrastructure leveraging standard IT virtualization technology that may support multiple use cases and fields of application simultaneously, as shown in Figure 8.59 [52, 53].

Table 8.17 provides a summary of potential functions for virtualization [54] and Figure 8.60 illustrates the NFV approach [55].

From an operator's perspective, the goals of SDN–NFV are as follows [51]:

Operational efficiencies
Elastic, scalable, network-wide capabilities
Automated OAM&P; limited human touch
Dynamic traffic steering and service chaining
Business transformation
Time-to-market improvements; elimination of point solutions
Agile service creation and rapid provisioning
Improved customer satisfaction.

The following are the key features of a network based on SDN and NFV [51]:

Separation of control and data plane.
Virtualization of network functions.
Programmatic control of network.
Programmatic control of computational resources using orchestration.
Standards-based configuration protocols.
A single mechanism for hardware resource management and allocation.
Automation of control, deployment, and business processes.
Automated resource orchestration in response to application/function needs.

Figure 8.59 NFVI supports a multitude of NFV use cases and fields of application.

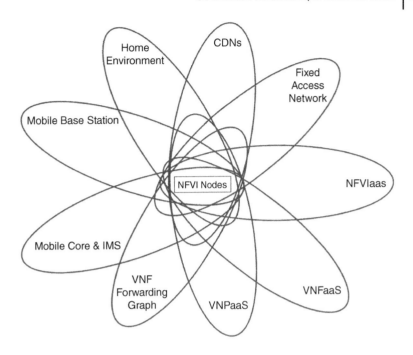

Table 8.17 Potential functions to be virtualized.

Network Element	Function
Switching elements	Broadband network gateways, carrier-grade network address translation (NAT), routers.
Mobile network nodes	Home location register/Home subscriber server, gateway, GPRS support node, radio network controller, various Node B functions.
Customer premises equipment	Home routers, set-top boxes.
Tunneling gateway elements	IPSec/SSL virtual private network gateways.
Traffic analysis	Deep Packet Inspection (DPI), quality of experience measurement.
Assurance	Service assurance, Service Level Agreement (SLA) monitoring, testing and diagnostics.
Signaling	Session border controllers, IP multimedia subsystem components.
Control plane/access functions	AAA servers, policy control, and charging platforms.
Application optimization	Content delivery networks, cache servers, load balancers, accelerators.
Security	Firewalls, virus scanners, intrusion detection systems, spam protection.

The ETSI NFV use cases are summarized as follows [56]:

NFV Infrastructure as a Service (NFVIaaS): NFVIaaS is analogous to a cloud IaaS that is capable of orchestrating virtual infrastructures that span a range of virtual and physical network, computing, and storage functions.

Virtual Network Functions as a Service (VNFaaS): Many enterprises are deploying numerous network service appliances at their branch offices: access routers, WAN optimization controllers, firewalls, and intrusion detection systems. Virtual Network Functions as a Service (VNFaaS) is an alternative solution for enterprise branch office networks, whereby VNFs are hosted on servers in the network service provider's access network PoP.

Virtualization of the Home Environment (VoHE): Virtualization of the home environment is analogous to VNFaaS. In this case, the residential gateway (RGW) and the set-top box (STB) are virtualized as VNFs residing on servers in the network service provider's PoP.

VNF Forwarding Graph (VNF FG): IT organizations need to be able to orchestrate and manage traffic flows between virtualized service platforms (e.g., VNFs) and physical devices in order to deliver a complete service to the end user. The VNF Forwarding Graph (VNF FG) is a service that provides flow mapping (a.k.a., service stacking or chaining) from a management and

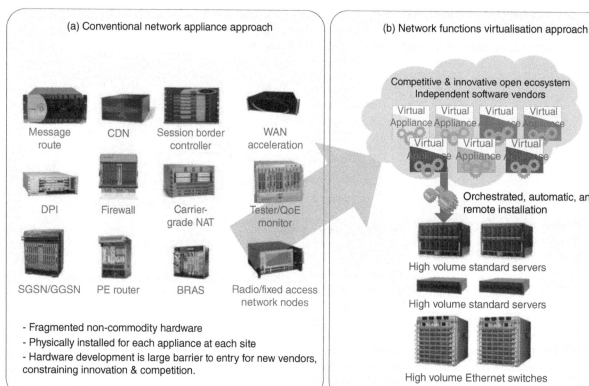

BRAS: broadband remote access server
CDN: content delivery network
DPI: deep packet inspection
GGSN: gateway GPRS (general packet radio service) support node
NAT: network address translation

PE: provider edge
QoE: quality of experience
SGSN: serving GPRS support node
WAN: wide area network

Figure 8.60 NFV approach.

orchestration system that may or may not be part of an SDN infrastructure.

Virtual Network Platform as a Service (VNPaaS): VNPaaS is similar to an NFVIaaS that includes VNFs as components of the virtual network infrastructure. The primary differences are the programmability and development tools of the VNPaaS that allow the subscriber to create and configure custom ETSI NFV-compliant VNFs to augment the catalog of VNFs offered by the service provider.

Virtualization of Mobile Core Network and IP Multimedia Subsystem: The 3GPP is the standards organization that defines the network architecture and specifications for network functions (NFs) in mobile and converged networks. Each NF typically is run on a dedicated appliance in the mobile network PoP. Running the NFs as VNFs on virtualized industry-standard servers is expected to bring a number of benefits in terms of CAPEX, OPEX, and flexible and dynamic scaling of the network to meet spikes in demand.

Virtualization of the Mobile Base Station: 3GPP LTE provides the Radio Access Network (RAN) for the Evolved Packet System (EPS). There is the possibility

that a number of RAN functions may be virtualized as VNFs running on industry-standard infrastructure.

Virtualization of Content Delivery Networks (CDNs): Some ISPs are deploying proprietary CDN cache nodes in their networks to improve delivery of video and other high-bandwidth services to their customers. Cache nodes typically run on dedicated appliances running on custom or industry-standard server platforms. CDN cache nodes and CDN control nodes can both, potentially, be virtualized.

Virtualization of Fixed Access Network Functions: NFV offers the potential to virtualize remote functions in the hybrid fiber/copper access network as well as PON fiber to the home and hybrid fiber/wireless access networks. In a DSL access network, some of the functions that may potentially be virtualized include the DSLAM and Message Display Unit (MDU) forwarding functions, while control functions remain centralized at the central office.

The ETSI NFV reference framework consists of three layers, as shown in Figure 8.61, and these may be described as follows:

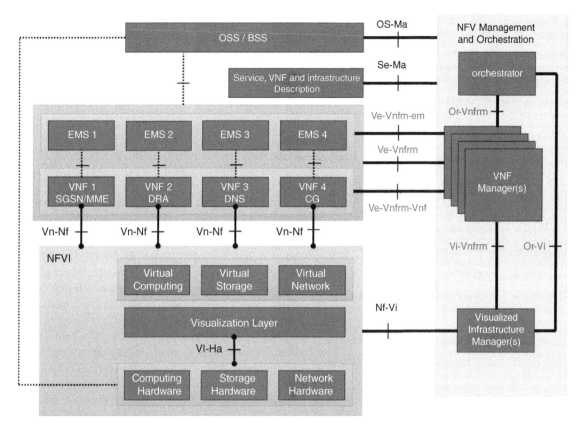

Figure 8.61 ETSI NFV model.

Figure 8.62 ETSI evolved reference framework.

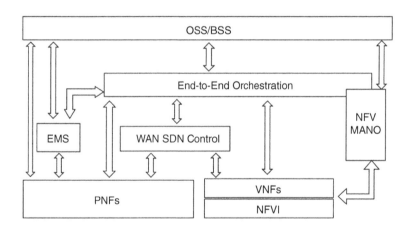

Service layer: Managing the requested service, guiding the application installation, expansion, and configuration.

Application layer: Containing all the applications and the relative management systems (EMSs + VNFMs) provided by each Telco vendor.

Infrastructure layer: Made up of hypervisors and operative systems allowing the allocation of HW resources to specific applications, providing computing power, storage and connectivity, and a management system for all virtualized resources.

The ETSI evolved reference framework is illustrated in Figure 8.62 [51].

Figure 8.62 illustrates the high-level management and control architecture. The figure shows a network infrastructure composed of virtualized and physical network functions. The Virtualized Network Functions (VNFs) run on the NFV Infrastructure (NFVI).

The management and control complex has three main building blocks:

NFV MANO: Manages the NFVI and has responsibility for lifecycle management of the VNFs. Key functions include, but are not limited to:

Allocation and release of NFVI resources (compute, storage, network connectivity, network bandwidth, memory, hardware accelerators).

Management of networking between virtual machines (VMs) and VNFs, i.e., data center SDN control.

Instantiation, scaling, healing, upgrade, and deletion of VNFs.

Alarm and performance monitoring related to the NFVI.

WAN SDN Control: Represents one or more logically centralized SDN controllers that manage connectivity services across multi-vendor and multi-technology domains. WAN SDN Control manages connectivity services across legacy networks and new PNFs, but can also control virtualized forwarding functions, such as virtualized provider edge routers (vPEs).

End-to-End Orchestration (EEO): Responsible for allocating, instantiating, and activating the network functions (resources) that are required for an end-to-end service. It interfaces with:

NFV MANO to request instantiation of VNFs.

WAN SDN Control to request connectivity through the WAN.

PNFs and VNFs for service provisioning and activation.

EEO and NFV MANO are shown as overlapping. This is because the ETSI NFV definition of MANO includes a Network Service Orchestration (NSO) function, which is responsible for a subset of functions that are required for end-to-end orchestration, as performed by EEO. NFV MANO knows whether a network function is virtualized without knowing what it does. WAN SDN Control knows what a network function does, without knowing whether it is virtualized [54].

Figure 8.63 shows the high-level architecture and identifies the functional blocks of Verizon's SDN-NFV architecture [51]. The architecture supports network function virtualization, software-defined networking, and service and network orchestration.

Verizon's SDN-NFV architecture may be summarized as follows:

Service orchestration is the customer-facing function responsible for providing the services catalog to the portal.

Portals include the customer portal, where a customer can order, modify, and monitor their services, and the ops portal.

End-to-end orchestration is responsible for lifecycle management of end-to-end network services.

The objective of EEO is to realize zero-touch provisioning; a service instantiation request – from the operations crew or from the customer, through a self-service portal – results in an automatically executed work flow that triggers VNF instantiation, connectivity establishment, and service activation.

The TM Forum distinguishes between service management and resource management and, in that context, uses the terms *customer-facing services* and *resource-facing services*. Service management and customer-facing services are related to the service as perceived by the end user. The customer-facing service aspects include order entry, billing, trouble ticketing, self-service portals, helpdesk support, etc. The resource-facing service consists of the network functions/resources required to deliver the customer-facing service. Resource management includes bandwidth allocation, QoS management, protocol handling, etc. Using the TM Forum definitions, EEO is responsible for resource management. All customer-facing service aspects are handled by other OSS and BSS systems.

The network services alluded to in the third point of the earlier list are what the TM Forum calls resource-facing

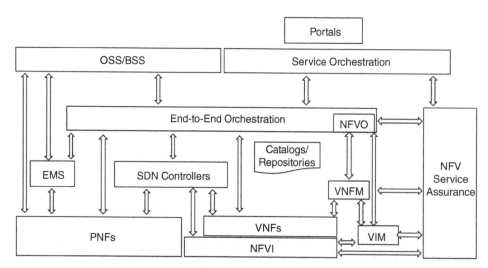

Figure 8.63 Verizon's SDN-NFV high-level architecture.

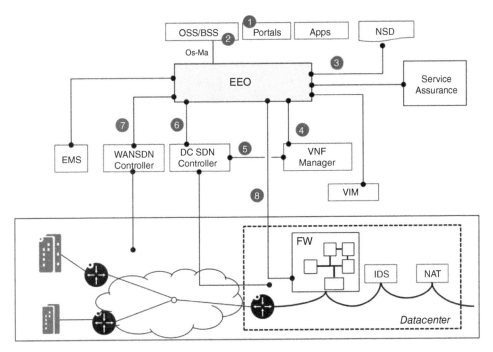

Figure 8.64 Management and control architecture with vCPE infrastructure.

services. Figure 8.64 demonstrates the call flow for the activation of an enterprise vCPE service [51].

The following steps describe the activation call flow for vCPE [51]:

The customer signs up via a Web portal. Via the portal, the customer can choose from services that are listed in the services catalog. Once it has selected a vCPE service, the customer can further customize the service to its needs.

An OSS system is responsible for orchestration of the customer-facing service aspects. For example, it interfaces with the billing system to register the user and its service choice. The OSS triggers EEO to line up all the network resources required for delivering the service.

EEO retrieves the End-to-End Network Service Descriptor information associated with the requested service. Based on this, it determines which resources are required and decides where to place new VNFs.

EEO interfaces with the VNFM and requests instantiation of the required VNFs.

For VNFs that consist of multiple VMs, the VNFM interfaces with the data center SDN (DC-SDN) controller to create connectivity between the VMs within that VNF.

EEO interfaces with the DC SDN controller to establish connectivity between VNFs. In the case of a vCPE, this entails creating a service chain that traverses the VNFs in the desired order.

EEO interfaces with the WAN SDN controller to extend the customer's VPN and connect to the newly established service chain in the data center.

EEO provisions the network elements in order to provide the end-to-end service requested by the customer. For example, the customer may have selected a specific security profile that needs to be enforced by the firewall. Therefore, EEO needs to make sure that the right profile is provisioned in the firewall.

As illustrated in Figure 8.64, no service orchestrator is involved in the vCPE activation. The EEO is facing all network functions including EMS, WAN SDN, DC SDN, VNFM, and VIM, and the customer-facing layer includes BSS, portals, and apps. So, Verizon uses mainly one orchestrator for all SDN–VNF applications. In the next section, we will introduce the concept of service orchestrators for multi-domain orchestration.

8.13.1 Convergence of IT and Telecom Domains

The adoption of NFV poses a number of other significant challenges that must be overcome in order to ensure the ability to continue to implement effective, end-to-end management. These challenges include [54]:

Dynamic relationships between software and hardware components. With NFV, software running on virtual machines (VMs) can be moved readily among physical servers or replicated to run on newly created VMs in order to maintain availability dynamically, expand/shrink capacity, or balance the load across physical resources.

Dynamic changes to physical/virtual device configurations. To accommodate the dynamic nature of

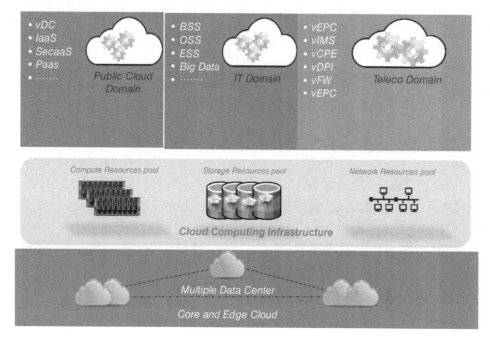

Figure 8.65 Multi-tenant converged common computing architecture.

virtualized networks, end-to-end management systems will need to be able to adjust the configuration of devices to react to changing conditions in the network.

Many-to-many relationships between network services and the underlying infrastructure. In a virtualized infrastructure, a network service can be supported by a number of VNFs, which may be running on one or several VMs. A single VNF may also support a number of distinct network services. In addition, the group of VNFs supporting a single network service could possibly be running on a number of distinct physical servers.

Hybrid physical/virtual infrastructures. As virtualization is gradually adopted, service providers will need to be able to integrate virtual environments into their existing end-to-end traditional/legacy monitoring infrastructures.

Network services spanning multiple service providers. Some of the VNFs comprising a virtualized network service may be hosted in the clouds of multiple collaborating providers.

IT and network operation collaboration. These organizations will need to cooperate effectively to establish new operational processes that meet the demands of end-to-end management of hybrid physical/virtual infrastructures.

The introduction of NFV and SDN will transform the legacy network of a service provider into a software-controlled network, and the significant overlap between the IT domain and the Telecom domain entails a new operating model. In addition, convergence between the IT cloud and the Telecom cloud is mandatory for efficient OPEX/CAPEX operating models. Therefore, the convergence of common computing infrastructure (computation, storage, and network) is mandatory for efficient resource utilization. Figure 8.65 demonstrates the concept of common computing infrastructure with multi-tenants including the IT domain, the Telecom domain, and the XaaS domain for cloud-managed services.

Figure 8.65 illustrates a target architecture for a service provider to serve multiple tenants on top of the common computing infrastructure. Figure 8.66 demonstrates detailed architecture for a multi-tenant approach for a service provider. As illustrated, NFVO is similar to EEO of Verizon, which manages the VNF domain. The cloud orchestrator for IT functions can be a separate private cloud or merged with the XaaS domain, where the service provider can offer cloud services to external organizations. An end-to-end service orchestrator manages the different domain orchestrators. In addition, the physical environment for legacy systems will be managed by the SO through the EMS. The customer-facing layer includes BSS/OSS, CRM, portals, etc.

8.14 Conclusion

3GPP started the 5G discussion in the RAN 5G Workshop in September 2015, and has made substantial progress since then. It has consolidated an accelerated timeline for early NR deployments and started normative work on the overall architecture in December 2016 and

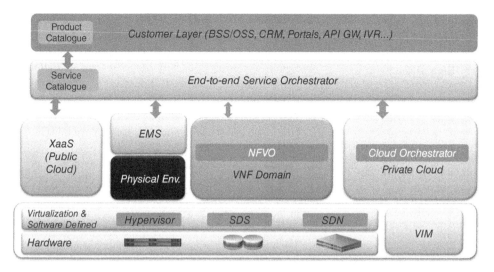

Figure 8.66 Detailed architecture for a multi-tenant approach.

on the new radio technology in March 2017. Defined to be a unified system covering a wide range of use cases, 5G is expected to cover a diverse deployment that is scalable and adaptable for different spectrum types: low and high bands. It is generally agreed by the industry that 5G standardization will take two main releases, spanning Release 15 and Release 16. The first phase will prioritize spectrum and waveforms up to 40 GHz for both eMBB and low-latency use cases, while the second phase will continue with waveforms above 40 GHz and will add mMTC and URLLC use cases.

At the same time, ITU-R extended the sub-6 GHz bandwidth for IMT at a nearly global scale in WRC-15, covering a new range of bands like the C-band, while considering the 24 to 33 GHz spectrum for mmWave technologies for the potential global harmonized band.

The envisioned market space for 5G technology is targeting a design that has the ability to unify the system needed by various use cases. In order to enable services for a wide range of users and industries with new requirements, the capabilities of 5G must extend beyond those of previous wireless access generations. These capabilities will include enhanced mobile broadband connectivity, massive system capacity for machine-type communications, very high data rates everywhere, very low latency, ultra-high reliability and availability, and low device cost.

4G has transformed the Internet, making it mobile and thus enabling countless services to flourish. 5G will leverage this further and make the Internet truly ubiquitous whilst enabling new, disruptive use cases even beyond what is feasible today. In order to accelerate next-generation mobile technology that is capable of meeting these requirements, co-existence with the LTE-A Pro network (eLTE in its Release 15 version) is expected to be a key enabler for 5G deployments.

References

1 TAICS (2015) "Taiwan 5G White Paper by MediaTek and ITRI," Technical Report, September.

2 3GPP RP-161901: "Revised work item proposal: Enhancements of NB-IoT," 3GPP TSG RAN Meeting #73, September 2016.

3 3GPP RP-161266: "5G architecture options," 3GPP RAN/SA meeting, June 2016.

4 ITU-R WP5D/TEMP/548-E: IMT Vision: "Framework and overall objectives of the future development of IMT for 2020 and beyond," September 2015.

5 Elnashar, A. *et al.* (2014) *Design, Deployment, and Performance of 4G-LTE Networks: A Practical Approach*, John Wiley & Sons.

6 Elnashar, A. and Elsaidny, M. (2014) "Extending the Battery Life of Smartphones and Tablets: A Practical Approach to Optimizing the LTE Network," *IEEE Vehicular Technology Magazine*, Issue 2, pp. 38–49, June.

7 Elnashar, A. and Elsaidny, M. (2013) "Looking at LTE in Practice: A Performance Analysis of the LTE System based on Field Test Results," *IEEE Vehicular Technology Magazine*, **8**(3), 81–92, September.

8 Tsai, L.-S., Zhuang, X. and Liao, P.-K. (2015) "Methods for Codeword Level Interference Cancellation with Network Assistance," MediaTek Patent, US20150146657 A1, *May* **28**, 2015.

9 Agilent (2013) *"Wireless LAN at 60 GHz – IEEE 802.11ad Explained,"* Application Note, May.

10 IEEE 802.11 working group: "IEEE 802.11ad, Amendment 3: Enhancements for Very High Throughput in the 60 GHz Band," December 2012.

11 IEEE 802.11 working group: "Wireless LAN Medium Access Control (MAC) and Physical Layer (PHY) Specifications," March 2012.

12 Rosenblum, E. S. (1961) "Atmospheric Absorption of 10–400 kMCPS Radiation: Summary and Bibliography to 1961," *Microwave Journal*, March.

13 3GPP TR 38.913 V14.0.0 (2016-10) "Study on Scenarios and Requirements for Next Generation Access Technologies."

14 3GPP TR 36.878 (2016-01) "Study on performance enhancements for high speed scenario in LTE."

15 3GPP TS 22.886 (2017-03) "Study on enhancement of 3GPP Support for 5G V2X Services."

16 3GPP TR 23.799 V14.0.0 (2016-12) "Study on Architecture for Next Generation System."

17 3GPP RP-160671 (2016-03) "New SID Proposal: Study on New Radio Access Technology."

18 3GPP TR 38.801 (2017-04) "Study on New Radio Access Technology; Radio Access Architecture and Interfaces."

19 3GPP TR 38.804 V1.0.0 (2017-03) "Radio Interface Protocol Aspects."

20 3GPP TS 23.401 (2017-09) "General Packet Radio Service (GPRS) enhancements for Evolved Universal Terrestrial Radio Access Network (E-UTRAN) access."

21 https://www.rohde-schwarz.com/us/solutions/wireless-communications/5g/5g-waveforms/5g-waveforms_230224.html

22 http://www.radio-electronics.com/info/cellulartelecomms/5g-mobile-cellular/modulation-waveforms.php

23 Zhang, X., Jia, M., Chen, L., Ma, J. and Qiu, J. (2015) "Filtered-OFDM – Enabler for Flexible Waveform in The 5th Generation Cellular Networks," arXiv:1508.07387 [cs, math], August 2015, arXiv: 1508.07387. Online: http://arxiv.org/abs/1508.07387

24 Schaich, F., Wild, T. and Chen, Y. (2014) "Waveform Contenders for 5G – Suitability for Short Packet and Low Latency Transmissions," in *79th IEEE Vehicular Technology Conference* (VTC Spring), *May* **2014**, pp. 1–5.

25 Dias, R. and Eschbach, B. (2013) *"A Comparison of OFDM with Cyclic Prefix and Unique Word Based on the Physical Layer of DVB-T."* Czech Technical University, Prague. Online: http://publications.rwth-aachen.de/record/226459

26 Schaich, F. and Wild, T. (2014) "Waveform contenders for 5G – OFDM vs. FBMC vs. UFMC," in *6th International Symposium on Communications, Control and Signal Processing* (ISCCSP), May 2014, pp. 457–460.

27 Berardinelli, G., Tavares, F., Sørensen, T., Mogensen, P. and Pajukoski, K. (2015) "On the Potential of Zero-Tail DFT-Spread-OFDM in 5G Networks," in *80th IEEE Vehicular Technology Conference* (VTC Fall), September 2014, pp. 1–6.

28 Gaspar, I., Mendes, L., Matthé, M., Michailow, N., Zhang, D., Alberti, A. and Fettweis, G. (2015)_ "GFDM – A Framework for Virtual PHY Services in 5G Networks," arXiv:1507.04608 [cs, math], July 2015, arXiv: 1507.04608. Online: http://arxiv.org/abs/1507.04608.

29 Elkourdi, M. *et al.* (2016) "Waveform Design Principles for 5G and Beyond," *17th annual IEEE Wireless and Microwave Technology Conference* (WAMICON).

30 Benvenuto, N., Tomasin, S. and Tomba, L. (2002) "Equalization methods in OFDM and FMT systems for broadband wireless communications," *IEEE Transactions on Communications*, **50**(9), 1413–1418, September.

31 Sahin, T. A. and Arslan, H. (2013) "Mask Compliant Precoder for OFDM Spectrum Shaping," *IEEE Communications Letters*, **17**(3), 447–450, March.

32 Berardinelli, G., Ruiz de Temino, L., Frattasi, S., Rahman, M. and Mogensen, P. (2008) "OFDMA vs. SC-FDMA: Performance comparison in local area IMT-A scenarios," *IEEE Wireless Communications*, **15**(5), 64–72, October.

33 Ankaral, Z., Karabacak, M. and Arslan, H. (2014) "Cyclic Feature Concealing CP Selection for Physical Layer Security," in *IEEE Military Communications Conference* (MILCOM), October 2014, pp. 485–489.

34 Guvenkaya, E., Tom, A. and Arslan, H. (2013) "Joint Sidelobe Suppression and PAPR Reduction in OFDM Using Partial Transmit Sequences," in *IEEE Military Communications Conference* (MILCOM), November 2013, pp. 95–100.

35 Future Mobile Communication Forum (2014) "5G White Paper – Rethink Mobile Communications for 2020+," Technical Report, November. Online: http://www.euchina-fire.eu/wp-content/uploads/2015/03/5G-SIG-white-paper-20141102-v0.compressed.pdf

36 Wunder, G. *et al.* (2014) "5GNOW: non-orthogonal, asynchronous waveforms for future mobile applications," *IEEE Communications Magazine*, **52**(2), 97–105, February.

37 NGMN Alliance (2015) "5G White Paper," Next Generation Mobile Networks, White Paper. Online: http://ngmn.org/uploads/media/NGMN_5G_White_Paper_V1_0.pdf

38 Huang, K.-C. and Wang, Z. (2011) *Millimeter Wave Communication Systems*, John Wiley & Sons.

39 Ho, K.-P., Cheng, S. and Liu, J. (2014) "MIMO Beamforming in Millimeter-Wave Directional Wi-Fi," arXiv:1403.7697 [cs, math], March 2014, arXiv: 1403.7697. Online: http://arxiv.org/abs/1403.7697.

40 Farhan, A., Marchetti, N., Figueiredo, F. and Miranada, J.-P. (2014) "Massive MIMO and Waveform Design for 5th Generation Wireless Communication Systems", 1st International Conference on 5G.

41 *5GNOW – 5th Generation Non-Orthogonal Waveforms for Asynchronous Signalling*. Online: http://www.5gnow.eu/

42 METIS (2014) Components of a New Air Interface – Building Blocks and Performance.

43 Saito, Y., Kishiyama, Y., Benjebbour, A. *et al.* (2013) "Non-orthogonal multiple access (NOMA) for cellular future radio access," *IEEE Vehicular Technology Conference* (VTC Spring), pp. 1–5.

44 Benjebbour, A., Saito, Y., Kishiyama, Y. *et al.* (2013) "Concept and practical considerations of non-orthogonal multiple access (NOMA) for future radio access," *IEEE International Symposium on Intelligent Signal Processing and Communications Systems* (ISPACS), pp. 770–774.

45 Nikopour, H., Yi, E., Bayesteh, A., Au, K. *et al.* (2014) "SCMA for downlink multiple access of 5G wireless networks," *IEEE Globecom'14*, pp. 3940–3945.

46 Nikopour, H. and Baligh, H. (2013) "Sparse code multiple access," *24th IEEE International Symposium on Personal Indoor and Mobile Radio Communications* (PIMRC), pp. 332–336.

47 Taherzadeh, M., Nikopour, H., Bayesteh, A. and Baligh, H. (2014) "SCMA codebook design," *IEEE Vehicular Technology Conference* (VTC Fall), pp. 1–5.

48 Nikopour, H. and Baligh, M. (2014) "System and Method for Sparse Code Multiple Access," U.S. Patent Application 2014/0140360, May 22.

49 Au, K., Zhang, L., Nikopour, H., Yi, E. *et al.* (2014) "Uplink contention-based SCMA for 5G radio access," *IEEE Globecom'13 Workshops*, pp. 900–905.

50 Nikopour, H., Yi, E., Bayesteh, A., Au, K. *et al.* (2014) "SCMA for downlink multiple access of 5G wireless networks," *IEEE Globecom'14*, pp. 3940–3945.

51 Verizon (2016) "SDN-NFV Reference Architecture," Version 1.0, White Paper, February 2016.

52 ETSI (2014) "NFVI Supports Multiplicity of NFV Use Cases and Fields of Application," White Paper presented at the *SDN & OpenFlow World Congress*, October 14–17, Dusseldorf, Germany. Online: http://portal.etsi.org/NFV/NFV_White_Paper3.pdf

53 ETSI NFV ISG Architectural Framework. Online: www.etsi.org/deliver/etsi_gs/ NFV/001_099/ 002/01.01.01_60/gs_NFV002v010101p.pdf

54 Metzler, J. (2015) *The 2015 Guide to SDN and NFV*, *webtorials*, January.

55 Obana, K., Kinoshita, T. and Shimano, K. (2014) "A Brief Overview of Network Function Virtualizations," *NTT Technical Review*, **12**(3), March.

56 ETSI GS NFV 001, Use Cases. Online: http://www.etsi.org/deliver/etsi_gs/NFV/001_099/001/01.01.01_60/gs_NFV001v010101p.pdf

Index

Practical Guide to LTE-A, VoLTE and IoT: Paving the Way Towards 5G, First Edition. Ayman Elnashar and Mohamed El-saidny.
© 2018 John Wiley & Sons Ltd. Published 2018 by John Wiley & Sons Ltd.